Molecular Theories of
Cell Life and Death

Molecular Theories of Cell Life and Death

Edited by
Sungchul Ji

RUTGERS UNIVERSITY PRESS ● NEW BRUNSWICK ● NEW JERSEY

Figure 11.4 reprinted by permission from
A. Müller, E. Cadenas, P. Graf and H. Sies,
Biochemical Pharmacology 33: 3235-3239 (1984). Pergamon PLC.

Figure 12.8 reprinted by permission from
H. de Groot and T. Noll,
Biochemical Pharmacology 35: 15-19 (1986). Pergamon Press PLC.

Copyright © 1991 by Rutgers, The State University

All rights reserved

No part of this book may be reproduced, stored in a retrieval system, or transmitted in any form or by any means, electronic, mechanical, photocopying, microfilming, recording, or otherwise, without written permission from the publisher.

Library of Congress Cataloging-in-Publication Data

Molecular theories of cell life and death edited by Sungchul Ji.
Includes bibliographical references, glossary and index.
ISBN 0-8135-1691-9.
1.Cytology—Congresses. 2. Molecular biology—Congresses.
3. Cell death—Congresses.
I. Ji, Sungchul
QH573.M64 1991   91-14289
574.87'6—dc20   CIP

British Cataloging-in-Publication information available

Manufactured in the United States of America

Dedicated to my parents,

Eung E Ji and Bok Nyo Keh,

my beloved family,

Dottie, Mia, and Doug,

and to the memory of my mentor,

Dr. David E. Green

CONTENTS

PREFACE · xiii

ACKNOWLEDGMENTS · xix

CONTRIBUTORS · xxi

INTRODUCTION · xxiii

I. THEORIES

Chapter 1
"Biocybernetics": A Machine Theory of Biology
Sungchul Ji

1.1.	Introduction	2
1.2.	A Bird's-Eye View of Biocybernetics	5
1.3.	The "Table Theory": A New Method in Theoretical Biology	8
1.4.	Cybernetics: A Theory of Machines	13
1.5.	Biopolymers as Molecular Machines: The McClare Machine	29
1.6.	Metabolic Pathways as Machines: The Holcombe-Welch Machines	44
1.7.	Subcellular Organelles as Machines	49
1.8.	The Cell as a Machine: The Bhopalator	62
1.9.	The Biocybernetics Model of the Human Body: The Piscatawaytor	141
1.10.	The Biocybernetics Model of the Human Society: "Newbrunswickator"	149
1.11.	A Biological Model of the Universe: The Shillongator	152
1.12.	Applications	163
1.13.	Concluding Remarks	209
	Acknowledgement	210
	References	211
	Appendix 1.A: The Law of Requisite Variety	221
	Appendix 1.B: Nucleic Acid and Protein Communication Channels	222
	Appendix 1.C: The Conformon-Based Model of the Origin of Life—The "Princetonator"	224
	Appendix 1.D: Cell Death Kinetics	225
	Appendix 1.E: The "Xenogene" Hypothesis	228
	Appendix 1.F: The Shillongator—the Abstract Version	230

Chapter 2 · 238
Schrödinger and the Riddle of Life
Ilya Prigogine

References · 242

Chapter 3
A Note on Chiral Symmetry Breaking and the Origin of Biomolecular Chirality
Dilip K. Kondepudi
                                                                                                                                243

3.1.	Introduction	243
3.2.	Chiral Symmetry Breaking in Non- Equilibrium Chemical Systems	244
3.3.	Selection of Biomolecular Chirality by Extremely Small Parity Non-Conserving Interactions	247
	References	248

Chapter 4
Mathematical Models of Cell Biochemistry
W. Mike L. Holcombe
                                                                                                                                250

4.1.	Introduction	250
4.2.	Windows on the Cell	251
4.3	A General Abstract Machine	259
4.4	The Topology of Death	261
	References	263

Chapter 5
Davydov's Soliton
Alwyn C. Scott
                                                                                                                                264

5.1.	Introduction	264
5.2.	Davydov's Soliton Theory	265
5.3.	Self-Trapping in Crystalline Acetanilide	268
5.4.	The Quantum Theory of Self-Trapping	271
5.5.	Biological Significance of Self-Trapping	280
	References	280

Chapter 6
Cytosociology: A Field-Theoretic View of Cell Metabolism
Harry A. Smith and G. Rickey Welch
                                                                                                                                282

6.1.	Introduction	282
6.2.	Cytosociology of Cell Metabolism	284
6.3.	Cytosociology of Enzyme Action in Vivo	285
6.4.	Enzymes: Field-Effect Elements	290
6.5.	A Primer of Field Theory	294
6.6.	Biological Space-Time	296
6.7.	Cell Metabolism: An Affine Geometry	298
6.8.	The Metabolic Microdomain: A Gauge Field	311
6.9.	Concluding Remarks	318
	Acknowledgements	320
	References	320

Contents

Chapter 7   324
Non-Dichotomous, Relative and Hierarchical Aspects of Life and Death
Jerome Rothstein

7.1.	Introduction and Summary	325
7.2.	On Whether Life and Death Are Mutually Exclusive and Exhaustive Logical Alternatives	327
7.3.	Why "Dead" and "Alive" Are Neither Molecular, Quantum Mechanical, Nor Thermodynamic States	331
7.4.	Toward a Definition of Living State	345
7.5.	Concluding Discussion	355
	References	358

Chapter 8   361
Emergent Properties in Neural Networks
Steven Finette

8.1.	Introduction	361
8.2.	The Netlet as a Paradigm for Neural Self-Organization	363
8.3.	Reliability and Cooperative Computation	370
8.4.	Conclusions	371
	References	372

II. EXPERIMENTS

Chapter 9
Molecular Mechanisms of Cell Death
Sten Orrenius

9.1.	Introduction	373
9.2.	Characteristics of Toxic Injury to Hepatocytes	374
9.3.	Mechanisms of Glutathione Depletion	374
9.4.	Role of Protein Thiol Modification	376
9.5.	Appearance of Surface Blebs in Hepatocytes Exposed to Toxic Agents	376
9.6.	Disruption of Intracellular Ca^{2+} Homeostasis by Toxic Agents	378
9.7.	Mechanisms of Ca^{2+}-Mediated Toxicity	379
9.8.	Future Perspectives	381
	References	383

Chapter 10   384
Damage to the Nucleus and Acute Cell Death Produced by Alkylating Hepatotoxins
Sidhartha D. Ray and George B. Corcoran

10.1.	Introduction	384
10.2.	Compartmentation of Molecular Damage Leading to Acute Cell Death in the Liver	386

10.3.	Cell Death and the Hepatocyte: Theories of Acute Cellular Necrosis	391
10.4.	The Nucleus as a Critical Site of Lethal Cell Injury	392
10.5.	Outlook	395
	Acknowledgements	396
	References	396

Chapter 11
Electronically Excited States in Cells and Organs: Relation to Prooxidant/Antioxidant Balance

Helmut Sies

401

11.1.	Introduction	401
11.2.	Low-Level Chemiluminescence	403
11.3.	Redox Cycling	404
11.4.	Antioxidant Activities	406
11.5.	DNA-Damage by Singlet Oxygen	408
11.6.	Excited Carbonyls	410
11.7.	Concluding Remarks	410
	Acknowledgements	410
	References	410

Chapter 12
Hypoxia and Molecular Mechanisms of Cell Death

414

Herbert de Groot and Thomas Noll

12.1.	Introduction	414
12.2.	Oxygen Partial Pressures in Mammalian Tissues	416
12.3.	Oxygen Dependence of Cell Functions	417
12.4.	Hypoxic Cell Death	418
12.5.	Pathological Cell Functions under Hypoxia	421
12.6.	Conclusions	425
	Abbreviations	426
	Acknowledgements	426
	References	426

Chapter 13
Population Dynamics of Tumors Attacked by Immunocompetent Killer Cells

430

Jacques Hiernaux, Patricia Meyers and René Lefever

13.1	Introduction	430
13.2	The Cytolytic Reaction and the Switching Parameter β	432
13.3	Conclusion	440
	Acknowledgements	440
	References	440

Contents

Chapter 14 — 442
Organization of Histone Genes and Regulation and Control of Their Expression in Early Development
Hans V. Westerhoff, Johanna G. Koster, and Olivier H. J. Destrée

14.1.	Introduction	443
14.2.	The Histones and the Organization of Eukaryotic Genes in Chromatin	444
14.3.	The Organization of Histone Genetic Information	448
14.4.	Kinetics of Histone Synthesis in *Xenopus Laevis*: The Race Against Time	454
14.5.	The Control of Histone-Gene Expression in *Xenopus Laevis*	459
14.6.	Concluding Remarks	467
	References	467

Chapter 15 — 473
How Fundamental Knowledge Can Help Solve Practical Problems in Toxicology
Robert A. Neal

15.1.	Introduction	473
15.2.	Mechanisms of Toxicity	475
15.3.	Toxic Response at Low Doses	476
15.4.	Validity of Animal Models	477
15.5.	Improved Methodology	477
15.6.	Measurements of Human Exposure	478
15.7.	Concluding Remarks	479

Glossary — 481

Index — 495

PREFACE

This book, *Molecular Theories of Cell Life and Death*, grows out of a conference held at the Robert Wood Johnson (then called Rutgers) Medical School in Piscataway on May 1 and 2, 1986, under the joint auspices of Rutgers University and the University of Medicine and Dentistry of New Jersey (UMDNJ). The primary goal of the meeting was to bring together a small group of internationally recognized experts from various fields of natural and mathematical sciences in an informal setting to discuss both the theoretical and experimental bases for judging what is alive and what is dead on the level of the cell, the "atom" of life. The adjective "molecular" in the title does not necessarily indicate that the contributors to the book subscribe to the reductionist philosophy of biology wherein life is thought to be completely explained in terms of molecules and molecular processes; rather the term "molecular" should be viewed simply as reflecting the belief that, whatever the eventual theories of life may turn out to be, those theories should be able to account for and be consistent with the vast amount of *molecular* biological data that have come to light during the past several decades.

Most of the chapters in this book were written on the basis of the lectures delivered at the conference, but there are a few (Chapters 10, 12 and 14) which were solicited long after the meeting, some as late as February, 1990 (Chapter 10). The contributors to this book vary widely in regard to their fields of specialization, which include mathematics, physics, chemistry, biology, and toxicology, as well as with respect to their professional accomplishments, ranging from a Nobel Laureate to deans of toxicology to assistant professors.

The book is organized into two parts, Part I consisting of 8 theoretical papers and Part II, 7 experimental ones. Unlike usual conference proceedings which are published within one or two years after the meeting, the present book has taken almost five years to appear, in part due to the late submission of some of the key chapters (received in mid-1989). This unusually long delay in publication might have rendered certain chapters "dated," a dreaded word in scientific publications, especially in the experimental arena, although I suspect that the highly respected scholarship of the authors involved would largely ameliorate this concern. On the positive side, the unintended delay in publication has provided some of the authors, including myself, with a longer incubation time than otherwise would have been possible in which to digest and contemplate the large amount of factual knowledge and thought-provoking theoretical ideas presented at the conference. Nevertheless, I, as the editor of the book, owe sincere apologies to those authors who, despite their busy schedules, kindly submitted their manuscripts on or near the original deadline (of September, 1986!)—especially S. Orrenius, H. Sies, and R. Neal.

Although the present book is seemingly the direct outcome of the colloquium held in 1986, the true beginning of the book may be traced back to the early 1970's when David E. Green and I began to develop a comprehensive biological theory which eventually led, in 1983, to the formulation of what appears to be the first comprehensive theoretical model of the living cell, a model called the Bhopalator. Since the colloquium and the genesis of this book are closely intertwined with the formulation of the Bhopalator, I may be allowed the privilege of briefly relating the events in the life of one scientist that played crucial roles in bringing about the 1986 Colloquium.

When I joined Rutgers in 1982, I had no intention of continuing my theoretical work on conformons (that is, the sequence-specific conformational strains of proteins and DNA postulated to carry free energy and Shannon information necessary and sufficient to drive all molecular machines) that I initiated in 1972 in Madison in collaboration with the late D. E. Green; rather I was bent on further developing and capitalizing the tissue photometric technique called the micro-light guide which I had developed while working with B. Chance in Philadelphia between 1974 and 1976. This novel experimental technique had proven to be useful in investigating regional metabolism and microcirculation in perfused organs. Robert Snyder, the chairman of our Department at Rutgers, hired me primarily on the basis of my experimental work in toxicology and tissue biophysics performed with micro-light guides and the complementary miniature oxygen electrodes while I was a visiting scientist in the laboratory of Manfred Kessler at the Max Planck Institute for Systems Physiology in Dortmund (1976 - 1979) and an assistant research professor in the laboratory of Ronald Thurman at the Department of Pharmacology, School of Medicine, University of North Carolina in Chapel Hill (1979 - 1982).

However, a few months after moving to Rutgers, I received a packet of old letters from Chapel Hill. One of these letters was from Professor R. K. Mishra of the All-India Institute of Medical Sciences in New Delhi, inviting me to give a lecture on conformons in an international meeting to be held in Bhopal, India, in November, 1983. Professor Mishra had sent the letter to Madison, Wisconsin, believing that I was still there, apparently based on the address appearing in the paper on conformons that Green and I wrote in 1972. From Madison, the letter was duly forwarded to Philadelphia, to Dortmund, West Germany, to Chapel Hill, and finally to Piscataway, retracing the string of the research institutions where I worked following my departure from Madison in 1974. The postal journey through the 4 international cities took about 3 months. The fact that this letter had reached me at all seemed to me miraculous—a living testimony to the care and thoughtful efficiencies of the secretaries at these scientific institutions as well as the fidelity of the postal services in the U.S. and Germany. What was amusing was that the letter from Mishra was so thin and inconspicuous (familiar to those who have received form letters from India) that I almost threw it away with the emptied manila envelope. But, by some instinct, I picked the envelope up again from the wastebasket (which I rarely do) and peered down into it to find the thin letter from India, which was to change the course of my research activities at Rutgers for the next seven years.

For about 6 months from the day I received the Mishra letter until I left for Bhopal, where the International Seminar on the Living State-II was held, I was completely absorbed in preparing for my two papers on conformons to be presented at the meeting. This meant that I had to go back to the literature on theoretical biology which I had not followed for almost 10 years since leaving Madison. The most striking revelation that came to me as the result of my reading the literature was that there had been a remarkable development in bioenergetics of which I was completely unaware—the theory of *solitons* (i.e., solitary waves propagating long distances without dissipation) formulated by the Russian physicist A. Davydov in 1973, which laid down the quantum mechanical basis for the conclusion that free energy may be stored in and transported through biopolymers. The Davydov solitons were later successfully simulated on a computer by A. Scott (see Chapter 5 of this book) and his coworkers in Los Alamos and Tucson. These developments represented a strong theoretical support for the validity of the concept of conformons that Green and I proposed in 1972. At present numerous workers, including A. Scott and R. K. Mishra, are viewing Davydov solitons and conformons as almost, if not completely, equivalent physical entities.

Another important development was the non-equilibrium thermodynamics theory of *dissipative structures* enunciated by I. Prigogine (see Chapter 2) and his school in Brussels and Austin, for which Prigogine was awarded the Nobel Prize in Chemistry in 1977. Encouraged by the unexpected theoretical support that the conformon concept received from soliton physics and imbued with a conviction that Prigogine's dissipative structures have a key role to play in biology, although I did not yet know how, I continued devoting all of my energy and waking hours to reading and thinking about theoretical problems in biology in preparation for my trip to Bhopal. During these months, I was greatly assisted by the almost daily luncheon meetings with Dr. Steven Finette (see Chapter 8) from the Department of Biomedical Engineering at Rutgers, now with the Naval Research Laboratory in Washington, D.C., who received his graduate training in biophysics from Syracuse University and had a keen interest in theoretical biology similar to mine. We used to spend hours discussing our favorite topics—the possibility that Prigogine's dissipative structures may provide the missing link between molecular biology and living phenomena on higher levels of biological organization.

The upshot of the long and intense preparation for the Bhopal meeting was my daring conclusion that it might be possible to formulate a theoretical model of the living cell, a task no one apparently had yet attempted up to that time, i.e., late 1983. When I left for the Bhopal meeting in early November, 1983, I had in my hand a tentative model of the living cell that consisted of a linear arrangement of about half a dozen arrows connecting various components of the cell, starting from DNA and ending not with proteins as is usually done in various biological publications but with the dissipative structures of Prigogine, also later called IDS (intracellular dissipative structures) or the Prigoginian form of genetic information (see Chapter 1). One of the novelties of this cell model was that dissipative structures were treated as *bona fide* cellular components for the first time on an equal footing with other familiar 3-dimensional intracellular structures such as enzymes, mitochondria, and the nucleus. In addition, dissipative structures were postulated to serve as the final form of the expression of genes.

One of the most important breakthroughs in the development of the cell model came early one morning in my hotel room in Bhopal, when I was going over my lecture material to be presented that day—I thought it might be better to connect Prigogine's dissipative structures in the cell with DNA (reminiscent of Kekulé's snake turning around and biting its own tail), thereby forming a circular arrangement of cell components rather than a linear arrangement as I had up to that moment. Having circularized the cell model, it was natural for me to *enclose* the whole arrangement within the cell membrane, and have the model act as a unit, receiving inputs from and producing outputs to its environment. In my mind's eye, this circular model seemed really *alive*—kicking, screaming, and self-replicating. This final insight leading to the circular model of the cell was undoubtedly one of the results of my participation in the Bhopal meeting; I decided to call the cell model the *Bhopalator* to express my gratitude to the organizers and supporters of the far-sighted and inspiring conference.

Upon my return from Bhopal, I informed Robert Snyder of the enthusiastic response accorded the Bhopalator model of the cell at the meeting. He suggested that I should discuss the cell model with Dr. Marion Anders of the University of Rochester who had long been interested in theoretical problems associated with mechanisms of cell death. During a meeting among three of us arranged by R. Snyder, it was decided that an international meeting should be organized to discuss theoretical models of the cell as related to molecular mechanisms of cell life and death. Marion Anders suggested that the conference title should include the word "life" as well as "death," Bob Snyder preferred the term "colloquium" over

Preface

"symposium," and I wanted to emphasize "molecular theories," thus the final title of the meeting, "International Colloquium on Molecular Theories of Cell Life and Death." The Organizing Committee consisted of M. Anders, R. Snyder, Robert L. Trelstad (Department of Pathology, Robert Wood Johnson Medical School, UMDNJ), S. Finette, and myself (chairman). With the hard work of the Organizing Committee, generous financial support from numerous organizations and offices, which are gratefully recognized in the Acknowledgments, and the unsparing support from R. Snyder and his staff, the meeting was successfully held on May 1 and 2, 1986 in Piscataway, with the scientific results recorded in the following pages.

ACKNOWLEDGMENTS

The major funding for the International Colloquium on Molecular Theories of Cell Life and Death from which this book originated were generously provided by the U. S. Environmental Protection Agency, and the U. S. Air Force Office of Scientific Research Life Sciences Directorate.

Partial funding were kindly provided by the Office of the President, the Office of the Dean of the College of Pharmacy, the Department of Pharmacology and Toxicology of the College of Pharmacy, and the Office of Research and Sponsored Programs, all of Rutgers-The State University of New Jersey, and from the Office of the Dean and the Department of Pathology, the Robert Wood Johnson (then called the Rutgers) Medical School, the University of Medicine and Dentistry of New Jersey (UMDNJ).

Thanks are due to the graduate students and postdocs of the Joint Graduate Program in Toxicology at Rutgers and UMDNJ R. W. Johnson Medical School who volunteered to help run the Colloquium and to Dr. C. Goldin and her staff from the Division of Continuing Education at Rutgers who so ably managed the day-to-day running of the Colloquium.

I owe special thanks to Drs. Bernard Goldstein, Robert Snyder, Marion Anders, Steven Finette, and Robert Trelstad for their advice and encouragement, to our department staff, Ms. Cathy Raymore, Bernadine Chmielowicz, Vivian Gallino and Judy Funari for their cheerful assistance, to my colleagues in the Department, Drs. Robert Guy, Mike Iba, Fred Kauffman, Debbie Laskin, and Carol Gardner, for their interest in my theoretical work, and to Dr. Alexander T. Pond, the Executive Vice President of the Rutgers University and Dr. John L. Colaizzi, Dean of the College of Pharmacy at Rutgers, for their financial support as well as for their role in creating the environment at Rutgers that has been conducive to carrying out abstract theoretical work such as mine.

Finally, I am grateful to my wife, Dottie, for the numerous hours spent editing the manuscript, despite her lower back pain, to Ms. Sue Konopack for her superb and careful word processing of the manuscript, and to Dr. Karen M. Reeds, the science editor of the Rutgers University Press, for her expert advice, patience, encouragement, and warm cooperation at all stages of the book preparation.

CONTRIBUTORS

George B. Corcoran, The Toxicology Program, College of Pharmacy, The University of New Mexico, Albuquerque, New Mexico 87131

Herbert de Groot, Institut für Physiologische Chemie I und Abteilung für Gastroenterologie, Universität Düsseldorf, Moorenstrasse 5, D-4000 Düsseldorf 1, Germany

Oliver H. J. Destrée, Hubrecht Laboratorium, KNAW, Utrecht, The Netherlands

Steven Finette[*], Department of Biomedical Engineering, College of Engineering, Rutgers University, Piscataway, New Jersey 08854

[*]Present address: Ocean Acoustics Branch, Naval Research Laboratory, Washington, DC 20375

Jacques R. Hiernaux, Service de Chimie-Physique II, Faculte de Sciences, Universite Libre de Bruxelles, Bruxelles, Belgium

W. Mike L. Holcombe, Department of Computer Science, University of Sheffield, England

Sungchul Ji, Department of Pharmacology and Toxicology, College of Pharmacy, Rutgers University, Piscataway, New Jersey 08855-0789

Dilip K. Kondepudi, Department of Chemistry, Wake Forest University, P.O. Box 7486, Winston-Salem, North Carolina 27109

Johanna G. Koster, Hubrecht Laboratorium, KNAW, Utrecht, The Netherlands

René Lefever, Service de Chimie Physique II, Faculte de Sciences Universite de Bruxelles, Bruxelles, Belgium

Patricia Meyers, Service de Chimie Physique II, Faculte de Sciences, Universite de Bruxelles, Bruxelles, Belgium

Robert A. Neal[*], Chemical Industry Institute of Toxicology, Research Triangle Park, North Carolina

[*]Present address: Department of Biochemistry, Vanderbilt University School of Medicine, Nashville, TN 37232-0146

Thomas Noll, Institut für Physiologische Chemie I, Uiversität Düsseldorf, Moorenstrasse 5, D-4000, Düsseldorf 1, Germany

Sten Orrenius, Department of Toxicology, Karolinska Institutet, P.O. Box 60400, S-104 01, Stockholm, Sweden

Ilya Prigogine, Department of Physical Chemistry, Free University of Brussels, Brussels, Belgium, and Department of Physics, University of Texas, Austin, TX 78712

Sidhartha D. Ray, The Toxicology Program, College of Pharmacy, The University of New Mexico, Albuquerque, New Mexico 87131

Jerome Rothstein, Department of Computer and Information Science, The Ohio State University, Columbus, Ohio 43210-1277

Alwyn C. Scott, Department of Mathematics, The University of Arizona, Tucson, Arizona 85721

Helmut Sies, Institut für Physiologische Chemie I, Universität Düsseldorf, Moorenstrasse 5, D-4000, Düsseldorf 1, Germany

Harry A. Smith, Department of Mathematics/Sciences, Tallahassee Community College, 444 Appleyard Drive, Tallahassee, Florida 32304-2895

G. Rickey Welch, Department of Biological Sciences, University of New Orleans, Lakefront, New Orleans, Louisiana 70148

Hans V. Westerhoff, Het Nederlands Kanker Instituut, Antoni van Leewenhoek Huis, Plesmanlaan 121, 1066 CX Amsterdam, The Netherlands

INTRODUCTION

Though philosophical and religious writings on the phenomena of life and death probably go back to thousands of years B.C., one of the first physical theories of life that had a wide influence in biology was not formulated until about fifty years ago, when Schrödinger wrote his famous little book entitled "What is Life? The Physical Aspect of the Living Cell." (Schrödinger, 1946). In Chapter 2 of these proceedings, Prigogine reemphasizes the significance of some of the key ideas contained in this book, despite the recent criticism on Schrödinger's work voiced by Perutz (1987).

Following in Schrödinger's footsteps, the present book, "Molecular Theories of Cell Life and Death," summarizes the results of both experimental and theoretical investigations by an interdisciplinary group of researchers that appear to shed new light on the phenomena of life and death of the cell, the fundamental unit of all living things. It takes account of both molecular and cellular experimental data and new theoretical concepts and frameworks that were not available in Schrödinger's time, such as far-from-equilibrium thermodynamics (Prigogine, 1961, 1977), cybernetics (Ashby, 1964), communication and information theories (Pierce, 1980), deterministic chaos (Tsonis and Tsonis, 1989), and the anthropic cosmological principle (Barrow and Tipler, 1986).

The transition of the cell from the living state to the death state is relatively sharp and can often be readily monitored using various experimental techniques such as enzyme leakage assays and light and electron microscopies. Just as in medicine where the study of the pathogenesis of various diseases has led to a deeper understanding of the structure and function of the human body in the normal state, so it is logical to assume that the rich experimental data on the intimate molecular details of cell death that are now available in toxicology would reveal the basic mechanisms of cell life.

I have summarized the major conclusions of each of the 15 chapters in Table I.1. The topics covered are highly heterogeneous, ranging from abstract theories to concrete experimental observations on the one hand, and from molecular mechanisms of cell death to the origins of life and of the universe on the other. To describe and analyze such a wide range of topics, the authors employ various specialized languages developed in the fields of evolutionary biology, toxicology, cellular and molecular biology, biochemistry, organic chemistry, physical chemistry, quantum physics, relativity theories, gauge field theories, statistical thermodynamics, irreversible thermodynamics, chemical kinetics, algebra of machines, logic, cybernetics, information theory, communication theory, and deterministic chaos. Such heterogeneity of discipline and discourse reflects the intrinsic complexities of living

Table I.1 The key conclusions of the contributors.

Author	Chapter	Conclusions
Ji	1	1. Conformons are the smallest, irreducible microscopic physical entities that can carry free energy and genetic information necessary and sufficient to drive all biomolecular processes in the living cell. 2. Conformons can be transduced into or generated from intracellular electrochemical gradients and waves called IDS (intracellular dissipative structures). 3. Conformons and IDS act as mediators of a novel force (called "cell force") that are responsible for all cellular processes. 4. The cybernetic theories of artificial machines can be generalized by reducing the size of machines to molecular dimensions and by introducing the concept of conformons as the ultimate driving force for all molecular machines. 5. A generalization of the concepts of conformons and IDS leads to the notion of gnergy, a physical entity resulting from a complementary fusion of energy and information. Gnergy symbolizes the information-energy duality which may subsume the wave-particle duality of quantum mechanics. 6. The application of the gnergy concept to cosmology suggests the possibility that this universe originated, not from the usual duality of energy and matter, but from the information-energy duality, which in turn leads to the notion of the "tetrahedrality" of energy, matter, information and life, viewing life as the packet of information and energy just as matter is regarded as the packet of energy.

(Table I.1 continued)

Prigogine	2	1.	Irreversibility is necessary for spatiotemporal organization in biology.
		2.	Chemistry encapsulates irreversible time into matter.
		3.	Schrödinger's aperiodic solids contain genetic information.
		4.	Nonequilibrium chemistry and physics provide mechanisms for long-range coherence.
Kondepudi	3	1.	The chiral asymmetry universally observed in biomolecules may arise from the parity violation in weak interactions amplified by nonequilibrium conditions on the evolutionary scales of space and time.
Holcombe	4	1.	Metabolic organization of a cell can be modelled using the mathematical language developed to describe artificial machines.
		2.	Cell death results when the index of successful intracellular communication at time t, $x(t)$, declines below a critical threshold.
Scott	5	1.	Biological macromolecules may store and transport free energy in the form of solitons.
		2.	The $C=O$ stretching vibration of the amide I bond of polypeptides may play a crucial role in the production of solitons in proteins.
Smith & Welch	6	1.	Cell metabolism, although composed of individual metabolic pathways, must be treated holistically as providing a unique microenvironment in which matter reveals its living qualities.
		2.	The holistic properties of cell metabolism can be described using the powerful language of general relativity theory (curved spacetime) and the local gauge field theories (local gauge symmetry).

(Table I.1 continued)

Rothstein	7	1.	Living systems are non-equilibrium and so complex in structure that the ordinary concept of states in thermodynamics and quantum mechanics are inadequate. The state concept developed in machine theories and computer science must be used in describing the "live" and "dead" states in biology.
		2.	Life is "meta-quantal" and "meta-molecular"; living phenomena cannot be completely described in terms of molecules and the laws of quantum mechanics. However, no special laws of nature need to be invoked to explain life.
		3.	The thermodynamic laws with the entropy concept suitably generalized to include information, organization, boundary condition, measurements, and computation provide the proper theoretical framework to account for life.
		4.	The notion of life and death cannot be discussed without taking into account the environment with which living systems interact; i.e., the phenomena of life and death are properties of the whole composed of system and environment and not those of either system alone or of its environment.
Finette	8	1.	Due to nonlinear interactions among neurons, emergent properties such as spatiotemporal ordering can be achieved by networks of neurons.
		2.	Emergent properties such as phase transition, hysteresis and limit cycles can be utilized to store information globally among cells rather than locally in a given cell.

Introduction

(Table I.1 continued)

Orrenius	9	1.	A sustained increase in cytosolic free Ca^{++} ion concentration induced by cytotoxic chemicals may provide a major mechanism of toxic cell killing.
		2.	Raised intracellular Ca^{++} ion concentrations may kill cells by activating Ca^{++}-dependent degradative enzymes.
Ray & Corcoran	10	1.	Many toxic chemicals injure cells by damaging cellular functions compartmentalized within the cell membrane, the endoplasmic reticulum, the cytoplasm, and the nucleus.
		2.	Many chemicals toxic to the liver cause acute liver cell injury by covalently modifying nuclear DNA.
Sies	11	1.	Experimental methods have been developed to measure electronically excited states of molecular oxygen and the carbonyl group in intact cells.
		2.	Electronically excited molecular species appear to play important roles in physiology as well as in pathophysiology.
de Groot & Noll	12	1.	Hypoxia (oxygen deficiency) can affect cell viability in two ways—(1) by diminishing ATP production, and (2) by decreasing the rate of electron abstraction from reactive radical intermediates of toxic chemicals.
		2.	Hypoxia increases misonidazole toxicity by prolonging the lifetime of the anion radical intermediate of misonidazole.
		3.	Hypoxia has dual and opposite effects on CCl_4 toxicity—(1) hypoxia increases the toxicity by disinhibiting the electron transfer from P-450 to CCl_4, and (2) hypoxia decreases CCl_4 toxicity by preventing the formation of lipid hydroperoxides, key toxic intermediates.

(Table I.1 continued)

Hiernaux, Meyers & Lefever	13	1.	Tumor growth can be approximated as a simple competition between proliferating malignant cells and immunocompetent killer cells.
		2.	Depending on the value of immune parameters characterizing effector-target binding and target lysis, various tumor states result, including tumor dormancy and immunostimulation.
Westerhoff, Koster & Destrée	14	1.	During early development, the regulation of histone gene expression consists of three parts—(1) translational regulation of mRNA stored in oocyte, (2) regulation of DNA replication, and (3) repression of histone gene expression at some pre-translational step.
Neal	15	1.	Toxicology is the science that defines limits of safety of chemical agents.
		2.	Toxicologists must be able, based on knowledge of mechanisms of action of chemicals, to make predictions on the potential hazards of chemicals to humans and environment.
		3.	The solution to the complex problems involving human health and environmental pollution may require the development of a new science of life that is constructed by integrating and extending the foundations of the existing scientific disciplines.

systems and processes that defy any description or analysis using a single, specialized language.

The recent debates concerning the role of intracellular Ca^{++} in chemically induced cell killing (Nicotera et al., 1990; D. Reed, 1990; J. Farber, 1990) highlight the urgent need for developing comprehensive theories of cell life and death that can provide a rational conceptual framework within which to evaluate the merits, or lack thereof, of various competing viewpoints. It is hoped that the rich collection of the theoretical ideas and principles presented in this book will contribute significantly to testing competing hypotheses and theories in cellular and molecular toxicology, including the intracellular Ca^{++} homeostasis hypothesis.

The epistemological underpinnings of the papers presented in this book may be grouped under two major headings—(1) <u>reductionism</u>, the belief that all living processes can be ultimately accounted for in terms of the known laws of physics and chemistry, and (2) <u>holism</u>, the doctrine that, although component structures and processes of living systems are ultimately chemical and physical in nature, the laws of physics and chemistry are insufficient to explain life.

It was N. Bohr (1933) who first pointed out that the phenomenon of life might not be completely accounted for in terms of the laws of physics and chemistry just as the structural stability of the hydrogen atom could not be explained in terms of classical mechanics, and that biologists might have to accept life as an irreducible fact of nature just as physicists accept quantum of action as a given. According to the principle of finite classes enunciated by Elsasser (1961, 1975, 1987), the laws of quantum mechanics cannot be applied to predict the behaviors of living objects because the mathematical derivations of these laws depend on the assumption that the classes of quantum objects are infinite in membership and homogeneous in properties, while the classes of living objects are finite in membership and heterogeneous in properties. These ideas led to the formulation of the so-called the "Bohr-Elsasser incompleteness theorem of physicochemical explanations of life", or simply the "Bohr-Elsasser incompleteness theorem" (Chapter 1).

To circumvent the Bohr-Elsasser incompleteness theorem, there appears to be two contrasting possibilities. (1) Fundamentally novel theories of nature, consistent with but fundamentally unrelated to the existing ones of physics and chemistry, must be developed in biology. The prediction of Elsasser (1961, 1975, 1987) that there may exist a set of what he calls "biotonic laws" supports this possibility. (2) Existing theories of physics and chemistry must be "extended", "generalized", or "modified" so as to accommodate living processes, in compliance with Bohr's correspondence principle. According to this principle, the generation of a new theory involves two steps--firstly the recognition and rejection of one of the presumptions of the old theory, and secondly its replacement by a new hypothesis and a new parameter which, when given a limiting value, enables the laws of the old theory to be derived as limiting cases of the laws of the new theory (Murdoch, 1987). Jerome Rothstein (1979) (also see Chapter 7 of this book) recently discussed how the concept of thermodynamic entropy can be generalized in such a way as to unite thermodynamics with the information theory of Shannon and Wiener and how such generalized entropy concept can provide a fundamental physical theory to account for biology, an approach clearly in line with the correspondence principle.

Another way to extend the fundamental theoretical framework of physics and chemistry so as to encompass the phenomenon of life may be to introduce the concept of conformons as a generalization of photons--photons carry free energy, while conformons, defined as conformational strains of biopolymers, carry both free energy and Shannon information (Chapter 1). The conformon postulate states that all macroscopic electrical machines are driven by photons (the mediators of electromagnetic interactions) while all molecular machines (enzymes, DNA and RNA) are driven by conformons. Given the concepts of conformons and molecular machines, all fundamental biological processes can be rationally accounted for, including the origin of life, enzymic catalysis, active transport, muscle contraction, oxidative phosphorylation, and gene expression (Chapter 1). Furthermore, the combination of the concept of conformons and that of dissipative structures of Prigogine (Chapter 2) leads to the formulation of a theoretical model of the cell.

In Chapter 6 Smith and Welch utilize general relativity and gauge field theories as analogies to develop a novel conceptual framework of cell biology. The deductions drawn from their "metabolic field" theory are basically in agreement with the results of Chapter 1.

In conclusion, it is hoped that the various novel concepts and approaches described herein will facilitate the development of the kind of science that R. Neal envisions in the final chapter of this book: " . . . only with the development of such a novel science of life, can we be equal to the difficult and challenging task of solving practical problems in toxicology and environmental health sciences in the coming decades and beyond."

REFERENCES

Ashby, W. R. (1964). An Introduction to Cybernetics. Metuhuen & Co., Ltd., London.
Barrow, J. D. and Tipler, F. J. 91986). The Anthropic Cosmological Principle. Oxford University Press, New York.
Bohr, N. (1933). "Light and Life." Nature 131: 421.
Elsasser, W. M. (1961). Quanta and the Concept of Organismic Law. J. theoret. Biol. 1: 27-58.
Elsasser. W. M. (1975). The Chief Abstractions of Biology. North-Holland Publishing Co., Amsterdam.
Elsasser, W. M. (1987). Reflections on a Theory of Organisms, Editions Orbis Publishing, Felighsburg, Quebec, Canada.
Farber, J. L. (1990). "The Role of Calcium in Lethal Cell Injury." Chem. Res. Toxicol. 3: 503-508.
Murdoch, D. (1987). Niels Bohr's Philosophy of Physics, Cambridge University Press, Cambridge, p. 243.
Nicotera, P., Bellomo, G., and Orrenius, S. (1990). "The Role of Ca^{2+} in Cell Killing." Chem. Res. Toxicol. 3: 484-494.
Perutz, M. F. (1987). Physics and the Riddle of Life. Nature 326: 555-558.
Pierce, J. R. (1980). Introduction to Information Theory. Second, Revised Edition. Dover Publications,

Inc., New York.
Prigogine, I. (1961). <u>Introduction to Thermodynamics of Irreversible Processes</u>. Second Edition. Interscience Publishers, New York.
Prigogine, I. (1977). Dissipative Structures and Biological Order. <u>Adv. Biol. Med. Phys.</u> <u>16</u>: 99-113.
Reed, D. J. (1990). "Review of the Current Status of Calcium and Thiols in Cell Injury." <u>Chem. Res. Toxicol.</u> <u>3</u>: 495-502.
Rothstein, J. (1979). "Generalized Entropy, Boundary Conditions, and Biology." In: <u>The Maximum Entropy Formalism</u> (R. D. Levine and M. Tribus, eds.), The MIT Press, Cambridge, pp. 423-468.
Schödinger, E. (1946). <u>What Is Life? The Physical Aspect of the Living Cell</u>. Cambridge University Press, Cambridge.
Tsonis, P. A. and Tsonis, A. A. (1989). Chaos: Principles and in Biology. <u>CABIOS</u> <u>5</u>(1): 27-32.

I. THEORIES

"The task of science is both to extend the range of our experience and to reduce it to order . . . . . "

Niels Bohr, 1929

Chapter 1

"BIOCYBERNETICS": A MACHINE THEORY OF BIOLOGY

Sungchul Ji

ABSTRACT

A comprehensive new theory of living processes has been formulated on the basis of the principles, concepts and analogies imported from physics, chemistry, and cybernetics (the study of machines). The most novel physical concept to emerge in the theory is that of "gnergy," a hybrid physical entity composed of free energy and Shannon information that is postulated to be ultimately responsible for driving all molecular machines. Discrete physical entities carrying gnergy are called "gnergons," and there are two examples of gnergons identified in biology—sequence-specific conformational strains of biopolymers called *conformons* and intracellular chemical gradients and waves called *IDS* (intracellular dissipative structures). Conformons and IDS are utilized to formulate what appears to be the first coherent theoretical model of the living cell called the Bhopalator, the suffix "-ator" indicating a self-organizing chemical reaction-diffusion system. In analogy to the strong force that is essential to explain the *structural stability* of atomic nuclei, it has been postulated that there exists a new force in the cell termed the *cell force* that maintains the *functional stability* of biopolymers in intracellular chemical milieu. Just as the strong force is mediated by gluons, it is suggested that the cell force is mediated by "cytons" composed of conformons and IDS. The biocybernetics theory has led to formulate theoretical models of the origin of life, the human body, and the human society; and the concept of gnergy as applied to the Big Bang has led to the formulation of a "biological" model of the universe. The utility of the various theoretical models in biology was then tested by applying them to solve practical problems, including mechanisms of oxidative phosphorylation, mechanisms of cell life and death, biological functions of cytochrome P-450 isozyme diversity, mechanisms of the P-450 isozyme induction by xenobiotics, the inflammatory response, and development of diseases including cancer, molecular mechanisms of aging, the significance of the brain size/body size ratio in human evolution, and computer-brain relationship.

1.1. INTRODUCTION

As G. G. Simpson (1964) pointed out, biology is the study of "phenomena to which all principles apply, "whereas physics is the search for "principles that apply to all phenomena." The proper relationship between biology and physics has long been debated among scholars (Bohr, 1933; Schrödinger, 1946; Perutz, 1987; Mayr, 1988; Elsasser, 1958, 1961, 1975, 1987). In the absence of any evidence to think otherwise, most physicists have tended to assume that both the scientific methods and theories developed in physics and chemistry would be necessary and sufficient to solve all fundamental problems in biology, one of the notable exceptions being N. Bohr (1933) (see also J. Rothstein, Chapter 7 of this book). As will become evident, the results to be presented in this chapter suggest quite an unexpected image of biology which seems alien to traditional physics but fits what Simpson might have envisioned in 1964.

Biology is like impressionistic painting, while physics is akin to photography; interestingly both methods of viewing nature developed within 20-30 years of each other in the middle of the nineteenth century in Europe (Pool, 1967). Impressionists endeavored to capture dynamic and fleeting impressions at the sacrifice of realistic details of objects in nature (Needham, 1988); in contrast, the photographer's camera can record objects in great detail, even beyond the capability of the human eye. This comparison between biology and physics conjures up other images: Biology vs. physics = dynamic and fleeting vs. static and permanent = deterministically chaotic vs. deterministic = living vs. non-living = qualitative vs. quantitative = non-mathematical vs. mathematical = horizonal (i.e., interdisciplinarity) vs. vertical (i.e., specialization) = direct observation of nature (e.g., behavioral biology) vs. indirect observation through measuring instruments = phenomena driving research vs. research methods (e.g., mathematical techniques) selecting phenomena to be investigated, etc. Because of these contrasting differences between biology and

"Biocybernetics": A Machine Theory of Biology

physics, it may be naive to expect that biological problems will be solved in the same way that physics or chemistry problems have been solved in the past.

The reason for this brief discourse on the interrelationship between biology and physics is by way of preparing the readers' mind for the somewhat unusual content of this chapter—a new theory of biology that has been constructed over the past two decades based on the principles, concepts and analogies drawn from practically every major scientific disciplines in physics, chemistry, and biology and covers an unusually wide range of topics, from enzymology to the living cell to the human society and the universe. In short, biocybernetics, the cybernetics of living systems, cannot be compared with any other physical theories presently known. I hope that this new biological theory will be accorded the same kind of attention befitting an impressionistic painting (with all the characteristic blurrings and imperfections and yet hopefully "beautiful" nevertheless) rather than the kind of scrutiny that might be given bubble chamber photographs of high-energy events by nuclear physicists.

The eminent evolutionary theorist Ernst Mayr (1988) divides biology into the "functional" biology and "evolutionary" biology and points out the mutual exclusivity between them. Functional biologists ask questions about "how" living processes occur or "how" certain living structures function. They attempt to answer these questions by studying the component parts of an organism in isolation, using basically the same experimental approach as used by chemists and physicists. Evolutionary biologists differ in both their method of investigation and the problems that they tackle. Their basic question is "why" living processes occur the way they do and not in some other ways. They are interested in knowing the causes for the existing characteristics and adaptive capabilities of organisms resulting from the biological evolution that has been in progress for more than 3 billion years. The method of research employed by evolutionary biologists is unique in that it, in general, does not need sophisticated measuring instruments, save the human brain.

This "how-why" or "function-evolution" duality in biology is reminiscent of the "wave-particle" duality of light. According to N. Bohr (1958), wave and particle characteristics of light are "complementary" in that which characteristics of light are actually observed in the laboratory depends on the nature of the experimental arrangement employed. To quote Bohr:

> Within the scope of classical physics, all characteristic properties of a given object can in principle be ascertained by a single experimental arrangement, although in practice various arrangements are often convenient for the study of different aspects of the phenomena. In fact, data obtained in such a way

simply *supplement* each other and can be combined into a consistent picture of the behavior of the object under investigation. In quantum physics, however, evidence about atomic objects obtained by different experimental arrangements exhibits a *novel* kind of complementary relationship. Indeed, it must be recognized that such evidence which appears *contradictory* when combination into a *single picture* is attempted, exhausts all conceivable knowledge about the object. Far from restricting our efforts to put questions to nature in the form of experiments, the notion of *complementarity* simply characterizes the answers we can receive by such inquiry, whenever the interaction between the measuring instruments and the objects forms an integral part of the phenomena.

Although Bohr formulated the complementarity idea within the context of the conflict between classical physics and quantum physics, the complementarity principle may have a generic character in that it can be applied to numerous other situations in nature, including those in biology, where two or more seemingly "irreconcilable" observations or viewpoints may arise. The "function-evolution" duality first clearly commented on by Mayr may be an example of such situations. As is well known, Bohr (1933) himself applied his complementarity principle to biology and suggested that:

the very existence of life must in biology be considered as an elementary fact, just as in atomic physics the existence of the quantum of action has to be taken as a basic fact that cannot be derived from ordinary mechanical physics. Indeed, the essential non-analyzability of atomic stability in mechanical terms presents a close analogy to the impossibility of a physical or chemical explanation of the peculiar functions characteristic of life.

As will become evident, the results presented in this chapter are completely consonant with Bohr's profound conclusion concerning the complementary relationship between biology and its sister sciences, physics and chemistry. But the present work goes beyond Bohr's general considerations; it specifies the nature of what may turn out to be the ultimate complementary conjugates—namely, *free energy* and *genetic information*.

The major aim of this chapter is to outline a new theory of living processes that is based on the fundamental postulate that living processes are driven ultimately by "gnergy," a physical entity characterized by a *complementary* union of free energy (represented by the Greek suffix "-ergy") and genetic information (indicated by the

"Biocybernetics": A Machine Theory of Biology

Greek prefix "gn-") (S. Ji, 1985a). The validity of the concept of "gnergy" rests on the logical consistency of the general theoretical framework of biology that results from its application. For lack of a better word, I elected to call this theory "biocybernetics," i.e., the machine theory of living systems. When N. Wiener (1948) defined "cybernetics" as the scientific study of control and communication in the animal and the machine, not much was apparently known about the fundamental differences between the mode of operation of animals and that of machines. But the experimental as well as theoretical advances that have been made during the recent decades have revealed not only some striking similarities but also fundamental differences between artificial machines and biological ones. For this reason, it was thought necessary to coin the new word, "biocybernetics," to represent the study of biological machines, which are best carried out separately from the study of artificial machines. Biocybernetics is, then, the theory of the phenomenon of life that is constructed upon the foundation of the complementarity principle of N. Bohr (1933) as reflected in the dualities of (1) function vs. evolution, (2) the deterministic (i.e., machine-like) vs. "deterministically chaotic" (i.e., life-like) behaviors of biological machines, and (3) Gibbs free energy vs. genetic information.

1.2. A BIRD'S-EYE VIEW OF BIOCYBERNETICS

There will be numerous new concepts and terms to be introduced as various aspects of biocybernetics are discussed in the following pages. To help guide the reader through the labyrinth of biocybernetics, I have prepared Table 1.1 as a mental map, wherein the key building blocks of the new theory are collected and compared with their counterparts in chemistry and physics. The main point of Table 1.1 is to bring out the interesting analogies that seem to exist among the three sets of the building blocks constituting chemistry, physics, and biology. New terms and concepts are explained in the extensive footnotes attached to the table. To recognize analogies (or similarities) among the building blocks, it is necessary to read across the table, one row at a time, keeping in mind the following common characteristics of each row:

Row 1 = The rule governing the organization (in space and time) of the material objects under study.
Row 2 = The material objects under study.
Row 3 = The fundamental constituents of the material objects under study.
Row 4 = The physical agents (i.e., field quanta or gauge bosons) that mediate fundamental forces.

Row 5 = The fundamental physical theories that are necessary to account for the static and dynamic properties of the material objects under study.

Row 6 = The fundamental forces that are responsible for the interactions within or in between the material objects under study.

The analogies are clearest between the first and the second column. The only unambiguous analogy that can be claimed in the third column is associated with the second row (i.e., organisms). Whether the analogies revealed in the first two columns can be extended to the remainder of the elements of the third column (i.e., Rows 1, 3, 4, 5 and 6) is an important question that cannot be answered unambiguously at present. However, it is interesting to note that the cited elements of the third column, namely the Bhopalator (Ji, 1985a,c), conformons (Green and Ji, 1972; Ji, 1974a, 1979, 1985b; Ji and Finette, 1985), IDS (Ji, 1985a), biocybernetics (this chapter) and the cell force (Section 8.9.), are all recent developments in theoretical biology that have taken place more or less independently of one another during the past two decades. It was only during the past year or so, while writing this chapter, that I began to recognize the possible interrelationships among these elements. It is almost certain that such interrelationships would have been impossible to recognize without utilizing the method of "analogical" thinking evident in Table 1.1 and other tables to follow. This realization prompted me to define rigorously the method of analogical thinking based on tables and in the form of what I call the "table theory" (see Section 1.3).

Table 1.1. A comparison of "biocybernetics" with quantum electrodynamics in chemistry and quantum chromodynamics in physics

	Chemistry	Physics	Biology
1.	Mendeleev periodic table	Gell-Mann quark model[1]	Bhopalator cell model[2]
2.	Chemical compounds	Subatomic particles	Organisms

(Table 1.1 continued)

3.	Elements (ca 10^2)	Quarks (6)	Molecules and ions (10^6 ?)
4.	Photons	Gluons[3]	Conformons[4] and IDS[5]
5.	QED[6] Statistical mechanics[7]	QCD[8] Electroweak theory[9]	Biocybernetics[10]
6.	Electromagnetic force[11]	Strong force[12] Electroweak force[13]	"Cell force"[14]

[1] The theory that hadrons are composed of quarks, there being 6 of them—u (up), d (down), c (charmed), s (strange), t (top) and b (bottom). The quark composition of the proton is uud and that of the neutron is udd.

[2] The comprehensive theoretical model of the living cell proposed in 1985 (Ji, 1985a). The key concepts of the cell model are "conformons" and IDS.

[3] Gluons are the physical agents that "glue" together quarks inside atomic nuclei.

[4] Conformons can be defined as functional domains of biopolymers (enzymes, DNA, RNA) arranged in nonequilibrium conformations or as sequence-specific conformational strains of biopolymers.

[5] IDS is the abbreviation for "intracellular dissipative structures" defined as nonequilibrium distributions of diffusible chemicals inside the cell, such as intracellular calcium waves and transmembrane ion gradients.

[6] QED stands for quantum electrodynamics—i.e., the quantum theory of electromagnetism.

[7] Statistical mechanics is the molecular theory of thermodynamics; i.e., the study of molecular motions and chemical reactions using the principles of mechanics and probability theories.

[8] QCD stands for quantum chromodynamics—i.e., the quantum theory of the strong force based on the assumption that quarks come in 3 "colors"—red, green or blue—and that only those combinations of quarks having the white color can be experimentally observed.

[9] Electroweak theory is the quantum theory that combines the electromagnetic force and weak force as different aspects of a common force, just as the electromagnetic force is the union of both the electric force and the magnetic force.

[10] Biocybernetics can be viewed as a molecular and cellular theory of information and free energy transductions in living systems.

1.3. THE "TABLE THEORY": A NEW METHOD IN THEORETICAL BIOLOGY

On numerous occasions, I have been struck by the absence of any high-powered mathematical equations in my publications in theoretical biology. I was aware that the fundamental breakthroughs in physics and chemistry have always been associated with elegant mathematical formalisms—e.g., the Maxwell equations of electromagnetism, the tensor algebra of Einstein's general relativity theory, the matrix algebra of Heisenberg's uncertainty principle, the Schrödinger wave equations for electrons and photons, and the Yang-Mills gauge theories of nuclear forces. My uneasiness was reinforced by David Hilbert's statement to the effect that, whenever scientific fields mature, they always come to mathematics for their final solutions.

I used to think that the absence of any sophisticated mathematical equations in my work was due to (1) my lack of mathematical skills, (2) the immaturity of the field of my research interest, or both. While these reasons may still be valid, I have recently begun to take the following alternative possibility seriously—that, at least for cell biology, the kind of mathematics that have been widely used in physics and chemistry is too simple to effectively handle the complex problems at hand. The natural question that now arises is, "Can fundamental problems in nature be solved without the help of mathematics?" The answer to this question, of course, will depend on the nature of the problems under consideration and the kind of solutions one envisions. However, this much is clear: Even without sophisticated mathematics, it is often possible to solve major problems in physics and chemistry. The best example is probably the discovery of the periodic table by Dmitri Mendeleev in 1896,

---

(Notes to Table 1.1 continued)

[11] Electromagnetic force is the force acting between all electrically and magnetically charged particles.

[12] Strong force is responsible for holding subnuclear particles together.

[13] Electroweak force that is responsible for radioactive decay of nuclei and for interactions among charged particles.

[14] "Cell force" is a hypothetical force thought to act within the living cell to hold together biopolymers in functional states. The cell force is postulated to be mediated by a combination of conformons and IDS called "cytons" (formed from "cyt-" meaning the cell and "-ons" meaning discrete entities), just as the strong force is mediated by gluons. The cell force should not be confused with the non-material, metaphysical "forces" that were invoked by vitalists in the 18th-19th centuries (E. Mayr, 1988, pp. 10-13). The cell force is distinctly physical. See Section 1.8.9 for more details.

"Biocybernetics": A Machine Theory of Biology

a full half a century before the advent of quantum mechanics (Kaku and Trainer, 1987). The discovery of the periodic table was one of the most important breakthroughs in the history of chemistry, because it introduced order into the chaos of the millions of chemical compounds known to nineteenth-century chemistry. Another example is the quark model of the strong force formulated by M. Gell-Mann in the 1960's, which led to a coherent systematization of hundreds of hadrons in terms of a set of just a few elementary constituents, long before the full mathematical theory of the strong force was formulated (Davies, 1986).

Emboldened by the non-mathematical reasoning that led to the discovery of the Mendeleev periodic table, I have decided to formalize what I have been practicing in my theoretical biology research for almost two decades, namely the use of *tables* to compare two or more sets of analogous objects, concepts, or theories, in order to discover new patterns hidden behind them. In analogy to the group theory in elementary particle physics, I elected to call the present method the "table" theory. The key components of the table theory are described below.

(1) The table theory is a logical system designed to reveal the similarities and analogies existing between two sets of objects, designated F (familiar) and U (unfamiliar), by arranging the elements of F and U in a table called the "analogy table."

(2) The relationship among the elements of F (called the "internal structure" of F) is better known than the relationship among the elements of U. The trend or property that is revealed by displaying the elements of F in a linear arrangement will be referred to as the "character" of that particular linear arrangement called an F vector. The internal structure of F may suggest a property, a concept, or a physical principle.

(3) A "table symmetry" (in analogy to gauge symmetry in particles physics (Adair, 1987)) is said to exist between F and U, if and only if the internal structure of F provides clear constraints for the linear arrangement of the elements of U and the F-directed U vector suggests a new relationship among the elements of U, leading to the recognition of a new property, concept, or principle.

(4) In the absence of any clear "table symmetries" apparent in a given analogy table, we may speak of the less stringent "table correlations," which are defined simply as any non-random pairings between the elements of an analogy table. Clearly there are two kinds of correlations due to the 2-dimensionality of a table—the "vertical" correlations found among the elements within a column, and the "horizontal" correlations found between the elements within a row of an analogy table.

(5) The reasons underlying table correlations can be divided into at least four

categories—(i) causality (i.e., the familiar cause-effect relationship of classical physics and chemistry), (ii) complementarity (resulting from the interactions between the measuring instruments and the phenomenon under observation, as in the particle-wave duality of light (Bohr, 1958)), (iii) deterministic chaos (see Section 1.8.11), and (iv) "historicity" (i.e., the sharing of a common historical background or experience, as in gene-directed phenomena in living systems like the correlated growth of two branches located on the opposite sides of a structurally symmetric tree).

It is important to remember, as Bohr repeatedly emphasized (1937, 1958), that the principle of causality has a limited applicability in physical sciences as exemplified by its exclusion in subatomic phenomena dominated by Planck's quantum of action.

(6) The maximum amount of the information (I) generated by a given analogy table consisting of a pair of F and U vectors may be quantified using a simplified version of Shannon's formula (Pierce, 1980; also discussed in Section 1.8.11);

$$I = \log_2 (W_0/W) \tag{1.1}$$

$$= \log_2 (N!) \tag{1.2}$$

where W_0 is the number of all possible U vectors, and W is the number of U vectors that are consistent with the internal structure of F, and N is the number of the elements of U. Clearly, I is maximal when $W = 1$, i.e., when there is only one U vector whose internal structure is compatible with that of F.

(7) The higher the value of I, the greater the confidence that can be placed on the validity of the internal structure of U. Therefore, I can be treated as a quantitative measure of the confidence that can be placed on the internal structure of U suggested in analogy to F.

(8) If there are M sets of objects to compare (where M is greater than 2), the objects can be displayed in a table with M columns and N rows. The maximum information generated by the M x N table can be calculated by using the following formula:

$$I = \sum_{1}^{M} \log_2 (N!/W_i) \tag{1.3}$$

where W_i is the number of all possible vectors of the ith unfamiliar set U_i that are

"Biocybernetics": A Machine Theory of Biology

compatible with F.

For example, the application of Equation (1.3) to Table 1.1 leads to

$$\begin{aligned} I &= \log_2 (6!) + \log_2 (6!) \\ &= 2 \log_2 (720) \\ &= \log_2 (720)^2 \\ &= \log_2 (5.184 \times 10^5) \\ &= 19 \text{ bits} \end{aligned} \quad (1.4)$$

In other words, the analogical comparison of the cell force to the electromagnetic force and the strong force "generated" 19 bits of information; or the probability that the concept of the cell force can arise randomly (i.e., U and F have a zero correlation) is less than one in a half million.

A highly relevant example to which the table theory summarized above can be applied is provided by the light-life analogy that Bohr enunciated in 1933. I have elaborated on Bohr's light/life analogy in the context of biocybernetics in Table 1.2.

As is well known, Bohr's light-life analogy originated from extending the unanalyzability of quantum phenomena in terms of Newtonian mechanics to the "suspected" un-analyzability of life in terms of chemistry and physics (see the top row and the second and third columns in Table 1.2) (Bohr, 1933). He implicated the particle-wave duality of light (second column in Table 1.2) as the basis for suggesting the analogy, but the logical steps which led him from the particle-wave duality to the light-life analogy were nowhere made clear. This logical deficiency in Bohr's argument may be remedied by supplementing Bohr's light-life analogy with Ernst Mayr's duality of evolutionary biology and functional biology, as shown in the third column of Table 1.2. I detect a reasonably convincing "table symmetry" between the first column (the gnergy vector) and the third column (the life vector), since genetic information (I) is the product of biological *evolution*, and the "how" questions of *functional* biology are primarily related to free energy (E) considerations, as are most of the problems in physics and chemistry (Mayr, 1988). The "light vector," in turn, shows some "table symmetry" with the gnergy vector, in that information (I) can be transmitted through space by modulating the frequency, amplitude or phase angle of *waves*, while energy (E) and matter (i.e., *particles*) are related through Einstein's equation, $E = mc^2$. Since the life vector and the light vector are separately "table symmetric" with the gnergy vector, perhaps the life and light vectors are also "table symmetric" to each other. A careful examination of the elements of the two vectors (including those elaborated within parentheses) indicates that this indeed may be true—e.g., *evolution* is impossible without the passage of geological *time*, and biological *functions* are impossible without dissipation of free *energy*. We may represent these "table symmetries" algebraically as follows:

$$V \text{ (life)} = V \text{ (gnergy)} \tag{1.5}$$

$$V \text{ (light)} = V \text{ (gnergy)} \tag{1.6}$$

Therefore, $\quad V \text{ (life)} = V \text{ (light)} \tag{1.7}$

where V signifies a vector and the equal sign signifies a table symmetry.

Table 1.2. The light and life analogy of N. Bohr

Gnergy	Light	Life
Information	Wave (or Field)	Evolution
(I)	(Position, Time)	(Adaptation, Behavior, Genetic information)
Energy	Particle (or Matter)	Function
(E)	(Momentum, Energy)	(Chemistry, Physics, Mechanisms, Free energy)

"Biocybernetics": A Machine Theory of Biology

The logical conclusion that Equation (1.7) may hold true is somewhat unexpected (Bohr's intuition notwithstanding) and suggests that there may be some deeper connection underlying these three vectors. One logical possibility is that the ultimate source of the correlations seen in Table 1.2 is "historicity," i.e., *the common descent of light and life from gnergy*, a conclusion consistent with the biological model of the universe called the Shillongator (see Section 1.11 and Appendix 1.F). The nice symmetries exhibited by the horizontal vectors, I and E, also support the notion that *genetic information* (I) and *free energy* (E) are *complementary* to each other as are particles and waves in quantum mechanics and not *supplementary* (Bohr 1958). These are very significant results, which probably would not have been readily recognized without the logical analysis afforded by the method of the table theory.

1.4. CYBERNETICS: A THEORY OF MACHINES

Since control, communication and computation cannot be carried out, whether in the physical world or in the abstract mathematical world, without some organized structures generally known as "machines," systems, or "automata," it follows that cybernetics can be viewed as a theory of machines. A common dictionary definition of a *machine* is "a structure consisting of a framework and various fixed and moving parts, for doing some kind of work" (Webster's New World Dictionary, College Edition, 1966). In the cybernetics literature, the term machines is used synonymously with *dynamical systems*, *general systems*, or *automata* (see Glossary). The examples of real-world machines are well known, e.g., coin-operated vending machines, automobiles, computers, etc. As an example of mathematical machines, the Turing machine may be cited which has been developed as a model for a universal computing machine (George, 1977; Hopcroft, 1984). A number of biologists have viewed enzymes as "molecular machines" (McClare, 1971; Ji, 1974a,b; Shaitan and Rubin, 1982; Lumry and Gregory, 1986; Welch and Kell, 1986).

1.4.1. The Four Causes of Machines

The concept of the machine can be applied to a wide variety of objects with dynamic internal structures or organizations, regardless of their size or material contents. In order to lay down a more rigorous definition of the machine that is compatible with all of the above attributes, I find it useful to employ the "four causes" formalism of Aristotelian epistemology. According to Aristotle (384-322 B.C.), all objects in nature possess four causes—the "material" cause ("What is it made out of?"), the "efficient" cause ("How does it work?"), the "formal" cause ("What is it?;" or "What is it that it represents?"), and the "final" cause ("What is it for?") (Lear, 1988).

A. <u>The Material Cause of Machines</u>: There are two fundamentally different kinds of machines—the first kind is constructed out of real material components as in coin-operated vending machines and computers, and the second kind consists of logical structures such as cellular automata (Codd, 1968; Farmer et al., 1983) and the Turing machine (George, 1977; Hopcroft and Ullman, 1979; Hopcroft, 1984). For convenience, we may refer to the former as the "concrete" or "material" machines and the latter the "abstract" or "logical" machines. Material machines, in turn, can be divided into "artificial" or "physical" machines and "evolution-made" or "biological" machines.

B. <u>The Efficient Cause of Machines</u>: The efficient cause of all abstract machines must be the human brain in the sense that the human brain produced them and implements and executes the logical steps embodied in abstract machines.

The efficient causes of the material machines of artificial origin, namely "human-made" machines, are either the human brain, if their production is to be explained, or the motions of the component parts dictated by the laws of nature as far as their operations are concerned. The efficient cause for biological machines is obviously the biological evolution from the point of view of their origin, or the molecular mechanisms of the various physical and chemical processes that are responsible for their operation.

The operation of all material machines are examples of irreversible processes and are accompanied by free energy dissipation and entropy production. The operation of machines must be coupled to some irreversible physical (e.g., electrical current flow, wind, expansion of steam) or chemical processes (e.g., combustion of gasoline in engine cylinders, ATP hydrolysis by molecular machines). To operate material machines without free energy dissipation is tantamount to violating the Second Law of Thermodynamics, which in effect states that all irreversible processes must be accompanied by entropy production (Prigogine, 1961).

C. <u>The Formal Cause of Machines</u>: All machines are constructed (or have evolved) to perform specific functions called for by appropriate input signals. In other words, all machines are constructed to receive some input signals and produce outputs that are evoked by the input signals. The identification of the input and output signals of a machine is an essential part of characterizing a machine. Therefore, the input and output signals of machines and their interactions within the machine may be identified as their formal causes.

D. <u>The Final Cause of Machines</u>: It is axiomatic that all machines perform

"Biocybernetics": A Machine Theory of Biology

certain functions to achieve specific purposes or goals. The final goals of a machine are in general not uniquely determined by the internal mechanisms of machines, since the same set of internal mechanisms of a machine can be coupled to other machines to accomplish quite different end results. For example, an electric motor may be utilized to operate a guillotine or an elevator; the mechanisms of cellular division may be essential for the normal development of an embryo or may be detrimental to an organism, if they occur in tumor tissues. In other words, the purpose of a machine may be determined only within the context of the environment with which the machine interacts.

1.4.2. Classification of Machines

Machines can be divided into simple machines that cannot be further reduced without destroying their machine characteristics and complex machines that are constructed by coupling two or more simple machines according to a certain set of rules.

All simple material machines are composed of moving parts whose motions in space and time are constrained so as to enable the machine to carry out specific functions (e.g., linear motions of pistons within engine cylinders that are constrained in such a way as to rotate the crankshaft of a car; vectorial movements and conformational changes of the catalytic groupings of intramembrane enzymes to translocate ions across biological membranes). Clearly, there exists a close correlation between the motions of internal parts of a machine and its functions or outputs. The moment-to-moment description of the positions of the internal parts is called the "internal states" of the machine (George, 1977). It is known that for a large class of machines (finite-state machines) there is a 1:1 correlation between their internal states and their outputs.

Mathematicians succinctly represent these various characteristics of a machine by using an algebraic expression called a "5-tuple," i.e., a vector with 5 components (Ehrig, 1974);

$$M = (I, S, O, d, l) \qquad (1.8)$$

where M = a machine, I = a set of inputs, S = a set of internal states, O = a set of outputs, d = a mathematical function relating the input of a signal and the change in the internal state, and l = a mathematical function relating the internal state to an output. These notations are schematically illustrated in Figure 1.1.

Table 1.3. Types of machines or automata (adopted from Ehrig, 1974).

I, S, O	d, l	Machine type
Sets	Deterministic[1]	Deterministic
Sets	Nondeterministic[2]	Nondeterministic
Sets	Stochastic[3]	Stochastic
Sets	Deterministically chaotic[4]	Deterministically chaotic[4]

[1] A one-to-one correspondence exists between the input signal and the internal state transition, or between the internal state and the output signal.

[2] A one-to-many correspondence exists; i.e., one input signal may lead to two or more internal state transitions, or one internal state may produce more than one outputs.

[3] The correlations between the input signal and the internal state transitions or between the internal state and the outputs are not unique but probabilistic.

[4] Moment-to-moment behavior is unpredictable but the global pattern observed over a period of time can be related to some deterministic rules (see Section 1.8.11.)

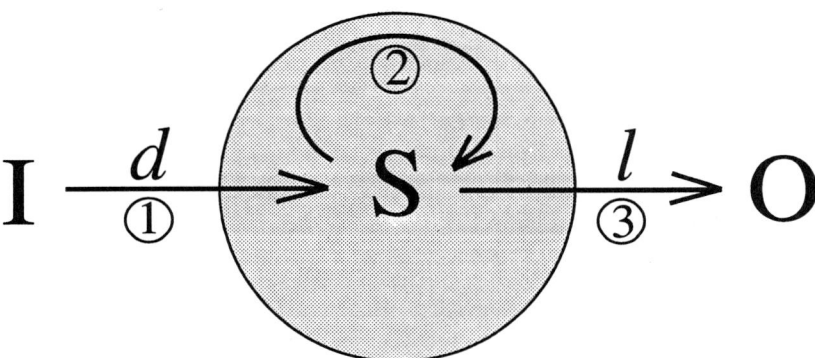

Figure 1.1. A schematic representation of a machine. 1 = input step; 2 = internal state transition; 3 = output step. I = the input set; S = the set of internal states of the machine; O = the output set; d = the function relating I to S; and l = the function relating S to O.

Depending on the nature of the elements of the 5-tuple, different classes of machines result as shown in Table 1.3. In addition, the sets I, S, O can be either finite or infinite. For our initial discussions, the simplest kind of machine called the deterministic "finite-state machine" will suffice, such a machine being characterized by the deterministic functions, d and l, and the finite sets, I, S and O (see Chapter 4).

1.4.3. Artificial vs. Biological Machines

As already pointed out, we can clearly distinguish two kinds of material machines—artificial or physical machines and evolution-made or biological machines. Some of the salient features of these two kinds of machines are listed in Table 1.4.

1.4.4. The Structural Hierarchy of Biological Machines

We can divide biological machines into at least seven groups as shown in Table 1.5. Higher level machines are constructed by coupling a set of lower-level machines, and the "purpose" of such a complexification of structures, we argue, is to "create" new properties. One of the desired properties that emerges with the structural complexification may be the "homeostasis" of living systems. In Section 1.5.3., we will discuss the theoretical connection between the increased structural complexity and homeostasis in biology in terms of the Law of Requisite Variety, a fundamental principle originating from cybernetics.

In Table 1.5, metabolic pathways are treated as complex machines. W. Holcombe (1982) appears to be one of the first (if not the first) to suggest this idea.

1.4.5. The Written Language as a Metaphor of Biological Complexity: The Principle of Emergent Properties

When faced with the immense complexity of biological organizations inherent in living stems ranging from enzymes to organs to whole organisms, we are forced to ask two questions: (1) "How did such complex structures arise?" and (2) "What is the meaning of such complex structures?" or "Why are they so complex?" A broad answer to the first question has been provided by the theory of biological evolution (Mayr, 1988; Gould, 1982). The answer to the second question, however, has not yet been fully explored, in my opinion.

Table 1.4. A comparison between artificial and biological machines

	Parameter	Artificial Machine	Biological Machine
1.	Constructor	Human	Biological evolution[1]
2.	Size	Macroscopic[2]	Microscopic[3]
3.	Thermal flexibility[4]	No	Yes
4.	Effects of temperature on machine efficiency[5]	Weak	Strong
5.	Behavior	Mostly deterministic	Often deterministically chaotic[6]

[1] All biological systems, from enzymes to man, are produced as the result of billions of years of biological evolution (Mayr, 1988).

[2] Visible to the naked eye; usually composed of large and rigid aggregates of atoms and molecules.

[3] Composed of individual deformable molecules.

[4] Artificial machines are designed to be so large and rigid as to withstand thermal motions of molecules.

[5] The performance of artificial machines are rather insensitive to lowering temperature, whereas biological machines critically depend on thermal motions for their function (Ji, 1974b; Welch, 1986).

[6] Due to the non-linear dependence on thermal fluctuations for functions, the behavior of individual biological machines are highly unpredictable and often reveal the phenomena of deterministic chaos (Markus and Hess, 1984). Hence biological machines can be appropriately called "deterministically chaotic" machines or "chaomachines."

Table 1.5. The structural hierarchy of biological machines

Level of Organization	Machine	Examples
I	Biomolecules	Receptors, enzymes, DNA, RNA
II	Metabolic	Krebs cycle, glycolytic pathway
III	Organelles	Mitochondria, nucleus, ER
IV	Cells	Neurons, muscle cells
V	Tissues	Hypothalamus, cerebral cortex
VI	Organs	Brain, heart, liver, kidneys
VII	Organisms	Man, mice, beetles, roses

I find it helpful to use the written language, say English, as a metaphor for understanding the connection between the complexity of an organized structure, whether material or semantic, and its function. Just as complex symbolic structures written in English are constructed by the process of concatenation (i.e., linking parts together into a chain) of letters into words, words into phrases, phrases into sentences, and sentences into texts, biological structures are constructed by coupling simpler structures to form more complex structures in space and time. As the complexity of a symbolic structure increases, new meanings appear; e.g., the meaning "beauty" belongs to the word itself and not to individual letters, b, e, a, u, t or y. The beauty of a great novel belongs to the whole book and cannot be reduced to the kinds and the frequency distribution of the individual words used by the author. Likewise, as biological structures increase in complexity, new functions can emerge; i.e., when parts are organized according to a certain set of rules new meanings or properties can emerge, which are totally absent in lower level organizations. Thus, structural complexity is a necessary condition for new properties, or "creativity."

The phenomena of new properties emerging from structurally "complexified" systems is not limited to the human language but appears to be universal, since such phenomena occur in artificial machines and in nature. Examples of the latter include (1) solitons—packets of waves that can transport energies over long distances without dissipation due to cooperative and nonlinear interactions between component particles in condensed media such as water (Olsen et al., 1984; Rebbi 1979), optical fibers (Mollenauer and Stolen, 1982) and biopolymers (Scott, 1985a,b; Ji, 1985b; Sobell et al., 1982; Polozov and Yakushevich, 1988), (2) superconductivity resulting from the interactions of electrons and lattice deformations in condensed media (Careri, 1984; Hunt, 1989), and (3) self-organizing chemical reaction systems exhibiting patterns of chemical concentrations organized in space and time due to nonlinear interactions between chemical reactions and diffusion processes (Epstein, 1987; Ross et al., 1988). Thus, we are here dealing with a novel law of nature which appears not to have been formalized into a clearly defined statement. As a start, I propose the following formulation of this law which may be called the "principle of emergent properties (PEP);"

> Whenever a complex system S is constructed out of subsystems s_1, s_2, s_3, . . . , s_n, according to a set R of rules, r_1, r_2, r_3, . . ., r_m, then a set P of new properties p_1, p_2, p_3, . . . , p_l can emerge that is not found in less complex systems composed of any subset of S.

The nature of the set of rules mentioned in PEP may be identified with a concept derived from concatenation, which is the most fundamental process of complexification in the written language. The concept of concatenation can be generalized from the one dimensional space into an N-dimensional one as follows;

> The N-dimensional concatenation is the process of producing complex systems from two or more component parts or subsystems, by coupling them in such a way that the behaviors or motion of the component subsystems are correlated in the N-dimensional space, where N is a natural number.

So defined, PEP and the process of N-dimensional concatenation can be seen to apply to all organized systems and structures, from linguistic and logical structures to biological machines.

Biology is replete with examples illustrating PEP and N-dimensional concatenations as shown in Table 1.6. The first column indicates the "dimensionality" of concatenations, N, which increases from 1 to 9. As N increases due to the coupling of two or more (N - 1)-dimensional components (see the second column) to form N-dimensional objects (the third column), new properties emerge (the fourth column), some specific examples of which are listed in the last column, which range from genes to Homo sapiens.

The content of Table 1.6 is largely self-explanatory. Only brief comments are added as footnotes. It is interesting to point out that the value of N exceeds 4—the dimensionality of spacetime—as biological structures become more complex than enzymes (see the rows with N equal to 5 or greater). In other words, what this table suggests is that, to completely describe the properties and behaviors of metabolic networks and beyond, it is necessary to invoke a "space" whose structure is more complex than can be specified in terms of the usual space (x,y,z) and time (t) coordinates of physics and chemistry. Such a space may be referred to as the "internal space" and the coordinates required to specify objects in the internal space—extra to spacetime—may be called the "internal coordinates." The concept of internal coordinates has been borrowed from particle physics where they are used to describe the properties of the fundamental particles. For example, according to the gauge theories (discussed in Section 1.8.7) of the fundamental forces, neutrons and protons represent the same kind of particles called nucleons, differing only with respect to their internal coordinates (Yang, 1977).

To describe the fundamental particles, it is often sufficient for physicists to use internal spaces with the dimensionality around 2. However, in order to completely describe living systems and processes, it seems necessary to utilize an internal space with a dimensionality of 5 (i.e., the difference between the N value of 9 in Table 1.6 and the dimensionality, 4, of spacetime). The existence of such a high-dimensional internal space in nature may not be as unrealistic as it might appear at first, because the superstring theorists who are interested in unifying the four fundamental (i.e., gravitational, electromagnetic, electroweak, and strong) forces of nature do invoke a 10-dimensional space (Green, 1985), which may be thought of being composed of the 4-dimensional spacetime and a 6-dimensional internal space. In fact, it may be considered that the multidimensionality of living systems and processes so clearly brought out in Table 1.6 is but the manifestation of the 10-dimensional physical world recently uncovered by superstring theorists. If this conjecture turns out to be true, then the connection between physics and biology may be much more intimate and profound than has been so far suspected to be the case by the majority of both physicists and biologists. The possible relationship between superstrings and the phenomenon of life will be discussed in a greater detail in Section 1.11 and Appendix 1.F (see Figures 1.15 and 1.A.5).

Table 1.6. Emergent properties of complex biological structures constructed by N-dimensional concatenations

N	Components	Complex Structures	Emergent Properties	Examples
1	Nucleotides	DNA	Genetic memory	Genes
2	Single-stranded DNA	DNA double helix	Intermolecular pairings	Watson-Crick base pairs
3	1-Dimensional proteins	Globular proteins	Molecular shape recognitions	Receptors, DNA-binding proteins
4	Globular proteins	Enzymes[1]	Catalysis	Chymotrypsin, Na^+/K^+ ATPase
5	Enzymes	Metabolic networks[2]	Metabolic regulations	Krebs cycle, Arachidonic acid cascade
6	Metabolic networks	Cells[3]	Intracellular communication[4]	Gene expression[4]
7	Cells	Organs	Intercellular communication[5]	Contact inhibition
8	Organs	Organisms	Increased adaptability to environment	Self-defense mechanisms[6]
9	Organisms	Species	Evolution[7]	*Homo sapiens*

1.4.6. Inputs, Outputs, and Internal States of Biological Machines

As was discussed in Section 1.4.2., machines can be characterized by their inputs, outputs and internal states. Table 1.7 lists examples of biological machines that belong to thirteen different levels of organization along with their inputs, outputs, and the nature of internal states. One of the useful results of applying the machine theory to biology is the notion that the so-called "internal state" of a biological machine can determine the nature of its outputs. Another useful results is that the machine concept provides a versatile theoretical framework for classifying the empirical observations of biology coherently. Such a coherent cataloguing of facts is enormously helpful in testing biological concepts and theories.

(Notes to Table 1.6)

[1] I am treating enzymes as more complex biopolymers than general globular proteins such as immunoglobulins and hemoglobin that do not possess full-blown enzymic activities. The distinction between enzymes and simple binding proteins may be that only the former possesses the ability to cause a series of time-symmetry breakings in molecular interactions. See Section 1.8.2. for related discussions.

[2] Metabolic networks are higher-order structures than enzymes, because the coordination of more than one kind of enzymes is required for the normal functioning of any metabolic network.

[3] A set of metabolic networks that are coupled to perform evolutionarily selected molecular work processes may be regarded as synonymous with the living cell.

[4] In order for the cell to function normally, it is axiomatic that the key components of the cell must interact with and influence (i.e., communicate with) one another. One of the most important intracellular communications is between the genome stored in DNA and the rest of the cell. In Section 1.8.11. the intracellular communication is discussed from the point of view of the communication in the time dimension.

[5] It is again axiomatic that no multi-cellular organisms can survive without efficient intercellular communications. The presence of the multitudes of hormones, cytokines, neurotransmitters, etc. in living tissues is an eloquent testimony to this fact.

[6] All organisms, whether unicellular or multicellular, have evolved efficient self-defense mechanisms against potential injuries from their environments. The efficient operation of various self-defense mechanisms in animals depends on the cooperation of almost all the organs in the body coordinated by the brain.

[7] It is a curious fact of biology that individual organisms cannot evolve and only a group of individuals can. This may be related to the fact that the evolutionary time scale is much longer than individual life spans.

Table 1.7. The inputs, internal states, and outputs of various biological machines.

Machine	Input	Internal state	Output
1) Receptors[1]	Hormones	Conformational states[2]	Second messengers[3]
2) Enzymes	Substrates	Conformational states[2]	Products
3) DNA/RNA Polymerase complex[4]	Sigma factor[5]	Conformational states[2]	Initiation of DNA transcription[6]
4) Krebs cycle	Acetyl-CoA NAD$^+$ ADP	States of "organized enzymes"[7]	NADH, CO_2 ATP
5) IP_3 pathway	Growth factors	States of "organized enzymes"[7]	Increase in intracellular Ca^{++}
6) Mitochondria	ADP	"Organelle state"[8]	ATP
7) Sarcomere	Ca^{++}	"Organelle state"[8]	Filament sliding
8) Immune cells	Concanavalin A	"Cell states"[9]	Cellular proliferation
9) Neutrophils	fMLP	"Cell state"[9]	Chemotaxis

"Biocybernetics": A Machine Theory of Biology

(Table 1.7 continued)

10) Hypothalamus	Hormones	"Tissue state"[10]	Releasing factors
11) Brain	Neurotransmitters	"Organ state"[11]	Feelings
12) Liver	Glucagon	"Organ state"[11]	Glucose
13) Human	Carcinogen	"Organismic state"[12]	Cancer[13]

---

[1] Including associated proteins and enzymes that act as a unit of cellular signal transduction, leading to the production of second messengers.

[2] The three-dimensional structures of biopolymers that can be transformed from one state to another without breaking or forming covalent bonds (i.e., only through rotating or stretching bonds). It is of fundamental theoretical importance to distinguish "conformation" from "configuration" of molecules, since configurational changes always involve breaking or forming covalent bonds through chemical reactions, whereas conformational changes involves only physical changes of molecules.

[3] Intracellular molecules (e.g., cAMP, IP_3, diacylglycerol) and ions (e.g., Ca^{++}) whose concentrations are altered as a direct consequence of hormone binding to its cell membrane receptor.

[4] The complex of proteins and the promotor region of a transcriptional unit of DNA ready to initiate transcription (i.e., DNA-directed RNA synthesis) upon binding a signal called a sigma factor.

[5] A bacterial protein of 85,000 daltons needed for the initiation of DNA transcription by RNA polymerase.

[6] DNA transcription appears to be what is usually meant when the term "gene expression" is used in molecular biology. However, whether or not this is the only possible meaning for "gene expression" remains an open question, since the realization of a phenotype involves more complex intracellular, intercellular and organismic-environmental interactions than just simple transcription of one or more genes.

[7] The set of enzymes constituting a given metabolic pathway can be organized in space (e.g. in mitochondria) and time (i.e., when and for how long they are to be synthesized through transcription and translation, or degraded); all metabolic pathways have spatiotemporal structures compatible with their biological functions and hence can be viewed as examples of a generalized machine.

[8] Each organelle state may be characterized by a unique set of values for the transmembrane gradients of H^+, Ca^{++}, K^+, Na^+, and other diffusible and actively transported molecules.

1.4.7. Naming of Machines: "City-ator"

It is a common practice in physical chemistry to name the theoretical or molecular mechanistic model of a self-organizing chemical reaction-diffusion system after the city where the related research was carried out, followed by the suffix -ator. The first well-known example is the Brusselator formulated by I. Prigogine and R. Lefever in Brussels in 1968 to model the spatio-temporal organization of chemical concentrations characteristic of the Belousov-Zhabotinskii (BZ) reaction. The BZ reaction was discovered in Russia in 1958 and is probably the most thoroughly investigated self-organizing chemical reaction-diffusion system known (Babloyantz, 1986). The chemical and kinetic properties of the BZ reaction were studied in detail by J. Field and R. M. Noyes in Oregon, leading to the formulation of their chemical model of the BZ reaction called the Oregonator. There are now hundreds of self-organizing chemical reactions known and more will undoubtedly be discovered. The study of such chemical reactions is a rapidly growing research field in chemistry and is referred to as "New Chemistry" (Babloyantz, 1986) (see Sections 1.8.2 - 1.8.5).

Since all biological machines are ultimately driven by chemical reactions and diffusions, they can be viewed as examples of self-organizing chemical reaction-diffusion systems. Therefore, it seems logical to name all biological machines according to the same procedure used in New Chemistry, namely by adding "-ator" at the end of the name of the city closely associated with research on the biological machine under consideration.

(Notes to Table 1.7 Continued)

[9] It is postulated that the living cell can exist in numerous states called "cell states" characterized by a set of spatiotemporally organized intracellular metabolic fluxes called "intracellular dissipative structures (IDS)" (Ji, 1987). Some IDS may manifest themselves as intracellular ion gradients or chemical waves (see Section 1.8.2.).

[10] Although it is not definitely known how best to characterize the "tissue states," it seems necessary that the state of microcirculation (blood flow through capillaries) (Kessler et al., 1976; Ji et al., 1978) and locally acting chemical messengers like prostaglandins, nitric oxide and cytokines must be included as a prominent parameter of the tissue state.

[11] "Organ states" must exist as evidenced by the well-known EEG and EKG waves whose patterns depend critically on the physiological states of the brain and the heart, respectively.

[12] "Organismic states" may be identified with the internal state of an organism that determine its behavior.

[13] Disease state characterized by uncontrolled proliferation of cells in an organism.

In fact the idea of calling self-organizing chemical reaction diffusion systems "-ators" can be extended to include not only biological machines but also any *organized processes* in nature, regardless of whether they are physical, chemical or biological in origin. Therefore, I have taken the liberty of naming all "organized processes" appearing in this chapter as "X-ators," where X is the name of a city. Readers not accustomed to using these names may find them awkward and even offensive (as I found the various names of enzymes such as DNA "gyrase," protein "foldase," adenine nucleotide "translocase," etc. very awkward at first); but unless and until a better method of systematically naming *organized processes* is devised, I see no alternative but to follow the Brusselator tradition. Since there are a lot of "organized processes" to catalogue in nature and since there are seemingly inexhaustible number of cities in the world, "organized processes" and "cities" seem to be "meant" for each other. For the convenience of the reader, I have collected in Table 1.8 all the "ators" that have been published so far. As I emphasized elsewhere (Ji, 1990), no deep significance should be attached to the names of the cities that have been "commandeered" to represent these various organized processes or machines. The association between the city names and the organized processes that they represent is mostly anecdotal or gratuitous rather than logical or historical. Gratuity, after all, has its place in biology as exemplified by the phenomenon of Monod's allostery (Monod, 1971).

Table 1.8. Examples of self-organizing processes in nature

	Organized Processes	Name	References
1	Belousov-Zhabotinskii reaction	Brusselator (mathematical model)	Babloyantz, 1986
2	"	Oregonator (chemical model)	"

(Table 1.8 continued)

3	Living cell	Bhopalator	Section 1.8
4	Human body	Piscatawaytor	Section 1.9
5	Universe	Shillongator	Section 1.11
6	Human society	Newbrunswickator	Section 1.10
8	Origin of life	Princetonator	Appendix 1.C
9	Oxidative phosphorylation	Madisonator	Section 1.7.3
10	Inflammatory response	Londonator	Section 1.12.10
11	Self-defense mechanisms of the body	Hanoverator	Section 1.12.11
12	Cancer	Valhallator	Section 1.12.12
13	Apoptosis	Edinburghator	Ji, 1990
14	Xenobiotic removing mechanisms in the cell	Oklahomacityator	"
15	Cellular differentiation	Philadelphiator	"
16	Cellular proliferation	Torontoator	"
17	Toxic cell death	Bethesda-ator	"

1.5. BIOPOLYMERS AS MOLECULAR MACHINES: THE McCLARE MACHINE

Biopolymers can be divided into two groups—what may be called "structural" biopolymers such as collagen that act mainly as structural supports and "molecular machines" that can utilize chemical free energy to perform some work functions such as translational and rotational motions and transmembrane ion pumping. The concept of "molecular machines" has been discussed qualitatively in the biological literature for at least two decades (McClare, 1971; Lumry and Gregory, 1986; Ji, 1974a, 1987). We can now express this idea more precisely using the algebraic formalism introduced in Equation (1.8), Section 1.4.2.

$$M^{bp} = (I^{bp}, S^{bp}, O^{bp}, d^{bp}, l^{bp}) \tag{1.9}$$

where M indicates a machine, the superscript "bp" denotes a biopolymer and can be replaced by e (enzymes), d (DNA) or r (RNA) and the other symbols are defined as in Equation (1.8).

In most cybernetics literature, it is commonly taken for granted that all machines are powered by electricity. However, when considering molecular machines, which are definitely not driven by electricity, we can no longer ignore the detailed mechanisms by which molecular machines are driven. In fact, the whole field of bioenergetics can be viewed as the study of the molecular mechanisms responsible for driving living processes, utilizing the free energy released from chemical reactions or light absorption (Nicholls, 1982; Kalckar, 1969). Despite much progress that has been made in recent decades in determining the structures of biomembranes and enzymes carrying out bioenergetic processes, the basic molecular mechanisms by which enzymes utilize the free energy of ATP to perform various vectorial and scalar work processes essential for cell life remain poorly understood. With most biopolymeric machines, our current knowledge is limited to the nature of the inputs and outputs, namely I^{bp} and O^{bp} in Equation (1.9). The greatest gap in our knowledge about molecular machines involves the nature of the set of internal states, S^{bp}, and the associated input (d^{bp}) and output (l^{bp}) functions, the essence of the "mechanisms" of the internal workings of biopolymers.

1.5.1. The "Molecularization" of Machines and the "Thermal Barrier"

Although many investigators have discussed the concept of machines to explain the functional properties of enzymes (McClare, 1971; Lumry and Gregory,

1986; Gavish, 1986), not much attention appears to have been given to the question, "Is there any limitation to extending the concept of machines from the macroscopic scale to the microscopic scale?." Quantum mechanics has taught us that "molecularization" of the Newtonian mechanics cannot be continued beyond Planck's quantum of action (Bohr, 1933). Is there any analogous limitation to the "molecularization" of macroscopic machines to the microscopic scale?

One of the most fundamental differences between macroscopic machines and molecular machines is the thermal rigidity of the former and the thermal flexibility of the latter (Ji, 1985a, 1987). Thermal motions are detrimental to macroscopic machines (e.g., short circuiting in computer chips when the diameter of "wires" become much smaller than a fraction of a micron), while they are essential for the functioning of enzymes (for a recent review, see Welch, 1986). This difference in the role of thermal fluctuations in the operation of macroscopic and microscopic machines, I believe, is fundamental and will be referred to as the "thermal barrier" between macroscopic and microscopic machines.

Two reasons may be advanced for the necessity of the thermal flexibility of molecular machines. (1) The energetic reasons: To release free energy from chemicals necessary for driving molecular machines, molecular machines must act as catalysts. It is becoming clear that enzymic actions depend on thermal energies to overcome high activation energy barriers and that the "utilization" of thermal energies in turn depends on the ability of enzymes to undergo thermal fluctuations (see Welch and Kell, 1986 for a recent review). Elsewhere, I proposed possible molecular mechanisms by which enzymes can "utilize" thermal energies for catalysis without violating the Second Law of Thermodynamics (Ji, 1974b). The critical element of the proposed mechanism is that the free-energy releasing chemical steps are catalyzed within the life-time of the thermally activated and conformationally strained local structures of an enzyme. (2) The information reason: In general, when enzymes catalyze a given biochemical reaction, they carry out not just one but a series of elementary catalytic acts that are organized in space and time. Such an organization requires information, namely molecular constraints, that "select" or "favor" a subset of elementary catalytic steps out of a large set of the elementary chemical reactions allowed for by the laws of quantum statistical mechanics. Molecular biology has now established that the information necessary for carrying out enzymic catalysis is stored in enzymes as their primary sequence of amino acids. But the question as to how enzymes retrieve the necessary information during a catalytic act has not been widely discussed. In 1974, I proposed a possible molecular mechanism for information retrieval that implicates thermal fluctuations of enzymes and the application of the Franck-Condon principle (see Glossary). This principle in effect states that,

"Biocybernetics": A Machine Theory of Biology

whenever an elementary chemical reaction occurs, the slower rearrangements of nuclei must precede the faster electronic rearrangements (Ji, 1974b). According to the proposed mechanism, thermal fluctuations provide mechanisms for enzymes to "scan" their stored information and to retrieve the catalytic information called for by the next elementary step. Therefore, the thermal flexibility of enzymes is *sine qua non* of their ability to carry out catalysis.

The thermal barrier separates macroscopic machines (whose frameworks are designed to resist thermal motions with energies given by kT where k is the Boltzmann constant and T the absolute temperature) from molecular machines (whose frameworks are small enough to be "agitated" by the impacts received from molecules moving about with kinetic energies in the range of kT). It is interesting to note that the barrier between Newtonian (or macroscopic) and quantum (or microscopic) mechanics is expressed in terms of the Planck constant, h, whereas the barrier between macroscopic and microscopic machines implicates the Boltzmann constant, k, which is embedded in the concept of conformons, conformational strains of biopolymers storing free energy and genetic information (Table 1.9). It is also interesting to point out that h is related to the minimum amount of energy associated with molecular motions, whereas k is connected with the minimum amount of entropy that must be produced in any measurement leading to one bit of information (Tribus and McIrvine, 1971; Pierce, 1980).

The key point of Table 1.9 is that, just as the transition from the macroscopic mechanics to the microscopic mechanics necessitated the introduction of the concept of the *quantum of action* and the birth of *quantum mechanics* in the early decades of this century, so the molecularization of machines may require inaugurating a similarly profound new concept, perhaps *conformons*, which may be viewed as the *quantum of communication*. As will be detailed in Section 1.8.11, communications play a fundamental role at all levels of biological organizations, from enzymes to metabolic networks to cells and beyond. The study of communications in living systems based on the fundamental assumption that all molecular communications are driven ultimately by conformons may be called *molecular communications theory*. Biocybernetics is a broader theory of biological communications in that it encompasses not only molecular communications but also communications at the cellular and supra-cellular levels.

1.5.2. Conformons and Molecular Machines

A. <u>Conformons as the Source of Free Energy</u>: The idea that conformational strains of enzymes may play a crucial role in free energy storage and transfer in

Table 1.9. The "Molecularization" of motions and machines.

Scale	Motions (Energy)	Machines (Energy & Information)
Macroscopic	Newtonian mechanics	Macroscopic machines
	↓ h^1	↓ conformons[2]
Microscopic	Quantum mechanics	Molecular machines

[1] h = Planck constant, 6.6252×10^{-27} erg sec.

[2] The sequence-specific conformational strains of biopolymers containing free energy and information needed to operate molecular machines, e.g. enzymes. The energy content of a conformon must be greater than kT, where k is the Boltzmann constant (1.3805×10^{-16} erg $°K$) and T is the absolute temperature. Therefore, at room temperatures (T = 298°K), the energy content of a conformon must be greater than $(1.3805 \times 10^{-16})(298) = 4.127 \times 10^{-14}$ ergs, or 0.594 Kcal/mole.

bioenergetic processes was first suggested by Lumry in the early 1960's (Lumry, 1974) and later applied to mechanisms of oxidative phosphorylation by numerous workers including Boyer (1977), Green (1974), Slater (1977), and Ji (1974a, 1976, 1977, 1979). In 1972, Green and Ji combined the conformational energy concept of Lumry with the energy "packet" concept borrowed from solid state physics and proposed that, during oxidative phosphorylation, the free energy released from redox reactions in electron transfer complexes is transferred to the ATP synthetase via "mobile" conformational strains of proteins. The ATP synthetase then utilizes the conformational free energy to drive the "endergonic" (i.e., free energy-requiring) phosphorylation of ADP to form ATP (see Figure 1.3, Section 1.7.3). Such conformational strains mediating free energy storage and transfer in biopolymeric systems were given the name "conformons," a term derived from "conform-" and "-on" indicating the "conformation" of a protein and a "mobile packet" or "discrete entity," respectively. Thus Green and Ji (1972) characterized conformons as follows:

It will be convenient to introduce the concept of the "conformon" which is defined as the free energy associated with a localized conformational strain in biological macromolecules. As a first approximation, we may assume that the conformon is to protein systems under physiological conditions what the phonon is to inorganic crystal lattice structures. We may list the following as the characteristic properties of the conformon. (a) The conformon is mobile. Conformon migration requires a relatively rigid protein framework such as α-helical structure. (b) The conformon differs from the generalized electromechanochemical free energy of protein conformational strain in the sense that the conformon has the property of a "packet of energy" associated with conformational strain localized within a relatively small volume compared with the size of the supermolecule (i.e., the macromolecular system composed of electron transfer complexes and the ATP synthetase of the mitochondrial inner membrane). (c) The path of conformon migration need not be rectilinear but will be dependent on the 3-dimensional arrangement of the linkage system. (d) The properties of the conformon are believed to be intimately tied in with the vibrational coupling between adjacent bonds in polypeptide chains.

It is truly remarkable that, about the same time, Volkenstein (1972; Volkenstein et al., 1972; Volkenstein, 1986) in Moscow coined the same term, the conformon, as a part of his quantum mechanical theory of enzymic catalysis:

> The displacement of an electron or of the electronic density in a macromolecule produces the deformation of the lattice, i.e., the conformational change. It can be treated as an excitation of the long-wave phonons and the system electron plus local deformation of macromolecule becomes like polaron. Let us call such a specific system the conformon...

Kemeny and Gorklany (1974) recognized that the conformon concepts formulated by Green and Ji and by Volkenstein are fundamentally related on the quantum mechanical level and pointed out that the entropic term in the free energy expression of conformons may play a dominating role.

A still another unexpected connection between conformons and a solid state physics idea surfaced in the 1970's following Davydov's suggestion (1973; 1981; Scott, 1985a,b) that free energy transfer and storage in proteins may be effectuated by localized vibrational events called "solitons." Numerical studies based on the quantum mechanical expression of solitons supported the possibility of transferring free energy in the form of localized conformational deformations of polypeptide chains at physiological temperatures (Hyman et al., 1981; Scott, 1982; Cruzeiro et al., 1988). Therefore, it may be concluded that the development of the soliton theory provides strong theoretical support for the existence of conformons of Green and Ji in biopolymers (Scott, 1985a,b).

B. <u>Conformons as the Source of Genetic Information</u>: Most of the discussions on conformons and solitons up to 1985 focussed on the need for storing and transferring free energy through protein (Green and Ji, 1972; Ji, 1974a, 1979) and DNA systems (Englander et al., 1980; Sobell, 1985; Sobell et al., 1982). Realizing that the availability of free energy is *necessary but not sufficient* to effectuate goal-directed molecular processes of the cell that can be transferred from one generation to the next, I felt it necessary to extend the concept of the conformon of Green and Ji by including genetic information as follows:

> Conformons are genetically determined local conformational strains of biological macromolecules, each endowed with specific biological functions including ligand binding, catalysis, free energy storage, and free energy transfer (Ji, 1985a,c).

The genetic information associated with a conformon was attributed to the collection of the spatiotemporally organized 5 - 10 amino acid residues that make physical contacts with substrates bound to the active sites of enzymes (Ji, 1990). Using various approximations, it was estimated that the free energy content of an average conformon was 5 - 10 Kcal/mole, and the information content 40 - 80 bits (Ji, 1990; also see Appendix 1.B). The ratio between the free energy and information contents of an average conformon, therefore, is $(5 \times 10^3 \text{ cal}/6.06226 \times 10^{23})/40$ bits $= 2.076 \times 10^{-22}$ cal/bit or 8.684×10^{-22} joules/bit which is about 20 orders of magnitude smaller than the most efficient information processing possible with macroscopic machines (Tribus and McIrvine, 1971); i.e., molecular machines are immensely more energy-efficient than artificial machines. Since it is unlikely that the efficiency of macroscopic machines can exceed that of molecular machines, primarily because of the thermal barrier discussed in Section 1.5.1., it may be predicted that the upper limit to the efficiency of artificial computing machines is approximately 9×10^{-22} joules/bit.

C. <u>Conformons as the Irreducible Microscopic Entities that Provide the Necessary and Sufficient Condition for Operating Molecular Machines</u>: It follows from the definition of conformons given above that the necessary and sufficient condition for realizing all genetically determined molecular processes (and hence heritable from one generation to the next) is simply to postulate that there exists a finite set of conformons in the living cell.

1.5.3. Why Are Enzymes So Complicated?

The fact that enzymes possess so complex structures as to defy any generalizations came as somewhat of a surprise to many X-ray crystallographers and enzymologists. If we assume that the average number of the conformational states accessible to each peptide residue that are separated from one another by free energy barriers much greater than the thermal energies, kT, is at least 2, then the total number of possible conformational states available to a polypeptide of say 200 amino acids would be at least 2^{200} or about 10^{60}, an astronomical number. If it takes 1 x 10^{-12} seconds for the polypeptide to undergo a conformational transition, the time required for the polypeptide to "scan" all its conformational states would be 10^{48} seconds or 3 x 10^{40} years, approximately 10^{30} times the age of this universe! It is obvious, therefore, that not all these theoretically possible conformational states are accessible to polypeptides and that there exist mechanisms by which polypeptides are constrained to populate only a small fraction of these. The most important such mechanism must be the process of folding polypeptides into 3-dimensional globules as they are synthesized on ribosomes and extruded into the aqueous environment of the cell. Thus, it is possible that the folding of nascent polypeptides into globules serves to "select," out of the almost infinite number of conformational states permitted by quantum statistical mechanics, a few that are essential for the biological function of a given polypeptide.

An experimental window to the vast world of these accessible conformational states has been provided by the temperature-dependent X-ray crystallographic studies of myoglobin carried out by Frauenfelder (1987) and others. Frauenfelder and his colleagues have demonstrated that the numerous conformational "substates" available to myoglobin could actually be "seen" spectrophotometrically, and the rates of transitions among them could be measured. The main results from his studies and similar studies by others (Artymiuk et al., 1979; Gurd and Rothgeb, 1979; Karplus, 1987) may be summarized as follows: (1) Globular proteins in their ground states possess a large number of "conformational substates" (CS) which they "visit" through the process of equilibrium fluctuations (EF), and (2) Only a small subset of these EF called "fims" (functionally important motions) are implicated in executing biological functions of globular proteins.

It is possible to view EF as the background out of which a globular protein "selects" fims for the purpose of carrying out a specific molecular work processes. Hence we can associate a finite amount of information (I) with fims which can be estimated by using the simplified form of Shannon's formula (Pierce, 1980; Ji, 1988).

$$I = \log_2 (N_{EF}/N_{fims}) \quad \text{bits} \tag{1.10}$$

where N_{EF} and N_{fims} designate the number of conformational states accessible by thermal fluctuations and the number of functionally important conformational transitions, respectively.

In addition to the information content of fims, we can also associate finite amounts of free energy with fims, since fims are non-equilibrium, local conformational strains, whose relaxation leads to the equilibration of the conformational states of myoglobin.

These considerations suggest that fims of Frauenfelder may be closely related to the conformons of Green and Ji (1972), Volkenstein (1972), and Kemeny and Gorklany (1974). Therefore, we may employ Equation (1.10) to estimate N_{EF}, since N_{fims} is estimated to be 4 (Frauenfelder, 1987) and the information content of fims may be assumed to be similar to that of an average conformon which has been estimated to be approximately 40 bits (Ji, 1990; see Appendix 1.B);

$$40 = \log_2 (N_{EF}/4) \tag{1.11}$$

which leads to $N_{EF} = 2.8 \times 10^{11}$, an approximate number of the equilibrium fluctuations available to myoglobin. If this numerical result turns out to be consistent with other temperature-dependent X-ray crystallographic data of myoglobin, the concept of conformons will have gained a further experimental support.

It should be pointed out, however, that conformons and fims differ in an important way, namely conformons are generated in globular proteins at physiological temperatures and pressures in response to biological input signals and perform specific biological functions, whereas fims are generated at low temperatures in response to flash photolysis of the myoglobin heme moiety, and therefore their biological functions are not clear at this time. Despite this reservation, it is likely that fims reflect some fundamental aspects of myoglobin function and that the concept of fims captures a property universal to globular proteins (Frauenfelder, 1987).

Before we attempt to provide a possible answer to the main question posed at the beginning of this Section, it would be helpful to summarize the essential points of the above discussion, using the matrix algebraic formalism. We can represent the basic characteristics of a biopolymer machine as a matrix (see below) where α, β, γ, δ . . . represent the internal states of a globular protein which may be identified with EF of Frauenfelder (1987), the arabic numerals indicate the input signals originating

		Internal States					
		α	β	γ	δ	...	
I n p u t	1	a	p	e	t	...	d
	2	b	q	f	u	...	e
	3	c	r	g	v	...	f
	4	d	s	h	w	...	g
S i g n a l s	.	.	.	.	.	...	.
	.	.	.	.	.	...	.
	.	.	.	.	.	...	.
	n	w	v	u	a	...	z

from the environment of the protein, and the English letters designate the new conformational states of the protein (to be called "outcomes") induced by the interactions with input signals. The only rule employed in writing down the "outcome" conformational states is that no two elements in a given column are identical. This means that the conformational transition of a protein is mandatory upon interacting with an environmental signal. A small number of these "outcome" conformational states of the protein are "target" or "desired" states and are therefore identifiable with fims or conformons.

Now we can apply the Law of Requisite Variety to enzyme machines. The Law of Requisite Variety is probably one of the most fundamental contributions made by cybernetics to science. The essential content of the law is that the complexity of a machine must be proportional to its ability to perform sophisticated tasks under varying environmental conditions (see Appendix 1.A). In order for the molecular machine to maintain the homeostasis of selecting a small number of target states under the dominating influence of the environment (i.e., input signals affect the internal states of the enzyme but not *vice versa*), it is necessary for molecular machines to increase the number of their internal states, since only the variety of the internal states of the machine can reduce the variety of the outcomes of the interactions between the machine and input signals. Thus, the Law of Requisite Variety provides a possible rationale for accounting for the existence of a large number of conformational substates of globular proteins. The fact that the number of the internal states of a biopolymer machine is large (e.g., in the order of 10^{11}) indicates that the molecular machine can absorb extraneous inputs and perturbations from its environment in order to "select" and "maintain" one or more functionally important conformational states accessible through thermal fluctuations. Therefore it may be concluded that the reason for biopolymeric machines to be structurally complex is because the environmental perturbations are so complex and, due to the Law of Requisite Variety, the only way to maintain the homeostasis of machine function is to increase the variety of its internal states (Appendix 1.A).

1.5.4 Conformons in DNA

The idea is gaining general acceptance that conformational strains of DNA (very likely RNA as well) may play a crucial role in DNA-directed polymerization of nucleic acids involved in self-replication and transcription (Ji, 1990; Polozov and Yakushevich, 1988; Sobell et al., 1982). An indirect support for such a possibility comes from the fact that cells possess enzymes (e.g., DNA gyrase, topoisomerases, helicases, etc.) that can increase or decrease the amount of mechanical strains embedded in DNA double helices, often leading to the generation or abolition of DNA supercoils, and these molecular work processes are driven by the free energy of ATP hydrolysis (Alberts et al., 1983; Wang, 1985). Space does not allow this topic to be discussed in detail here; but it is clear that the machine theoretic principles discussed so far, primarily as applied to enzymes, may apply to DNA with an equal force. Conformons embedded in DNA-protein complexes may provide both the free energy and genetic information necessary (and sufficient) to carry out the delicate and precise mechanical maneuvers, organized in space and time, which are required for gene expression and self-replication of DNA.

Although the concept of conformons was originally invoked to explain the properties of enzymes, the direct experimental proof for the existence of conformons (i.e., sequence-specific conformational strains) in fact may ultimately come from studies of DNA, because there is reasons to believe that the conformons embedded in DNA molecules are large enough to be visualized by electron microscopy. Preliminary calculations (see Appendix 1.B) based on the channel capacity equation of Shannon (Pierce, 1980) indicate that the free energy content of conformons in DNA (and hence the associated structural perturbations) may be up to 10^7 times as large as that of conformons embedded in enzymes. The basic assumptions underlying these calculations are (1) that the maximum free energy content of a conformon embedded in enzymes is the same as the free energy of hydrolysis of one molecule of ATP in the cytosol (ca. 16 Kcal/mol), and (2) that the rate of information retrieval from the DNA system in bits per symbol is the same as the rate of the information flow through the enzymes mediating DNA-directed protein synthesis. The resulting large difference between the free energy contents of protein conformons and DNA conformons appears to be due to the fact that the DNA system employs only 4 different symbols to encode genetic information, whereas the protein system utilizes 20 symbols. In other words, the band width of the nucleic acid communication channel is much narrower than the band width of the protein communication channel.

1.5.5. Conformons and the Origin of Life

One of the most important concepts to diffuse into biology from solid-state physics in recent years is that of "frustrations." Frustrations result when a large number of "particles" (e.g., electrons, atoms, molecules, monomeric units within polymers, etc.) are forced to interact among themselves and yet no matter how they are arranged in space there are always some particles that must go into thermodynamically unfavorable (hence "frustrated") configurations. In order for such a situation to develop, it is necessary that these particles can exist in at least two different states and that at least one of the possible ways with which a particle can interact with its neighbor is more stable (hence "favored") than the other (Frauenfelder, 1987).

Using the concept of frustrations, P.W. Anderson and his colleagues (1983, 1987; Anderson and Stein, 1984) have developed an ingenious and convincing model of the origin of biological information. The most significant conclusion suggested by their model is that primordial biopolymers that contained frustrations had a greater probability of self-reproduction than frustration-free counterparts. Therefore, according to the Anderson model, the nucleotide sequence that have survived the evolutionary selection are those that have led to the generation of "frustrations." By definition, frustrations are localized conformational strains and therefore can be identified with conformons. In effect, Anderson and his colleagues have proposed a molecular mechanism of the origin of life that utilizes the sequence information associated with conformons of primitive biopolymers but not the reservoir of free energy contained in conformons. In my personal discussion with P. W. Anderson and in my letter to him dated December 10, 1984, I suggested that the model could be improved by utilizing the free energy available in conformons to drive the polymerization reactions needed in self-replication and that this would eliminate the need for postulating that the primordial environment somehow had available a pool of high-energy nucleoside triphosphates ready to be utilized for polymerization, which may be highly improbable. The model of the origin of life proposed by Anderson and his group in Princeton as modified to fit the conformon concept has been referred to as the "Princetonator" and is described in Appendix 1.C. Furthermore, I propose that the conformational strains postulated to have played a key role in the genesis of biological information in the prebiotic environment be called the "Anderson conformon," in order to emphasize the theoretical connection that exists between the Anderson model of the origin of life which was motivated by the theory of frustrations and the concept of conformons derived from enzymology (Table 1.10). The conformon-based model of the origin of life represented by the Princetonator is in principle similar to the recently proposed model of the origin of life based on the soliton concept (Careri and Wyman, 1985).

1.5.6 Conformons as the Unifying Concept in Molecular Biology

The discussions presented above demonstrate that the most characteristic feature of molecular machines is that they are driven by conformons. This conclusion can be highlighted by asserting that conformons are the irreducible microscopic entities that are necessary and sufficient to drive all molecular machines, in analogy to photons that mediate all molecular interactions. Since the number of amino acid residues constituting a conformon in enzymes is estimated to be 9 (Appendix 1.B), the maximum number of different conformons is $20^9 = 5.12 \times 10^{12}$. These conformons can be divided into 8 groups depending on their biological functions (Table 1.10). It is possible that two or more families of conformons may merge as we learn more about them in the future.

It is satisfying to note that, through the Anderson model of the origin of biological information, conformons make contact, in a realistic manner, with the possible mechanisms of the origin of life. In a sense, the Anderson model suggests that the present-day conformons are but evolutionarily selected frustrations of primordial biopolymers. These frustrations in turn are but the electromagnetic quanta originating from the sun. Thus, it may be concluded that conformons are

"Biocybernetics": A Machine Theory of Biology

evolutionarily selected photons. The origin of life problem is intimately linked to all the molecular processes that are going on in living systems today, since it is very likely that most, if not all, of the molecular processes that gave rise to the first self-replicating molecular systems more than 3 billion years ago may be still operating in living systems. The present-day molecular biologic processes may be coupled to the OL (from origin of life) processes, just as they are coupled to macroscopic processes performed by all organisms; interestingly, the former coupling involves primarily the time dimension and the latter primarily the spatial dimension, and hence we may speak of the temporal coupling (phylogeny) and the spatial coupling (ontogeny and physiology) in biology. It seems logical to demand of any molecular theories of biology to provide reasonable mechanisms for explaining not only the molecular processes of immediate interests but also the origin of life. Not to demand the latter explanation would be equivalent to viewing the spatial coupling as more important than the temporal coupling, while both are obviously equally essential for all living phenomena. The conformon theory of molecular biology summarized in Table 1.10 clearly satisfies both of these requirements.

Table 1.10. Classification of conformons according to their biological functions

Conformational strains	Biological functions	References
1. Anderson conformons[1]	Origin of biological information	Anderson (1983, 1987)
2. Frauenfelder-Lumry conformons[2]	Background thermal fluctuations of proteins	(Frauenfelder, 1987; Lumry and Gregory, 1986)
3. Volkenstein-Jencks conformons[3]	Substrate & product binding by proteins (specificity & recognition)	(Volkenstein, 1972, 1986; Jencks, 1975; Kemeny & Gorklany, 1974)

(Table 1.10 continued)

4. Franck-Condon conformons[4]	Determination of the free energy level of the transition state of the ES complexes (catalysis)	(Ji, 1974b, 1979, 1985, 1990; Shaitan & Rubin, 1982)
5. Green-Ji-Davydov conformons[5]	Free energy transfer through biopolymers	(Green & Ji, 1972; Ji, 1974a, 1979; Davydov, 1973; Scott, 1985)
6. Kehkls conformon[6]	DNA transcription & self-replication	(Englander et al., 1980; Sobel, 1985; Polozov and Yakushevich, 1988; Sobell et al., 1982)
7. Klonowski-Klonowska conformons[7]	Timing in enzymes	(Klonowski & Klonowska, 1982)
8. Gedda conformons[8]	Timing in DNA	(Gedda & Brenci, 1978)

---

[1] What Anderson calls "frustrations" in his model of origin of biological information (1983, 1987) can be viewed as examples of conformons.

[2] R. Lumry is widely regarded as the father of conformational strains. Clear experimental data supporting the role of conformational fluctuations in biopolymer functions are supplied by H. Frauenfelder (1987) and others.

[3] This name is recommended in order to recognize the facts (1) that Volkenstein independently coined the term, conformons, (2) that his seminal work on the quantum mechanical formalism of enzymic catalysis based on the conformon concept probably contributed significantly to the wide acceptance of the conformon concept among theoreticians (Shaitan and Rubin, 1982; Zgierski, 1975), and (3) that the role of the substrate-binding free energy in enzymic catalysis was clearly elucidated by Jencks (1975). The Volkenstein-Jencks conformons are postulated to be involved in *specific recognition* by biopolymers of their ligands including enzymes, immunoglobulins, receptors, and regulatory sequences in DNA.

"Biocybernetics": A Machine Theory of Biology 43

(Notes to Table 1.10 continued)

[4] The activation free energy of an enzyme-catalyzed chemical reaction is determined by the difference between the free energy levels of the enzyme-substrate (ES) complex at the ground state and the transition state, also known as the Franck-Condon state (Reynolds and Lumry, 1968; Ji, 1974b). The Franck-Condon conformons are thought to be responsible for the *catalytic* function (in contrast to the *recognition* function carried out by the Volkenstein-Jencks conformons) of enzymes and other biopolymers such as ribozymes (McSwiggen and Cech, 1989). Unlike classical enzymes which seem to utilize both the Volkenstein-Jencks conformons and the Franck-Condon conformons, certain enzymes including cytochrome P-450 isozymes may have the capacity to "uncouple" these two types of conformons so that they can perform "nonspecific catalysis." Some workers estimate that there are about 200 different P-450 isozymes, each of which can metabolize 10^2 - 10^3 different xenobiotics (foreign compounds) (Coon, 1990). The ability of P-450 isozymes to "uncouple" the Volkenstein-Jenck conformons and the Franck-Condon conformons may be essential for cells to remove evolutionarily unfamiliar chemicals that invade cellular interiors (Ji, 1990). We may speak of those enzymes in which only the Franck-Condon conformons are active as "partial enzymes" and those in which both the Volkenstein-Jencks conformons and the Franck-Condon conformons act as "complete enzymes" (see Sections 1.12.6 and 1.12.7).

[5] Although R. Lumry, P. Boyer and others had previously entertained the possibility of storing and transmitting free energy in proteins as mechanical strains, Green and Ji (1972) and Ji (1974a, 1976, 1979, 1985, 1990) appear to be the first to rigorously formalize this idea using the concepts borrowed from condensed matter physics. Independently M. Volkenstein (1972) proposed the concept of the conformon in the context of enzyme catalysis in general but denied the possibility of conformons transferring free energy in proteins due to lack of structural periodicity. However, later work by A.S. Davydov (1981), A. Scott (1982, 1985a,b) and others on solitons strongly supported the idea of storing and transporting free energy in proteins as conformational strains.

[6] "Kehkls" is an acronym derived from the first letters of the names of the authors, Kallenbach, Englander, Heeger, Krumhansl, Litwin, and Sobell, who discussed for the first time the possible role of conformational strains in DNA replication and transcription (Englander et al., 1980; Sobell, 1985).

[7] Klonowski and Klonowska (1982) suggested that the relaxation of the conformational strains introduced during the synthesis of certain proteins may serve as a biological clock that determines the lifetimes of these proteins and the associated cellular functions. Thus it seems appropriate to refer to the subset of conformons in proteins whose function it is to determine the time evolution of cellular functions as Klonowski-Klonowska conformons.

[8] Based on the study of the contemporaneity of disease states in human monozygotic twins, L. Gedda in Rome came to conclude that the human genome is 4-dimensional; i.e., genes contain not only spatial information dictating the nature of amino acid sequences of proteins but also temporal information determining the timing and duration of the expression of genes (Gedda and Brenci, 1978). He suggests a picturesque metaphor of the human genome; a birthday cake with many lit candles which differ in both color (representing spatial information) and lengths (representing the temporal information) so that different candles go out at different times. To implement the 4-dimensional genes of Gedda, all one has to invoke is the existence of a set of conformons embedded in DNA or DNA-protein complexes that act similarly to the Klonowski-Klonowska conformons in proteins defined above. Such conformons are termed the Gedda conformons (Ji, 1985a).

1.6. METABOLIC PATHWAYS AS MACHINES: THE HOLCOMBE-WELCH MACHINES

1.6.1. "Metabolic Machines"

During the past half a century, biochemists have discovered numerous intracellular metabolic pathways such as the glycolytic pathway, the Krebs cycle, the fatty acid oxidation cycle, the mitochondrial electron transport chain, the urea cycle, the arachidonic acid cascade, the inositol triphosphate pathway, etc. A metabolic pathway consists of a set of enzyme-catalyzed steps that are connected in parallel or in series. If we designate the number of all the metabolic pathways found in the cell as N, the average number of enzyme-catalyzed steps per metabolic pathway as n, the average number of copies of an enzyme as m, then the total number of enzyme molecules present in a cell would be mnN. An order-of-magnitude approximation suggests that $m = 10^5$, $n = 30$, and $N = 10^2$, leading to $mnN = 3 \times 10^8$.

There is increasing experimental evidence that the multifarious metabolic pathways and the associated enzymes inside the cell are not randomly distributed but are highly organized both in space and in time (see Smith and Welch, Chapter 6). Thus it is not surprising that Welch and Kaleti (1981) speak of "cytosociology," the study of the community of enzymes organized to accomplish common cellular goals.

To describe the goal-directed, machine-like properties of metabolic pathways such as the Krebs cycle, M. Holcombe (1982) utilizes the algebraic formalism borrowed from cybernetics (see Chapter 4). I believe that the algebraic language discussed by Holcombe will be of enormous help in rigorously developing new ideas such as "cytosociology" and in analyzing complex experimental data in the field of cell metabolism. For this reason, I suggest that we refer to all intracellular metabolic pathways as "Holcombe-Welch machines," for convenience.

Following Holcombe (Chapter 4), we can treat a given intracellular metabolic pathway as a finite-state machine as a first approximation;

$$M^{mp} = (I^{mp}, S^{mp}, O^{mp}, d^{mp}, l^{mp}) \tag{1.12}$$

where the symbols are as defined in Equation (1.8), except that the superscripts "mp" stand for "metabolic pathways."

For any given metabolic pathway, the nature of the input and output sets, I^{mp} and O^{mp}, are relatively easy to define. Using the Krebs cycle as an example, we have I^{mp} = (acetyl-CoA, NAD^+, CoASH, GDP, FAD, H_2O), and O^{mp} = (CoASH, NADH, CO_2, GTP, $FADH_2$). What is not well known is the nature of the internal

"Biocybernetics": A Machine Theory of Biology

states of the metabolic machine, S^{mp}, and the functional relationships between the inputs and the internal state transitions, d^{mp}, and between the internal states and outputs, l^{mp}.

In trying to characterize S^{mp}, it may be helpful to use the digital computer as a model. The internal state of a digital computer reflects the dynamic state of the activities of its hardware and not the static structure. The maximum number of the internal states in which a computer can exist is constrained by the hardware construction of the computer, but the actual internal states in which a computer exists from moment to moment is dictated by a software extrinsic to the computer. According to the Law of Requisite Variety (Appendix 1.A), the larger the variety of the internal states available to a computer, the greater is its computing capacity. Just as the hardware of a computer is organized into multiple levels, from transistors, to flip flops and gates, and to registers and data paths (Hofstadter, 1980), so individual enzymes are organized (or "chunked" in the computer parlance (Hofstadter, 1980)) into functional units. Flip flops and gates are constructed by spatially organizing the right kind and the right number of transistors on a silicon chip. Similarly, we can envision that a metabolic pathway is constructed by organizing the right kind and the right number of enzymes in the right locations inside the cell. The temporal organization of the activities of the components of a computer are controlled by computer programs extrinsic to the computer as pointed out above. Similarly, the temporal regulation of metabolic pathways may be effectuated by "metabolic programs" (herein identified with d^{mp} and l^{mp} in Equation (1.12)) which may be extrinsic to the metabolic pathway and reside in the cell as a whole. One of the key elements of the "metabolic programs" may be the selection and timing of the activation of genes, leading to the synthesis of particular proteins. Since gene expression implicates the whole cell and not just DNA, plasma membrane receptors, or the translational or transcriptional machineries alone, the conclusion that the ultimate locus of the "metabolic programs" is the whole cell and not just any of its parts seems to be justified.

1.6.2. The Organizational Hierarchy of Intracellular Metabolic Pathways

Again using the digital computer as an analogy, wherein the multitudes of the possible states of activities of the hardware components are organized into more readily recognized and manageable "chunks" (Hofstadter, 1980), the various metabolic pathways themselves may be "chunked" into higher-order functional units. Hofstadter lists 3 distinct levels of "chunking" in the hardware design of modern-day digital computers; (1) transistors, (2) flip flops and gates, and (3) registers and data paths as already mentioned. These "chunks" are manipulated further to produce higher levels of organization of hardware activities by using softwares with increasing complexities: (1) machine instructions, (2) compiler or interpreters, (3) LISP (list of production rules), (4) embedded pattern matcher, and (5) intelligent programs.

The various metabolic machines discussed in the above section may be compared to registers and data paths of a computer and are further organized in space and time by "metabolic programs" extrinsic to them in order to produce myriads of cellular functions. The number of levels into which the classical metabolic pathways are further organized is not yet known. Designating this number as L (with the definition that classical metabolic pathways have L = 1), I hereby assert that L is a whole number much greater than 2. Pursuing the computer analogy literally, L may be 5.

Any metabolic organizations beyond the level of the classical metabolic pathways may be called "supra-" or "super-metabolic pathways." The level of organization of a given "supra-metabolic pathway" may be defined as the number of the classical metabolic pathways that are coupled to produce the supra-metabolic pathway under consideration; thus, we will speak of "level-L" metabolic machines referring to the metabolic machines composed of L classical metabolic pathways. Clearly it may not be always possible to "isolate" and identify distinct metabolic pathways, leading to the ambiguity in the assignment of the value of L. Some examples of these postulated supra-metabolic pathways are listed in Table 1.11.

Table 1.11. Hierarchical organization of intracellular metabolic pathways

L	Metabolic organization	Machine-type
1	Krebs cycle	Level-1 metabolic machine (Krebs machine)[1]
1	Glycolytic pathway	"
1	Oxidative phosphorylation	"

"Biocybernetics": A Machine Theory of Biology

(Table 1.11 continued)

1	Hormone-sensitive second messenger production	"
2	Pasteur effect[2]	Level-2 metabolic machine (Pasteur-Crabtree machine)[4]
2	Crabtree effect[3]	"
2	Protein kinase reactions[5]	"
3	DNA Transcription[6]	Level-3 metabolic machine
4	RNA Synthesis[7]	Level-4 metabolic machine
5	RNA Translation[8]	Level-5 metabolic machine
6	Protein folding[9]	Level-6 metabolic machine
7	Gene-directed enzymic activities[10] (e.g., cell division)	Level-7 metabolic machine[10]

---

[1]Since the Krebs cycle is one of the first well-established classical metabolic pathways, we may refer to all metabolic pathways belonging to level 1 as the Krebs machines. We will reserve the term "Holcombe-Welch machines" to indicate the set of all metabolic machines, regardless of their degree of organizations.

[2]Inhibition of glycolysis by enhanced mitochondrial respiration.

[3]Inhibition of mitochondrial respiration by enhanced glycolysis.

[4]Whenever convenient, we will refer to the supra-metabolic pathways with $L = 2$ as the Pasteur and Crabtree machines, since the Pasteur and the Crabtree effects are the first clear examples of the manifestations of complex metabolic machines that are composed of two classical metabolic pathways, namely glycolytic pathway and mitochondrial respiration.

(Notes to Table 1.11 continued)

[5] The numerous intracellular protein kinase reactions triggered by various second messengers, e.g., cAMP and Ca^{++}, may be considered to be the examples of the "second level" metabolic machines (or "level 2" metabolic machines), since they result from coupling two level 1 machines, namely the hormone-sensitive receptor complexes and the second-messenger-sensitive protein kinases.

[6] Certain hormone-induced phosphorylation of cytoplasmic proteins can activate the transcription of genes in the nucleus. Hence, the hormone-directed DNA transcription must implicate at least one more metabolic pathway than level-2 metabolic machines.

[7] The synthesis of mRNA implicates a post-transcriptional modification of the nascent nuclear RNA before mRNA can be exported from the nucleus to the cytosol. The metabolic pathway responsible for the intranuclear RNA modification is additional to DNA transcriptional machinery and hence belongs to the level-4 metabolic pathway.

[8] The translation of the information carried by mRNA into a polypeptide depends on the whole complex of the enzymic machinery associated with active ribosomes and hence belongs to the metabolic organization one level higher than level-4.

[9] The translation of mRNA into a polypeptide on ribosomes do not guarantee the production of an active enzyme. To produce an active enzyme from a mRNA, the nascent polypeptide must be guided into a correctly folded 3-dimensional protein, most likely assisted by a set of special enzymes. Therefore, the genesis of active enzyme requires a higher-order metabolic organization than level-5, the level of the synthesis of linear sequences of amino acids.

[10] Gene-directed enzymic activities leading to cell functions can be compared to the moment-to-moment activities of a computer performing a task under the direction of a computer program. Just as the physical manifestation of a computer in action is the spatiotemporally organized voltage values (i.e., on or off states) of the transistors on silicon chips constituting the hardware of the computer, so the physical manifestations of a cell performing gene-directed enzymic activities are the spatiotemporally organized distributions of various material entities (including both small- and large-molecular-weight species) within and around the cell. Such spatiotemporally organized material distributions associated with some cell functions can be viewed as examples of "dissipative structures" investigated by I. Prigogine (1978; Prigogine and Stengers, 1984) and hence were called "intracellular dissipative structures (IDS)" (Ji, 1985a; Ji and Finette, 1985). IDS have been also called the "Prigoginian form of genetic information" (Ji, 1988), "intracellular chemical waves (ICCW)," or "intracellular ion gradients (ICIG)," but all these different terms share the same basic meaning—the organization of matter and energy in space and time (see Section 1.8.2.). Gene-directed enzymic activities are synonymous with gene-directed cell functions such as mitosis, which implicates at least 18 distinct steps (Pardee, 1989).

"Biocybernetics": A Machine Theory of Biology

Because of the fact that biopolymers in general are dynamic molecular entities whose functional properties are sensitively influenced by IDS, we have a situation where the final product of gene expression, i.e., IDS, folds back onto itself in the sense that IDS can influence the properties of genes themselves. This is an example of a feedback control and is reminiscent of Kekule's dream about "a snake biting its own tail," which led to the discovery of the cyclic structure of benzene. The cyclical (or feedback) nature of the intracellular metabolism was first explicitly invoked in 1983 in the theoretical model of the living cell called the Bhopalator (Ji, 1985a,c). It is possible that this cyclicity of metabolic control is a prerequisite for effectuating self-motility, self-thinking, and self-reproduction of the cell.

1.7. SUBCELLULAR ORGANELLES AS MACHINES

In principle, the coupling of basic metabolic pathways (i.e., Krebs machines) to produce suprametabolic pathways (i.e., Holcombe-Welch machines) can occur in either a homogeneous, isotropic and symmetric environment (such as in a well-stirred test tube) or a highly heterogeneous, anisotropic and asymmetric environment.

It is possible that as living cells evolved, the "structure" of intracellular metabolic machines underwent transitions from symmetric states to asymmetric ones, most likely to increase the efficiency of coupling (or communication among) different metabolic machines. Closely related to this idea is the notion of "metabolic channeling" which forms a cornerstone in "cytosociology" (Welch and Kell, 1986). Equivalently, it may be stated that as the biological function of the cell became more diversified, its metabolic machines had to acquire a greater degree of asymmetry in space and time in order to increase the number of the possible internal states (i.e., S^{mp} of Equation (1.12)) of the cell, which in turn could have led to an increase in the adaptability of the living cell in accordance with the Law of Requisite Variety (Appendix 1.A). In other words, the greater the degree of asymmetry in the structure of intracellular metabolic machines, the greater would be the versatility of the cell to perform its functions, a conclusion that follows immediately from the application of the Law of Requisite Variety to Holcombe-Welch machines.

1.7.1. Metabolic Symmetry Breakings in Space and Time

The homogeneity of the intracellular metabolism can be broken in two ways—in space or in time. A metabolic symmetry breaking in space simply means that given metabolic pathway is activated "here" and not "there" inside the cell. This can happen, for example, when the set of enzymes catalyzing a metabolic pathway are physically constrained to exist within an intracellular organelle such as mitochondria, peroxisomes, lysosomes, etc. A metabolic symmetry breaking in time signifies that a certain metabolic pathway is activated only during a certain time period and neither always nor randomly. The activation of a cell with appropriate signals to enter the cell cycle provides a clear example of a temporally ordered sequence of metabolic events, which can be viewed as a sequence of time-symmetry breakings in metabolism. Another example would be sustained oscillations of intracellular metabolic variables such as the intracellular concentration of Ca^{++}, ATP or NADH (Hess et al., 1975).

In some cases, it may be impossible to separate spatial symmetry breakings from temporal symmetry breakings in metabolism. A case in point is the intracellular trafficking of organelles. Depending on the polarity of the intracellular microtubules, "cargos" such as mitochondria and ER (endoplasmic reticulum) bound to microtubules are transported either toward the periphery or the center of the cell. The thermodynamic force responsible for the organellar movement derives from the hydrolysis of ATP catalyzed by "microtubule motors." If the hydrolysis of ATP is triggered only when a cargo to be transported bind to the microtubule motor, the force-generating metabolism will occur asymmetrically not only in space but also in time. Thus we are forced to consider space and time connected through metabolism; i.e., space and time are intricately weaved together as a result of the peculiarities of the metabolism catalyzed by the microtubule motor. It is possible that we are dealing with a rather novel phenomena of "metabolic space-time," a term introduced by Welch in analogy to the general theory of relativity (Smith and Welch, Chapter 6).

In the final analysis, the fusion of space and time in metabolism may result from the coupling of the electronic transitions (i.e., quantum jumps of electrons) attending enzyme-catalyzed chemical reactions and the nuclear rearrangements associated with the conformational alignments of the catalytic groupings in the active site of enzymes. The coupling of the *electronic transitions* and the *nuclear rearrangements* may be dictated by the so-called generalized Franck-Condon principle first formulated in qualitative terms in 1974 (Ji, 1974b, 1979) and restated in the next section. The generalized Franck-Condon principle in turn may be thought of as resulting from applying the Franck-Condon principle of quantum mechanics to the molecular mechanism of chemical reactions catalyzed by enzymes whose functions critically depend on thermal fluctuations. To the extent that the *nuclear rearrangements* of catalytic residues in enzymes are mediated by thermal fluctuations (Ji, 1974b; see also Welch and Kell, 1986), the space-time coupling intrinsic to enzyme-catalyzed processes must be in part dictated by the fluctuation-dissipation theorem of statistical mechanics (Kubo, 1966). According to this theorem, equilibrium and non-equilibrium properties of matter are intimately related. The first example obeying the fluctuation-dissipation theorem was provided by Einstein in 1905, who connected the diffusion coefficient D (associated with equilibrium thermal fluctuations of diffusible particles) and the friction coefficient f (associated with a irreversible displacement of particles in space and hence with a dissipation of free energy) in the simple formula, D = kT/f, where k is the Boltzmann constant and T the absolute temperature. Frauenfelder (1987) recently discussed the fluctuation-dissipation theorem in connection with the role of equilibrium fluctuations (EF) of myoglobin and the functionally important molecular motions (fims) of the hemoprotein. However, he did not mention the Franck-Condon principle, probably because the electronic-nuclear interactions are not so clearly manifest in non-enzymic proteins such as myoglobin.

It is evident that the generalized Franck-Condon principle contains both the quantum mechanical and statistical mechanical contents, which can be expressed algebraically as:

$$\begin{matrix} \text{Generalized} \\ \text{Franck-Condon} \\ \text{Principle} \end{matrix} = \begin{matrix} \text{Franck-} \\ \text{Condon} \\ \text{Principle} \end{matrix} + \begin{matrix} \text{Fluctuation-} \\ \text{Dissipation} \\ \text{Theorem} \end{matrix} \qquad (1.13)$$

Equation (1.13) succinctly summarizes the essence of the quantum statistical mechanical theory of molecular machines or enzymes, the Franck-Condon principle governing the chemical reaction (i.e., electronic transitions), the source of *free energy*, and the fluctuation-dissipation theorem regulating the mechanical motions of enzymes (i.e., nuclear rearrangements of catalytic residues) under the constraints imposed by the *genetic information* encoded in the linear sequence of enzymes. Since enzyme-catalyzed processes must be both *driven* by free energy and *controlled* by genetic information, it is logical to conclude that any fundamental theory of enzymic catalysis must incorporate both quantum mechanical and statistical mechanical elements in ways consistent with the Franck-Condon principle and the fluctuation-dissipation theorem as indicated in Equation (1.13). To the best of my knowledge, the conformon theory of enzymic catalysis proposed in 1974 (Ji, 1974b, 1979) appears to be the first enzymatic theory that has been formulated in accordance with Equation (1.13).

Most, if not all, of the phenomena found to obey the fluctuation-dissipation theorem in statistical mechanics so far implicate non-equilibrium (or irreversible) processes that are driven by free energy, *external* to individual molecular (or more generally microscopic) species (e.g., concentration gradients as in batteries, temperature gradients as in combustion engines, electrical potential energy as in electrical and electronic machines, etc.). However, most enzyme-catalyzed irreversible processes are driven by free energy *internal* to the enzyme-substrate complex in the sense that the requisite free energy derives from the very exergonic chemical reactions that are catalyzed by enzymes. Therefore, we may distinguish two kinds of free energies—(i) the external free energy, the usual form of free energy studied in non-enzymic chemical reactions, and (ii) the internal free energy unique to enzyme-catalyzed processes. Depending on which kind of free energies provide the driving force, we may in turn divide machines into two categories—(i) "macroscopic machines" driven by external free energy, and (ii) "microscopic" or "molecular machines" driven by internal free energy. The former may operate via what McClare (1971) called "constrained equilibrium" mechanism, and the latter via "molecular energy" mechanism. Future studies may reveal deep connections between the theory of molecular machines represented by Equation (1.13) and the theory of molecular machines advocated by McClare about two decades ago before his creative career came to an abrupt cessation due to his untimely death. To honor his far-sighted and profound contributions made to the field of molecular bioenergetics, I recommend that we name all molecular machines as the *McClare machines*, in analogy to the Turing machine in computer science (Hopcroft, 1984).

1.7.2. Spatio-Temporal Scaling in Biology: The Connection between Cytology (Cellular Morphology) and Biochemistry

Since the discovery of the cell, the basic building block of all living systems, more than one-and-a-half centuries ago, cytologists have documented at least thirty distinct microscopic structures in the living cell (Novikoff and Holtzman, 1976). As somewhat latecomers to cell biology, biochemists have independently discovered a large number of intracellular metabolic pathways, some of which later turned out to be closely associated with subcellular organelles. For example, when D. E. Green discovered in the 1940's that the set of enzymes catalyzing the Krebs cycle activities and oxidative phosphorylation could not be easily separated and acted as a unit, he designated this unusual complex of enzymes as "cyclophorase" (Tzagoloff, 1982). Cyclophorase is one of the first well-established examples of what later came to be known as "multienzyme complexes" or "multienzyme systems" (Welch, 1977). Subsequently, using electron microscopy, investigations revealed that Green's cyclophorase was identical with the cytologist's mitochondria (Tzagoloff, 1982). Of course, we now know that mitochondria contain in addition other metabolic pathways such as the fatty acid oxidation pathway, the urea cycle, the calcium/proton pumps, and the mitochondrial DNA, etc. The biological significance of these metabolic pathways being localized within the mitochondrion is not yet clear.

How are the intracellular structures that have been revealed by cytological investigations related to the metabolic pathways established in biochemistry? Is there any fundamental physical principle connecting the two aspects of the cell? If so, what is the nature of this principle? I think it is possible to answer the first two questions in the affirmative, because I believe that there is a physical principle that will allow us to connect cytology to biochemistry. This physical principle, I suggest, is what I called the "generalized Franck-Condon Principle" (Ji, 1974b). To make this principle more readily applicable to a wide variety of biological situations, I now formulate a new expression of the principle as follows:

> Whenever an observable process, P, results from the coupling of two partial processes, one slow (S) and the other fast (F), with F proceeding faster than S by a factor of at least 10^2, then S must precede F.

This statement is simply a qualitative generalization of the Franck-Condon Principle well known in quantum mechanics (Reynolds and Lumry, 1966). A more detailed discussion of this topic is omitted here since it was given elsewhere (Ji, 1974b).

I further claim that this principle is general in its applicability to rate processes in biology, from enzymic catalysis to biological evolution, as indicated in Table 1.12.

A kind of "symmetry" is evident in the Table 1.12 in the sense that the same principle of slow and fast processes was applied to all levels of biological organizations, from electronic rearrangements of enzymic reactions to biological evolution. Thus, we may regard this table as reflecting what may be called the "Principle of Slow and Fast Processes."

One of the major assets of the above principle is that it provides a plausible mechanistic framework for constructing "temporal structures" of living processes in contrast to spatial structures. Modern molecular biology tells us how enzymes, often using nucleic acids as templates, construct various 3-dimensional molecular structures in the cell. Less well understood is the mechanisms by which the cell coordinates the various intracellular rate processes in time that can differ in rates maximally by a factor of 10^{19}, from picoseconds (10^{-12} s) to years (10^7 s).

Table 1.12. The application of the generalized Franck-Condon Principle to various biological rate processes.

Overall Process (P)	Partial Processes	
	Fast (F)	Slow (S)
1. Enzymic catalysis	Covalent bond rearrangements (i.e., electronic transitions)	Conformational rearrangements of catalytic groups (i.e., nuclear rearrangements)
2. Gene expression	Enzymic reactions	Conformational rearrangements of double-stranded DNA
3. Memory	Input of signals to neurons	Rearrangements of genes in DNA(?)
4. Morphogenesis	Gene expression	Rearrangements of the connections between cells and between cells and extracellular matrix(?)
5. Evolution	Events in individual organisms	Rearrangements of physical and social environments of organisms

Based on the Principle of Slow and Fast Processes I hereby postulate that:

The intracellular temporal structures are constructed by decomposing the overall process, P, into N consecutive processes, p_i, where $i = 1$ (fastest) through N (slowest), in such a way that $v(p_i)/v(p_{i+1})$, is always approximately 10^2 where $v(p_i)$ is the rate of the ith process.

This postulate leads to;

$$v(p_1)/v(p_N) = (10^2)^N = 10^{2N}, \text{ or} \tag{1.14}$$

$$\log (v(p_1)/v(p_N)) = 2N \tag{1.15}$$

where p_1 and p_N are the fastest and the slowest partial processes, respectively, of the overall process, P.

For example, if a certain unicellular organism exhibits temporal organization over a time span of say months (i.e., 10^6 seconds) driven by metabolic processes whose fastest partial process has a time constant of about 10^{-3} seconds, then Equation (1.15) predicts that there would be $\log (10^6)/(10^{-3}) = \log 10^9 = 9 = 2N$, or $N = 4.5$, or about 4 - 5 component processes underlying the temporal structure.

It is not yet clear how the number of the component processes predicted by Equation (1.15) is related to the various levels of organization of the intracellular metabolic pathways discussed in Table 1.11. As a working hypothesis, it may be postulated that the level of metabolic organization, L, in Table 1.11 is approximately determined by the number of the components of the temporal organization, N, given by Equation (1.15);

$$L = N \tag{1.16}$$

Equation (1.16) seems to make sense, since it is likely that myriad intracellular metabolic pathways may be "chunked" into functional units (i.e., Holcombe-Welch machines) according to their kinetic similarities. Just as molecules with complementary 3-dimensional shapes universally attract one another, perhaps metabolic pathways (viewed as 4-dimensional structures) with "complementary" kinetic behaviors may "attract" one another as well.

Another way of describing the relationship between cytology and biochemistry (or metabolic pathways) would be to view the cell as a 4-dimensional structure as already alluded to and cytology and biochemistry primarily as the images of the 4-

dimensional cell projected onto the 3-dimensional space of structures and the 1-dimensional space of time, respectively.

As Prigogine points out (see Chapter 2), life is essentially a dissipative structure; i.e., the quintessence of the living cell is the spatiotemporally organized intracellular chemical reactions and physical processes called cellular metabolism. Cellular metabolism implicates spatial structures that range in size from 10^{-10} cm (diameter of the proton) to 10^{-5} cm (diameter of a cell), spanning 5 orders of magnitude, and in time from 10^{-9} s (electron transfer reactions) to 10^6 s (cellular differentiation), spanning 15 orders of magnitude. To organize the intracellular processes that span such wide ranges of space and time, these processes may have to be "chunked" into manageable functional units, to each of which the generalized Franck-Condon Principle may apply. Various intracellular structures from DNA to enzymes to subcellular organelles may be viewed as a part of the cell's tactics for subdividing the spatial and temporal scales into optimal sizes for efficient control and regulation. We will refer to this point of view as the "Spatiotemporal Scaling Hypothesis of Biology." The number of the levels of the spatiotemporal "chunking" of cellular metabolism may be N given by Equation (1.15). This line of thinking provides a possible rationale for the existence of the multiple levels of metabolic organization so characteristic of the intracellular processes, including gene expression and cellular replication (see Table 1.11).

1.7.3. The Mitochondrion: A Subcellular Organelle Machine (the "Madisonator")

(1) Mitochondria are subcellular organelles that are composed of the outer membrane and the inner membrane, about 1 micron in diameter and tens of microns long. They are not static structures but dynamic and highly organized in the cytosol. Tzagoloff (1982) gives an accurate description of mitochondrial dynamics:

> . . . far from being the static structures observed in thin sections, mitochondria are extremely dynamic organelles capable of profound changes in size, form, and location. Mitochondria from different parts of the cell can stream through what appear to be cytoplasmic channels where they often meet and fuse into larger organelles. Such compound mitochondria can, at a later time, break up into a series of smaller particles which again become dispersed throughout the cell. Mitochondrial plasticity is also evidenced by recent studies of yeast in which three-dimensional reconstructions from serial sections have shown that in certain stages of growth, the cell contains a single

mitochondrion presumed to arise from a coalescence of a large number of smaller mitochondria. In addition to fission and fusion, mitochondria are constantly contracting, expanding, and undergoing the same kind of shape changes seen in moving amoebas. The purpose and cause of this wild motion are not known at present.

(2) Mitochondria are known to carry out numerous biological functions inside the cell (Tzagoloff, 1982): (i) synthesis of ATP from ADP and inorganic phosphate driven by the free energy of the oxidation of NADH (see Equations (1.41) and (1.42) in Section 1.12.4), (ii) active uptake of Ca^{++} and Na^+ and active extrusion of H^+, (iii) reverse electron flow (i.e., electron flow away from cytochromes aa_3 and toward NAD^+), (iii) energized transhydrogenation (i.e., transfer of the hydride ion from NADH to $NADP^+$), (iv) export of reducing equivalents to the cytosol (Thurman and Kauffman, 1980), (v) oxidation of pyruvate to produce NADH via the Krebs cycle, (vi) ß-oxidation of fatty acids, and (vii) mitochondrial DNA-directed protein synthesis (Grivell, 1983).

It is clear that these different processes cannot occur randomly in mitochondria but must be coordinated in space and time, both within individual mitochondria and in interaction with the rest of the cell. Such a metabolic coordination most likely requires sophisticated regulatory mechanisms operative in mitochondria (see below). Therefore mitochondria should be viewed as an intracellular machine performing some essential regulatory functions in addition to serving as the "powerhouse" of the cell. In fact, the co-existence in mitochondria of the postulated intracellular regulatory functions and the function as the intracellular energy source may have evolved because it provided a survival advantage to the cell due to a greater efficiency in regulating cellular activities. This may be analogous to the situation in a capitalistic society where the control of money provides one of the most efficient ways of influencing human social and economic activities.

(3) To focus on the possible molecular mechanisms of energy coupling in mitochondria, only the first two of the mitochondrial processes listed above are singled out and depicted schematically in Figure 1.2. The main purpose of this figure is to clarify the topology of the energy-coupling enzymes located in the mitochondrial inner membrane. Notice that the enzymes are embedded in the phospholipid bilayer which forms an elongated tube (whose long axis is perpendicular to the plane of the page). The inner membrane has two asymmetric surface—the M-side facing the matrix space where the Krebs cycle enzymes are located, and the C-side facing the cytoplasm of the cell through the intracristal space entrapped in between the inner membrane and the outer membranes (not shown).

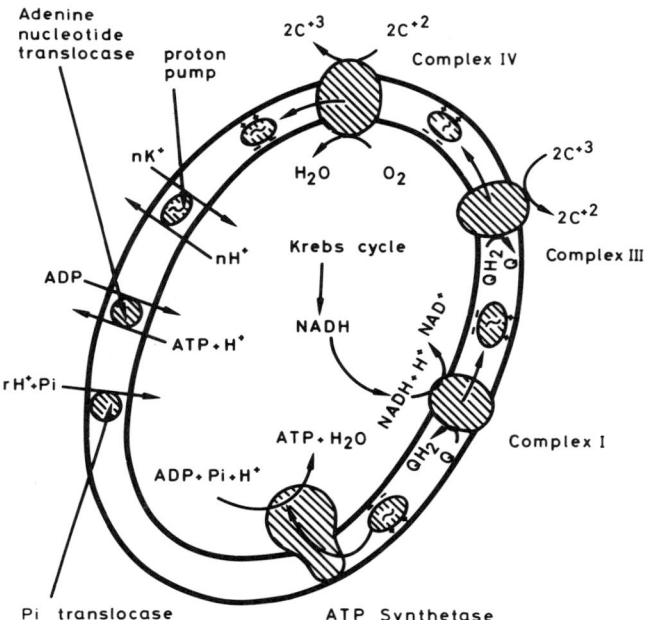

Figure 1.2. A schematic representation of the ATP synthesis and active transport processes in mitochondria. Notice that only the inner mitochondrial membrane is shown for simplicity.

(A) <u>Mechanisms of oxidative phosphorylation</u>—(i) The Krebs cycle produces NADH, which is oxidized by the system of respiratory enzymes located in the inner membrane called Complex I, resulting in the formation of QH_2 from Q and an electrically polarized "proton transfer complex (PTC)" postulated to exist in the hydrophobic phase of the inner membrane (Ji, 1974a, 1976); see the "oblong" protein with 2 negative charges on the matrix side and two positive charges on the intracristal side of the inner membrane; the "squiggle" sign indicates the presence of conformationally strained regions within PTC (i.e., conformons). (ii) Complex III oxidizes QH_2 and reduces 2 molecules of cytochrome c generating another molecule of electrically polarized PTC (see the 2 o'clock direction). (iii) Two molecules of reduced cytochrome c are oxidized and one atom of oxygen reduced to form water by Complex IV leading to the production of a third molecule of polarized PTC (see the 12 o'clock direction). (iv) One molecule of the polarized PTC "collides" (Ji, 1976, 1977) with the ATP synthetase and transfers its free energy to the ATP synthetase to drive the synthesis of ATP from ADP and Pi (see the 6 o'clock direction). The molecular mechanisms responsible for the electrical polarization of PTC and the free energy transfer from PTC to the ATP synthetase most likely implicate asymmetric protonation and deprotonation reactions as will be discussed in Figure 1.3 below (Ji, 1976, 1977).

(B) <u>Mechanisms of active transport</u>—(i) The electrically polarized PTC can depolarize itself by binding two protons (or metal cation equivalents) on the matrix side and releasing two protons on the intracristal side (see the 10 o'clock direction), thus effectively carrying out the so-called proton/proton or proton/metal ion antiport of P. Mitchell (Nicholls, 1982). (ii) The electrochemical gradient of protons so generated can drive the secondary active transport of Pi and adenine nucleotides in appropriate directions.

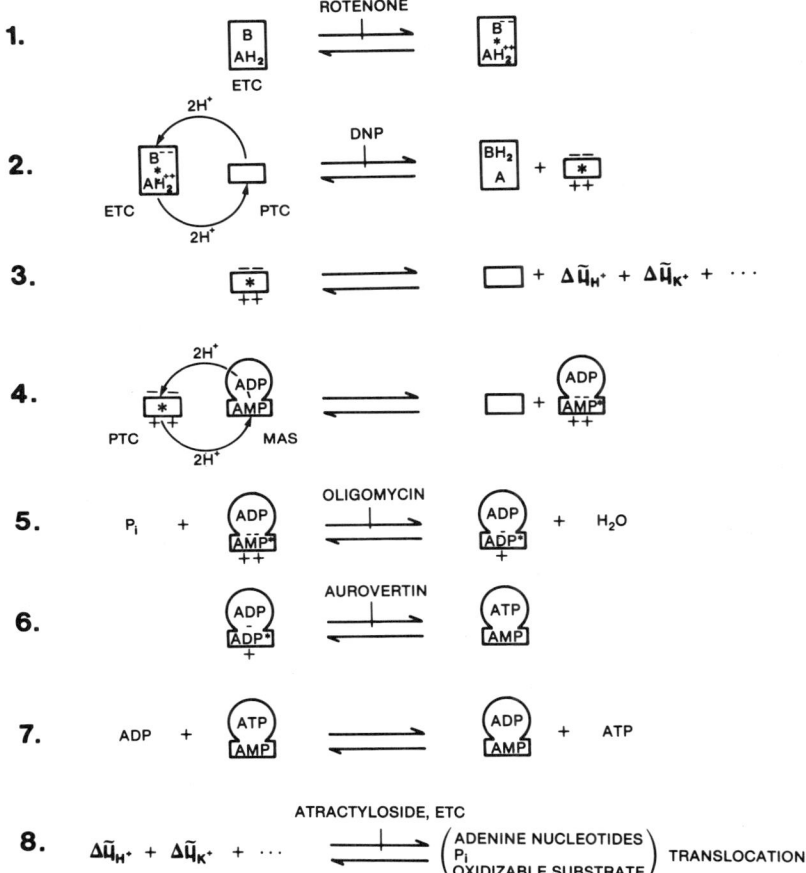

Figure 1.3. A proposed mechanism of oxidative phosphorylation and active transport in mitochondria (the Madisonator). Full accounts of this mechanism have been presented on several occasions (Ji, 1974a, 1976, 1977 and 1979). ETC (electron transfer complex) with substrate AH_2 bound on its intracristal side and substrate B bound on its matrix side undergoes thermal fluctuations leading to AH_2^{++} and B^{--}. In the charge-separated state of ETC, specific regions within ETC are thought to "trap" conformational strains (i.e., conformons) indicated with the symbol *. The electrically polarized ETC undergoes Brownian motions within the inner membrane and collide with PTC (proton transfer complex) to exchange protons asymmetrically, resulting in the depolarization of ETC and the polarization of PTC (see step 2). Conformons stored in PTC can drive either the active transport of protons and metal ions to generate the electrochemical gradients of these ions across the inner mitochondrial membrane (step 3), or the energization of MAS (mitochondrial ATP synthetase) (see the curved arrows in step 4), both by exchanging protons asymmetrically. The conformons stored in MAS are partially utilized to phosphorylate the bound AMP to ADP (step 5). A part of the remaining conformons in MAS drive the second transphosorylation reaction from ADP to another bound ADP, generating ATP and AMP (see step 6). The newly synthesized ATP still bound to MAS is displaced by ADP from the matrix space (step 7). Finally, the ATP in the matrix is actively transported out into the cytosol in exchange for Pi and ADP driven by the electrochemical gradient generated in step 2 (see step 8).

(4) A more detailed description of the molecular mechanisms underlying oxidative phosphorylation and respiration-driven active transport in mitochondria is given in Figure 1.3. To the best of my knowledge, this eight-step mechanism of oxidative phosphorylation rationally accounts for most, if not all, of the important experimental data published in the literature during the past several decades (Green and Ji, 1972; Ji, 1974a, 1976, 1979) and, in addition, makes testable predictions (Ji, 1976, 1979).

(5) It is well known that mitochondria can catalyze spatiotemporally organized metabolic processes. B. Hess and his associates (1975, 1978) have demonstrated that under certain experimental conditions the following variables of mitochondria can exhibit temporal oscillations—NADH fluorescence intensity, respiratory rate, cytochrome b redox state, H^+ uptake, K^+ uptake, mitochondrial volume, and the ATP level, all with a period of about 1 minute but with different phases. Although no computer models have yet been built to simulate the observed data quantitatively, it appears that the mechanism shown in Figure 1.3 can provide at least qualitative explanations for all of the observed data on mitochondrial oscillations. In analogy to the Brusselator representing the Belousov-Zhabotinskii reaction, I am taking the liberty of calling the mechanism shown in Figure 1.3 the "Madisonator" to indicate the profound influence that my collaboration with David E. Green in Madison during the years 1970 and 1974 had on the development of the proposed model.

(6) The chemosmotic hypothesis of oxidative phosphorylation proposed by P. Mitchell in 1961 (Nicholls, 1982; Skulachev and Hinkle, 1981) postulates that ATP synthesis is driven by the electrochemical gradient of protons across the inner mitochondrial membrane which is set up by respiration; in other words, in Mitchell's model the generation of the transmembrane electrochemical gradient of protons driven by respiration precedes the phosphorylation reaction. However, according to the Madisonator, the respiration-driven generation of the transmembrane proton gradient (step 3 in Figure 1.3) and the phosphorylation reaction (steps 5 and 6) are parallel events that are driven by a common free energy precursor generated from respiration, namely conformons (see steps 1 and 2). The primary biological role of the active transport in mitochondria is thought to be not the synthesis of ATP as P. Mitchell assumes but most likely the communication between mitochondria and metabolic events going on in the cytosol. If this interpretation turns out to be correct, then the phenomenon of the proton gradient-driven ATP synthesis, well-known in the literature, may have no general biological significance in mitochondria, except perhaps that such a process may contribute to the survival of cells under anoxic conditions when the cytosolic pH drops due to the accumulation of lactate produced

by anaerobic glycolysis and that the resulting transmembrane proton gradient may drive ATP synthesis according to the chemosmotic mechanism of P. Mitchell.

(7) The Madisonator involves two distinct kinds of protons—those protons whose movements are confined within the inner membrane as first invoked by R. J. P. Williams (1961) (see steps 1, 2, and 4) and those protons that move in and out of the inner mitochondrial membrane as postulated by P. Mitchell (1961) (see step 3). For convenience, we may designate the first kind of protons as the intramembrane protons or the "Williams protons" and the second kind as the extramembrane protons or the "Mitchellian protons." As pointed out in (6), according to the Madisonator, the primary role of the Williams protons is to couple respiration to ATP synthesis or active transport, while the primary role of the Mitchellian protons is to enable mitochondria to communicate with the metabolic events occurring in the cytosol.

(8) It is interesting to note that during the past half a decade, numerous investigators in bioenergetics came to the conclusion that newly accumulating experimental observations cannot be readily accounted for by the chemosmotic hypothesis as originally formulated and therefore that Mitchell's hypothesis should be modified (Westerhoff et al., 1984; Slater et al., 1985). The modified versions proposed by these workers all seem to agree with the mechanistic framework embodied in the Madisonator (see related discussions in Section 1.12.4). In addition, the soliton-driven Turning Wheel model of biological free energy transduction recently put forward by Careri and Wyman (1984) is very similar to the conformon theory of free energy coupling embedded in the Madisonator, the only major difference being that the soliton hypothesis is based on energy storage in the vibrationally excited amide-I bond in polypeptides whereas the conformon hypothesis implicates energy storage primarily in local conformational strains in ground vibrational states, although these two modes of storing free energy in proteins may be coupled under certain metabolic conditions.

(9) Finally, the significance of the Madisonator goes beyond its ability to provide coherent mechanistic explanations for mitochondrial data; it is the desire to understand the mitochondrial phenomenon in molecular terms that led to the notion of the conformon in 1972. Just as the black body radiation data were essential in the formulation of the concept of the *quantum of molecular actions* by M. Planck in 1900 (Polkinghorne, 1984; Kuhn, 1978), perhaps it may be said that mitochondrial data were indispensable for formulating the idea of the conformon, the *quantum of molecular machine actions* or the *quantum of molecular communications* (see Section 1.8.11). Therefore, mitochondria may provide an important experimental model to test the validity of the conformon concept. The availability of the Madisonator may facilitate this testing.

1.8. THE CELL AS A MACHINE: THE BHOPALATOR

1.8.1. The Cell as the Atom of Biology

Before we consider the machine aspects of the cell, it is instructive to compare the living cell with the atom in physics. Novikoff and Holtzman (1976) provide an apt comparison:

> Cells and atoms are units. Each is composed of simpler components which are integrated into a whole that exhibits special properties not found in any of the parts or in random mixtures of the parts. Both exhibit considerable variations in properties, based on different arrangements of components; the number of variations far exceeds the number of major components. Both serve as basic building blocks for more complex structures. However, the analogy cannot be pressed too far; cells can reproduce themselves, whereas atoms cannot. . . . .

Although viewing the cell as the "atom of life" is well accepted by biologists as the above quotation shows, the fundamental similarities and differences between the atom and the cell appear not to have been exhaustively documented. I suggest that the atom and the cell can be compared at minimally twelve different levels as shown in Table 1.13. Of these twelve levels, five represent well-known facts requiring little explanations (see rows 1, 7, 9, 11 and 12). The remaining seven levels of comparison (i.e., rows 2, 3, 4, 5, 6, 8 and 10) in contrast have been motivated by the new molecular theory of the living cell presented in this chapter. Since the contents of these comparisons are either already given in previous sections (e.g., rows 2 and 5 in Section 1.6) or will be discussed in detail in the following sections of this chapter as indicated in the footnotes, it is not necessary to go into detailed explanations of Table 1.13 at this time.

The main purpose of Table 1.13 is to give an overview of the interrelationships among the various building blocks of the new theory of the living cell that are listed in the last column of the table. Probably the most important new insight emerging from the atom/cell analogy table is the idea that the basic components of the cell (i.e., rows 2 and 5) are not individual molecules (as might have been thought) but rather metabolic processes (or metabolic machines as discussed in Section 1.6.1), each involving a set of a large number of molecules whose mechanical motions and chemical transformations are organized in space and time to form what we know as "metabolic pathways." We may represent this idea

"Biocybernetics": A Machine Theory of Biology

algebraically as;

$$V_{atom}(e, p, n) = V_{cell}(\text{metabolic machines}) \quad (1.17)$$

where e, p, and n represent electrons, protons and neutrons, respectively. The notation, $V_s(c_1, c_2, \ldots, c_n)$, indicates the vector consisting of n components, c_1, c_2, etc., embedded in the system, s. The vector, V, can be interpreted as a structure in the Euclidean space (requiring x, y, and z to specify), spacetime (requiring x, y, z, and t for specification), or a greater-than-four-dimensional space composed of the four-dimensional spacetime and a one or higher dimensional "internal space," an abstract space essential for the complete specification of physical objects, entities, or events (see Section 1.8.7. and Figure 1.10 for related discussions). The equal sign is used to indicate not necessarily "equality" but an "analogical relationship" in general.

Table 1.13. A comparison of the cell with the atom of physics

	Parameter	Atom	Cell
1	Size (cm)	10^{-8}	10^{-3}
2	Number of components	3^1	$30 - 50^2(?)$
3	Structural dimensionality	3^3	4^4 or greater
4	Thermodynamic state	Equilibrium[5]	Far from equilibrium[6]
5	Substructures	Electronic orbitals[7]	"Supra-metabolic pathways"[8]
6	Rules of construction	Quantum mechanics[9]	"Biocybernetics"[10]

(Table 1.13 continued)

7	First theoretical model	Bohr atom[11]	Bhopalator[12]
8	Uncertainty principle	$(\Delta q)(\Delta p) \geq h$[13] or $(\Delta E)(\Delta t) \geq h$	$(\Delta G)(\Delta I) \geq k$[14]
9	Parent physical theories	Newtonian mechanics[15]	Quantum mechanics Statistical mechanics Thermodynamics Cybernetics[16]
10	New concept	Planck's quantum of action[17]	Gnergons[18]
11	Self-reproduction[19]	no	yes
12	Goal-directed self-motion[20]	no	yes

---

[1] Electrons, protons and neutrons. The latter two can be further broken down into quarks, but quarks do not directly affect any biological processes.

[2] It is postulated that cellular components can be identified with the basic metabolic pathways (i.e. Krebs machines) discussed in Table 1.11 and Section 1.6.

[3] The Euclidean space of x, y and z.

[4] Cells are dynamic systems driven by chemical reactions, thus requiring time (t) in addition to the spatial coordinates x, y, and z to specify them. Therefore cells are at least 4-dimensional (see Sections 1.8.8 and 1.12.1 for related discussions).

[5] No energy dissipation is required to maintain its structure.

[6] Cells are in nonequilibrium states in the sense that continuous dissipation of free energy is required to maintain their structures (e.g., cellular morphology).

(Notes to Table 1.13 continued)

[7] The chemical properties of an atom are determined by the spatial distribution of the electrons around the nucleus of the atom which can be calculated using the Schrödinger equation and the position operators (Polkinghorne, 1984).

[8] All of the cellular properties may be viewed as resulting from the spatiotemporal organization of the activities of the basic metabolic pathways (Krebs machines) to form a large number of "supra-metabolic pathways" (Holcombe-Welch machines).

[9] Quantum mechanics whose formal structure was completed by the mid-1920's is necessary and sufficient to account for all atomic structures.

[10] It is likely that the rules of quantum mechanics (which determine the energetics involved) and statistical mechanics, although necessary, are not sufficient to account for cellular behaviors in molecular terms. The additional ingredient needed has been suggested to be the genetic information underlying living processes. Thus, the necessary and sufficient condition for describing all living processes is thought to be a higher dimensional entity called "gnergy" formed by combining free energy and Shannon information (Ji, 1985a,c, 1990; see Section 1.1). The study of gnergy in living systems are called "biognergetics" which is synonymous with "biocybernetics," a machine theory of biology, since gnergy is necessary and sufficient for driving all biological machines.

[11] The famous "planetary" model of the atom proposed by Bohr in 1913 (Murdoch, 1989).

[12] The Bhopalator proposed at the International Seminar on the Living State-II, held in Bhopal, India, in November, 1983, appears to be the first comprehensive theoretical model of the living cell (Ji, 1985a,c) (also see Sections 1.8.2).

[13] The Heisenberg uncertainty principle; it is impossible to determine neither the position (q) and momentum (p) of a particle nor the transmitted energy (E) and the time of the transmission (t) simultaneously with an arbitrarily small uncertainty. h is Planck's constant.

[14] A hypothetical relation which states that it is impossible to determine both the energetics (G) of a biological process and its biological significance (I) simultaneously with an arbitrarily small uncertainty. k is the Boltzamnn constant. A detailed derivation is given in Section 1.8.10.

[15] Quantum mechanics was derived from Newtonian mechanics by introducing discreteness into the motions of physical bodies, namely Planck's quantum of action (Murdoch, 1989).

[16] The physical theory capable of accounting for all living processes and systems, from enzymes to man, in a logically coherent manner can be constructed by combining a set of the well-established physical theories listed. Cybernetics includes the information or communication theory of Shannon (Pierce, 1980).

[17] To explain the black body radiation data of Rayleigh and Jeans, M. Planck was forced to assume in 1900 that thermal energy was absorbed or emitted not continuously but in discrete quantities which he called

The fundamental difference between electrons, protons and neutrons on the one hand and metabolic machines on the other is that the former are stable 3-dimensional objects, while the latter are dynamic entities that require for their complete description not only the 4-dimensional spacetime (to specify the mechanical motions of molecules in terms of x, y, z and t) but also an internal space for characterizing the chemical transformations among different molecules). The dynamic entities, structures, objects or events that cannot be completely described in the 3-dimensional Euclidean space but require for their complete specification physical spaces with four or more dimensions may be identified with what Prigogine termed "dissipative structures" about two decades ago (Babloyantz, 1986). Therefore, a firm grasp of the concept of dissipative structures inhabiting the four-dimensional spacetime (to be called 4-structures for short) in contrast to those existing in the usual 3-dimensional space (to be called 3-structures) appears to be an essential prerequisite for understanding the fundamental characteristics of the living cell. The conceptual jump demanded by the transition from 3-structures to 4-structures may be analogous to the expansion of the world view afforded by the transition from the 3-space of the Euclidean geometry to the 4-space of Riemannian geometry in general relativity theory (Gribbin, 1983; Misner et al., 1973). Therefore, it is possible that both biology and Einstein's general relativity theory are embedded in the common matrix of the 4-space, namely spacetime, and that Einstein's general relativity theory is concerned with the cosmological scale of the 4-space, while biology deals with the other, microscopic extreme of the 4-space. Just as the Riemannian geometry provided the mathematical language needed to develop general relativity theory, so

---

(Notes to Table 1.13 continued)

"quanta" (Polkinghorne, 1984; Murdoch, 1989; Kuhn, 1978).

[18] Discrete physical entities containing both free energy and genetic information are called "gnergons" (Ji, 1985a, 1990). There are two kinds of gnergons in living systems—conformons (see Section 1.5.2) in biopolymers and intracellular dissipative structures (IDS) (see Section 1.8.2) in the liquid phase of the cell.

[19] Self-reproduction is absolutely unique to the cell. In fact, the cell may be the smallest self-reproducible physical entity in the universe. Although computer scientists frequently discuss the possibility of creating self-reproducible physical (as contrast to abstract) machines, no man-made self-reproducible machine exists as of 1991. Even when they finally succeed in constructing such machines, it is unlikely that their size can approach the cellular dimensions.

[20] Certain atoms may be considered to exhibit a primitive form of self-motility in the sense that they can undergo random radioactive decomposition. But such motions cannot be harnessed to do non-random work, such as the chemotaxis of the cell.

"Biocybernetics": A Machine Theory of Biology

the fiber bundle geometry discussed in Section 1.8.7 may prove to be similarly essential in constructing the fundamental theories of living processes, the microscopic elaborations of the 4-space.

1.8.2. Dissipative Structures of Prigogine

Although it had been known for a long time that living systems must obey the laws of thermodynamics derived from the study of non-living systems such as the steam engine discovered in the 1700's, the first clear exposition of the connection between thermodynamics and biology appears to have been made in the theory of dissipative structures developed by I. Prigogine and his school in Brussels and Austin in the late 1960's and early 1970's (Glansdorff and Prigogine, 1971; Prigogine, 1977, 1978, 1980; Schieve and Allen, 1982; Prigogine and Stengers, 1984). This theory contains two fundamental notions: (1) There exist in nature dynamic states of matter called "dissipative structures" that are characterized by unique distributions of matter in space and time, and (2) Those material systems exhibiting spatio-temporal organizations exist "far from equilibrium" associated with nonlinear differential equations.

The first component of the theory may be viewed as the extension of the traditional concept of "structures" from the 3-space (of x, y and z) to the 4-space (of x, y, z, and t). By definition, dissipative structures are those material distributions in space and time that can be maintained only if appropriate "dissipation" of free energy accompanies them, as for example in the flame of a candle. Depending on whether the second component of the theory is regarded as the necessary and sufficient condition for dissipative structures or only as a necessary condition, we may have either the "strong" or the "weak" version, respectively, of the dissipative structure theory.

Since we now have numerous examples of far-from-equilibrium systems that do not necessarily show any spatio-temporal organization of matter (e.g., self-organizing chemical reactions discussed below usually do not show any chemical organizations, unless the reaction conditions are carefully controlled; uncoupled mitochondria can respire more rapidly and generate more heat than intact mitochondria performing constructive roles such as the synthesis of ATP), it may be concluded that only the weak version of the dissipative structure theory is valid; i.e., the "far-from-equilibrium" condition alone cannot guarantee the formation of dissipative structures. Then, what is the missing requirement for dissipative structures? According to the "biognergetics" theory proposed recently (Ji, 1985a, 1990), the necessary and sufficient condition for organizing matter in space and time is the combination of both free energy (resulting from being far from equilibrium) and Shannon information. Shannon information simply refers to the past act of selection leading to biases or preferred actions; whenever an object, either static or dynamic, is selected out of a set containing more than two similar objects, the chosen object can be said to possess Shannon information. In other words, the biognergetics theory may be said to supplement the dissipative structure theory of Prigogine by introducing Shannon information as the missing requirement for dissipative structures. Therefore, it can be concluded that the necessary and sufficient condition for creating dissipative structures or 4-structures is gnergy.

A. <u>4-Structures in Physics</u>: By 4-structures in physics, I mean those spatio-temporal organizations of matter that occur in purely physical systems devoid of any chemical reactions. The thermodynamic force behind these structures can derive from the flow of heat or electricity through macroscopic devices or paths. One of the simplest and most frequently cited examples of the dissipative structures in physics is the so-called Bérnard instability, which was elegantly described by Prigogine and Stengers (1984):

> .....The instability is due to a vertical temperature gradient set up in a horizontal liquid layer. The lower surface of the latter is heated to a given temperature, which is higher than that of the upper surface. As a result of these boundary conditions, a permanent heat flux is set up, moving from the bottom to the top. When the imposed gradient reaches a threshold value, the fluid's state of rest—the stationary state in which heat is conveyed by conduction alone, without convection—becomes unstable. A convection corresponding to the coherent motion of ensembles of molecules is produced, increasing the rate of heat transfer. Therefore, for given values of the constraints (the gradient of temperature), the entropy production of the system is increased; this contrasts with the theorem of minimum entropy production. The Bérnard instability is a spectacular phenomenon. The convection motion produced actually consists of the complex spatial organization of the system. Millions of molecules move coherently, forming hexagonal convection cells of a characteristic size.

A computer, uncoupled from its electrical source, is a static structure requiring only the x, y and z axes to be unambiguously specified and hence is a 3-structure. In contrast, the computer in action, driven by a software program and dissipation of electrical energy, is a 4-structure, because the complete description of the moment-to-moment state of the internal components of the machine requires not only the x, y and z axes but additionally the time axis. Notice that no chemical changes take place within the machine during its operation, only the flow of electricity through its circuitry.

B. <u>4-Structures in Chemistry</u>: The first well-characterized chemical reaction capable of exhibiting spatio-temporal organizations of chemicals was discovered in Russia in 1958 by B. P. Belousov. This reaction was later extended by A. M. Zhabotinskii and hence is called the Belousov-Zhabotinskii (BZ) reaction (Epstein et al., 1987). The BZ reaction mixture usually contains (1) bromate, BrO_3^-, (2) a strong acid such as sulfuric acid, (3) malonic acid, $CH_2(CO_2H)_2$, and (4) a metal ion redox couple such as Ce^{+3}/Ce^{+4}, or Fe^{+2}/Fe^{+3}. As the reaction proceeds through a set of coupled intermediate steps, whose detailed characteristics have been well characterized (Field, 1985), the various intermediates (e.g., Br^-, reduced or oxidized metal ions, and CO_2) undergo changes in their concentrations that vary in space and in time, generating spatiotemporal patterns of chemical concentrations that are visible to the naked eye (Müller et al., 1985). The characteristic distances across which visible concentration gradients are formed is in the range of 10-10^2 microns, and the time constants of the gradient formation is in the neighborhood of tens of seconds or longer (Müller et al., 1985).

C. <u>4-Structures in Biology</u>: The spatiotemporal organization of chemicals as illustrated by the Belousov-Zhabotinskii reaction and literally the hundreds of other so-called self-organizing chemical reactions that have now been identified (see the series of articles in J. Chem. Educ., Volume 66, 1989) are not confined to purely chemical systems. The experimental evidence available firmly establishes the notion that spatiotemporal organizations of chemicals can occur inside the cell, giving rise to what has been called "intracellular dissipative structures (IDS's)" (Ji, 1985a,c, 1987, 1988, 1990), or the 4-structures that are confined within the cell. It is convenient to discuss the existing data under three separate headings: (1) cell-free systems, (2) biopolymeric systems, and (3) cellular systems.

(1)   4-Structures in Cell-Free Biochemical Systems

During the past two decades, B. Hess and his school in Dortmund have investigated numerous self-organizing biochemical reactions in cell-free systems such as glycolytic enzymes and mitochondria (Hess et al., 1975; Hess, 1983; Boiteux et al., 1980). Their investigations have clearly demonstrated that, under appropriate reaction conditions, cell-free biochemical systems can show temporal variations in the concentrations of glucose, NADH, ATP, FAD, reduced cytochrome b, K^+, and H^+. In addition, they have shown that propagating waves of NADH concentration can be generated. Therefore, we can conclude that cell-free biochemical systems can generate 4-structures; i.e., they can be constrained in space and time to form dissipative structures of Prigogine (1980; Ji, 1985a; 1988).

(2)   4-Structures in Intracellular Metabolism

During the International Seminar on the Living State held in Bhopal, India in November, 1983, I postulated that the dissipative structures composed of diffusible biochemicals can exist inside the cell. I later called these "intracellular dissipative structures (IDS's) (Ji, 1985a). I further proposed that there may exist two kinds of IDS's, namely space-dependent and time-dependent IDS's (see Figure 1.4). These conjectures appear to have been fully validated by the experimental measurements of intracellular calcium activities reported by numerous investigators using the calcium-sensitive fluorescent dyes, Quin2 and fura-2, developed by R. Tsien and his colleagues (Sawyer et al., 1985; Wier et al., 1987; Tank et al., 1988; Connor et al., 1988; Hernández-Cruz et al., 1990).

The first experimental evidence supporting the concept of IDS's that came to my attention was reported by Sawyer et al. (1985). Using Quin2, a dye whose fluorescence efficiency increases upon binding Ca^{++}, these investigators demonstrated that the intracellular calcium activity in a neutrophil increased several folds in the front half of the chemotaxing (i.e., moving toward some chemical signal) cell with little or no change in the tail half. They also showed that the intracellular calcium activity was much greater around the plasma membrane surrounding a phagocytosed (i.e., engulfed) particle. These observations clearly demonstrate that intracellular ion gradients can exist without any membrane barrier. One possible mechanism for creating such a membrane-free cytosolic ion gradient may involve two steps—(i) a chemical-mediated increase in the calcium conductance of the plasma membrane of the lamellipodium (i.e. the advancing edge of the cell), leading to an influx of extracellular calcium, and (ii) a voltage-dependent activation of the ATP-driven calcium pump located in the plasma membrane of the tail end of the cell, resulting in a rapid active transport of intracellular calcium out into the extracellular space. The combined actions of (i) and (ii) should lead to a steady-state intracellular calcium gradient of the correct polarity.

Using fura-2, another dye whose fluorescence efficiency increases with Ca^{++} binding, Wier et al. (1987) demonstrated that, in spontaneously contracting single rat heart cells, the spatial uniformity of the intracellular calcium activity can be interrupted periodically by spontaneously propagating waves of high intracellular calcium activity. These waves often originate as a single wave created in the middle

"Biocybernetics": A Machine Theory of Biology

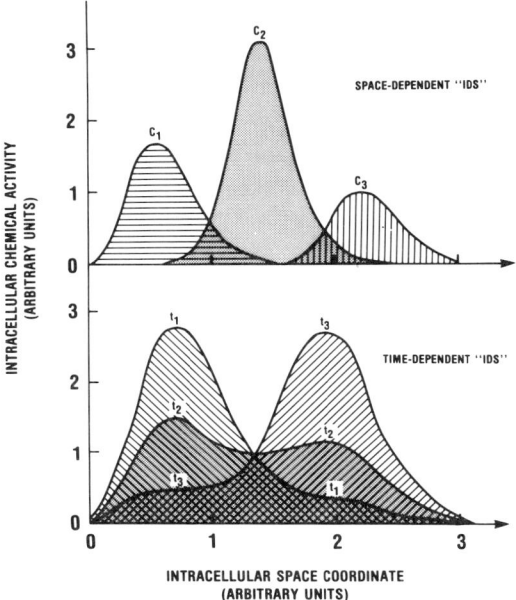

Figure 1.4. Hypothetical shapes of intracellular dissipative structures (IDS). The y-axis measures the concentration (or activity) of various chemicals, C_1, C_2, and C_3, and the x-axis represents the intracellular locations. IDS can also vary as a function of time (t_1, t_2, t_3). We can envision two types of IDS—the space-dependent and time-dependent IDS's. One possible mechanism by which space-dependent IDS's could be generated would be to have various enzymes responsible for producing associated products located at specified loci within the cell, for example, on specific sites on the cytoskeleton. Then the local concentration profiles of the products will be determined by the relative rates of product formation, the rate of their thermal diffusion and their lifetimes (see top panel). To create time-dependent IDS, one possibility is to distribute an enzyme in two loci, again possibly anchored on two different sites on the cytoskeleton, and to couple these two enzymic systems (electrically or mechanically through microfilaments, microtubles or membrane surfaces) in such a way that the activation of one enzyme is associated with the inhibition of the other. Such a coupled system of identical enzymes could give rise to chemical waves as schematically shown in the lower panel.

of the cell which then breaks up into two waves, each moving in the opposite directions. These moving waves provide unambiguous examples of the 4-structures of calcium ions present in the cytosol.

One of the interesting properties predicted by the dissipative structure theory of Prigogine is that systems existing far from equilibrium can, under a right set of experimental conditions, exhibit the phenomenon of "successive bifurcations" when the control parameter (which determines the distance from equilibrium) is varied consecutively (Prigogine and Stenger, 1984; Ji et al., 1986). This principle is illustrated schematically in Figure 1.5.

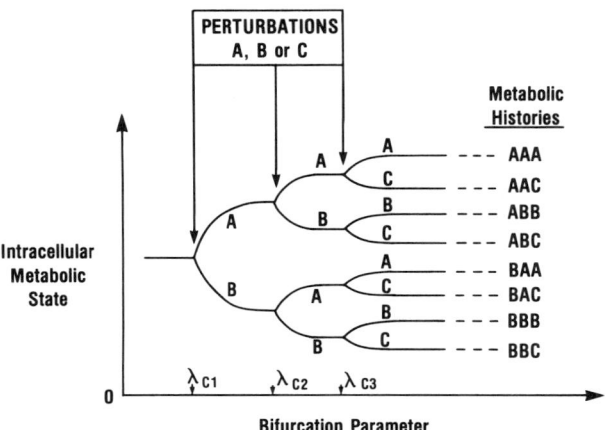

Figure 1.5. The bifurcation tree for cell metabolism. The y-axis represents the intracellular concentration of metabolites such as NADH or 7-hydroxycoumarin. The x-axis indicates the extent of the metabolic perturbations induced by substrate additions as measured by the increased free energy of the system relative. The points where metabolism bifurcate are called bifurcation points (λ_1, λ_2, λ_3). Depending upon the nature of perturbants (A or B) cellular metabolism can undergo successive bifurcations to reach new metabolic states represented by the symbols AAA, BAA, etc. These symbols reflect the order of perturbations and consequently the metabolic history of cells.

Such a "bifurcation tree," in turn, suggests the possibility that a pair of consecutive perturbations of a metabolic system may lead to two different end states, depending on the order of the perturbations, thus imparting to the system the ability to "memorize" its past history; i.e., the metabolic system's current state reflects its past history, an example of temporal symmetry breaking and a support for IDS.

We tested this idea experimentally by measuring the effects of adding two different substrates, ethanol and 7-hydroxycoumarin, to the isolated perfused rat liver on the fluorescence changes of the liver tissue measured with the tissue photometric technique developed by B. Chance (Ji et al., 1986). As shown in Figure 1.6, the final level of the tissue fluorescence was found to be much greater when ethanol was added first followed by 7-hydroxycoumarin than when the order of the substrate addition was reversed.

The biochemical mechanisms underlying these fluorescence changes, which are most likely very complex, need not concern us here, since these mechanistic considerations are not essential in arriving at the conclusion we desire; i.e., the final states of metabolism that the perfused rat liver assumes following two consecutive

perturbations are not identical and depend upon the order of perturbations. We have called this phenomenon the "non-commutativity of the order of substrate addition" (Ji et al., 1986), in analogy to the non-commutativity of the order of applying the position and momentum operators to the Schrödinger equation in quantum mechanics (Pauling and Wilson, 1963). For convenience, we may refer to this phenomenon simply as the "metabolic non-commutativity." The significance of the metabolic non-cummutativity in the liver observed in Figure 1.6 is that this supports the idea of liver metabolism being far from equilibrium, thus satisfying one of the two requirements for the existence of IDS in liver cells.

A more direct evidence for the "metabolic non-commutativity" has been reported by Medina and Númez de Castro (1988). These authors measured the changes in the redox state of the mitochondrial electron transport chain in Ehrich ascites tumor cells following addition of glucose (5 mM) and glutamine (0.5 mM). It was found that cytochromes c, b and $(a+a_3)$ exhibited different redox states (10-40% variations), depending on the order of the substrate addition, although the redox state of cytochrome c was found to be unaffected. They concluded that the Ehrich ascites tumor cell respiratory chain functions far from equilibrium and thus is capable of supporting IDS's.

(3)   Biopolymeric 4-Structures

It is customary to think of biopolymers (proteins, DNA and RNA) only as 3-structures, perhaps influenced by their elegant (and yet static) X-ray crystallographic structures. A little reflection would reveal that biopolymers can also exist as 4-structures, since they are endowed with the ability to perform a series of molecular motions in space (e.g., chromosomal rearrangements during mitosis and meiosis, the sliding of the thin and thick filaments of the actomyosin system during muscle contraction, and the "evolutionarily choreographed" motions of active site residues of enzymes during catalysis (Ji, 1974b, 1979), etc.) and in time (e.g., microtubular oscillations (Ueda et al., 1986)). Metaphorically, it may be said that the 3-structures of biopolymers are to their 4-structures, what dancers at rest are to them dancing (This may answer W. B. Yeats' question, "How can we know the dancer from the dance?"), or what the inactive state of a digital computer is to its active state. In relativity theory, 4-structures are referred to as events or spacetime (Wald, 1977). Thus, biopolymeric 4-structures can be viewed as biopolymeric events or catalytic events.

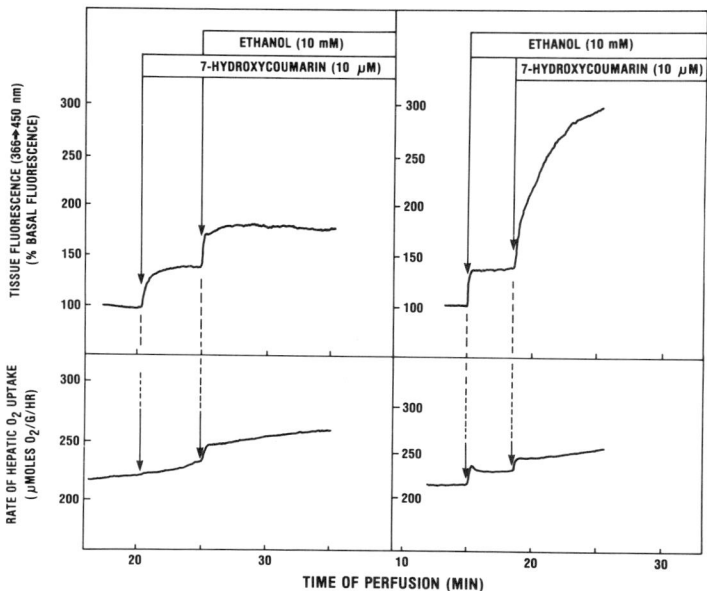

Figure 1.6. The metabolic non-commutativity. The rat liver was isolated under pentobarbital anesthesia and perfused at 37°C with the Krebs-Henseleit bicarbonate buffer equilibrated with 95% O_2 and 5% CO_2. The perfusate was pumped through the liver at a constant flow rate (3 to 4 mls/min/g liver) via a cannula inserted into the portal vein. The liver tissue was illuminated with the near-UV light peaking around 360 nm and the resulting fluorescence was measured at 450 nm using a bifurcated quartz lightguide with 2 mm tips. The signal was detected by an EMI photomultiplier and further amplified using a Johnson Foundation (Univ. of Pennsylvania) fluorometer. The 450 nm fluorescence reflects the tissue contents of both reduced pyridine nucleotides and 7-hydroxycourmarin. 7-Hydroxycoumarin was dissolved in the Krebs-Henseleit buffer and was infused into the liver by switching the perfusate source from the control to the 7-hydroxycoumarin containing buffer via a three-way stopcock. Ethanol was introduced through the portal vein by an infusion pump. The rate of oxygen uptake was measured with a Clark-type electrode.

To cause a biopolymer to undergo a transition from a 3-structure state to a 4-structure state requires an input of Gibbs free energy to the biopolymeric machine;

$$M^{bp}(\text{3-state}) + \Delta G \rightarrow M^{bp}(\text{4-state}) + \text{heat} \quad (1.18)$$

where M^{bp}(3-state) and M^{bp}(4-state) indicate biopolymers in a 3-structure state and a 4-structure state, respectively, and ΔG is the Gibbs free energy change accompanying the biopolymeric state transitions. In another respect, M^{bp} (3-state) can be regarded as the ground state of a biopolymer, and M^{bp} (4-state) as its "excited," "energized," or "activated" state. Equation (1.18) can be incorporated into Equation (1.9) discussed in Section 1.5, by assuming that the input set I^{bp} includes ΔG, while the output set, O^{bp}, includes heat. In the traditional cybernetics literature discussing macroscopic machines, ΔG and heat are not explicitly mentioned, because almost all machines are driven by the same source of energy, namely electricity. But in discussing biopolymer machines, we cannot leave out ΔG, because the mechanism of ΔG utilization may be intimately linked to the mechanisms of biopolymer functions.

The fundamental consequences of activating biopolymers via Equation (1.18) are the breaking of symmetries in the distribution of chemicals both in space and time through their catalytic cations. The concept of symmetry breaking is borrowed from particle physics and statistical mechanics (Prigogine, 1977, 1978, 1980; Anderson and Stein, 1984) and is here employed to indicate the most elementary processes underlying all organizations of matter in space and time. Depending on whether the effects of the spatiotemporal symmetry breakings are confined mainly within or outside the biopolymeric machine involved, we can recognize "intramolecular" and "extramolecular" spatiotemporal symmetry-breakings.

(a) <u>Intramolecular spatiotemporal Symmetry-Breakings by the 4-Structures of Biopolymers</u>: When soluble enzymes are activated through substrate binding under thermally fluctuating conditions, enzymes catalyze specific chemical steps with specific rate constants (Ji, 1974b, 1979); i.e., they carry out what may be called a "catalytic program," defined as a linear sequence of ordered pairs, each pair consisting of the substrate(s) and the associated (or "conjugate") time duration (or time delay) elapsed between substrate binding and the time of the catalytic act (i.e., electronic rearrangements) specific for the bound substrate. The execution of a catalytic program is synonymous with a series of time symmetry breakings. Therefore, enzymes can be viewed as "time symmetry-breaking molecular machines" driven by the chemical reactions that they catalyze.

Accompanying these temporal symmetry breakings are the "spatial symmetry-breakings" that the Frank-Condon principle requires for the electronic transitions underlying the catalysis of specific chemical steps (Ji, 1974b). Clearly, these spatial symmetry-breakings are confined within the active site of enzyme molecules, and the release of products resulting from the catalysis occurs more or less isotropically, unless the enzymes are anchored in an anisotropic environment.

(b)   **Extramolecular Spatiotemporal Symmetry Breakings by the 4-Structures of Biopolymers**: One of the most dramatic examples of the spatiotemporal symmetry-breaking induced by the activation of biopolymers has been reported recently by B. Hess's group in Dortmund. They have demonstrated that a homogenous solution of tubulin molecules can generate waves of microtubule polymerization that can propagate with a speed of 0.015 mm/sec (Mandelkow et al., 1989). Since the microtubule formation is driven by the hydrolysis of GTP into GDP and inorganic phosphate catalyzed by individual tubulin monomers, we can regard the spatiotemporally organized polymerization—depolymerization reaction as a sort of an amplification of the spatiotemporal organization occurring inside the catalytic sites of individual tubulin enzymes. That is, the extramolecular spatiotemporal symmetry-breakings are supported by intramolecular spatiotemporal symmetry-breakings. It may be that *no extramolecular spatiotemporal symmetry-breakings can occur in the cell without the associated intramolecular spatiotemporal symmetry-breakings driving them.*

1.8.3. Conformons-IDS Interactions

The act of symmetry breakings, whether in space or time, can be regarded as the most fundamental process underlying spatiotemporal organizations of matter, regardless of whether the symmetry breakings occur on the microscopic or the macroscopic scale. Symmetry breakings are synonymous with "selecting" one out of at least two available choices ("this" over "that," "now" over "then," and "here" over "there," etc.) and hence associated with at least one bit of Shannon information (Pierce, 1980). In addition, the laws of thermodynamics dictates that no act of selection of any kind can be carried out without appropriate dissipation of free energy (Prigogine, 1977). Therefore, it follows that no symmetry breakings of any kind are possible without (1) at least one bit of Shannon information, and (2) the dissipation of the requisite amount of free energy. According to the biognergetics theory (Ji, 1985a, 1990), these two requirements are met by conformons in biopolymers and IDS in the cytoplasmic phase.

We can recognize four types of spatiotemporal symmetry breakings inside the cell:

"Type I" = conformon-driven symmetry breakings (e.g., enzymic catalysis, effects of binding proteins on DNA transcription, etc.).

"Type II" = IDS-driven symmetry breakings (e.g., action potentials, MPF

"Biocybernetics": A Machine Theory of Biology

(maturation promoting factor)-driven mitosis (Murray and Kirschner, 1989)).

"Type III" = the conformon-induced IDS (i.e., all IDS driven by enzyme-catalyzed chemical reactions).

"Type IV" = the IDS-induced conformons (e.g., DNA conformational transitions driven by changes in pH or ionic strength or membrane receptor activation-induced DNA covalent modifications (Holliday, 1989)).

Because of these multiple mechanisms, it is possible for the symmetry breaking influences to propagate from the interior of enzymes to the limits of cell boundaries and beyond (Type III), or from extracellular space to DNA or enzymes in the cytosol (Type IV). In other words, the distance scale over which various symmetry-breaking influences can be exerted can extend from the molecular dimensions (i.e., 1Å or 10^{-10}m) to the cellular dimensions (i.e., 10 microns or 10^{-5}m) (or *vice versa*), representing an amplification (or contraction) of distance by a factor of about 10^5. The net consequence is that the cell can convert the genetic information stored in DNA into its dynamic functions, and *vice versa* (see Section 1.12.5).

1.8.4. The Bhopalator: A Theoretical Model of the Cell

Two related ideas are essential in formulating the coherent and comprehensive theoretical model of the living cell that I named the Bhopalator: (1) It is possible to encode rate constants in the primary amino acid sequence of enzymes (and hence in the genome), and the sequence information of enzymes can be transduced into specific rate processes in the cell through the conformon mechanism of enzymic catalysis (Ji, 1974b), and (2) Prigogine's dissipative structures can act as the link between the sequence information of biopolymers and cellular functions (Ji, 1985a,c; 1988).

Using the terminologies and concepts introduced earlier in this chapter, we may summarize the essential features of the Bhopalator as follows:

(1) The cell consists of two major classes of machines—biopolymeric (i.e., DNA, RNA, proteins), and metabolic machines (i.e., basic metabolic and suprametabolic pathways (see Equation 1.12 and Table 1.11)).

(2) Both biopolymers and metabolic pathways can exist in either the "resting" or the "activated" states, the latter requiring an input of Gibbs free energy. The resting states are examples of 3-structures and the activated states represent 4-structures, or events.

(3) The activated state of a biopolymeric machine can be characterized in

terms of its unique set of conformons whose movement in space and time determines the spatiotemporal structures (or 4-structures) of biopolymers and hence their biological functions, just as the spatiotemporal distribution of the voltage values over the transistors in a computer chip determines its computing functions.

(4) The activated state of a metabolic machine is associated with a unique distribution of a set of diffusible chemicals around it, like a field around electrically charged bodies. A set of activated metabolic machines in the cell generates the intracellular gradients of molecular species called IDS; and IDS serve as the immediate (i.e., proximate) causes for all cellular functions, just as conformons are the immediate causes for all biopolymeric functions.

(5) Conformons and IDS can be interpreted as representing the internal states of biopolymeric machines (i.e., S^{bp} in Equation (1.9)) and metabolic machines (i.e., S^{mp} in Equation (1.12)), respectively.

(6) Conformons and IDS contain both Gibbs free energy and genetic information and hence are two different forms of gnergons, hybrid entities derived from the fusion of ergons (carrying free energy) and gnons (carrying information). Biopolymers and metabolic pathways in the resting states are examples of gnons while their activated states represent gnergons.

(7) Gibbs free energy and Shannon information can be exchanged between conformons and IDS's. The spatial amplification (or dilation), namely spreading of the spatiotemporal symmetry-breaking influences from the enzyme interior to the cytosol requires the conversion of conformons to IDS (i.e., Type III spatiotemporal symmetry-breakings), while the spatial contraction, namely the "focusing" of the spatiotemporal symmetry-breaking influences originating from the bulk cytosolic phase onto biopolymers, requires the conversion of IDS into conformons (i.e., Type IV spatiotemporal symmetry breakings).

(8) Through the interconversions and couplings possible between conformons and IDS, the living cell can transmit, receive, or transduce information both within the cell (e.g., from DNA to enzymes) and between the cell and its environment (e.g., hormone-induced activation of gene expression), a prerequisite for ontogeny (development of individual organisms) and phylogeny (evolutionary development of species) of living systems.

1.8.5. The Bhopalator in the Molecular and Algebraic Representations

The cell model incorporating all of the features listed in Section 1.8.4. is schematically shown in Figure 1.7.

The Bhopalator divides the overall cellular processes into 20 major steps:

"Biocybernetics": A Machine Theory of Biology 79

DNA replication necessary for cell division, differentiation and evolution (step 1); DNA transcription and RNA splicing whereby genetic information is transferred from DNA to mRNA (step 2); translation in which the genetic information of mRNA is further transferred to the linear sequence of amino acids in polypeptides (step 3); expression (Monod, 1971) whereby polypeptide chains synthesized on ribosomes fold into catalytically active 3-dimensional proteins influenced by intracellular microenvironment (step 4); substrate binding in which specific amino acid side chains (labeled w, x, and y) and substrates interact to form an enzyme-substrate complex (step 5); activation where the thermal fluctuations of the enzyme-substrate complex create a transient conformational strain at the active site called the Franck-Condon conformon represented by an alignment of catalytic side chains w, x, y and z which is intermediate between the substrate-binding (w, x and y) and the product-binding (x, y and z) conformational strains, the Franck-Condon conformon facilitating the conversion of the substrate to an intermediate, I, by quantum mechanical tunneling (step 6); deactivation whereby the highly strained conformation of the active site in the transition state relaxes into the product-binding conformation, forcing the conversion of I into the product, P (step 7); product release (step 8) by which the newly formed P dissociates from the enzyme active site and contributes to the creation of intracellular dissipative structures (IDS), with the enzyme returning to its initial conformation ready for the next cycle of catalysis (step 9); and mutation and recombinations (step 10), leading to the production of new genes under the influence of local environment determined by IDS (e.g., local electrical field gradients, pH gradient, mechanical stress gradients in DNA, etc.). Steps 11 through 18 (represented by dotted lines) indicate feedback control interactions between IDS's and conformons of biopolymers, thus forming negative or positive feedback control loops. Finally, steps 19 and 20 depict the reception by the cell of information originating from its environment (including neighboring cells) and the output of information by the cell to its environment, respectively. Although not explicitly indicated, various receptors located in the plasma membrane, the cytosol, or the nucleus can be treated as pseudo-enzymes and members of biopolymer machines (see Table 1.10, footnote 4).

The living cell depicted in Figure 1.7 can be viewed as a complex machine and thus is amenable to description by the algebraic formalism used in cybernetics (see M. Holcombe, Chapter 4). The following equation encapsulates the essential properties of the living cell;

$$M^c = (I^c, S^c, O^c, d^c, l^c) \qquad (1.19)$$

where M^c indicates the cell (c) machine; I^c is the set of inputs into the cell, including Gibbs free energy in the form of exergonic (i.e., free energy-releasing) chemical reactants and photons from the sun, and chemical messengers; S^c is the set of the internal states of the cell (called the "cell states" in analogy to the electronic, vibrational and rotational states of atoms and molecules (Ji et. al., 1986)) characterized by the spatiotemporal distributions of conformons and IDS; O^c is the set of outputs by the cell, including heat, chemical products, and mechanical work on its environment such as cell division (leading to the control of the Euclidean space), chemotaxis, and phagocytosis; d^c is the relationship (or mechanisms of interactions) between the inputs and the internal states of the cell (to be called the "input program of the cell"); and finally l^c is the relation (or mechanisms of interactions) between the internal states of the cell to its outputs (to be called the "output program of the cell"). Clearly, this algebraic representation of the cell fits the universal scheme representing machines in general depicted in Figure 1.1.

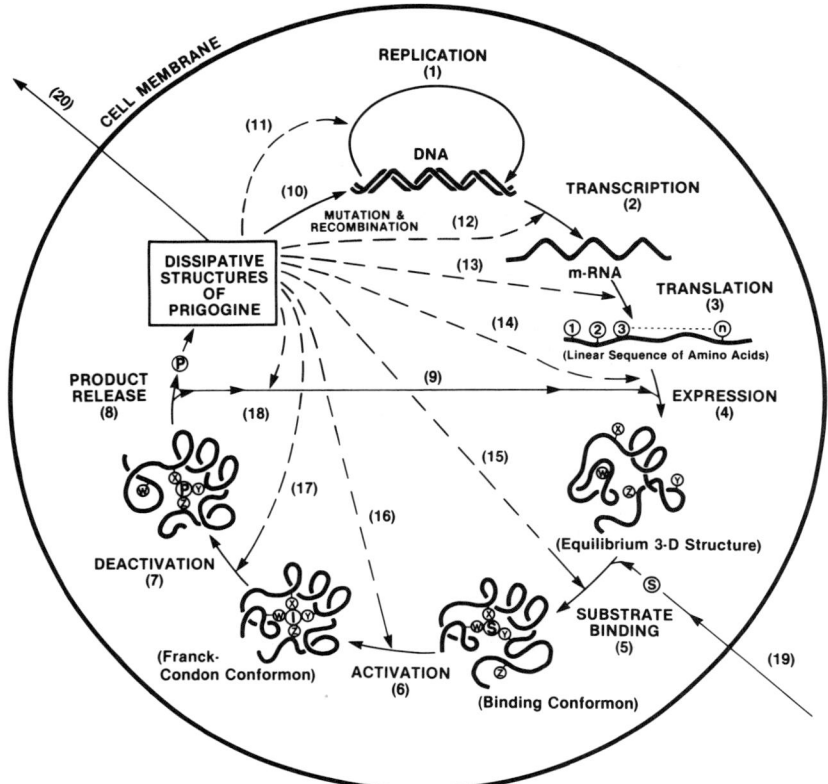

Figure 1.7. A molecular model of the living cell. It was first proposed at the Second International Seminar on the Living State -II, held in Bhopal, India on November 13-20, 1983.

"Biocybernetics": A Machine Theory of Biology

1.8.6. The Cell as a Molecular Computer

The cell as modeled by the Bhopalator reveals striking similarities to the digital computer (see Table 1.14). Just as a computer is designed to transform inputs to outputs according to computer programs, so the cell can be viewed as a physical system of a microscopic dimension that has evolved to transduce (via S^c in Equation (1.19)) inputs (I^c) to outputs (O^c), according to the "cell programs" (d^c and l^c) stored in it.

The concept of the "cell program" emerges naturally from the Bhopalator. This concept is related to, but not identical with, the genetic programs stored in DNA. The cell programs presuppose the existence of genetic information in the cell in two forms—(1) the traditional, static sequence information stored in biopolymers, and (2) the new form of information stored in dynamic, spatiotemporally organized distributions of diffusible molecular species, namely IDS. The former is called the "Watson-Crick form" of genetic information, and the latter the "Prigoginian form," and biopolymers are postulated to interconvert these two forms utilizing conformons (Ji, 1988) (also see Section 1.8.3).

Table 1.14. A Comparison between the Cell and the Computer

Parameter	Cell	Computer
Current carrier[1]	Electrons	Electrons
Current conductors[2]	Chemicals	Wires
Current controller[3]	Enzymes	Transistors
Mechanisms of control[4]	Conformons	Photons

(Table 1.14 continued)

Electron source	High-energy chemical	External voltage
Electron sink	Low-energy chemicals	Ground
Memory	Biopolymers IDS	Flip-flops Capacitors
Structural rigidity[5]	Thermally fluctuating	Thermally robust
Size	Microscopic (ca. 10^{-3} cm)	Macroscopic (ca. 10^2 cm)
Program location	Biopolymers	Softwares
Self-reproductibility[6]	Yes	No
Model[7]	Bhopalator	Turing machine

---

[1] Chemical reactions are not usually discussed in terms of electron flows, but I believe it is legitimate to view all chemical reactions (and not just redox reactions) as resulting from shifts in electron densities within the molecules participating in electronic rearrangements, leading to the conclusion that the only difference between electric currents through conductors and chemical reactions is that the former involves "continuous" electron flows in the macroscopic scale ($\geq$ 50-100 nm ?) and the latter involves discrete electronic jumps in the microscopic scale (probably $\leq$ 0.1 nm), a scale difference of at least 10^2-10^3 fold.

[2] Electron movements in chemical reactions can be divided into two parts—(1) the electronic density shifts associated with the electronic rearrangements occurring at the transition state of chemical reactions, and (2) the diffusion of chemicals that precedes or follows the electronic rearrangements. Therefore, the current flow associated with chemical reactions is spatially random due to the thermal fluctuations essential for diffusion, while the current flow in computers is deterministic due to the constraints imposed upon it by the macroscopic wires of the computer circuitry.

[3] Enzymes regulate the time evolution of enzyme-catalyzed chemical reactions through spatiotemporal symmetry-breakings driven by conformons as discussed in connection with Equation (1.18). Since the progress of chemical reactions is analogous to current flow in electronic circuitry, enzymes can be compared

"Biocybernetics": A Machine Theory of Biology

The concept of the Prigoginian form of genetic information requires that there exist mechanisms for transmitting the information stored in IDS from the mother cell to daughter cells. Since IDS can control enzymic activities (e.g., intracellular pH and Ca^{++} affecting metabolism) and since "maintenance methylase" can methylate DNA, it is possible that the cytosolic IDS information can be transduced into the information encoded in the pattern of DNA methylation, which can then be transmitted to daughter cells (Holliday, 1989).

It was postulated that the reason for the existence of these two forms of genetic information is that the Watson-Crick form is necessary for transmitting genetic information in the time dimension, while the Prigoginian form is required for transmitting genetic information in the spatial dimension, just as musical notes and sound waves are needed for transmitting musical information in time and in space, respectively (Ji, 1988). Such a clean separation of the two forms of genetic information may be possible only over the time and length scales that are greater than those accessible to molecular species through their thermal fluctuations, because in biopolymers information transmission in space and time appears to be coupled in the form of conformons.

1.8.7. The Bhopalator in a Field Theoretic Representation

A. <u>What is a Field?</u>: In physics, the concept of a field began with the discovery of the electric and magnetic fields (Adair, 1987). A field in general refers

---

(Notes to Table 1.14 continued)

to transistors which control electron flow in computer circuity.

[4] Just as the actions of transistors can be described in terms of the spatiotemporal distributions of photons, energy packets mediating all electromagnetic interactions, enzymic actions may be described in terms of the spatiotemporal distributions of conformons, the ultimate physical objects responsible for the biological actions of biopolymeric machines (see Section 1.5).

[5] Thermal fluctuations are essential for all enzymic catalyses, whereas thermal fluctuations of the circuit components of silicon chips are generally detrimental to computer functions (see 5.1.). This may be one of the most fundamental differences between computers and biological machines and serve as the ultimate barrier to the miniaturization of computers.

[6] Including not only the central processing unit but also input and output devices.

[7] The mathematical model of the universal computer was formulated by Turing in the 1940's (Hopcroft, 1984). The Bhopalator, although expressible in an algebraic form, has not yet led to a rigorous mathematical model.

to a quantity defined at each point throughout some region of space and time ('t Hooft, 1980). There are two kinds of fields. A "scalar field" has only a magnitude associated with each point (e.g., the distribution of temperature values of each point on a frying pan). A "vector field" associates each point with a magnitude and direction (e.g., the distribution of the velocity of each molecule in a fluid).

B. <u>The Intracellular Metabolic Velocity Field (IMVF)</u>: The concept of vector field is useful in visualizing the instantaneous metabolic state of the cell. This can be represented schematically as a collection of arrows distributed inside the cell as shown in Figure 1.8. Each arrow conveys three kinds of information—(1) the origin of each arrow coincides with the center of mass of an enzyme or an enzymic system located in the cell at a given time point, (2) the magnitude of an arrow indicates the instantaneous rate of the chemical reaction catalyzed by the enzyme, and (3) the orientation (not in the Euclidean space as shown but in an abstract "internal coordinate" space; see Section 1.8.1. and Equation 1.17) reflects the nature of the chemical reaction involved. All these three quantities of the vectors can vary with the location of enzymes within the cell and with time; i.e., each vector is a function of space and time. For convenience, we will call such a vector field the "intracellular metabolic velocity field (IMVF)."

The IMVF depicted in Figure 1.8 conjures up an image of a living cell in which all the arrows undergo constant fluctuations and motions, showing changes in their positions, directions, and sizes. As some enzymes get degraded and others newly synthesized, certain arrows may disappear altogether (annihilation of an IMV vector) from the scene and some others may appear suddenly out of nowhere (creation of an IMVF vector). Still other arrows may collide and thereby undergo changes. Collisions are not the only mechanisms by which arrows can interact; they can exchange diffusible chemicals as a means of interactions. All these mental images are reminiscent of particle physics, wherein the interactions among numerous high-energy particles have been successfully characterized using field theories (Adair, 1987).

C. <u>The Biopolymeric Force Field (BFF)</u>: The mechanism underlying each arrow in Figure 1.8 can be analyzed in terms of the interaction between the catalytic properties of an enzyme and the local concentration of chemical substrates and cofactors for that enzyme. For this purpose, it is necessary to introduce two more fields—the biopolymeric force field (BFF), and the intracellular chemical concentration field (ICCF).

"Biocybernetics": A Machine Theory of Biology

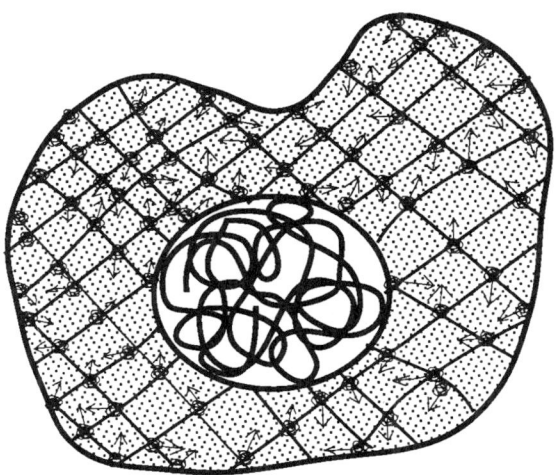

Figure 1.8. The Intracellular Metabolic Velocity Field (IMVF). The lines represent the cytoskeletal network, highly stylized, the little wiggly "particles" or "globules" positioned at the intersections are enzymes, and the small dots signify diffusible chemical species. Each enzyme possesses a vector, whose magnitude is proportional to the rate of the chemical reaction catalyzed by the enzyme at any given time and whose direction indicates the kind of the chemical reactions being catalyzed.

The arrows in Figure 1.9 schematically depict an instantaneous distribution of the force vectors of the catalytic sidechains of the enzymic active site. Of course these force vectors are determined by the vector sum of all of the other force vectors generated all along the polypeptide chain, but these are not shown for simplicity. Again, it is not difficult to imagine how the time evolution of these active site vectors would look like as the enzyme involved carries out its catalytic function. The spatiotemporal distribution and the changes in size and direction of the force vectors would look similar to the dynamics of the arrows in IMVF but much more constrained due to the linear linkage of the arrows in the polypeptide chain. We will call this vector field the "biopolymeric force field" to be designated with the symbol $\vec{F}$. Note that the collection of the force vectors is identical with a conformon located at the active site at the same time point; i.e., conformons can be viewed as the quanta of the biopolymeric force field, just as photons are the quanta of the electromagnetic force field.

D. <u>The Intracellular Chemical Concentration Field (ICCF)</u>: The living cell can be viewed as a collection of a finite number of material particles whose concentration inside the cell varies in both space and time. The intracellular material particles can be conveniently divided into two broad categories— (i) l-particles (l from "light") composed of small-molecular-weight chemical species such as protons, metal ions, inorganic phosphate, ATP, etc., and (ii) h-particles (h from "heavy") composed of biological macromolecules such as nucleic acids, proteins and carbohydrates. The most important property that differentiates l-particles and h-particles is probably their diffusibilities under physiological conditions of temperature and pressure. The diffusion coefficient (D) is defined as the number of particles per second crossing unit

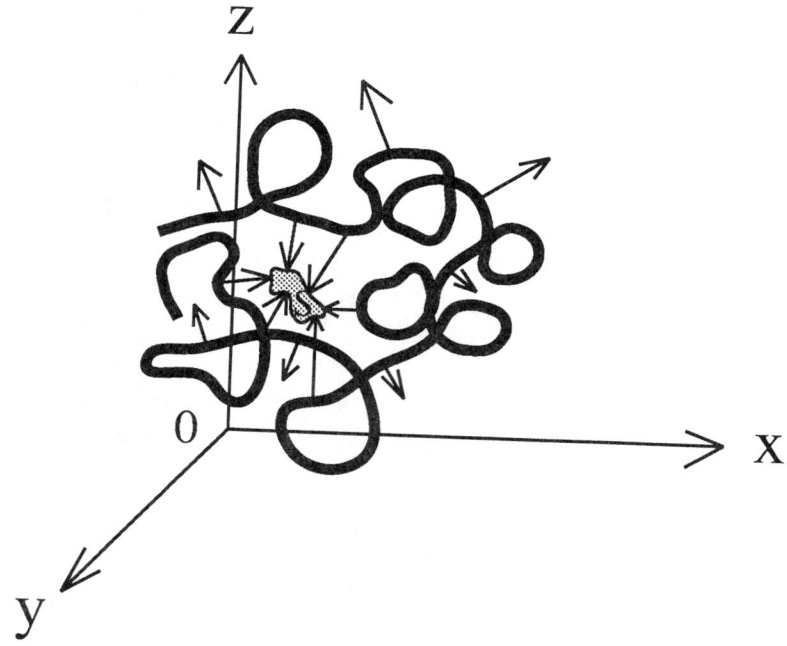

Figure 1.9. Biopolymeric Force Field (BFF). The thick curves indicate the 3-dimensional arrangement of the protein backbone, side chains having been excluded for clarity. The arrows represent the instantaneous forces acting along the representative points of the backbone. The substrate (S) is being acted upon by a set of 6 force vectors at time t. Unlike the arrows in Figure 1.8, those in Figure 1.9 represents the direction of the force vectors in the Euclidean space.

area under unit concentration gradient and is inversely proportional to the square root of the molecular weight (M) of the particle; i.e., $D \propto (M)^{-1/2}$. Assuming that the average molecular weight of l-particles is 10^2 daltons and that of h-particles 10^6 daltons, we can conclude that the diffusion coefficient of l-particles is greater than that of h-particles by a factor of about 10^2.

To completely describe the dynamic (i.e., quantum statistical mechanical) states of all these intracellular particles, it is necessary to have a mathematical or logical space composed of the 4-dimensional physical space of x, y, z, and t (to describe the time-dependent positions of all the particles inside the cell) and an N-dimensional "internal space" (to specify the chemical transformations of the particles, where N may be identified with the variety of the chemical species participating in the intracellular metabolism). Such a (N + 4)-dimensional space will be referred to as the *intracellular chemical concentration field (ICCF)* (see the dots (l-particles) and wiggly globules (h-particles) in Figure 1.8). So defined, any objects describable in ICCF can be identified with intracellular dissipative structures (IDS), discussed in Section 1.8.2. and Figure 1.4. In other words, IDS can be viewed as subsets of the elements of (or topological objects in) ICCF that have been selected by biological evolution owing to their essential roles in the survival of the cell. ICCF is to IDS what the ocean is to oceanic waves, what the field of mass densities in air is to sound waves, what the electromagnetic field is to radio signals in electrical communication systems (see Section 1.8.11. for related discussions), and finally what marble (i.e., the material cause of Aristotle) is to marble statutes (i.e., the formal cause). In the parlance of electrical communications theory, ICCF may be equated with the communication channel and IDS with the signals carrying messages through that channel (Pierce, 1980). It was for these reasons that I pointed out several years ago that IDS might represent a new form of genetic information, alternative to the DNA sequence information, and suggested that we recognize two distinct forms of genetic information—the Watson-Crick form (analogous to musical notes) and the Prigoginian form (analogous to sound waves) (Ji, 1988).

E. <u>A Field Representation of the Living Cell</u>: Since the intracellular concentration of any chemical species would be determined by the balance of the rates of its formation and removal at any given time and location inside the cell, ICCF is a function of IMVF. At the same time, the rates of chemical transformations of any given chemical species are influenced by both its intracellular concentrations and the activities of the enzymes catalyzing the reactions involved; therefore, IMVF is clearly a function of ICCF and BFF. However, ICCF in general cannot directly affect IMVF, because most if not all of intracellular chemical reactions do not proceed until and unless catalyzed by appropriate enzymes. This contrasts with nonenzymic

chemical reactions where chemical concentrations directly influence the rates of their chemical transformations according to the laws of chemical kinetics, a hybrid discipline of quantum mechanics and statistical mechanics (Frost and Pearson, 1961). It appears as though biological evolution has selected for cellular metabolism only those chemical reactions that do not proceed in the absence of enzymes, a condition without which evolution cannot control or regulate cellular metabolism and cell functions through the modulation of enzyme structures. Therefore, enzymes must carry genetic information—i.e., proteins must act as a communication channel to transmit genetic information (see Appendix 1.B for related discussions). All of these complex interactions among IMVF, ICCF and BFF and their relations to cell functions may be schematically represented as shown in Equation (1.20).
where the arrows indicate the exertion of unidirectional influences or controls.

Step 4 in Equation (1.20) represents those metabolic processes in which the mechanical state of biopolymers directly influences the cytoplasmic concentrations of chemical species. Examples of such processes include the set of metabolic reactions generally known as "energy-coupled" processes (Ji, 1974a, 1979) such as oxidative phosphorylation, active transport, and muscle contraction, where the free energy stored in the form of the conformational strains of proteins (i.e., conformons) are thought to drive the endergonic (i.e., free-energy-consuming) partial processes (e.g., ADP phosphorylation, transmembrane transport of ions, sliding of the thin filament relative to the thick filament, etc.). In addition, such conformon-driven processes may be implicated in the structural transformations of DNA underlying, for example, (i) recombinations that may involve long-range protein-protein interactions through DNA looping (Heichman and Johnson, 1990) and (ii) the enhanced binding by a repressor at a low-affinity site in the presence of a high affinity site.

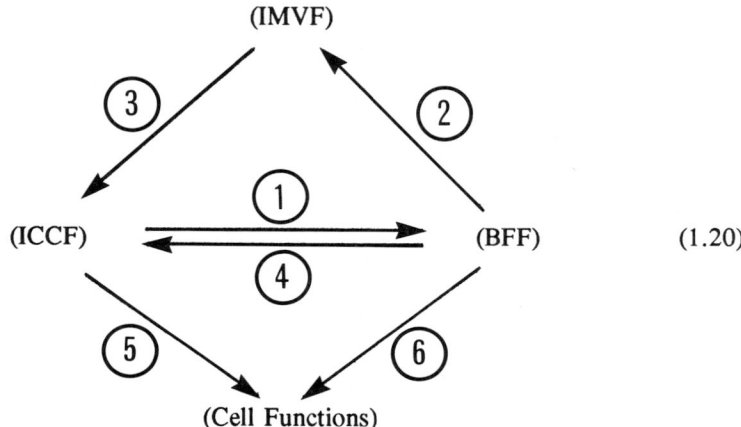

(1.20)

Just as IDS can be viewed as the signals transmitting genetic information through the communication channel composed of ICCF, so conformons can be considered as the signals transmitting genetic information through the communication channels composed of biopolymers, namely nucleic acids and proteins. Again just as IDS represent a new form of genetic information as indicated above, so conformons can be treated as still another form of genetic information, which may be called the "conformon form" of genetic information (see Table 1 in Ji, 1988). It is interesting to note that the Prigoginian form of genetic information is effectuated (i.e., converted into cellular actions) through step 5 and the conformon form through step 6 in Equation (1.20).

Two features about Equation (1.20) seem to stand out—(i) The unidirectionality of the interactions between BFF and IMVF on the one hand (step 2) and IMVF and ICCF on the other (step 3). (ii) The duality of the mechanisms of cell functions, namely through ICCF and BFF (steps 5 and 6). The unidirectional interaction between IMVF and ICCF has been already explained above as the result of the fact that biological evolution has selected for cellular metabolism only those chemical reactions that absolutely depend on enzymes for their progression. The unidirectionality of step 2 may be considered as the inevitable consequence of the same fact. The cell functions that depend on ICCF (step 5) include all the secretory or excretory functions of cells such as endocrine cells secreting hormones and metabolic cells releasing nutrients and structural and transport molecules into circulation. The cell functions mediated by BFF (step 6) should include cellular motility underlying cell shape changes, cell division, and chemotaxis. In general, the ICCF-mediated cell functions have a larger domain of influences than the BFF-mediated cell functions; i.e., the ICCF-mediated cell functions are long-ranged, while the BFF-mediated cell functions short-ranged, just as the electrical communications are effective over much greater range of distances than the communications utilizing sound waves. The evolution of these two kinds of cellular effector mechanisms must have provided the cell with a greater degree of control over its space and time, leading to a greater adaptability to its changing environment.

Based on the above discussions, it can be concluded that Equation (1.20) is a field-theoretic representation of the living cell. In another sense, Equation (1.20) can be viewed as describing what may be called the "spacetime-elaborated" or "spacetime-modulated" chemical concentration field conducive to cell life. By "spacetime-modulated," I simply mean that the concentrations of the chemical species inside the cell vary, above and beyond the variations allowed for by nonenzymic reactions alone, as functions of space (x, y, and z), time (t), and genetic information (I) encoded in enzymes. This may be analogous to the general relativity theory, wherein the gravitational force field can be equivalently represented by the set of numbers called "curvatures" that depend on space (x, y, and z), time (t), and the distribution of matter (m); i.e., the presence of matter curves spacetime, and matter moves along the shortest route (i.e., the geodesic) on the curved spacetime (for related discussions see Sections 1.8.9 and 1.12.1 of this chapter, and Smith and Welch in Chapter 6). Just as the *gravitational force* can be thought of as "elaborating" or "manifolding" (see Section 1.8.8) the curvature of spacetime from the uniform and flat one to a heterogeneous and "bumpy" one with peaks and valleys (Figure 6.1 in Chapter 6), so the *cell force* elaborates the structure of the intracellular chemical concentration field (ICCF) at every point in space and time to produce highly heterogenous and complex structures called IDS; i.e., IDS can be regarded as an indication of the existence of a *new force* in the cell called the cell force, just as the bending of light around the sun is an indication of the existence of the gravitational force around the sun. Furthermore, *gravitational force* is "caused" by the presence of *matter*; similarly, the *cell force* can be thought of as being "caused" by the presence of the *genetic information* encoded in biopolymers.

1.8.8. Manifolds, Fractals, Fiber Bundles, and Gauge Fields: The Role of Genetic Information in the Theory of the Cell Force

As evident in the previous section, the formulation of a physical theory of the living cell may be greatly facilitated by the availability of the concepts and terminologies developed in Einstein's general relativity theory. This may not be a happenstance, if our conjecture is correct that the domain of the validity of the general relativity theory is the spacetime on the cosmological scale, while the domain of the validity of cell biology is the spacetime on the microscopic scale, as already pointed out in Section 1.8.1. What I mean by this statement is simply this: To the extent that we can regard the "curving" of spacetime by matter as the "*elaboration*" and "*complexification*" of the structure of spacetime *by matter* on the cosmological scale, so we can view intracellular metabolism as the *elaboration* and *complexification* of spacetime *by the cell* on the microscopic scale. The physical motion implicated in the former is the *accelerations* of moving bodies relative to a frame of reference, while the physical motions underlying the elaboration of spacetime by the cell is *chemical transformations of matter*. The result of the macroscopic elaboration of spacetime by matter is the curved spacetime, or equivalently the genesis of the "gravitational force," and the consequence of the microscopic elaboration of spacetime by the cell is the "curved" intracellular chemical reactions (giving rise to IDS), or equivalently the birth of the "cell force."

The conclusion that the connection between general relativity and cell biology may be more than just an analogy but is a natural consequence of the underlying multifaceted properties of *spacetime* justifies our curiosity to know more about the connections between the general relativity theory and the other theories of the fundamental forces that have been established in physics during the past half a century. It is now an established fact that there indeed exist deep connections between the general relativity theory (which is in effect the theory of the gravitational force) and the other theories of the fundamental forces of nature, namely the electromagnetic, electroweak and strong forces (Davies, 1984, 1986; Kaku and Trainer, 1987; Lopes, 1981; Moriyasu, 1985; 't Hooft, 1980; Yang, 1977; Zee, 1986). It has taken theoretical physicists more than a half century to uncover the underlying unity among the fundamental forces, and this has been possible through the developments of gauge field theories (Moriyasu, 1985) and superstring theories (Green, 1985).

Since physicists have already demonstrated that gauge field theories provide the proper language to describe the connection between general relativity theory and other theories of fundamental forces, perhaps it is logical to ask the question whether or not gauge field theories would prove equally useful in analyzing the possible connections between the cell force and the gravitational force on the one hand and the cell force and strong force on the other. These and related topics are discussed below.

A. <u>Manifolds</u>: A manifold is a mathematical concept denoting a set of points or elements with characteristic properties such that they can be grouped into subsets. A plane is a two-dimensional manifold of points, because it is a set of all points, each of which being specified by two coordinates, (x,y), and contains lines as its subsets. Euclidean space is a 3-dimensional manifold of points, each point requiring three coordinates, (x,y,z), for its specification and contains subsets of lines and planes. In general, an n-dimensional manifold is a collection of n values assigned to any n variables $x_1, x_2, \ldots, x_n$ (Levi-Civita, 1977). The study of the invariant properties of the structure of manifolds is called topology, a generalized geometry. The most important manifold in physics is the one involving the spatial and time coordinates, namely the set of 4 values assigned to x, y, z, and t. This manifold is referred to as spacetime (Misner et al., 1973; Wald, 1977; Taylor and Wheeler, 1966).

One of the most useful and fundamental properties of a manifold is that each point in it can be replaced with another manifold; so you can have a manifold of manifolds of manifolds of . . . etc. Such a mathematical tool is essential to represent complex physical systems or situations which cannot be analyzed at once but only in steps, due to the limited "focusing" power of the measuring instruments (e.g., light microscope or electron microscope) or the observer (recall that the human eye cannot be focussed on the "forest" and the "tree" simultaneously).

The manifolds composed of a large number of smaller manifolds may be called "nested manifolds." The Russian doll that contains a half dozen smaller replicas of itself, each nesting within the next larger, can be regarded as a simple example of nested manifolds. In a sense, the human body can be viewed as a nested manifold, since the body is a 4-dimensional manifold of cells, each cell is a 4-dimensional manifold of molecules, and each molecule is a 4-dimensional manifold of atoms—4-dimensional, because the elements of each manifold are arranged not only in space but in time as well (e.g., the properties of the cells constituting our body vary depending on their locations in and the age of the body). Therefore, the mathematical concept of manifolds can be clearly related to the physical objects that we call "machines" in this chapter, in the sense that a large machine can be constructed by properly coupling a set of smaller machines, each one of which in turn can be constructed by coupling a set of even smaller machines, etc.

The mathematical objects that possess fundamentally similar (or self-similar) features at different levels of magnification (i.e., at different size scales) are called "fractals" (Barnsely et al., 1987; McNamee, 1990; Jürgens et al., 1990). Fractals are so called because of the fact that they have fractional dimensions unlike the integral dimensions of familiar figures of the Euclidean geometry. The self-similarity characteristic of fractals is inherent in nested manifolds and biological machines. Hence, we can assert that all biological machines, above the level of molecular machines, can be treated as *fractal objects* or *fractals*. Thus, we can anticipate that both the topology of manifolds and fractal mathematics will play important roles in the study of living systems and processes viewed as machines. It is encouraging that the utility of the fractal geometry has already been demonstrated in some branches of biology and medicine (Glass and Mackey, 1979; Jürgens et al., 1990; West, 1990).

B. Fiber Bundles: Fiber bundles are a special class of manifolds that have the following structures: (i) the basespace, (ii) the fiber space erected at each point of the basespace, and (iii) the connection which correlates the points of one fiber space to those of another (Moriyasu, 1985). The combination (or union) of the base space and the fiber space is called the "fiber bundle space" or "total space." These various components of the total space are depicted in Figure 1.10.

Although the fiber space in this figure is represented as a line, the dimensionality of the fiber space can be increased to 2 (circle), 3 (sphere), 4 (4-dimensional space), etc. The concept of the connection is a generalization of the concept of the curvature of a 2-dimensional surface such as the surface of the earth, and its formulation was due to H. Weyl who in 1917 attempted to develop a unified

"Biocybernetics": A Machine Theory of Biology

theory of gravitation and electromagnetism (Bernstein and Phillips, 1981; Moriyasu, 1985; Kaku and Trainer, 1987).

The fiber bundle geometry provides us with a convenient method to visualize the concepts of the manifold and the internal space. To construct an (m + n)-dimensional manifold from an m-dimensional manifold (where m and n are whole numbers), it is necessary and sufficient to carry out two operations: (i) Represent the m-dimensional manifold as a basespace, and (ii) Replace each point of the m-dimensional basespace with n-dimensional fiber space, which is equivalent to the "internal space" discussed in Section 1.8.1. Gauge theories discussed below can be also represented in terms of the fiber bundle geometry by associating gauge fields with the fiber space (or the internal space) erected at each point of spacetime which is treated as a basespace.

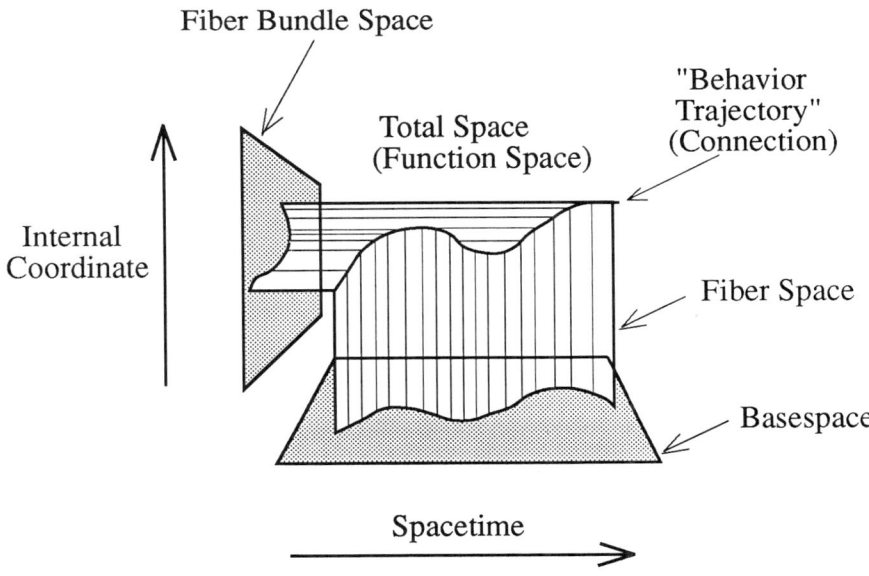

Figure 1.10. A schematic representation of the fiber bundle space (adapted from K. Moriyasu, 1985). The total space is divided into the fiber space (represented as the vertical plane on the left) and the basespace (represented as the horizontal plane). The basespace is identical with the spacetime in physics, and the fiber space refers to the internal space erected at each point of spacetime that specifies the nature of the events going on in spacetime. The total space (also called the fiber bundle space) is useful in depicting the function or the behavior of a machine; therefore the total space may be also called the "function space." The rules that allow one to compare the internal coordinates of one event with those of another are called "connections." (See footnote #2 in Table 1.15).

I believe that the fiber bundle geometry is valuable for biology, because it enables us to represent *genetic information* in terms of the *fiber space*, just as the *curvature* of spacetime can be so depicted. In this sense, the genetic information of cell biology may be equivalent to the curvature of spacetime in general relativity (see below). In view of the conjecture presented in Section 1.8.8 that cell biology elaborates spacetime on the microscopic scale while general relativity elaborates it on the cosmological scale, it may now be stated that the curvature of general relativity indicates the structure of spacetime on the cosmological scale, whereas genetic information of cell biology reflects the spacetime structure on the microscopic scale (Table 1.15).

Table 1.15. The unification of the fundamental forces of nature by gauge field theories

Force (1)	Interacting particles (2)	Gauge bosons (3)	Base space (4)	Fiber space[1] (5)	Connection[2] (6)
1 Gravitational	Masses	Gravitons	Spacetime	Curvature[3]	Gravitational potential field[4]
2 Electromagnetic	Charged particles	Photons	Spacetime	Phase angle[5]	Vector potential field[6]
3 Weak	Quarks & leptons[7]	W^+, W^-, Z^{08}	Spacetime	-	Weak vector field
4 Strong	Quarks[9]	Gluons[10]	Spacetime	Isotopic spin[11]	Isotpic spin potential field[12]

"Biocybernetics": A Machine Theory of Biology

(Table 1.15 continued)

| 5 | Cell[13] | Biopoly-mers[14] | Cytons[15] | Spacetime | Genetic information[16] | Intra-cellular concentration field (ICCF)[17] |

---

[1] Also called "internal degrees of freedom," "internal coordinates," "internal space," "internal symmetry space," or "local variable" (Moriyasu, 1985; Adair, 1987). As particles move from one point in spacetime to another, their coordinates in this internal space will vary according to some rule, if spacetime and the fiber space are "coupled" through the presence of the gauge field or the connection.

[2] The physical field that connects or couples the motions of particles in spacetime to the changes in their internal coordinates. Alternatively, it is called the gauge field and the associated field quanta are called gauge bosons.

[3] Newton postulated that two masses separated in space attracted each other through the agency of the "gravitational force" that acted instantaneously. Such an "action at a distance" is disallowed by the special theory of relativity, according to which no interaction in vacuum can proceed faster than the speed of light; i.e., any instantaneous interactions violate the principle of the special theory of relativity. The correct theory of the gravitational force was successfully formulated by Einstein in 1925 in the form of the general theory of relativity, wherein the motions of material bodies driven by the "gravitational force" were replaced by the "geodesics" (i.e., the paths of the shortest distance between two points) on a "curved spacetime." In other words, Einstein replaced the gravitational force of Newton with the "curvature" of spacetime (Adair, 1987).

[4] A unique set of numbers representing the curvature of a spacetime point determined by the presence of other masses in the neighborhood.

[5] Quantum mechanics describes particles and fields in terms of wave functions. Any wave motion has three characteristics—frequency, amplitude, and phase angle. It is the phase angle of the waves of charged particles that provides the internal degrees of freedom which exhibits the property of "local gauge symmetry" or "local gauge invariance."

[6] Also called the electromagnetic potential ($A\mu$). This is distinct from the ordinary electromagnetic field, since it can exist in the absence of the magnetic field as demonstrated in the Bohm-Aharonov experiment (Moriyasu, 1985; Adair, 1987).

[7] Leptons are spin 1/2 particles that do not respond to the strong force. Six different leptons are known—electron, mu particle, and tau particle, with massless neutrino for each.

[8] The gauge bosons of the weak force predicted by the Weinberg-Salam-Glashow theory of electroweak interactions and experimentally observed in 1983 (Adair, 1987).

C. Gauge Theories: In recent years, physicists have made great strides in their attempt to develop a unified theory of the gravitational, electromagnetic, weak and strong forces (Davies, 1986). The gravitational force is important primarily for describing interactions between material bodies at large scales (e.g. macroscopic machines, planetary mechanics, cosmology), the weak and strong forces are effective only at very short distances ($< 10\text{-}1^5$ m) and hence operate within the nuclei of atoms, and the electromagnetic forces are mainly responsible for the physical and chemical processes essential for life, ranging in scale from the microscopic ($\sim 10^{-10}$m) to the macroscopic (~ 10 m). The physical theory that many believe will eventually

---

(Notes to Table 1.15 continued)

[10] The agents (i.e., field quanta) mediating strong interactions.

[11] The internal degrees of freedom of nucleons that determine whether a nucleon is a proton or a neutron.

[12] Also called the Yang-Mills potential.

[13] It is postulated that there exists a new kind of force in nature called the *cell force* that "holds" together h-particles (i.e., biopolymers) and l-particles (i.e., small-molecular-weight chemicals) of the cell together in the living state against environmental perturbations, just as the *strong force* holds nucleons together in atomic nuclei against electrostatic repulsions.

[14] The living state of the cell can be thought of as arising from the interactions among biopolymers (primarily between nucleic acids and proteins) mediated by the cell force, just as the structural stability of atomic nuclei can be viewed as originating from the interactions among nucleons mediated by the strong force. The interactions among biopolymers may be direct (if mediated by conformons) or indirect (if mediated by IDS). The conformon-mediated direct interactions may be regarded as the short-range actions of the cell force, and the IDS-mediated indirect interactions as the long-range actions.

[15] The gauge bosons mediating the cell force will be called "cytons," a term constructed from "cyt-" meaning the cell and "-ons" meaning discrete entities. Cytons are composed of conformons and IDS (see Section 1.8.9 for more details).

[16] The genetic information encoded in biopolymers is postulated to represent an abstract internal space of biopolymers that is analogous to the matter-induced curvature of spacetime, the phase angle of matter waves, and the isotopic spins of nucleons. This is the space whose structure is elaborated by biological evolution in contrast to the other internal spaces in Table 1.15 which are elaborated by the cosmological evolution that began with the Big Bang.

[17] The identification of the component of the cell force corresponding to connection was least obvious. As evident in the text, ICCF has been assigned to this category more or less by default. Therefore, if future investigations produce more rigorous logical reasons to support this assignment, to that extent the concept of the cell force will be further strengthened.

lead to the unification of these forces is known variably as "local gauge field theories," "gauge field theories," or simply "gauge theories" ('t Hooft, 1980; Moriyasu, 1985; Adair, 1987). Some of the excitement of the physics community engendered by gauge theories is summarized by Moriyasu (1985):

> The discovery of gauge theory rivals in importance the development of both relativity and quantum mechanics. In contrast to the situation less than 10 years ago, gauge theory now dominates nearly all phases of elementary particle physics today. Even the reasons for performing the new experiments are now judged by their relevance for testing the predictions of gauge theory.9Spin 1/2 particles that respond to the strong force and the fundamental constituents of the hadron (i.e., neutron, proton, and meson). Six different quarks are known—up, down, strange, charmed, bottom, and top quarks.

Essential to understanding gauge theories is the notion of gauge symmetries. I quote P. Davies (1986):

> ... A simple example of a gauge symmetry concerns lifting a weight. The work done in raising a weight from A to B depends only on the height *difference* between A and B, not on their absolute height. It wouldn't matter whether we measured height from ground level or sea level, for instance. Thus, we may *regauge* the zero level of height (i.e., the gravitational potential) without in any way altering the physics of the lifting process. In other words, the system is symmetric under gauge transformations of height.

So, a "gauge symmetry" is associated with the existence of a measurable property of a physical system whose magnitude remains invariant (symmetric) when the measuring scale is changed (regauged).

Another important element in gauge theories is the idea of "local gauge symmetry," in contrast to "global gauge symmetry" which the example cited by Davies above is. A local gauge symmetry refers to the existence of a physical property associated with every point in spacetime whose magnitude remains invariant when the measuring scale is changed from place to place and from moment to moment (Zee, 1986). The all-important physical quantity whose value is kept invariant in local gauge transformations is called the "action," which has the dimension of a product of energy and time (Lopes, 1983). The importance of action in gauge theories can be traced to Hamilton's principle in classical mechanics (also known as the principle of least action), which states that a system composed of many interacting particles undergo changes in the positions of the particles only in such a way as to keep the cumulative magnitude of the action of the system minimal (Goldstein, 1980).

We can now state the basic content of what is called the "gauge principle" (Zee, 1986) qualitatively as follows:

> Whenever it is possible to perform local coordinate transformations in such a way as to maintain the action of the system invariant, there exists at least one new force field that counteracts the effects of changing the local coordinates.

The new field so identified is called the "gauge field" and its quanta (i.e., the agent that mediates interactions in gauge fields) are known as "gauge bosons" ('t Hooft, 1980).

D. A Qualitative Comparison of the Cell Force with the Fundamental Forces of Nature: Utilizing the concepts and terminologies of the gauge field theories and the fiber bundle theory introduced in the previous subsections, we can now compare the cell force with other fundamental forces in nature, namely the gravitational, electromagnetic, weak, and strong forces (Table 1.15). If the cell force indeed exists, the cell force might reveal a set of common physical characteristics shared by the other fundamental forces.

The purpose of Table 1.15 is two fold: (i) To describe the qualitative characteristics of the four fundamental forces (rows 1, 2, 3 and 4) using the concepts and terms derived from gauge field theories (columns 2 and 3) and fiber bundle geometry (columns 4, 5 and 6), and (ii) To compare the cell force (below the dotted line) with the common characteristics of the four fundamental forces (above the dotted line) under five different categories (columns 2 through 6). The technical terms and concepts appearing in the table are explained in the extensive footnotes attached.

On the basis of the "table theory" introduced in Section 1.3, it is clear that Table 1.15 contains four F vectors (F from "familiar;" i.e., rows 1 through 4) and one U vector (U from "unfamiliar;" i.e., row 5). The four F vectors establish the common characteristics of the row vectors under five different categories—(i) interacting particles, (ii) gauge bosons, (iii) basespace, (iv) fiber space, and (v) connection. The validity of the U vector (i.e., row 5 and the concept of the cell force) will largely depend on the "goodness of fit" between the components of the U vector identified under these five categories and the corresponding components of the F vectors.

The first and the third components of the cell force (i.e., biopolymers, and spacetime) are self-explanatory and reasonable. The term "cyton" appearing as the second component of the U vector is a new word coined to indicate the "gauge boson" mediating the cell force (see footnote #15 for the derivation of the word). Unlike the gauge bosons of particle physics, which have no internal structures, the cyton is postulated to be composed of conformons and IDS, one of the most fundamental assumptions of the cell theory advanced in this chapter.

It is probably safe to say that what distinguishes *life* from *nonlife* is *genetic information*. Since the *cell force* is postulated to be also fundamental to life, the question naturally arises as to the nature of the theoretical relationship between the cell force and genetic information. Table 1.15 suggests two possible interrelationships, genetic information represents (i) the fiber space, or (ii) connection of the fiber bundle space describing the living cell (see the L-space in Figure 1.11). Primarily based on the assumed similarity between genetic information and the curvature of spacetime as already alluded to above (Sections 1.8.7 and 1.8.8), I have been led to assign genetic information to the category of fiber space (see row 5 and column 5 in Table 1.15); i.e., I am assuming that *genetic information in cell biology is analogous to the curvature of spacetime in general relativity*. This leaves only one component of the cell force undefined, namely the component belonging to the connection category (row 5 and column 6). The best candidate for this component, I think, is the intracellular chemical concentration field (ICCF) discussed in Section 1.8.7. According to this assignment, ICCF is analogous to the gravitational potential field in general relativity on the one hand and to the vector potential field of electromagnetism on the other. It is possible that the rise or fall of the concept of the cell force will depend critically on future theoretical investigations on the validity of the last two assignments, namely genetic information as the fiber space and ICCF as the connection (see Figure 1.11 for a related discussion).

It is interesting to point out in Table 1.15 that the coordinates to which local gauge transformations are applied are different for different forces (see Fiber space)—the spacetime coordinates themselves for the gravitational force, the phase angle of the electron wave function for the electromagnetic force, and the isotopic spin states of the nucleons (i.e., protons and neutrons) for the strong force ('t Hooft, 1980). The spacetime coordinates are *external* to material particles, while the phase angle of the electron wave function and the isotopic spin states of nucleons are *internal* to the particles involved. For this reason the latter two are examples of what is called the "internal coordinates" (see Section 1.8.1).

This idea of "internal coordinate" is interesting from two perspectives: (i) It is related to the fiber space of the fiber bundle theory, and (ii) The genetic information encoded in biopolymers (DNA, RNA and enzymes) and the cell (i.e., "cell programs") may be viewed as the biological analogue of the internal coordinates of subatomic particles as already mentioned.

Just as the development of Einstein's theory of general relativity heavily depended on the Riemannian geometry, many believe that the ultimate unification of the fundamental forces through gauge theories will be facilitated by the theory of fiber bundles. If the theory of fiber bundles is useful in unifying the fundamental interactions in nature, it may also be useful in developing fundamental theories of living processes. By expressing biological theories in the same language as used in describing fundamental interactions in physics, albeit qualitatively, it may be possible to obtain useful hints and guidelines from gauge theories. The "fundamental particles" approach of physics may be applicable to biology in the sense that we may begin to view chemicals (l-particles), biopolymers (h-particles), cells (c-particles) and multicellular organisms (o-particles) as "fundamental" particles at different levels of biological organization, despite the fact that these are immensely more complex in their internal structures compared to elementary particles studied in physics. In defense of the this conclusion, I present the following argument:

(i) In contrast to particle physicists whose direction of progress has been from the less detailed to the more detailed picture of elementary particles (e.g., from nucleons to quarks and leptons (Davies, 1986)), perhaps biologists may benefit from approaching biological interactions from quite the opposite direction—from the detailed to the less detailed; e.g., away from the detailed 3-dimensional and static X-ray crystallographic enzymes or even from the 4-dimensional "screaming and kicking" NMR-determined enzymes in solution to a much simpler view of enzymes as small, deformable particles with "internal degrees" of freedom, called genetic information.

(ii) The "direction of decreasing details" may not be as heretical as it may sound, if we take into account the fact that the "depth of focus" of the human mind may be intrinsically limited, so that the detailed picture of what may be called the "bio-particles" (namely, l-, h-, c-, and o-particles defined above) at one "focal plane" may obscure the picture on the adjacent planes of focus. In other words, it is possible that we can be confronted with too detailed pictures of bio-particles all at once, preventing us from "seeing the forest for the trees." We may formulate the following statement to be referred to metaphorically as the "forest-for-trees" principle of the human mind:

> The depth of field of the human mind is much narrower than the thickness of the objects of nature, so that the full description of nature requires viewing

objects at more than one focal planes.

In the spirit of this metaphor, we may speak of two opposite approaches to biological research, namely the "from-forest-to-trees" approach, and the "from-trees-to-forest" approach. I suggest that in many areas of contemporary biological researches, we may have been too preoccupied with detailed description of trees, losing sight of the forest. More and more, it may now be necessary for us to keep some distance away from the enchanting trees of the molecular biology and try to discover the majestic beauty of the forest of life, namely cell and organismic biology.

(iii) We have in manifold theories and fractal mathematics tailor-made mathematical languages with which to describe bio-particles. o-Particles are manifolds of c-particles, which are in turn manifolds of l-particles, and finally h-particles are manifolds of l-particles. All bio-particles are self-similar, and the self-similarity derives from the fact that they are all biological machines.

1.8.9. The Cell Force—The Link between Genotype and Phenotype

Whenever physical objects, A and B, influence each other in the sense that the motion of A causes that of B to change or *vice versa*, a force is said to act between A and B, or A is said to exert a force on B or *vice versa*. The gravitational force is the first force in nature to be mathematically defined in the form of Newton's second law;

$$F = m \alpha \qquad (1.21)$$

where F is force, m is the mass of the falling body due to the gravitational force, and α is the acceleration (i.e., the rate of change of the velocity of the falling body).

Physicists tell us that all physical interactions in nature are ultimately reducible to four fundamental forces—gravitational, electromagnetic, weak and strong (see Table 1.15) (Davies, 1986). Furthermore, these four fundamental forces can be expressed using the common mathematical language of gauge theories (Section 1.8.8.)(Moriyasu, 1985). This raises the question as to whether or not living processes (e.g., cell division, embryogenesis, etc.) can also be completely explained in terms of the four forces. In other words, are the four fundamental forces of physics sufficient to account for the phenomenon of life? Based on the results of my theoretical work since the early 1970's, I have recently come to the conclusion that the four fundamental forces of physics are *necessary* but *not sufficient* to give rise to the phenomenon of life on the cellular level. To completely account for cell life, it seems to me to be necessary to postulate the existence of a new force acting in the cell, which I have called the *cell force*. The justification for invoking the cell force is: (1) There are "beautiful" analogies that can be detected between the cell force and three of the four fundamental forces—i.e., the electromagnetic (Tables 1.1 and 1.15), the strong (Tables 1.1 and 1.16), and the gravitational forces (Table 1.17). (2) The concept of the cell force provides a natural way of integrating the varied structural elements, concepts and theories essential for describing the living cell that have accumulated over the past several decades. The ultimate test for the validity of the cell force concept is the logical consistency and the elegance that this concept imparts to the general theory of living processes built upon it.

One of the most fundamental problems facing the contemporary biology is the question as to how genotype (i.e., the genetic constitution of an individual) determines phenotype (i.e., a category or group to which an individual belongs based on its observable characteristics). During the past century or so, many fundamental ideas and theories have been developed (e.g., the theory of biological evolution, the universality of the cell, etc.), and numerous important experimental breakthroughs have occurred (e.g., the discoveries of DNA double helix, genetic codes and metabolic pathways, etc.) that have revolutionized our understanding of the mechanisms underlying the phenomenon of genotype-phenotype coupling. In general, these ideas and observational data may be grouped into two broad categories referred to by Mayr (1988) as "functional biology" and "evolutionary biology" (see Section 1.1). These breathtaking advances in biology notwithstanding, no physical theory yet exists that can provide a coherent and comprehensive explanation for the genotype-phenotype coupling in fundamental molecular, cellular, organismic and societal terms.

The cell force concept elaborated in Table 1.15 appears to have the capacity to couple not only genotype to phenotype but also participate in coupling genotype to biological evolution itself. These global ideas may be expressed in the form of algebraic equations:

$$\text{Genotype} \xrightarrow{\text{Cell Force}} \text{Phenotype} \tag{1.22}$$

$$\text{Evolution} \xrightarrow[\text{Natural Selection}]{\text{Cell Force}} \text{Genotype} \tag{1.23}$$

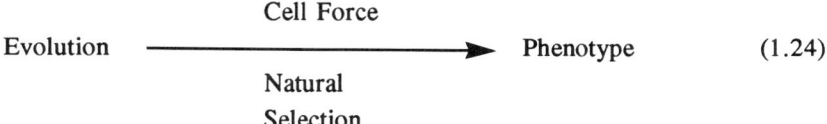

$$\text{Evolution} \xrightarrow[\substack{\text{Natural} \\ \text{Selection}}]{\text{Cell Force}} \text{Phenotype} \quad (1.24)$$

Equation (1.22) indicates that, through the agency of the cell force, genotype controls phenotype. In another sense, Equation (1.22) can be viewed as an operational definition of the cell force in contrast to Table 1.15 which characterizes the cell force in mechanistic terms; i.e., *the cell force is the totality of the molecular and cellular mechanisms that are both necessary and sufficient to effectuate the genotype to phenotype coupling.*

Similarly, Equation (1.23) indicates that the cell force is essential for the evolution of genotype, which must be true since *no biological evolution would have been possible without the cell.* This last statement, although rarely so expressed, seems to be fundamental to all biological thoughts. Hence this statement deserves to be explicitly recognized as an important biological principle by giving it a special appellation—I suggest that we call the above italicized statement the "Cell Principle of Biological Evolution." Unlike Equation (1.22), the realization of Equation (1.23) requires not only the cell force but also natural selection dictated by organism-environment interactions (Gould, 1979, 1982; Mayr, 1988). Equations (1.22) and (1.23) may be "added" algebraically to obtain Equation (1.24), which in effect reiterates the well-accepted biological tenet that phenotypes are the consequences of biological evolution.

The fact that the cell force appears in both Equations (1.22) and (1.23) is consistent with the notion expressed in Section 1.5.6 that whatever molecular mechanisms underlying the genotype-phenotype coupling in individual cells must have played equally essential roles in the origin of biological information some 2-3 billion years ago (see Appendix 1.C). As indicated in Section 1.5.6 and Table 1.10, conformons are postulated to be responsible for driving all molecular machines in the cell (Ji, 1974, 1979, 1988; 1990) and in addition are thought to have participated in the mechanisms of the origin of biological information (Anderson, 1983, 1987; Ji, 1990). Since conformons constitute a key component of the cell force (see cytons in Table 1.15 and 1.16), it is logically self-consistent to conclude that the cell force is essential for both cell life and the origin of life, thus supporting Equations (1.22) and (1.23).

It is important to recognize (i) that the characteristic turnover time of Equation (1.22) is typically less than 10^2 years (in humans) while that of Equation (1.23) is much longer, 10^3 - 10^9 years, and (ii) that Equation (1.22) occurs within individual cells or organisms, whereas Equation (1.23) implicates populations of large numbers of cells or organisms. Therefore, the mechanisms underlying Equations (1.22) and (1.23) are quite different in both temporal and spatial scales.

It is interesting to ask the question as to which of the three equations considered above corresponds to functional biology and which to evolutionary biology discussed by Mayr (1988). If we interpret Equation (1.22) to represent those biological researches whose ultimate aims are to correlate biological structures and functions to underlying genes, this equation could be equated with functional biology including biochemistry, molecular genetics, molecular biology, molecular cell biology, molecular physiology, etc. In contrast, it seems reasonable to interpret Equation (1.24) as representing the research fields concerned with accounting for biological structures and functions in terms of the biological evolution as reflected in fossil records and variations of biological forms as a function of the environment of organisms. Thus, Equation (1.24) may reasonably be associated with the evolutionary biology as defined by Mayr (1988). This leaves Equation (1.23) uncharacterized, and the question naturally arises as to which branch of biology, if any, Equation (1.23) represents. In my opinion, this equation is best interpreted as reflecting those fields of biology concerned with the molecular mechanisms of the origin of life and with molecular evolution—as exemplified by the studies published by Crick et al. (1976), Kuhn (1983), Eigen (1986, 1987), Prigogine et al. (1972), Anderson (1983, 1987), Kimura (1979), Lowenstein (1985), Cairns-Smith (1982, 1985), Kondepudi and Nelson (1985; see also Chapter 3 of this book), Ferris (1984), and Wilson et al. (1985), just to cite a few.

The gap between functional biology and evolutionary biology justifiably deplored by Mayr (1988) and others may now find a natural explanation. According to Equation (1.24), evolutionary biology is a much broader field of study than functional biology, being composed of functional biology (i.e., Equation (1.22)) *and* another branch of biology represented by Equation (1.23) and yet to be properly named. Therefore, the domain of evolutionary biology cannot be completely covered by the domain of functional biology, leaving certain areas of evolutionary biology that cannot be accounted for by the results of functional biological researches.

The classification of biology into functional biology and evolutionary biology suggested by Mayr (1988) appears incomplete in view of Equations (1.22), (1.23), and (1.24). These equations suggest another, perhaps a more logical, way of classifying biology; i.e., biology may be first divided into two major branches—(i) "macroevolutionary biology" (i.e., the biology of "macroevolution") encapsulated in Equation (1.24) , and (ii) "microevolutionary biology" (i.e., the biology of "microevolution"). The microevolutionary biology is then further divided into what

"Biocybernetics": A Machine Theory of Biology

may be called the "user biology" (Equation (1.22)) and the "sender biology" (Equation (1.23)), the terms "user" and "sender" originating from communications theory to be discussed below (see Section 1.8.11 and Figure 1.12). According to this classification scheme, then, Mayr's functional biology and evolutionary biology become user biology and macroevolutionary biology, respectively.

The use of the prefixes, micro- and macro-, in the above discussion was motivated by their well-known uses in physics, but with a slight "twist" in their meanings. The chief concerns of physics and chemistry are to understand the behaviors of macroscopic objects (e.g., magnets, computer chips, chemical reactions, etc.) by "extending" or "extrapolating" the knowledge of the behaviors of microscopic objects (e.g., electrons, atoms, molecules, etc.) on the *spatial dimension*—e.g., from atomic scale to human body scale, which differ from each other by a factor of about 10^9. In contrast, the main objective of biology is to understand macroevolution (i.e, Equation (1.24)) by extending our knowledge of microevolution (i.e., primarily Equation (1.22)) on both the *spatial dimension* (i.e., from genes to organisms) and *time dimension* (i.e., from the present to the past). In other words, physics is concerned mainly with the correlations between *the small* and *the large* (i.e., spatial correlations), whereas biology deals with the correlations between *the small* and *the large* (spatial correlations) as well as between *the past* and *the present* (temporal correlations). Furthermore, it appears that the spatial and temporal correlations observed in living systems are made possible because of the unique capabilities of biopolymers that can mediate *spatiotemporal* correlations, namely correlations between *events* as discussed in more detail in Section 1.8.11.

On the basis of these considerations, we can rephrase the contents of Equations (1.22), (1.23) and (1.24) as follows:

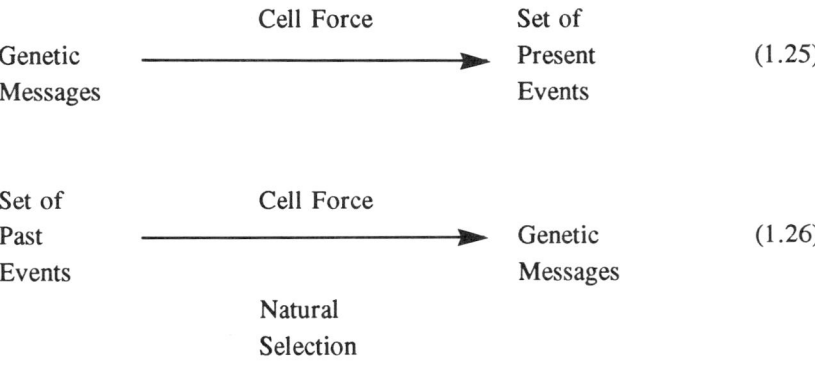

$$\text{Set of Past Events} \xrightarrow[\text{Natural Selection}]{\text{Cell Force}} \text{Set of Present Events} \qquad (1.27)$$

where I have treated evolution as equivalent to a set of events (i.e., organism-environment interactions) that have transpired in *the past*, phenotype as equivalent to a set of events proceeding in *the present*, and genotype as equivalent to a set of *genetic messages* encoded or stored in DNA of the cell.

Equation (1.27) can be viewed as still another operational definition of the cell force—i.e., *the cell force is the totality of the molecular and cellular mechanisms essential for past events to exert influences on present events in organisms*. The quintessence of macroevolutionary biology as represented by Equation (1.27) is the fact that events in living systems that transpired in the evolutionary past are influencing the events that are happening in living systems right now, through the agency of the cell force. In the language of communication theory (Pierce, 1980; also see Section 1.8.11), this is equivalent to the statement that past events are *sending* genetic messages to contemporary organisms, which receive and *use* them (i.e., Equation (1.26)). This is the rationale for the suggestion made above that the branch of biology represented by Equation (1.22) (or equivalently Equation (1.25)) be called the "user biology," and that represented by Equation (1.23) (or equivalently Equation (1.26)) the "sender biology."

Since genetic messages are carried by DNA, and biological events are all mediated by the cell, the set of three equations already displayed twice above may be rewritten in still another way in order to highlight these points:

$$\text{DNA} \xrightarrow{\text{Cell Force}} \text{Cell Functions} \qquad (1.28)$$

$$\text{Evolution} \xrightarrow[\text{Natural Selection}]{\text{Cell Force}} \text{DNA} \qquad (1.29)$$

$$\text{Evolution} \xrightarrow[\text{Natural Selection}]{\text{Cell Force}} \text{Cell Functions} \qquad (1.30)$$

Written in this highly abbreviated form, these three equations clearly reveal (i) the messenger role of DNA, (ii) evolution as the source of genetic information, and (iii) the cell as the user of genetic information.

It is one of the crowning achievements of the twentieth-century molecular biology to have deciphered genetic messages into nucleotide sequences—we now know the precise linear sequences consisting of hundreds and thousands of the nucleotides that encode the structure of hundreds of proteins. In addition we know a lot about the nucleotide sequences of DNA that regulate gene expression as well as their associated DNA-binding proteins (Alberts et al., 1983; Watson et al., 1987; Lewin, 1987). However, in spite of the rapidly accumulating experimental knowledge in molecular genetics, the feeling is growing among biologists that detailed molecular data on genes and their accessory molecular apparatuses alone may not be sufficient to unravel the mysteries of gene-directed cellular functions. This feeling of alienation from the ultimate molecular processes underlying cellular functions may not be remedied by a continued accumulation of new and better empirical data, because it may have a deeper cause. It is possible that this unsatisfactory state of affairs arises from the fact that the theoretical framework currently available in molecular biology, namely the laws of chemistry and physics as embodied in quantum mechanics, statistical mechanics, and chemical kinetics, although *necessary*, are *not sufficient* to account for the rich experimental observations that have been unearthed during the recent decades.

The inadequacy of the current theoretical framework employed in molecular biology can be analyzed using Equations (1.28) through (1.30). As pointed out above, the primary goal of functional biology may be equated with the understanding of cell functions in terms of genes, namely Equation (1.28)—that is, to *explain* cell functions in terms of the nucleotide sequences of DNA, or to prove that cell functions are *caused* by DNA sequences. It is probably fair to say that most molecular biologists hold the optimistic view that, given detailed enough experimental data on genes and associated molecules inside the cell, the application of the existing laws of physics and chemistry would be sufficient to account for all of the fundamental properties of the living cell such as mitosis and cellular differentiations. They seem comfortable in assuming that no new rules or principles, other than those already manifest in physics and chemistry, would be required to explain functional biology

(Mayr, 1988). Such a viewpoint may be referred to as *biological reductionism*. Alternatively, it is possible that the existing laws of chemistry and physics are insufficient to completely account for Equation (1.28). In other words, it may turn out that the rigorous application of the laws of physics and chemistry to DNA structures under the realistic conditions of the interior of the living cell fails to reproduce cellular functions, just as the laws of Newtonian mechanics failed to explain the stability of atoms and the laws of special relativity cannot explain the bending of light in a gravitational field nor the existence of black holes. In fact, I subscribe to this alternative viewpoint, which is consistent with the so-called Bohr-Elsasser Incompleteness Theorem of Physicochemical Explanations of Life formulated below and the conclusions of more recent workers such as Smith and Welch (see Chapter 6) and Rothstein (see Chapter 7).

Even if one succeeds in completely correlating DNA structures and cell functions, this would not answer the "why" question about, or the biological "meanings" of, the DNA-cell function correlations. As I make this statement, I am thinking of an analogy (motivated by the parable of the Chinese room told by J. Searle (1984)) involving a powerful supercomputer that is programmed to translate Chinese to English following an elaborate algorithm and a Chinese scholar. Suppose that both are assigned the task of translating a Chinese poem into an English version, which they duly execute flawlessly. Clearly, the English poem would have elicited different sets of internal states or responses in these two "translators"—the most significant difference being that the computer completely missed the "meaning" of the Chinese poem that the Chinese scholar could not escape but perceive in his mind along with the rich mental imagery to go with it. Analogously, I assert that the laws of chemistry and physics can only provide the DNA-cell function correlations and cannot reveal the biological "meanings" of the DNA-induced cell functions, because these meanings are in addition to the laws of physics and chemistry and reflect the genetic messages imparted by natural selection, i.e., Equation (1.29).

Using the grammar of a written language as an analogy, it may be said that DNA has two distinct sets of attributes—(i) syntax (i.e., the arrangement of words as elements in a sentence to show their relationship; sentence structure), and (ii) semantics (i.e., the study of the relations between signs or symbols and what they represent). The laws of physics and chemistry governing the physicochemical behaviors of DNA (i.e., Equation (1.28)) indicates the syntax of DNA, while the evolutionary events leading to the generation of DNA (i.e., Equation (1.29)) determine the semantics of DNA. That is, *the application of the laws of physics and chemistry reveals the syntax of DNA (i.e., its predictable physicochemical behaviors in harmony with those exhibited by man-made polymers with comparable structures and sizes) but not its semantics (i.e., the biological functions or "fitness value" of DNA-dependent cell functions)*. The semantics of DNA may be defined as the set of the correlations between the evolutionary events (Equation (1.29)) and cell functions (Equation (1.28)) that DNA effectuates. That is, the semantics of DNA is embodied in Equation (1.30) and not in Equation (1.28) nor Equation (1.29). The last two equations reveal only the syntax of DNA. One of the major conclusions of biocybernetics presented in this chapter is that the electromagnetic force underlying all physical and chemical interactions in nature can only reveal the syntax of DNA but not its semantics. To uncover both the syntax and the semantics of DNA, it is necessary to invoke a new force, the cell force.

As will be discussed in Section 1.8.11, the cell can be viewed as a communication system in the sense used in electrical communications engineering (Pierce, 1980; Gagliardi, 1988; Ji, 1988). Unlike the artificial electrical communication systems which are specifically designed to transmit information in the spatial dimension only (e.g., a telephone call or a telegram from L.A. to New York City), the biological communication systems are designed to transmit information in either space (e.g., endocrinology, allostery, etc.) or time (e.g., biological evolution, memory, cell cycle, etc.). The electromagnetic force (mediated by photons) is necessary and sufficient to operate artificial electrical communication systems. However, I claim that the electromagnetic force, although necessary, is not sufficient for biological communication systems to transmit information in space and time and that a necessary condition for biological communications is the cell force (mediated by cytons which are in turn composed of conformons and IDS). In other words, the cell force may be thought of as a new kind of force in nature that can act not only in space but also in time in contrast to all the other fundamental forces (gravitational, electromagnetic, weak and strong) which can act only through space. Therefore, we may distinguish two kinds of forces—(i) S-forces acting in space (meaning that its strength changes with distance), and (ii) T-forces that act in time (meaning that its strength changes as a function of time). Since any change can be either + (increase) or - (decrease), it is logical to predict that there will be four classes of forces—(+)S, (-)S, (+)T and (-)T-forces. The cell force may be the first clear example of a T-force, and there may be other T-forces in nature yet to be discovered. It seems possible to this author that the weak force responsible for apparently random radioactive decays of atomic nuclei may represent a (+)T-force and the strong force that confines quarks in atomic nuclei a (-)S-force (Nambu, 1976). Aging processes in living systems may be controlled in a *dual* manner, namely by both the (+)T- and (-)T-forces in order to ensure the fidelity of aging processes (see Section 1.12.13).

Based on the assumption that the principle of complementarity deduced from atomic physics can be analogically applied to biology, Bohr (1933) more than a half century ago suggested that the phenomenon of life may never be completely explained in terms of the laws of chemistry and physics. This idea was further extended and elaborated on by W.M. Elsasser (1958, 1961, 1975, 1987) during the past three decades. The notion that the laws of physics and chemistry are *necessary but not sufficient* to account for the phenomenon of life will be referred to as the "Bohr-Elsasser Incompleteness Theorem of Physicochemical Explanations of Life," or simply the "Bohr-Elsasser theorem." The cell force concept, which cannot be derived from the fundamental forces of physics, is consistent with the Bohr-Elsasser theorem and provides molecular and cellular explanations as to why the laws of chemistry and physics are insufficient to explain living processes; i.e., chemical and physical processes are driven by free energies, but living cells are driven by gnergy, a higher dimensional entity formed by a *complementary union* of genetic information and free energy as in conformons and IDS (Ji, 1985a,c, 1990).

A. The Analogy between the Cell Force and the Electromagnetic Force: The strength of the analogy between the cell force and the electromagnetic forces derives in part from the fact that the Bhopalator model of the cell can be favorably compared to the periodic table in chemistry (Table 1.1). Just as the electromagnetic force is intimately associated with a set of objects such as the Mendeleev periodic table, elements, chemical compounds, photons and QED, so the cell force can be viewed as an integral part of a coherent theoretical structure consisting of the Bhopalator, organisms, conformons, IDS, and biocybernetics. Therefore, when I assert that there exists a new force, the cell force, I mean that there exists a coherent physical theory consisting of a set of building blocks, i.e., the Bhopalator, conformons, IDS and biocybernetics, that can "explain" living organisms, just as the Mendeleev periodic table can explain millions of chemical compounds. The term "cell force" should be regarded as a mnemonic device for indicating the complex theory of interactions among chemicals and biopolymers inside the cell according to the rules of biocybernetics.

B. The Analogy between the Cell Force and the Strong Force: The thought that the cell may possess a new force of its own first occurred to me when I was thinking about how physicists arrived at the conclusion that there was the strong force in atomic nuclei. In 1911, physicists discovered that atoms contained nuclei and that these consisted of neutrons and protons confined in a small volume, about 10^{-15} m in radius. It was immediately clear to them that the stability of atomic nuclei could not be explained in terms of the then known fundamental forces, namely the electromagnetic and the gravitational forces. The electromagnetic force should cause atomic nuclei to practically "explode" due to the electrostatic repulsion between

"Biocybernetics": A Machine Theory of Biology

positively charged protons. Besides, the gravitational force was known to be too weak to have any significant effects. In order to account for the structural stability of atomic nuclei, physicists were therefore forced to conclude that there must exist a new force with an effective range limited to the interior of atomic nuclei that is strong enough to "hold together" the nucleons (i.e., protons, neutrons and mesons) against electrostatic repulsion. They called this new force the "strong force."

In my opinion, biologists trying to explain the living processes of the cell in terms only of the electromagnetic force (i.e., the principles of physics and chemistry) are akin to physicists attempting to explain the stability of atomic nuclei without invoking the strong force. This mode of thinking sensitized my mind to entertain the idea of the *cell force* around late 1989. But a more compelling reason for postulating the cell force in earnest came to mind when I noticed that the cell force concept contained all the qualitative theoretical ingredients required by the gauge theory of Yang and Mills (Yang, 1977), the mathematical theory of the strong force. In other words, the cell force seems to satisfy qualitatively the same mathematical theory obeyed by the strong force. This analogy between the cell force and the strong force is displayed in detail in Table 1.16 along with a set of extensive footnotes.

Table 1.16. Toward a gauge theory of the cell force

Building blocks	Strong force	Cell force
1. Gauge potential field[1]	Isotopic spin potential field[2]	Intracellular chemical concentration field (ICCF)[3]
2. Conserved quantity[4]	Isotopic spin[5]	"Fitness spin"[6]
3. Physical particles[7]	Hadrons[8]	Biopolymers[9]

(Table 1.16 continued)

4. Substructure of particles[10]	Quarks[11]	Functional domains[12]
5. Gauge bosons[13]	Gluons[14]	"Cytons"[15]
6. Local system[16]	Nuclei[17]	Cells[18]
7. Global system[19]	Atoms & Molecules[20]	Organisms[21]

---

[1] Explained in Footnote #2 in Table 1.15.

[2] The space in which to specify the internal coordinates of nucleons.

[3] The physical field through which the cell force exerts its influence. See Footnote #17, Table 1.15.

[4] A function of the internal coordinates of particles that remains invariant upon local gauge transformations (see Section 1.8.7.).

[5] The internal parameter of a nucleon whose value determines whether that nucleon is a proton or neutron. The proton-ness or neutron-ness of a nucleon is mutually exclusive because the number of neutrons and protons in an atomic nucleus must remain constant, although the proton-ness or neutron-ness of individual nucleons can freely change (local gauge transformation).

[6] The internal parameter of a biopolymer whose value determines whether or not a given biopolymer is "fit" (i.e., makes an essential contribution to the *living state* of the cell at a given point in space and time) or "unfit" (i.e., does not make any essential contribution to maintaining the *living state* of the cell). The *fit* and *unfit* states of a biopolymer are not identical with its *active* and *inactive* states, since the activities of biopolymers alone do not guarantee the living state of the cell; only the spatiotemporal coordination of their activities does. The fitness spins of biopolymers inside the cell may exist in at least two states—f in which biopolymers are fit enzymologically or otherwise, and u in which biopolymers are unfit. As the fitness spins of the biopolymers in the cell change (e.g., "rotate" say from 0 to 360° in the abstract fitness space of biopolymers), the enzymatic activities of different biopolymers may be turned on or off, but the total "vector" sum of the fitness spins of biopolymers in a cell must always add up to a value greater than a critical threshold so that the cell as a whole is *alive*, just as the total number of neutrons and protons in an atomic nucleus remains invariant despite the seemingly arbitrary changes in the position of the individual isotopic spins. What guarantees the conservation of the isotopic spins of the nucleons is the presence of the *strong force*. Analogously, I suggest that what guarantees the living state of the cell, despite seemingly arbitrary on and off states of individual biopolymers, is the presence of a new force inside the cell, the *cell force*. IDS and conformons are the microscopic entities that mediate the cell force and therefore can be regarded as

(Notes to Table 1.16 continued)

constituting the gauge bosons of the cell called "cytons."

[7] Physical particles whose motions in space and time are influenced by a given force.

[8] The subnuclear particles (neutrons, protons, mesons, etc.) that respond to the strong force. Notice that other subnuclear particles such as W^+, W^- and Z^0 particles and leptons are insensitive to the strong force.

[9] It is postulated that biopolymers such as DNA, RNA and proteins, in addition to their usual response to the electromagnetic force, are the only molecules in the cell that respond to the cell force.

[10] Particles responding to the strong force or the cell force have internal structures.

[11] Quarks are the ultimate material entities (i.e. do not have internal structures) that respond to the strong force.

[12] It is postulated that a given biopolymer is composed of two or more functional domains that interact with each other by exchanging "cytons," namely conformons and IDS.

[13] The field quanta that mediate the interactions between particles in a given force field.

[14] The quanta of the strong field that "glue" together quarks to form hadrons.

[15] The physical entities that mediate the cell force, namely conformons and IDS.

[16] The physical structure formed through the action of the force under consideration.

[17] The strong force is not only responsible for the formation of hadrons but also the formation of atomic nuclei as a whole due to the residual strong force left over after binding quarks.

[18] The cell force "holds together" biopolymers (h-particles) and biochemicals (l-particles) inside the cell and creates the local "dissipative structure" called the living cell.

[19] The structure larger than the structure in which the immediate action of the force under consideration is confined. The force under consideration is a necessary condition for this larger structure.

[20] Since atoms and molecules cannot exist without stable nuclei, the strong force is a necessary condition for the existence of atoms and molecules.

[21] Since organisms cannot exist without the living cell, the cell force is a necessary condition for all living systems.

One of the most important results of invoking the existence of the cell force is that the new force allows one to go from the smallest building blocks of the cell, namely biochemicals and biopolymers, to the biological functions of the cell; i.e., the cell force "connects" genotypes and phenotypes as discussed earlier, just as the strong force connects quarks to the atomic structure. Without the strong force to use as a analogical model, it would have been very difficult and perhaps even impossible to connect, even qualitatively, the functional properties of the cell to its structural constituents. In other words, the identification of the appropriate analogies (i.e., a table symmetry; see Section 1.3) may be absolutely necessary to formulate a coherent physical theory of the living cell.

C. The Analogy Between the Cell Force and the Gravitational Force: When Newton first invoked the existence of the gravitational force over two hundred years ago, he thought of this force somehow acting between material bodies instantaneously. However, through his special and general theories of relativity, Einstein demonstrated that the "mechanism" of action of the gravitational force is not as simple as Newton once thought but implicates "curved" spacetime and "geodesic" (i.e., the path of the shortest distance between two points on a curved surface such as that of the earth) motions of material bodies in it (Misner et al., 1973; Moriyasu, 1985). In other words, the term "gravitational force" can be viewed as a mnemonic device for representing the whole complex of physical theories and concepts that provide theoretically consistent explanations for all "gravitational" interactions. In the same manner, the "cell force" should be regarded as a short-hand label for the complex structure composed of the physical theories and concepts (as revealed in Tables 1.1, 1.15, 1.16, 1.17) that provide logically coherent explanations for all living structures and processes.

We have already compared the cell force with the gravitational force in Table 1.15 under five distinct headings and found interesting analogical relationships among the five pairs of the parameters that characterize these two forces. The purpose of this section is to bring out the cell force/gravitation force analogy in another way— through the consideration of the following statement:

"Chemical reactions proceeding inside the cell is to those occurring in a test tube what general relativity is to special relativity."

This qualitative statement may be expressed in terms of a set of what may be called "analogical" equations:

"Biocybernetics": A Machine Theory of Biology 115

General Relativity	=	Special Relativity	+	Acceleration due to Gravity	(1.31)
Curved Spacetime	=	Flat Spacetime	+	Curvature due to Matter	(1.32)
"Cell" Metabolism	=	"Test-Tube" Metabolism	+	Life due to the Cell Force	(1.33)
"Curved" Chemistry	=	"Flat" Chemistry	+	Spacetime-Specificity due to Biopolymers	(1.34)

Equation (1.31) encapsulates the difference between special relativity whose validity is restricted to describing the motions of material bodies in an inertial frame (i.e., the frame of reference moving with a constant speed) and general relativity that governs the motions of bodies in an accelerating frame (e.g., the movement of a test particle in a freely falling elevator). To successfully describe the motions of bodies in an accelerating frame, it is necessary to view spacetime as "curved" or "warped" due to the presence of matter (Misner et al., 1973; Moriyasu, 1985; Ridley, 1984; Adair, 1987), and the rules of the ordinary flat geometry (i.e., Euclidean geometry) must be replaced by the rules of the curved spacetime (i.e., Riemannian geometry), as indicated in Equation (1.32).

Just as there are two sets of rules governing the motions of material bodies in spacetime (i.e., general relativity and special relativity), depending on whether material bodies are observed in an accelerating frame or an inertial frame, so I suggest that we recognize two sets of rules governing chemical reactions (designated "cell" metabolism and "test-tube" metabolism), depending on whether enzymic reactions occur inside the living cell or in test tubes (Equation (1.33)). Again just as the gravitational force is responsible for the difference between special relativity and general relativity, so the cell force is claimed to be responsible for the difference between "cell" metabolism and "test-tube" metabolism.

Biologists commonly assume that enzymic reactions occurring inside the cell are identical to those proceeding inside a test tube. This may be analogous to physicists who do not distinguish between the motions of test particles observed in an *inertial frame* and those observed in an *accelerating frame*, thereby denying the existence of the gravitational force. Without the gravitational force, there are many motions of bodies in nature that cannot be explained rationally, including the bending of light by the gravitational field of the sun and the phenomenon of black holes (Adiar, 1987). In other words, physicists cannot explain motions of test particles observed in an accelerating frame in terms of the rules governing the motions of test particles in an inertial frame. Similarly, it is possible that biologists will not be able to completely account for the behaviors of the metabolism occurring in the living cell in terms of the rules derived from enzymic reactions investigated in test tubes.

By "curved" chemistry, I mean those characteristics of the chemical reactions occurring inside the cell that cannot be completely accounted for by the laws of chemistry and physics—i.e., by the laws of "flat" chemistry. In this sense, the existence of "curved" chemistry may be said to have been already presaged by Bohr more than half a century ago (Bohr, 1933, 1958), since he conjectured, on the basis of an analogy to quantum mechanical complementarity principle as already pointed out, that the laws of chemistry and physics might be insufficient to completely account for life. "Curved" chemistry is characterized by heterogenous chemical reactions whose rates are regulated "locally" in space and time. In contrast, "flat" chemistry involves chemical reactions that obey the rules of chemistry and physics that apply "globally" in space and time. In other words, "curved" chemistry is spacetime-specific, and this is thought to arise from two facts: (i) The chemical reactions giving rise to "curved" chemistry do not occur spontaneously under physiological conditions and require enzymic catalysis. (ii) The distribution of the requisite enzymes in space and time is constrained by genetic information encoded in cellular DNA (see Equation (1.34)). This statement is equivalent to saying that intracellular enzymes are subject to not only the electromagnetic force but also the cell force, just as nucleons are subject to the electromagnetic force and the strong force in atomic nuclei (see Tables 1.15 and 1.16).

The key result of the above considerations may be summarized in one sentence: Just as the gravitational force modifies the rules of special relativity to those of general relativity and the presence of matter introduces curvature into spacetime, it may be stated that the cell force transforms the rules of ordinary chemistry to those uniquely applicable to the chemistry proceeding inside the cell catalyzed by biopolymers. In view of the discussions already presented in connection with Table 1.15, we may rephrase this summarizing statement in the form of a metaphor:

"Biocybernetics": A Machine Theory of Biology

"Just as matter tells spacetime how to curve and curved spacetime tells matter how to move (Misner et al., 1973), so biopolymers tell chemistry how, when and where to occur and spatiotemporally organized intracellular chemical concentrations (i.e., IDS) tell biopolymers how to move."

This statement, which may be called the "cell force/gravitational force analogy" for future reference, serves as a convenient mental model to envision how intracellular biopolymers (i.e., DNA, RNA and proteins) interact with intracellular chemical concentration gradients and waves (i.e., IDS) to effectuate cell life, in analogy to the interactions between matter and spacetime giving rise to gravity.

Probably the most significant result derived from the analogical comparison of the cell force with the gravitational force is the idea of bifurcating chemistry into the "curved" chemistry and "flat" chemistry as shown in Equation (1.34). Just as the presence of matter transforms flat spacetime into curved spacetime, it has been suggested that the presence of biopolymers (or molecular machines) is responsible for the conversion of flat chemistry into curved chemistry. This bifurcation of chemistry seems to fall out naturally from a slightly different approach—by applying gauge theories to *chemical reactions* in contradistinction to *physical motions* in spacetime. Since there are two distinct kinds of gauge invariances—*global* and *local* gauge invariances (see below), we can set up a simple 2x2 table based on two pairs of dichotomies, namely the dichotomy of global vs. local on the one hand and the dichotomy of physical motions vs. chemical reactions on the other, as shown in Table 1.17. Special relativity theory can be viewed as representing the rules of physical motions obeying the principle of global gauge invariance, while general relativity theory reflects the rules of physical motions of bodies obeying the principle of local gauge invariance (see row #1 in Table 1.17). Analogically, flat chemistry can be treated as the chemistry wherein the principle of global gauge invariance holds (i.e., chemical reactions obey the laws of chemistry and physics globally at any place or at any time), whereas curved chemistry represent the set of chemical reactions wherein the principle of local gauge invariance holds true (i.e., the presence of biopolymers alters the rate of chemical reactions in space and time in an unpredictable manner and yet in such a way as to maintain cell life invariant) (see row #2 in Table 1.17).

In Section 1.8.8, a brief discussion was given on the difference between "global" and "local" gauge invariances (or symmetries). A simple example of a *global gauge invariance* is provided by the process of lifting a weight from A to B in a gravitational field. The work required to perform the weight lifting depends only on the height difference between A and B and not on the absolute heights of A and

Table 1.17. The analogy between the cell force and the gravitational force

	Global[2]	Local[3]
Physical reactions[1]	Special relativity theory[4] (Mass-free spacetime)	General relativity theory[5] (Mass-induced curved spacetime = Gravitational force)
Chemical reactions	"Flat chemistry[6] (Cell-free metabolism)	"Curved chemistry[7] (Cell-directed metabolism = cell force)

[1] Interactions between material bodies that do not lead to any chemical transformations, i.e., physical motions of material bodies.

[2] The influences affecting physical or chemical interactions are transmitted uniformly in spacetime.

[3] The influences affecting physical or chemical interactions vary from one point to another in space and time.

[4] The theory of motions in the flat spacetime—i.e., the space with the same curvature everywhere and all times.

[5] The theory of the motions of material bodies in the spacetime whose curvatures differ from one locality to another and from one point in time to another. Such heterogeneous spacetime curvature is due to an inhomogeneous distribution of matter in spacetime.

[6] Chemical reaction rates are homogeneously affected in space and time.

[7] Chemical reaction rates are individually regulated in every point in space and time in the cell so as to maintain the living state of the cell. Such a spacetime-elaborated chemistry is equivalent to the cell force, just as the curved spacetime is equivalent to the gravitational force.

B. Therefore, the work of lifting the weight remains *invariant* upon applying coordinate transformations *globally* to all the points in spacetime (Davies, 1988); local coordinate transformations (e.g., changing position A but not B), however, will not leave the work of the weight lifting invariant because this would lead to a change in the height difference between A and B. The essential idea behind the concept of a *local gauge invariance* is that the measuring scale (i.e., gauge) can be changed from place to place and moment to moment (i.e., "locally") without altering some fundamental properties of the physical system under consideration. In the case of gravitational interactions, Einstein's general theory of relativity has shown that there exists a set of numbers called "metric tensors" associated with each point in spacetime that can be varied from place to place and moment to moment without violating the laws of physics. The crucial question that now arises is "What constitutes the local gauge invarience in the living cell?" In view of the analogy already presented between the curvature of spacetime (as measured by metric tensors) and the genetic information of cell metabolism (see Table 1.15), it seems logical to conclude that what varies from place to place and from moment to moment in the cell is the nature of the DNA sequences (i.e., genetic information) that are actually activated and expressed and that what is kept invariant, despite the seemingly random turnings on and off of various cellular genes, is the ability of the cell to respond to environmental signals in such a way as to enhance its gene-directed functions.

1.8.10. The Bhopalator and the "Cellular Uncertainty Principle (CUP)"

According to the Bhopalator, all cellular functions are determined by its cell states, i.e., S^c in Equation (1.17). In other words, S^c contains all the information about the molecular processes underlying cellular behaviors and therefore can be identified with the set of "immediate" or "proximal" causes for all cell functions. However, because of the facts (1) that the internal states of the cell are composed of IDS's, and (2) that IDS's are dynamic 4-structures consisting of intracellular gradients of diffusible molecular species which can be easily disrupted whenever cells are subject to the procedures that are necessary for experimental measurements, we can reasonably conclude that it would be impossible in general to determine unambiguously the immediate molecular causes underlying cell functions (Ji, 1988). This conclusion reminds us of the Heisenberg Uncertainty Principle (HUP) which delimits the accuracy with which certain pairs of quantum mechanical observables (canonical conjugates) can be measured simultaneously (Pauling and Wilson, 1963). In analogy to HUP, therefore, the uncertainty associated with cellular processes was referred to as the "Biological Uncertainty Principle (BUP)" (Ji, 1988, 1990).

However, I now propose to rename the above statement as the "Cellular Uncertainty Principle (CUP)" and reserve BUP for a more general uncertainty principle that may be discovered in the future, involving biological structures and processes more complex than cells.

In Table 1.13, the cellular uncertainty relation analogous to HUP was suggested to be

$$(\Delta G)(\Delta I) \geq kT \tag{1.35}$$

where ΔG is the uncertainty attending the measurement of the Gibbs free energy changes associated with cellular processes, ΔI is the uncertainty about the significance of the cellular processes under study with respect to their contribution to the living state of the cell, namely the uncertainty about the "fitness" value of the cellular processes involved, k is the Boltzmann constant (1.3805×10^{-16} erg/°K), and T is the absolute temperature. Equation (1.35) simply states that there is a limit beyond which it is impossible to determine both the Gibbs free energy change underlying a given cellular process and its contribution to the fitness of the cell simultaneously with an arbitrary accuracy; i.e., beyond certain limit, the more accurately one determines the Gibbs free energy change, the less accurately can one determine the associated biological significance.

Equation (1.35) can be derived using the fiber bundle geometry introduced in Section 1.8.8. Following Moriyasu (1985), we can represent the structure and function relationship of the living cell as shown in Figure 1.11. The total space composed of spacetime and the genetic information space provides a geometric means to depict the behavior of all kinds of machines executing associated programs, including biopolymers and the living cell. Hence the total space in Figure 1.11 will be called the "function" space when it is used to describe artificial machines and the "L-space" (the letter L from life), when it is used to investigate the behavior of biological machines. The behavior of a living cell will trace out a thick trajectory R (from "river," a symbol of life) composed of N sub-trajectories called "streams," where N is the number of biopolymers inside the cell. Each stream represents the behavior of one biopolymer inside the cell. Clearly, then, the uncertainty about the behavior of the cell cannot be less than the uncertainty about the behavior of one of the N biopolymers.

To estimate the uncertainty about the behavior of a biopolymer inside the living cell, the following steps are required:

(1) There is a finite amount of uncertainty (ΔG) that is associated with the determination of the Gibbs free energy change underlying a given intracellular process catalyzed by a biopolymer. Since to drive any net biological process it is

"Biocybernetics": A Machine Theory of Biology

necessary to dissipate Gibbs free energy at least as large as thermal energies, kT, it follows that the smallest uncertainty about the measurement of Gibbs free energy changes attending a biopolymer-catalyzed process inside the cell can be estimated to be:

$$\Delta G \geq kT \quad (1.36)$$

(2) Due to ΔG, the cross-section of the behavior trajectory of the biopolymer possesses a finite size. This leads to an uncertainty about the internal coordinate (i.e. genetic information) of the biopolymer, since there are at least two internal coordinates that can be accommodated within the cross-section of the behavior trajectory (see 1', 1, and 1'' in Figure 1.11). Therefore, the uncertainty about the genetic information associated with the biopolymer behavior is at least one bit;

$$\Delta I \geq 1 \text{ bit} \quad (1.37)$$

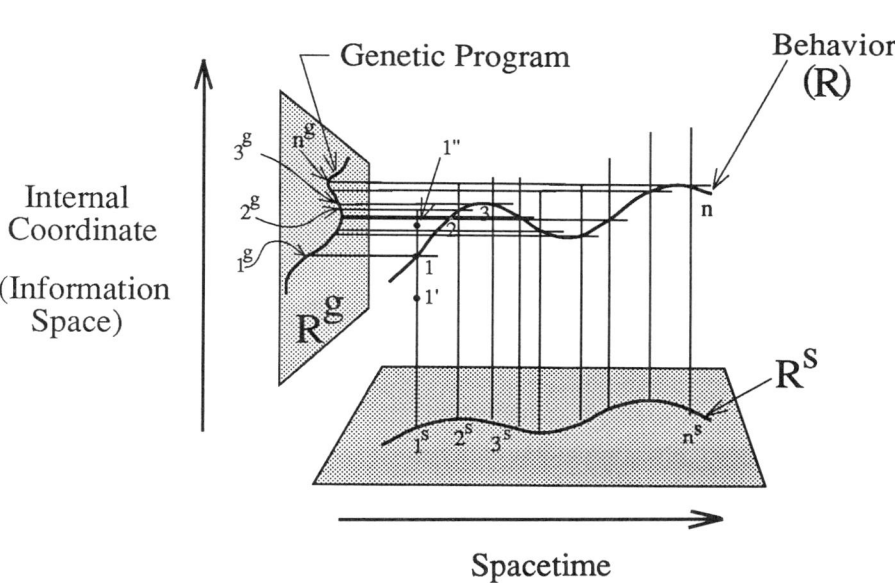

Figure 1.11. The derivation of the "Cellular Uncertainty Principle (CUP)" based on the fiber bundle theoretic representation of the living cell. The cell behavior is depicted as a curvy line (see R) in the total space. The genetic program responsible for the cell behavior, R, is indicated as the projection of R onto the internal coordinate space (see the vertical plane on the left) to generate an image labeled R^G, the superscript G indicating "genetic information." Since the internal coordinate space is where genetic information can be specified, we can call this space the genetic information space. Also, since the genetic information space is very closely related to the "sequence space" recently described by M. Eigen (1987; Eigen et al., 1988, 1989), it may be called the Eigen space in honor of his contribution to the information theory of living processes. The spacetime trajectory of R is labelled R^S. Since all living systems can be depicted in the total space composed of the Eigen space and spacetime, we will refer to the total space as the "L-space," L meaning life.

(3) Equations (1.36) and (1.37) can be combined by multiplication to obtain Equation (1.35), the Cellular Uncertainty Principle.

Just as Heisenberg's Uncertainty Principle can be viewed as defining the smallest packet of energy, namely Plank's quantum of action, h, having the value 6.6252×10^{-27} ergs sec, so the Cellular Uncertainty Principle as expressed in Equation (1.35) may be interpreted as defining the smallest packet of gnergy (to be designated with the symbol "l," meaning life) having the value

$$ l = 1.3805 \times 10^{-16} \quad \text{erg bit} \tag{1.38}$$

In analogy to h, the quantum of action, we may call l the "quantum of biological communication," since no life would be possible without communications within or between cells. Biological communications are discussed in more detail in the next sub-section. Equation (1.35) can now be written as

$$(\Delta G)(\Delta I) \geq l \tag{1.39}$$

1.8.11. The Cell and Biological Communications: The Cellular Communication Theory

The living cell can be viewed as an essential component of the biological communication system capable of transmitting information in space and time. We may define biological communications (or "bio-communications," for short) as the phenomena of exchanging of information and control influences mediated by living cells. Unlike electrical communication systems whose main purpose is to transmit information through space, bio-communication systems (consisting of one or more cells) are capable of transmitting information not only through space (e.g., endocrinology) but also through time (e.g., evolution, embryogenesis) (Ji, 1988). The major aim of this sub-section is to apply the mathematical expressions and the conceptual framework of electrical communication systems (ECS) to the description and analysis of bio-communication systems (BCS).

(1) There appears to be no universally accepted scheme for representing electrical communication systems diagrammatically. For our purpose, it is sufficient to depict ECS as shown in the upper portion of Figure 1.12, which resembles closely the scheme used by Gagliardi (1988). ECS is divided into two subsystems called "sender" and "user" that are connected by "channel." Sender is in turn composed of (i) source of message (e.g., a book or man speaking), (ii) encoder (which transduces original messages into sequences of signals that can be transmitted through the communication channel), and (iii) transmitter (which emits signals), while user is

"Biocybernetics": A Machine Theory of Biology

composed of (iv) receiver, (v) decoder (which converts sequences of received signals back to original messages), and (vi) received messages. The purpose of ECS is to transmit messages *through space*, regenerating original messages accurately for the benefit of the user. Information transmission through time is not the concern of ECS.

It is interesting to note that practically the same scheme used to depict ECS can be used to represent bio-communication systems, with only a minor relabeling of two of the six components, namely receptor and effector (see the lower portion of Figure 1.12). In contrast to ECS, the purpose of BCS is to "effectuate" the content of received messages rather than simply regenerating them. To effectuate (i.e., actualize or reify) messages, molecular mechanisms (collectively called effector) are needed that can perform work processes called for by received messages. Another fundamental difference between ECS and BCS is that the latter has the ability to transmit information through space as well as through time as already pointed out.

(2) C. Shannon, one of the foremost pioneers of the communication theory (also called the information theory (Pierce, 1980)), derived two important mathematical expressions, one defining the entropy (H) of the message (or information) source, and the other characterizing the channel capacity (C). In passing, it may be mentioned that H (also called the Shannon, information-theoretic, or informational entropy) is not identical with the thermodynamic (or physical) entropy,S, as is often claimed to be by some physicists and biologists. To avoid possible confusions, Wicker (1987) recently recommended that H in information theory be called "complexity" (of the message source), reserving the term entropy to denote S in thermodynamics. However, in the following discussions, we will continue referring to H as entropy but in the sense of "complexity" as pointed out by Wicker (1987).

(i) <u>The Entropy (H) of a Message Source</u>

$$H = - \sum_{i}^{n} p_i \log_2 p_i \quad \text{bits/symbol} \tag{1.40}$$

where n is the number of symbols in the message source (a message may be composed of one or more symbols), p_i is the probability of the ith symbol being selected as a part of a message. If a message source produces n symbols per second and if each symbol carries H bits of information, then the message source can emit information at the rate of nH bits/second. Therefore, H can be expressed in the units of either bits/symbol of bits/second. The physical meaning of H is that it represents

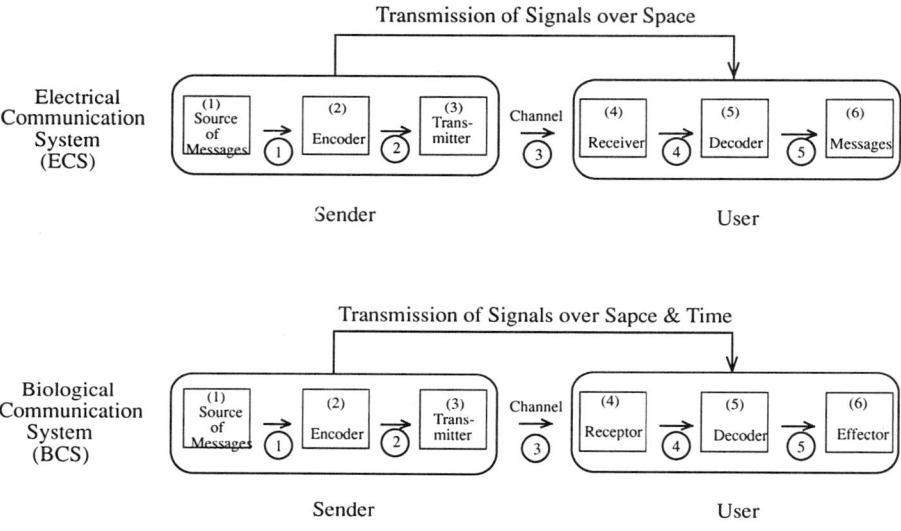

Figure 1.12. Schematic representations of the electrical communication system (ECS) and the biological communication system (BCS). Both communication systems are divided into 6 major structural components. Functional homologies between the two sets of components are only partial. The source of messages in BCS may also be called the "generator," since numerous biological messengers are generated by enzymes.

the complexity of the message source (Wicker, 1987) or the uncertainty concerning which of the symbols in the message source will be selected for transmission (Pierce, 1980). The greater the value of H, the greater the amount of information carried by the chosen symbol. If a message consists of N symbols, the amount of information carried by the message is NH bits. Hence, the information content of a message is not completely determined by the properties of the message alone (e.g., the length of a message) but profoundly influenced by the statistical properties of the message source. For example, a symbol transmitted by a message source containing 10

"Biocybernetics": A Machine Theory of Biology

equally probable symbols will carry more information (i.e., 3.333 bits, calculated using Equation (1.41) below) than a symbol originating from a message source containing only 5 equi-probable symbols (i.e., 2.333 bits).

If the probabilities of choosing symbols from the message source are equal, then Equation (1.40) reduces to:

$$H = \log_2 n \qquad (1.41)$$

which was referred to as the "simplified" version of Shannon's formula in Appendix 1.B.

(ii) <u>The Channel Capacity (C)</u>: The channel capacity is a quantitative measure of the maximum speed with which information can be transmitted through a communication channel and is given by the following expression;

$$C = W \log_2 (1 + P/N) \qquad (1.42)$$

where W is the bandwidth (i.e., the width of a band of frequencies used to encode messages, or a measure of the degree of freedom of choosing symbols), P is the power of signals (i.e., the rate of dissipation of energy in transmitting signals), and N is the power of noise. According to Equation (1.42), there are two ways in which the speed of information transmission through a given channel can be increased—(a) by broadening the bandwidth (e.g., by increasing the complexity or variety of the alphabet of the message source, an example of which may be the increase in the number of the letters from 4 nucleotides in the nucleic acid alphabet to 20 amino acids in the protein alphabet), and (b) by increasing the power or amplitude of transmitted signals (e.g., to compensate for the smaller bandwidth of the nucleic acid communication channel, the free energy content of conformons embedded in DNA (i.e., P) may be increased relative to the free energy content of conformons embedded in proteins by a factor of 10^7 (see Appendix 1.B)).

The relationship between *symbols* and *signals* is that the former is static while the latter is dynamic. Signals are usually generated by modulating "carrier waves" such as the electromagnetic waves which propagate through space. A given symbols can be encoded in more than one ways, depending on the properties of the communication channel available.

One of the most fundamental theorems of the communication theory of Shannon is that, in ECS, C must be always greater than or equal to H (Pierce, op cit., p. 98);

$$C \geq H \tag{1.43}$$

which may be referred to as *Shannon's inequality*, for convenience. At least two corollaries maybe derived from this inequality: (a) It is impossible to transmit information from source to user without a communication channel (Since no communication channel is equivalent to $C = 0$, and since the entropy of a message source cannot be negative, transmitting information without a channel is tantamount to violating Equation (1.43)), and (b) In order to transmit information of any kind, a message source must be coupled to a communication channel with a capacity greater than or equal to the entropy of the source.

Of the two mathematical expressions presented above, only H is utilized in most of the discussions on biological information (e.g., the information content of DNA, nucleotide triplets carrying $\log_2 4^3$ or 6 bits of information per triplet, etc.). The concept of the channel capacity, although as fundamental as that of the entropy of the message source, is rarely mentioned in connection with biological communications. It is possible that there are numerous cellular processes now being intensely investigated (e.g., receptor-ligand interactions, intracellular signal transductions via cAMP, Ca^{++}, and the inositol triphosphate cascade, regulation of transcription by DNA-protein interactions, etc.) that will not be completely understood until and unless the concept of the channel capacity is fully taken into account.

(3) We can clearly recognize two kinds of bio-communications—(i) the *intercellular communication*, and (ii) the *intracellular communication*. Of course, these two types of communications must interact with each other. It is conjectured that the primary role of intercellular communications (e.g., endocrinology) is to transmit information in the spatial dimension in multicellular organisms and that one of the major biological functions of intracellular communications (e.g., from the nucleus to the cytosol to the extracellular space) is to transmit information in the time dimension as in all gene-directed cellular processes.

Applying the communication system diagram given in Figure 1.12 to both the intercellular and intracellular communications, I have tentatively identified the various components of these two communication systems as shown in Table 1.18. The identification of the system components suggested in this table is not rigorous. This may be primarily because of the complex nature of the bio-communication systems under consideration and not necessarily because Shannon's communication theory cannot be extended to the description of information transmission through time. Despite certain degree of arbitrariness and seeming contrivance, the versatility of the communication theory in accommodating bio-communication systems is evident in Figure 1.18. I know of no other physical theory which shows a similar versatility and power of organization as the communication theory.

"Biocybernetics": A Machine Theory of Biology

Table 1.18 Comparison of intercellular and intracellular communication systems in biology

System component	Intercellular communication system	Intracellular communication system
Sender	Cell #1	Biological evolution[2]
User	Cell #2	Cell[3]
Channel	Diffusible chemicals in aqueous media[1]	Variable nucleotide sequences of DNA or RNA[4]
Source of messages[5]	Metabolism of Cell #1	Mutation, recommendation and natural selection[6]
Transmitter	Mechanisms for transporting molecules across cell membrane of Cell #1	Cell division[7]
Channel	Diffusible chemicals	Linear sequences of symbols in biopolymers[8]
Receptor	Membrane or cytosolic receptors of Cell #2	DNA-binding proteins[9]

(Table 1.18 continued)

Decoder	Signal-transducing enzymes of Cell #2	Sequence-specific conformational strains of biopolymers[10]
Effector	Hormone-induced metabolic changes in Cell #2	Conformons[11] and IDS[12]

---

[1] Life cannot exist without water—this may be primarily because diffusible chemicals dissolved in aqueous media serve as the channel for intercellular communication, just as the electromagnetic waves provide the channel for electrical communications. The concentration gradients, the molecular configurations (i.e., the covalent linkages of atoms) and the conformational states (i.e., the noncovalent linkages of atoms) of diffusible chemicals may be analogous to the amplitude, the frequency and the phase angle of electromagnetic waves carrying electrical signals.

[2] Biological evolution can be regarded as a set of events that transpired in the evolutionary past. Treating biological evolution as the sender of biological information is consistent with the requirement that the sender and the user be separated in time so as to transmit information in the time dimension (Ji, 1988).

[3] The cell can be viewed as a set of events realizable at the present and future, *utilizing* the genetic information received from the past. This is consistent with the idea that the intracellular communication involves information transmission in the time dimension.

[4] The linear sequences of nucleotides in DNA and RNA serve as the communication channel operating over the time dimension, just as written languages serve this role for human society. The signals transmitting genetic information through the biopolymer channel are written in the letters of 4 nucleotides. The 4 letter alphabet of the nucleic acid language contrasts with the 26 letters in the English alphabet and 24 in the Korean alphabet.

[5] It is interesting to think that the immediate source of the information (i.e., physical and chemical constraints) that led to the origin of life on this planet 2-3 billion years ago was the environment of the earth surface in the prebiotic era and that the ultimate source of the information contained in the earth surface in turn was the Big Bang that occurred 15 billion years ago. It is generally believed among cosmologists that, because of the particular manner in which this universe began at the time of the Big Bang, the evolution of life on this planet was inevitable and pre-ordained (Barrow and Tipler, 1986); i.e., whatever gave rise to the Big Bang, that entity might have already contained in it the "blue print" of life. I speculated (somewhat metaphysically) that the entity that caused the Big Bang was what I called the "gnergy tetrahedrality" (in analogy to the wave-particle duality), defined as the complementary union of four primordial elements—*energy*, *matter*, *information*, and *life* (see Section 1.11).

(4) It is clear from the above discussion that there exists in nature a communication system centered around the living cell, which we will call the "cellular communication system." We may define the *cellular communication theory (CCT)* as the *theory of the living cell viewed as the embodiment of the physical mechanisms for storing, transducing and transferring information in space and time driven by chemical reactions*. The Bhopalator model of the living cell described in Section 1.8.4 provides the ultimate molecular and cellular basis for all intercellular and intracellular communications. The key postulates underlying CCT are described below:

(i) The quantitative expressions for the amount of information (Equation (1.40)), the channel capacity (Equation (1.41)), and the Law of Requisite Variety (Appendix 1.A) apply to the cellular communication system.

(Notes to Table 1.18 continued)

[6] Since the variation of genes through mutation and recombination and the subsequent transmission of select genotypes through natural selection determines the nature of the transmitted genes, the totality of these complex processes can be viewed as the encoder of genetic information for the purpose of transmitting it through time.

[7] Cell division is probably the most fundamental property of the living cell, and I maintain that this is because cell division serves as the mechanism of transmitting genetic information on the evolutionary time scale.

[8] There may be intracellular communication channels other than the DNA channel that participate in the transmission of genetic information from the nucleus (the past) to the cell as a whole (the present). It is possible that the additional channels include the RNA channel, the protein channel, and the IDS channel.

[9] It appears that the genetic information stored in nucleic acids cannot be expressed without these macromolecules first interacting with sequence-specific DNA-binding proteins, whether they are regulatory or enzymatic.

[10] According to the conformon theory of gene expression (see Section 1.12.5.), no information retrieval from DNA is possible without dissipation of free energy in the form of conformons. Conformons are postulated to be introduced into DNA via DNA sequence-specific ATPases. Such ATPases may be conveniently called "conformon generators" or "conformon-generases," the enzymes that catalyze the free-energy transfer from ATP to sequence-specific regions in DNA.

[11] Conformons can directly drive cell functions such as cellular motility and cell shape changes, since conformons are necessary and sufficient for muscle contraction (Ji, 1974a).

[12] Certain cell functions may require further transduction of conformons into intracellular chemical concentration gradients and waves (i.e., IDS), as in the maintenance of membrane potentials and cell volume.

(ii) The Law of Requisite Variety (LRV), namely $V_O \geq V_E/V_M$, states that, in order for a machine to maintain a constant variety of outputs (V_O) (e.g., mitosis, cellular motility, etc.) as the variety of environmental inputs (V_E) (e.g., thermal fluctuations, chemicals, particles, microorganisms, etc.) increases, the variety of the internal states of the machine (V_M) (i.e., different spatiotemporal patterns of distributions of intracellular ions, small molecules, and biopolymers) must increase.

CCT postulates that LRV is ultimately responsible for the complexity of all organisms. As the variety of extracellular environment increases, cells are thought to increase the variety of their internal states through five distinct mechanisms—(a) chromosomal rearrangements (e.g., pericentric inversions, Delbrück, 1986), (b) gene recombinations, (c) post-transcriptional modifications (e.g., RNA splicing), (d) post-translational *covalent* modifications (e.g., protein phosphorylation, formylation, methylation, protonation, etc.), and (e) post-translational *noncovalent* modifications (e.g., specific and nonspecific ion-induced conformational transitions of proteins, DNA and RNA; non-covalent interactions between proteins and proteins, between proteins and DNA, and between proteins and RNA).

The increased variety of the internal states of the cell so produced may serve two major biological functions—(a) to allow cells to transmit more information per unit time by increasing the bandwidth of the cellular communication channel (i.e., the W term in Equation (1.42)), and (b) to allow cells to increase the *fidelity of cellular communications* by using more than one channels for transmitting identical messages. The latter idea was referred to as the "multiple channel hypothesis of cellular (or biological) communication" and was found to provide logical explanations for the phenomenon of the multi-stage chemical carcinogenesis and the power-law dependency of cancer incidence rate on age (see Section 1.12.12).

Applying the multiple channel hypothesis to cellular signal transduction, it may be predicted that most, if not all, extracellular messengers (e.g., growth hormones) may trigger two or more intracellular signal transduction cascades (e.g., cAMP-dependent and IP_3-dependent pathways) to activate or inhibit a given cellular function such as gene expression, cell division or chemotaxis. The principle of increasing the fidelity of information transmission using redundancy is well established in electrical communications engineering and in linguistics (Campbell, 1982), and the same principle appears to operate in the cellular communication system.

(iii) CCT views endocrinology and morphogenesis as two variations of the same theme, namely messenger-cell interactions (consummated through ligand-protein interactions; see (viii) below) in that endocrinology involves two immobile (or fixed) cells connected by mobile messengers (e.g., ACTH connecting the pituitary cells to adrenocortical cells), while morphogenesis implicates mobile cells communicating through immobile messengers (e.g., cells depositing extracellular matrix proteins to which the same or other cells subsequently respond). Messengers can be small-molecular-weight species or small segments of DNA, RNA or proteins, and the cells connected by messengers can be located at different anatomical loci (endocrinology) or different time points (morphogenesis).

(iv) CCT assumes that biological evolution has utilized all possible variations of the ligand-protein interactions in the cellular communication system within the constraints imposed by the laws of physics and chemistry, so long as these variations provided survival advantages to organisms. Ligands can be small-molecular-weight species, other proteins, DNA, or RNA.

(v) The Law of Requisite Variety (Appendix 1.A) may provide the fundamental rationale for biological evolution. Thus, the structure of organisms may continuously increase in complexity in order to maintain the functional homeostasis of organisms in the face of increasingly diversifying environmental conditions, namely the surface of the earth. The unidirectional increase in the complexity of the earth's surface is, in turn, connected to the evolution of the solar system and the universe itself according to the Second Law of Thermodynamics. The Second Law dictates that the entropy of the universe increases with time.

(vi) Consistent with (iii), CCT recognizes at least three kinds of proteins participating in protein-ligand interactions—(a) familiar receptors which act as the receivers of messengers (e.g., the receptor portion of receptor kinases, steroid receptors), (b) "generators" which synthesize and release messengers for transmission (e.g., adenylate cyclase producing cAMP, guanylate cyclase, IP_3 synthetase, ion and other pumps of biological membranes, etc.), and (c) "receptor-generators" which produce messengers upon receiving other messengers (e.g., receptor kinases such as PDGF receptors, cell adhesion receptors (?)).

Since no communication system is complete without the source of messages (Figure 1.12), it is logical to think that the cellular communication system should possess the message source, or "generators of messages." The nitric oxide (NO) synthase recently isolated and purified by Solomon Snyder and his coworkers (the lecture given at the conference entitled "Advances in Receptor-Ligand Interactions" held at the Robert Wood Johnson Medical School-UMDNJ, Piscataway, N.J., on September 26-28, 1990) may provide an interesting example of a generator. The receptor for NO has been found to be guanylate cyclase (and certain other Fe-containing enzymes) whose enzymic activity increases when NO binds to the iron ion embedded in the cyclase. I agree with S. Snyder that the NO synthase-NO system does not fit nicely into the traditional conceptual framework of receptor-ligand interactions, although NO can clearly be viewed as a messenger. The generator concept proposed herein may resolve this dilemma.

In some cases a receptor and a generator may be covalently linked so as to form two separate domains of a single receptor-generator (or receptor-effector) protein, such as autophosphorylating receptor kinases. Certain G-proteins may act as receptor-generators if they receive information from the receptor subunit and transfer it to the catalytic subunit of a receptor complex such as the ß-adrenergic receptor.

(vii) CCT predicts the presence of much greater varieties of messengers than are currently conceived of by researchers in cellular signal transduction. According to CCT, messengers can be classified in terms of three basic criteria—(a) stability (stable (s) and unstable (u)), (b) specificity (specific (s) and nonspecific (n)), and (c) mobility (mobile or diffusible (m) and fixed (f)). Therefore, in principle, there can be eight distinct classes of messengers operating in the cellular communication system, and these will be designated as "ssm" (stable-specific-mobile), "ssf" (stable-specific-fixed), "snm" (stable-nonspecific-mobile), "snf" (stable-nonspecific-fixed), "usm" (unstable-specific-mobile), "usf" (unstable-specific-fixed), "unm" (unstable-nonspecific-mobile), and "unf" (unstable-nonspecific-fixed) messengers.

Most of the hormones studied in classical endocrinology and neurotransmitters clearly belong to the ssm class of messengers, since they are chemically stable (unless degraded by specific enzymes), interact with specific receptors and diffusible. Prostaglandins playing important roles in the inflammatory response belong to the usm class, since they are unstable with half-lives measured in minutes, specific, and mobile. Nitric oxide has a half-life of only 5 seconds, binds specifically to Fe-containing enzymes such as guanylate cyclase and mobile and hence can be classified also as an usm messenger.

The ligand for the osmoreceptor of bacterial cell membrane may turn out to be water molecules themselves (why not?), in which case H_2O will be the first example of a snm messenger, since H_2O is stable, nonspecific, and mobile. One prediction based on this hypothesis is that the functional capacity of the bacterial osmoreceptor would be relatively insensitive to point mutations (unlike many specific hormone receptors) so long as the amino acid substitutions conserve the hydrophobicity of the receptor.

Another example of messengers that belong to the nonspecific class may be the mechanical strains induced by stresses applied to collagen lattice infiltrated with living fibroblasts, which has been demonstrated to affect cell growth and protein synthesis (Jain et al., 1990). I suggest that the "macroscopic" stresses applied to the collagen lattice induce a set of "microscopic" conformational strains (i.e., conformons, Section 1.5.6; see (viii) and Figure 1.13) confined in proteins constituting the extracellular matrix, cell membrane receptors, cytoskeletons, or nuclear scaffoldings, and some of these conformational strains may act as messengers of the unf kind, since they will be unstable (mechanical strains will disappear when stress is removed), nonspecific (because a large number of different kinds of conformons may be activated simultaneously) and fixed (because they cannot "get out" of biopolymers, although they can "rattle around" or migrate from one site to another within them through thermal fluctuations). In contrast, the conformational strains generated in receptors by binding specific hormones may act as messengers of the usf kind (since they would be unstable, specific, and confined in biopolymers). Such a novel kind of messengers may be utilized (or generated) by membrane proteins such as G-proteins and the receptors that bind extracellular matrix proteins including fibronectin and laminin (Aebi and Engle, 1989).

It is possible that hormone- or mechanical stress-induced conformational strains can be trapped in sequence-specific regions of biopolymers where the free energy stored in conformational strains provides the thermodynamic force for initiating certain nuclear events, such as DNA transcription and replication. Although thermal fluctuations alone can initiate DNA transcription or replication, it may be that such thermally initiated processes cannot be organized in time without violating the Second Law of Thermodynamics, since any form of control processes absolutely demands dissipation of requisite free energy (Hess, 1975). This may be called the "principle of the free energy cost for control." Ligand- or stress-induced conformons may provide the appropriate free energy required for controlling the timing of nuclear events, and here may lie in part the fundamental significance of the binding of multiple protein factors essential for certain DNA transcription and replication and RNA splicing.

Examples of messengers belonging to the fixed class may include the cell receptor-binding segments of extracellular matrix proteins that now number in the hundreds. These messengers can be stable, unstable, specific, or nonspecific. Therefore it can be predicted that there will be 4 distinct classes of what may be logically called "extracellular fixed messengers (EFM)"; namely, ssf, snf, usf, and unf. EFM may play important roles in morphogenesis, angiogenesis, cellular differentiation, and metastasis.

Accumulating evidence indicates that there are great similarities between cytoskeletal and extracellular matrix proteins, so that the influence of metabolic events occurring in the cytosol is not confined within the space delimited by the cell membrane but spills over into what may be called the "pericellular space", the extracellular sphere of influence of the living cell. This conjures up an image of the living cell as the atom of life endowed with a "halo of cellular influences" around it, just as organic molecules are surrounded by the Van der Waals force that plays a critical role in molecular recognition. This "halo of cellular influences" may be intimately related to the cell force discussed in Section 1.8.9, just as certain properties of atoms and molecules can be traced to the strong force (see footnotes 19 and 20 in Table 1.16). This expanded vision of the cellular influence beyond the cell volume raises the question of the role of the cellular membrane anew. Why is it necessary for cells to have plasma membrane? What is the fundamental distinction between the intracellular and the extracellular spaces?

I agree with B. Olsen (October, 1990, personal communication) that one of the fundamental roles of the cell membrane is the retention of certain metal ions (K^+, Mg^{++}, etc.) and the exclusion of still others (Ca^{++}, Na^+, etc.) from the cytosol so as to maintain various ion gradients across the cell membrane. I suggest that, in accordance with the Law of Requisite Variety, such membrane-supported ion gradients are important for the cell because they provide novel mechanisms for increasing the variety of the internal states of the cell (i.e., cell states), thus allowing cells to perform their essential functions under increasingly more complex extracellular environment. In other words, the cell membrane-supported ion gradients contribute to increasing the variety of the cell states (i.e., the value of V_M in the equation discussed in (ii)). The molecular mechanisms responsible for ion gradient-induced increase in cell states can be traced ultimately to ion-induced conformational changes of biopolymers, which was invoked as an important component of the post-translational noncovalent modifications discussed in (ii).

The above discussion makes it evident that ions can act as messengers, since they can cause conformational transitions of biopolymers just as classical hormones do. We can recognize two kinds of ionic messengers—(a) *specific* ionic messengers such as Ca^{++} and Mg^{++} acting through specific receptors (e.g., calmodulins, Mg^{++}-dependent enzymes), and (b) *nonspecific* ionic messengers as in the case of the ions whose influence is exerted through changing the ionic strength of the medium in which biopolymers are dissolved. The first kind will belong to the ssm class of messengers, and the second kind will belong to the snm class.

(viii) One of the fundamental assumptions underlying CCT is that ligand-protein interactions provide the ultimate molecular mechanisms enabling cellular communications. As pointed out above, ligands can be metal ions, small molecules (e.g., ATP, inorganic phosphate, IP_3, etc.), other proteins, DNA or RNA. The focus is on proteins because proteins have evolved to perform two major functions—(a) the storage of information in the form of the primary, secondary, tertiary and quaternary structures, and (b) the utilization of chemical free energy to perform molecular work functions through their ability to catalyze chemical reactions.

In other words, proteins are important in ligand-protein interactions because they can act as molecular machines. To emphasize the machine aspect of proteins, we may view protein molecules as molecular analogs of familiar coin-operated vending machines (COVM). Two features of COVM are useful in this analogy—(a) specific inputs (e.g., dimes, quarters, dollar bills, or tokens), and (b) specific outputs (e.g., candies, coffee, milk, music, cash, photographs, shoe-shining, etc.). Just as a COVM requires dissipation of free energy (e.g., the pulling of knobs by the arm muscle of a customer, or the rotation of carousels by electric motors, etc.) to perform their designed functions, so proteins must dissipate free energy in order to perform molecular work processes. In most cases the requisite free energy can be supplied by the free energy of the interaction between ligands and proteins. The collection of these ideas and concepts will be referred to as the "ligand-operated molecular vending machine (LOMVM)" hypothesis of ligand-protein interactions. This hypothesis is schematically illustrated in Figure 1.13. The most significant feature of this figure is that it enables us to visualize how molecular machines can utilize both the biological information stored in their structures and the free energy generated from ligand-protein interactions in order to perform various molecular work processes essential for cellular functions. Depending on the structural characteristics of the effector, ligand-protein interactions can lead to oxidative phosphorylation, active transport, DNA transcription and translation, RNA splicing, chromosomal rearrangements, mitosis, meiosis, etc.

It is possible that the a to d transition shown in Figure 1.13 can be driven by external forces such as the mechanical stress applied to the fibroblasts infiltrating the collagen lattice which increased protein synthesis in these cells, presumably mediated by conformationally strained membrane receptors and other proteins (Jain et al., 1990).

(ix) CCT also provides possible rationales for the complex molecular assemblies necessary for effectuating DNA transcription and replication and RNA splicing. Some replisomes contain 20 or more proteins (A. Kornberg and R. Kornberg, the Fourteenth Annual CIBA-Geigy/Drew Symposium on Biomedical Research, "The Double Helix: Replication, Recombination and Transcription," held at Drew University, Madison, N.J., on October 16, 1990), and splicesomes may contain up to 100 protein factors (X. D. Fu, "Identification and Nuclear Localization of Functional Components of Mammalian Splicesomes," a seminar given at the Department of Molecular Biology and Biochemistry, Rutgers University, Piscataway, N.J., on October 18, 1990). As indicated in (iv), LRV may be the key to understanding the complexity of such molecular apparatuses; these enzymic

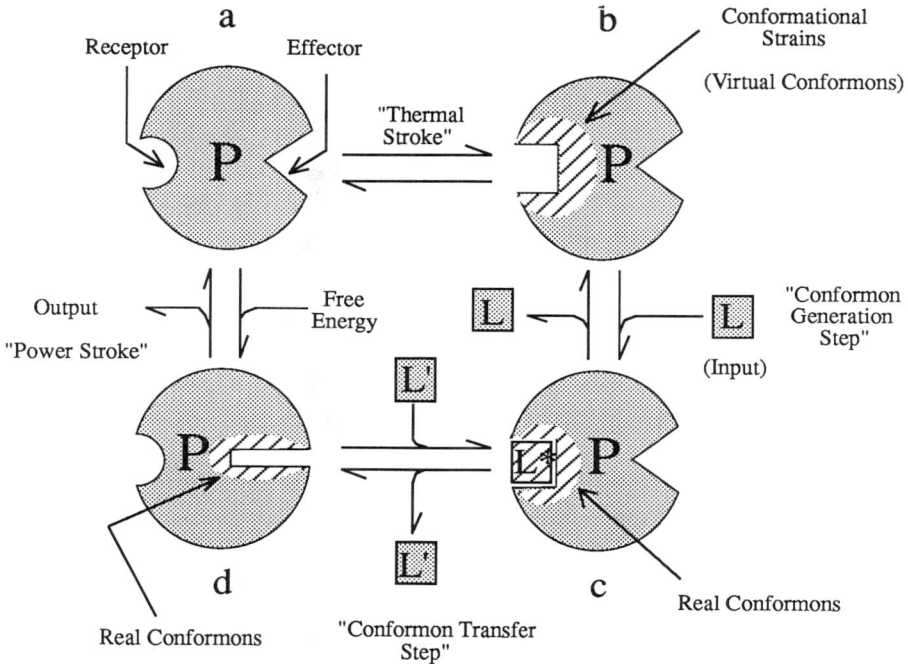

Figure 1.13. The "ligand-operated molecular vending machine (LOMVM)" hypothesis of ligand-protein interactions. The protein (P) possess two functional domains—the receptor and the effector domains. Each domain can exist in two (relaxed and strained) conformational states represented by the partial figures of a circle, a square, a triangle, and a rectangle. The conformationally strained regions of P are indicated with hatched lines. The model consists of four major steps.

(I) The *thermal stroke* (see a to b): Through thermal fluctuations, P can exist in either conformationally relaxed state, a, or conformationally strained state, b. The conformational strain energy localized in the receptor domain is thought to be derived from the thermal environment of P and hence called "virtual conformons" in analogy to the virtual quanta in the relativistic quantum field theory (S. Ji, 1985), because such thermally derived energy cannot be utilized to do any useful work without violating the Second Law of Thermodynamics.

(II) The *conformon generation* step (see b to c): The lifetime of the conformational strains around the receptor in b is thought to be too short to do any useful molecular work processes (including its directional migration to the effector domain; i.e., the b to d transition is prohibited) but long enough to bind the ligand (L) that has diffused toward the binding groups of the receptor domain, leading to the ligand-protein interactions, either covalent or noncovalent. To be consistent with the principle of microscopic reversibility, it is necessary to invoke at least three structural states of the ligand—the reactant ligand (L), the structurally strained bound ligand (L^*), and the product ligand (L') that has undergone either covalent (i.e., electronic)

"Biocybernetics": A Machine Theory of Biology 137

machineries must possess a large variety of internal states (as expressed in terms of both conformational substates of individual proteins and the combinatorial variety of subunit compositions) in order to maintain the homeostasis of DNA replication and transcription and RNA splicing under a wide variety of input conditions, both external and internal (see below); only variety can destroy (or reduce) variety (Equation 1.A1 in Appendix 1.A).

(x) CCT admits of two kinds of causes (anything producing an effect or result)—(a) external causes, and (b) "internal causes". External causes refer to messengers arriving at the cell from outside (e.g., chemoattractants causing a directional movement of phagocytes), while "internal causes" are due to those messengers generated from within the cell, without any immediate inputs from outside (e.g., spontaneous growth and division of cells in plant seeds given right humidity and temperature, presumably caused by the molecular clock embedded in nuclear

---

(Notes to Figure 1.13 continued)

alterations through interaction with P or spatial rearrangements (e.g., passive diffusion across biological membranes or conformational transitions). It is mandatory that in order to convert the virtual conformons in b to the real conformons in c, the requisite free energy must be provided by the L to L' transition; i.e., the L to L' transformation must be exergonic.

(III) The *conformon transfer* step (see c to d): Conformons localized around the receptor domain in c are postulated to have the ability to disengage from the receptor domain and migrate to the effector domain as depicted in the c to d transition, without appreciable dissipation of free energy probably due to evolutionarily prepared special structural characteristics of P. The migration of conformational strains in polypeptides has been predicted theoretically using the soliton equations of condensed matter physics (Davydov, 1981; Scott, 1985a,b; Section 1.5.2) and seems to play a fundamental role in DNA transcription and replication as evidenced by the requirement of these processes for DNA supercoils, clear examples of conformons.

(IV) The *power stroke* (see d to a): Once conformons are localized in the effector domain, the free energy and control information available in them can be utilized to drive any goal-directed molecular work processes underlying cell functions, be they specific covalent modifications of the effector domain itself or other molecules such as ADP (during oxidative phosphorylation), active transport, the opening or closing of membrane ion channels, the separation of DNA double helices for replication and transcription, the directional movement of replisomes, splicesomes and elongation complexes, and the microtubule-mediated trafficking of intracellular cargoes such as endoplasmic reticulum and mitochondria and possibly the chromosomal "choreography" in the nucleus during mitosis and meiosis. It is possible that under certain conditions, the information stored in the conformons localized in the effector domain (see d) can be carried away by diffusible molecules generated by the conformationally strained effector, just as the binding of a ligand to the receptor domain produces conformons during the b to c transition. This notion will be designated as the *conformon-messenger transduction hypothesis*, where messengers can be diffusible or fixed molecules or even other conformons. The LOMVM hypothesis described here can be viewed as a molecular realization of what was called by J. Rothstein a "well informed heat engine (WHE)" (1971; also see Chapter 7 of this book).

DNA). External causes are familiar in physics and chemistry—so much so that they are often thought to be the "only" kind of causes acceptable to science. Strongly influenced by physicists and chemists, most, if not all, biologists apparently have accepted, uncritically in my opinion, the epistemological precept of physics and chemistry that *nothing can move without being moved*. This precept is embodied in the third law of Newtonian mechanics, namely the law of the equality of action and reaction (Goldstein, 1980). However, there is no *a priori* reason to think that this law derived from studying motions of nonliving objects should hold with an equal force for all motions of living systems. The existence of these two kinds of causes is clear to any conscious human being—our bodily motions can be triggered either by *external* causes such as sights, sounds, touch, and tastes or by *internal* causes (i.e., by our own free will) in the absolute absence of any external stimuli. I am now sitting in front of my IBM PC, word-processing; but I can get up at any moment and go out for a nice walk around the campus, on my own free will. This simple example demonstrates that my body is capable of performing motions that violate Newton's third law; i.e., this law does not always apply to the motions of my body if they are initiated by my own free will. I now postulate that Newton's third law does not always apply to the living cell, either. We can therefore define those causes of cellular behaviors that obey Newton's third law as external causes and those outside the domain of validity of Newton's third law as "internal causes."

The notion of internal causes described here may be closely related to the concept of "unmoved mover" that Aristotle enunciated more than two millennia ago. M. Delbrück (1976) recently revived this concept as the physical principle underlying the biological role of DNA (Campbell, 1982). CCT is consistent with Delbrück's interpretation of Aristotle's doctrine of unmoved mover in light of the modern molecular biology. Not only, CCT identifies cell's ability to transmit information in time (e.g., DNA-directed timing of intracellular events) as the ultimate mechanistic basis for the phenomenon of "unmoved moving" such as the timing of mitosis and differentiation and intentional or creative actions of the human mind. The ability of cells to transmit information in time, in turn, depends on the capacity of proteins to undergo thermal fluctuations (i.e., the thermal stroke in Figure 1.13) and generate conformons, utilizing the free energy of ligand-protein interactions (see the conformon generation step in Figure 1.13).

Finally, astute readers may have been wondering what the relationship is between CCT and biocybernetics. Biocybernetics can be viewed as the study of information and free energy transactions in living systems, whereas CCT is the theory of information storage, transduction and transfer in space and time mediated by living cells. So defined there is clearly a considerable overlap between biocybernetics and

CCT. However, biocybernetics is a much broader field of study than CCT, just as quantum mechanics in physics subsumes the molecular orbital theory in organic chemistry.

1.8.12. The Principle of Deterministic Chaos and the Living Cell

A. <u>What is Deterministic Chaos?</u> The concept of "deterministic chaos" was first clearly formulated by E. N. Lorenz of MIT in 1963. He accidentally discovered that an identical numerical input into a set of nonlinear differential equations gave unidentical results, which was contrary to the common belief held up to that time by mathematicians. During the past 2-3 decades, the concept of deterministic chaos has been well established in various fields of physical sciences such as atmospheric science, hydrodynamics, classical mechanics and statistical mechanics (Shaw, 1981), biology (May, 1976; Tsonis and Tsonis, 1989), physiology (West and Goldberger, 1987), and medicine (Andrey, 1989).

Deterministic chaos can be defined as *randomness* generated by dynamical systems obeying *deterministic* rules. Dynamical systems are either mathematical objects composed of sets of differential or difference equations or physical systems composed of dynamically interacting components whose evolution from some initial state to a final state can be described by deterministic rules (Tsonis and Tsonis, 1989). An example of a simple dynamical system capable of exhibiting chaos is a fish population in a pond (May, 1976), whose growth can be modeled by the so-called logistic equation,

$$X_{T+1} = aX_T (1 - X_T) \qquad (1.44)$$

where X_{T+1} is the fish population at the $(T + 1)^{th}$ year, X_T the fish population at the Tth year, and a is the "non-linear" or "control" parameter (i.e., the number that characterizes the temporal behavior of the equation, without itself undergoing change). The term "non-linear" refers to the fact that the non-zero value of a always leads to a non-linear term in Equation (1.44), namely a $(X_T)^2$. Calculating the value of X as a function of time, T, using a computer and plotting X against T reveal four different patterns depending on the value of the control parameter, a: (i) a *steady state* when a = 2.707, (ii) an *oscillatory* state with period 2 (i.e., the same population number is repeated every two years) when a = 3.35, (iii) an oscillatory state with period 4 (i.e., the same population number is repeated every 4 years) when a = 3.5, and (iv) a *chaotic state* (i.e., the population number varies unpredictably and chaotically with no periodicity) when a = 3.829. The temporal behavior of the fish population undergoes a transition from a periodic state to a chaotic state when a becomes greater than 3.750.

B. <u>The Cell as a Deterministically Chaotic Machine ("Chaomachine")</u>: As illustrated above using the logistic equation, dynamical systems can exist in a steady state, an oscillatory state, or a chaotic state, depending on the value of the non-linear parameter of the system. There is no doubt that the living cell can exist in various steady states (as evidenced by the phenomena of intracellular homeostasis with respect to the concentrations of H^+, K^+, Na^+, Ca^{++}, etc.) (Hess, 1983; Boiteux et al., 1980) as well as in oscillatory states exemplified by the periodic changes in intracellular metabolites and contractile motions of amoeba, Physarium plasmodium (Ueda et al., 1986). Although there is as yet no direct experimental evidence, I now postulate that the cell possesses the capability to exist in deterministically chaotic (DC) states and that the DC states of the cell are essential for cellular evolution and function. This idea may be referred to as the *Postulate of the Deterministically Chaotic Cell*, which has the following essential elements.

(1) The cell is the product of 2 - 3 billion years of biological evolution. Biological evolution, in turn, consists of two fundamental processes; (i) the variation of the gene pool of organisms, and (ii) the natural selection of the fittest out of a population of organisms (Mayr, 1988). Therefore, biological evolution is accompanied by a continuous increase in the amount of the genetic information transmitted from one generation to the next, since W_b is always greater than W_a in Equation (1.45).

$$I = \log_2 (W_b/W_a) \qquad (1.45)$$

where I is the amount of the genetic information stored in DNA, W_b is the variety of organisms *before* natural selection, and W_a is the variety of organisms *after* natural selection.

(2) Naturally selected organisms have on average a greater I value in their genome than those not so selected, because this greater value of I is associated with a greater adaptability of the organisms selected (in part due to the Law of Requisite Variety discussed in Appendix 1.A); i.e., the greater the I value, the greater the survivability of organisms.

(3) W_b in Equation (1.45) is determined predominantly by the biological characteristics of organisms, while W_a is determined primarily by the natural environment in which organisms exist.

(4) Species whose individual cells can exist in deterministically chaotic states may have a greater probability of survival because the DC states contribute to

increasing W_b and hence I under a given environmental condition.

In other words, it is here postulated that the principle of deterministic chaos (PDC) has played an essential role in biological evolution; PDC determined W_b in Equation (1.45) and environmental conditions determined W_a. For a recent discussion on chaos from the perspective of the information theory, see Shaw (1981).

1.9. THE BIOCYBERNETICS MODEL OF THE HUMAN BODY: THE PISCATAWAYTOR

The human body is the most complex material system known to us. It consists of 10^{12} -10^{14} cells that are organized in space and time (i.e., different parts of the body have different cells and cells at a given site in the body change their properties with time). We described the cell as a system of "curved" spacetime-elaborated chemical reactions and physical processes that perform a set of biological functions (Table 1.17). Likewise the human body can be viewed as a spatiotemporally organized system of chemical reactions and physical processes that proceed under the influence of genetic programs (internal causes) and environmental inputs (external causes) to accomplish a set of biological functions. Following the tradition established in New Chemistry (Babloyantz, 1986), i.e., the field of study dealing with spatiotemporally organized chemical reaction-diffusion systems where any spatiotemporally organized chemical reaction-diffusion systems are named after the city where the research is done (see Section 1.4.7), I elected to call the model of the human body described below the *Piscatawaytor*. This name also reflects my indebtedness to the students, faculty and colleagues at the College of Pharmacy, Rutgers University, and the Robert Wood Johnson Medical School of the University of Medicine and Dentistry of New Jersey (UMDNJ) in Piscataway, who made direct or indirect contributions to the development of the *Piscatawaytor* during the academic years, 1985-1989.

1.9.1. The Physical Principles of the Human Body

It is fair to say that there is no major disagreement among practicing life scientists with respect to the notions (1) that our body is composed of a set of well-known inorganic, organic, and biochemical molecular species of finite structural complexities, and (2) that all of our bodily functions, including thinking and feeling (Searle, 1984), result from complex and highly organized interactions among a set of straightforward physical, chemical and biochemical processes and reactions. The mystery about the human body is not in the nature of component structures and processes, which are well known, but in the manner in which these are organized or coupled in space and time. We all accept that there exists some kind of correlations between the *macroscopic* (i.e., visible to the naked eye) structures and functions of our body and the associated *microscopic* (i.e., molecular) structures and processes. It is important to note that, although macroscopic structures and processes are directly observable without using any measuring instruments, microscopic structures and processes can be deduced only from experimental data obtained through the use of sophisticated measuring instruments such as spectrophotometers and electron microscopes applied to appropriately "prepared" biological samples. Sample preparation must be regarded as a part of measuring procedures. As already pointed out by Bohr (1933), these measuring procedures can profoundly alter the very living phenomena that we wish to investigate.

According to the table theory described in Section 1.3, there are at least four kinds of correlations in nature—causal (i.e., related to internal causes of Section 1.8.11), complementary, deterministically chaotic, and historical. I believe that the microscopic-macroscopic correlations observed in our body involve all of these four kinds of correlations. Causal correlations may be observed if and only if all the cells implicated in correlations exist in linear domains (so that there are 1:1 causal correlations between cellular genome and cell behaviors) and the measuring procedures employed do not perturb the microscopic phenomena under observation. Complementary correlations (i.e., there are no causal links between the microscopic and the macroscopic so that you cannot explain the macroscopic in terms of the microscopic or *vice versa*) are expected, (1) if one or more cells implicated in correlations exist in deterministically chaotic regimes (see Section 1.8.11) or (2) if the microscopic phenomena under consideration satisfy Bohr's complementarity condition (see Section 1.1). In other words, it is postulated that the human body can exist in either the linear regime or the nonlinear regime and that the macroscopic-microscopic correlations are determined by both the linear or nonlinear cell states involved and the experimental measuring procedures employed. Therefore, two basic physical principles seem to dominate the macroscopic-microscopic correlations of the human body—the Principle of Deterministic Chaos, and Bohr's Complementarity Principle. In other words, the human body is a "deterministically chaotic machine" driven by microscopic processes that obey to Bohr's complementarity principle. This conclusion seems in harmony with our earlier conclusion that the cell is a deterministically chaotic (DC) machine; since the human body is made out of trillions of DC machines, the body cannot be deterministic. This conclusion suggests that the following statement may be true: "It is impossible to construct a deterministic machine by coupling two or more deterministically chaotic machines."

"Biocybernetics": A Machine Theory of Biology 143

1.9.2. The Four Control Systems of the Piscatawaytor

In order for the trillions of cells in our body to work together to accomplish the common goal of healthy existence and reproduction under unpredictable and often hostile environmental conditions (e.g., bleeding due to accidental bodily injuries, infections by pathogenic microorganisms, etc.), all the cells in the body must specialize and coordinate their activities in space and time. The coordination of cellular activities throughout the body entails elaborate and effective communication and control, the central theme of cybernetics. As a first approximation, it appears reasonable to assume that the communication and control systems of the human body obey the same set of rules obeyed by artificial communication and control systems such as robots, computers and telecommunication systems.

During the academic years, 1985-1987, I had the privilege of teaching a pathophysiology course to the third-year students at the College of Pharmacy at Rutgers. The textbook adopted for the course was over one thousand pages long, divided into over 50 chapters, covering topics ranging from cells and tissues to congenital heart diseases and disorders of cerebral functions. To present the material without organizing the topics according to some rational scheme was thought to be too confusing and educationally unsound. My effort to discover some pattern around which to organize the 50-odd chapters was eventually paid off when I noticed that most, if not all, of the topics in the textbook could be fit into four "control" systems—the nervous (N), circulatory (C), endocrine (E), and immune (I) systems—all working together to effectuate macroscopic voluntary bodily motions and intentional microscopic processes in the brain (M). These systems are represented in the form of a tetrahedron, whose four apices are occupied by N, C, E and I and whose center is occupied by M as shown in Figure 1.14, a geometric representation of the Piscatawaytor.

The tetrahedron is known as the simplex of the 3-dimensional space (i.e., the simplest geometric figure lying in a 3-dimensional space and not in a space of lower dimension) (Aleksandrov et al., 1984). It is possible that the reason for the "fourness" of the control system (and not three or five) is because of the tetrahedron being the 3-simplex and that only in the tetrahedron are the apices in simultaneous contacts with one another or are equivalent (a form of geodesic?). If there is any physical meaning to this geometric conjecture, perhaps the most important property suggested by it may be the notion of what may be called the "simultaneous contacts" among all the control systems, that is, the suggestion that the perturbation of one control system in the human body leads to the perturbation of all the others sooner or later, or that it is impossible to perturb one control system of the human body

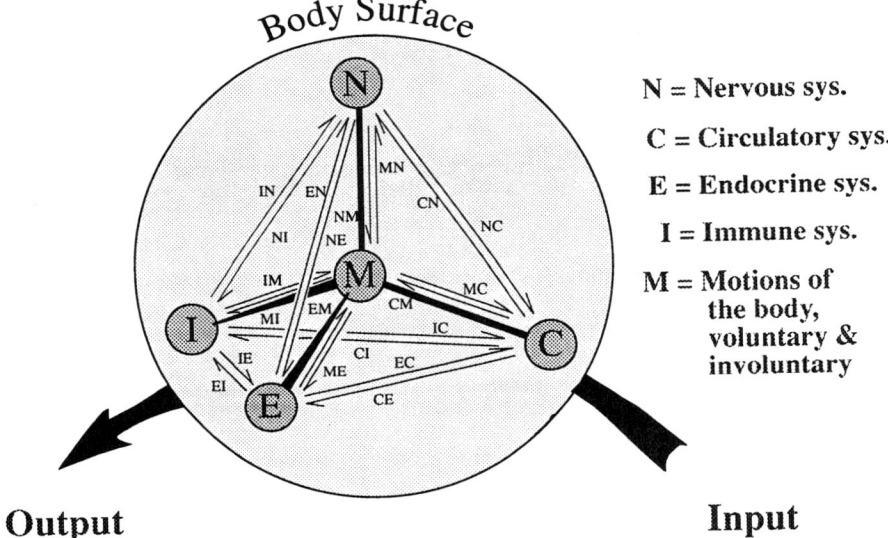

Figure 1.14. The Piscatawaytor: A theoretical model of the human body. The trillions of cells constituting our body can be divided roughly into 5 groups; the nervous (N), circulatory (C), endocrine (E), immune (I) and voluntary muscular (M) systems. The N, C, E and I systems are thought to act as control systems that cooperate to enable the human body to carry out voluntary motions, M, both macroscopic (i.e. bodily movements) and microscopic (i.e. thought processes). For details, see text.

without also perturbing all the other control systems. The fact that the M system must be relegated to the center of the tetrahedron in order to effectuate the "simultaneous contacts" suggests the possibility that the most important biological function of the human body is *voluntary bodily motions*, including thought processes. This conclusion places voluntary bodily motions, which we all too readily take for granted, at the center of our biological being. Is it possible that there is some deep philosophical significance to this conclusion? Have we underestimated the fundamental biological and evolutionary significance of our voluntary bodily motions?

1.9.3. The Piscatawaytor as the Periodic Table of Physiology and Medicine

In order for the human body to function normally, the trillions of cells constituting the human body must communicate with one another as already mentioned. There are now hundreds of chemicals that are known to mediate communications between cells, variously called hormones, neurotransmitters, autacoids, growth factors, cytokines, lymphokines, interleukins, etc. The naming of these chemical messengers is mostly historical or operational, largely devoid of any theoretical rationale.

"Biocybernetics": A Machine Theory of Biology

The Piscatawaytor may provide a theoretical framework for organizing all the varied intercellular chemical messengers operating in the human body that have accumulated in the literature during the past century, just as Mendeleev's periodic table formulated in 1869 introduced order into the confusing collection of millions of chemicals found in nature. The Piscatawaytor shown in Figure 1.14 immediately suggests three broad classes of intercellular messengers: (i) the "XY messengers," carrying information in the direction from control system X to control system Y, (ii) the "YX messengers," carrying information in the reverse directions from Y to X, and (iii) the "XX messengers," transporting information from one cell to another within a given control system, X. The topological features of the Piscatawaytor dictate that there exist exactly 25 classes of intercellular messengers (Table 1.19).

Table 1.19. A rational classification of the intercellular messengers in the human body according to the Piscatawaytor

Class	Information[1]		Notation	Examples
	Source[2]	Destination[3]		
1	Motion	Motion	MM	Lactic acid (?)
2	Motion	Nervous	MN	CO_2
3	Motion	Circulation	MC	O_2 Deficiency
4	Motion	Endocrine	ME	?
5	Motion	Immune	MI	Endorphine
6	Nervous	Nervous	NN	Neurotransmitters
7	Nervous	Motion	NM	Acetylcholine
8	Nervous	Circulation	NC	Epinephrine

(Table 1.19 continued)

9	Nervous	Endocrine	NE	Hypothalamic releasing factors
10	Nervous	Immune	NI	Endorphine
11	Circulation	Circulation	CC	PGE_2, TXA_2, PGI_2, NO
12	Circulation	Motion	CM	?
13	Circulation	Nervous	CN	?
14	Circulation	Endocrine	CE	?
15	Circulation	Immune	CI	Prostaglandins
16	Endocrine	Endocrine	EE	TSH, ACTH, FSH
17	Endocrine	Motion	EM	Testosterone
18	Endocrine	Nervous	EN	Sex hormones
19	Endocrine	Circulation	EC	Catecholamines
20	Endocrine	Immune	EI	Sex hormones, GH
21	Immune	Immune	II	Interleukins
22	Immune	Motion	IM	?
23	Immune	Nervous	IN	Interleukin-2

"Biocybernetics": A Machine Theory of Biology

(Table 1.19 continued)

| 24 | Immune | Circulation | IC | NO (?) |
| 25 | Immune | Endocrine | IE | Interferon |

---

[1] Something that reduces uncertainty. The reception of one bit of information leads to the removal of the uncertainty associated with a choice between two. Two bits of information removes the uncertainty associated with making a correct choice out of $2^2 = 4$ possibilities, etc. The information content of a messenger is determined by the variety of the messengers present in the information source out of which a messenger was chosen. The maximum amount of the information (I) carried by a messenger is given by Shannon's formula (Pierce, 1980), $I = \log_2 W_0$, where W_0 is the number of possible messengers in the information source.

[2] The cell emitting a chemical messenger.

[3] The cell that receives a chemical messenger.

Several features stand out in Table 1.19: (1) Same messengers often appear in more than one categories. This does not necessarily mean that there will be "cross talks," because different messages can still be transmitted through using an identical messenger, if the messenger is transmitted in concentration waves having different amplitudes, frequencies or phases. (2) Some messengers carry information in both directions, namely from X to Y and from Y to X. (3) There are many question marks in the table, which may be regarded as "predictions" as was the case with the original periodic table of Mendeleev or as indications of the author's ignorance which may be easily removed by further literature search.

1.9.4. Sociobiological Debates on Human Nature and the Piscatawaytor

According to the Piscatawaytor model of the human body (including the functioning of the brain), there are two distinct components to the human nature—(1) those largely determined by genes (to be called the "deterministic attributes"), and (2) those causally dissociated from genes (to be called the "deterministically chaotic attributes"). Deterministic attributes are those aspects of humankind that can be predicted, just as much of the behaviors of lower animals such as insects, birds and reptiles can be predicted. The theory of humans viewed primarily as a deterministic animal controlled by genes is called the "sociobiological theory of man" by E. O. Wilson (1982). In contrast to Wilson's theory, the Piscatawaytor introduces a second fundamental nature of humans: The "deterministically chaotic (DC)" one. The DC nature of humans may be the result of cells operating in nonlinear regimes so that the cellular behaviors are now "disengaged" from their genome, just as the strange "attractors" are causally separated from the deterministic equation underlying them. The term "strange attractor" refers to the collection of the solutions to a nonlinear differential or difference equation that reveals some readily recognizable *patterns* when graphically displayed in an appropriate mathematical space, as though the solutions "attract" one another. The adjective "strange" reflects some unusual mathematical properties possessed by these attractors, such as the property that no two points (i.e., solutions) within a strange attractor overlap (Cvitanovic, 1984). It is important to bear in mind that, although the individual points in a strange attractor cannot be predicted from the associated deterministic equation and input data, the "collective behavior" of all the points generated by the deterministic equation and input data is predictable; i.e., the collective behaviors of the solutions reflect the characteristics of the underlying deterministic equation. This is analogous to the human genome determining the collective (or global) anatomical and behavioral characteristics of *Homo sapiens* (vis-à-vis those of apes, say) but not individual (or local) characteristics; i.e., all human beings have a nose and two eyes, etc. but no two noses or pairs of eyes are exactly the same, even between two monozygotic twins! In other words, *the Piscatawaytor suggests a novel view of Homo sapiens, wherein humans are "globally" constrained by the human genome but "locally" left free*. This conclusion seems to agree with our current knowledge about genetic and non-genetic factors determining human behavior (Plomin, 1990).

The model of humankind as embodied in the Piscatawaytor suggests an interesting possibility of resolving the age-old debate concerning nature vs. nurture in shaping human social behaviors. I agree with E. O. Wilson that genes play a crucial role in determining certain social behaviors of humankind, just as they do in controlling social behaviors of insects (Wilson, 1982). I also agree with his opponents such as A. Montagu (1980) who uphold the view that not all human behaviors are determined by genes and that cultural influences are of prime importance in the shaping of some, if not all, human behaviors. I believe that we can now begin to unite such seemingly "irreconcilable" opinions into a harmonious and synthetic view of humankind based on the biocybernetics model of humans, namely the Piscatawaytor. Wilson's sociobiologic view of man may be regarded as reflecting the "deterministic" aspects of human behaviors, while the more traditional anthropological and sociological viewpoints of humanity championed by his opponents mirror primarily the "deterministically chaotic" attributes of human nature. According to the Bhopalator model of the cell (see Section 1.8), these deterministically chaotic behaviors of cells are revealed only when cells operate in far

"Biocybernetics": A Machine Theory of Biology

from equilibrium (and hence nonlinear) conditions, and under these conditions cells become "hypersensitive" to their environmental conditions due to their ability to amplify minute signals to macroscopic dimensions (Prigogine and Stengers, 1984; Ji, 1987). Therefore the deterministically chaotic state of humans would make it possible for our development to be controlled mainly by cultural influences, largely unaffected by individual genetic differences. By the same token, the Piscatawaytor suggests that, depending upon the biological state of the critical cell populations controlling the body, individuals can exhibit deterministic behaviors (e.g., egotism, altruism, sex drive, hate, love, etc.), irrespective of educational or cultural background.

1.10. THE BIOCYBERNETICS MODEL OF THE HUMAN SOCIETY: "NEWBRUNSWICKATOR"

The germ of the idea for constructing a theoretical model of the human society based predominantly on recently developed theoretical biology concepts (discussed in this Chapter) originated when I taught an honors seminar course to about 10 Rutgers College students in the fall of 1986, entitled "From Cells to Societies: A Biological View of Man and his Universe." I decided to name the model the Newbrunswickator, not only because it was born in the New Brunswick campus of Rutgers University but more significantly because I wanted to build the model using the same principle of dissipative structures formulated by I. Prigogine (1980) that already gave rise to the Bhopalator, a theoretical model of the cell (see Sections 1.4 and 1.8). Unlike the Bhopalator, however, in formulating the Newbrunswickator the principle of dissipative structures is used only in an *analogical* sense (and not in its physical context) as explained in Table 1.20.

I. Prigogine and his school have been emphasizing the "constructive" role of irreversible processes in nature in general (Glansdorff and Prigogine, 1971; Prigogine and Stengers, 1984). I first had the idea of connecting the irreversibility of human lives (no one can live forever) and the structure of the human society after hearing Professor Prigogine's lecture at Rutgers delivered on October 2, 1984, entitled "Constructive Role of Irreversible Processes." Can human deaths that inexorably follow all human lives serve as the primary driving force for the basic structure of human society? Perhaps not human death itself but our ability to be consciously aware of our forthcoming death, can this be the ultimate driving force for all human social structures? Lower animals also die, but they have not produced any culture or civilization. This led me to compare the cell and the human society at several different levels. Table 1.20 appears to reveal reasonably good "symmetries" between the Bhopalator and the Newbrunswickator. It also exposes the multiplicity of possible

Table 1.20. The "Newbrunswickator"—a theoretical model of the human society constructed in analogy to the Bhopalator, a model of the living cell.

	Parameters	Bhopalator	Newbrunswickator
1.	Interacting particles	Molecules	Human bodies[1]
2.	Irreversible processes	Chemical reactions	Human lives[2]
3.	Dissipative structures	Spatiotemporally organized chemical concentrations	Human *cultural* activities[3]
4.	Equilibrium structures	Molecular structures in 3-dimensions	Human *biological* activities[4]
5.	Genetic information	DNA	Literature[5], etc.
6.	Essential function	Self-reproduction	Civilization[6]
7.	Mechanism of genesis	Genetic variation Natural selection	Social contract[7](?)
8.	Extinction	99.99%	Common
9.	Purpose	None	To increase self-knowledge[8]
10.	Physical principle[9]	$dG/dt \leq 0$	$dI/dt \geq 0$ (?)
11.	Force	Cell force	"Society force"[10]

(Notes to Table 1.20)

[1] Human society can be viewed as a set of human bodies that interact with one another through direct physical contacts or through various indirect means of communication.

[2] Professor Prigogine informed me in 1984 that other workers have also considered the irreversibility of human lives as a possible source of constructive processes on the human societal level.

[3] It is interesting to view human cultural activities (e.g., education, basic research, concerts, plays, art exhibitions, religious activities, festivals, etc.) as analogous to dissipative structures in physical sciences, because such an analogy emphasizes how "ephemeral" human cultural activities can be, just as how easy it is to extinguish the flame of a candle, a canonical example of dissipative structures.

[4] Dissipative structures cannot exist without the stable equilibrium structures that provide proper boundary conditions, e.g., the petri dish with a right size and surface characteristics in order to observe chemical waves of the Belousov-Zhabotinskii reaction (Prigogine and Stengers, 1984). Similarly, the cultural activities of human societies cannot exist without satisfying the basic biological needs of the human body such as eating, excreting, sleeping, reproducing, etc.

[5] In the cell, DNA serves as the hard copy of genetic information that is stable enough to be conserved during cell divisions. There are other ways of storing genetic information in the cell (e.g., intracellular ion gradients) but these are too fragile to survive cell division processes. The written language as exemplified by the literature similarly serves as a stable means for transmitting cultural information from one human generation to the next.

[6] The essential function of the cell is clear; without its ability to self-replicate, no life is possible. Is there any analogous function of human society? That is, is there any function exhibited by human society, without which its very existence is endangered? I tentatively identify this essential function of human society to be the creation of "civilization," the sum total of all the human artifacts without which human society cannot exist.

[7] We now know how living systems came about, through biological evolution—i.e., through variations of the genome of individual organisms and the natural selection acting on these organisms (Mayr, 1988). How did human society originate? The symmetry of Table 1.20 would suggest that the ultimate driving force for the existence of human society is the desire on the part of individuals to be a member of the society because this membership provides them with certain desirable consequences not obtainable by isolated individuals. Human society cannot exist without such a cohesive force that brings individual human beings together into a group. We may call such a "force" the "society force," a complex system of multiple factors that all work together to bring about human society. It is possible that one of the most important components of the "society force" is the innate propensity of the human being to pursue self-interest.

[8] Knowledge is here defined as meaningful information. "Meaningful" in turn signifies "referent-revealing" in the postmodern sense of Jean-Francois Lyotard.

[9] All irreversible processes in living systems are driven by decreases in Gibbs free energy. Analogously, perhaps all cultural activities supported by the irreversible passage of individual human lives may be associated with a net increase in the potential for human communication. In other words, cultural activities of the human

factors contributing to the cohesive societal structures as noted in Footnote #10.

Another unique feature of the Newbrunswickator is that it shares the same machine-theoretic framework as the Piscatawaytor (the model of the human body) and the Bhopalator (the model of the living cell), without both of which human society cannot exist. Since the cell and the human body are both thought to be "deterministically chaotic machines," it is probably true to conclude that human society must also behave like a deterministically chaotic machine. It is possible that biological activities (driven by instincts) underlie the deterministic aspects of the human society, while cultural activities (driven by the "society force" which may ultimately be connected to the death-consciousness of the human mind) are predominantly responsible for the deterministically chaotic aspects.

1.11. A BIOLOGICAL MODEL OF THE UNIVERSE: THE SHILLONGATOR

I regard the concept of "gnergy" as one of the most important results of my theoretical investigations in biology over the past two decades. Gnergy is defined as a hybrid physical entity or object that contains both information (gn-) and energy (-ergy) (see Section 1.1). Discrete entities carrying gnergy are called "gnergons." Gnergons in turn consist of gnons (carrying only information) and ergons (carrying only energy). We now know of two concrete examples of gnergons in biology: (1) conformons (see Section 1.5.2) and (2) IDS (intracellular dissipative structures (see Section 1.8.2), both of which are thought to constitute cytons that mediate the cell force (See Section 1.8.9).

(Notes to Table 1.20 continued)

society may contribute, on average, to the creation of the societal environment possessing a greater "communication potential" so that individuals who wish to do so can communicate with one another more efficiently, i.e., the channel capacity (the rate of information transmission, dI/dt; Pierce, 1980) may increase with human history.

[10]Just as the cell force concept (see Section 1.8.9) represents the complex system of the elements, both material and theoretical, that are needed to account for the functional integrity of the living cell, so I am suggesting that it is legitimate to talk about the "society force" as a short-hand notation for the complex aggregate of the numerous factors that are essential for the functional integrity of human society. I have already indicated that the irreversibility of human deaths constitutes a major component of the "society force," but there must be other components just as essential, including the human inclination to seek pleasure and avoid pain, both physical and mental.

What is very unusual about the concept of gnergy is that energy and information exist as a "fused" microscopic entity in all molecular machines, whereas these two maintain independent existences in macroscopic machines; i.e., gnons and ergons are "fused" in molecular machines, while they are separated in macroscopic artificial machines. For example, gasoline (ergons) and an automobile (gnons, i.e., structural information) are separate entities, while all active molecular machines such as the Na^+/K^+ ATPase and actomyosin systems must undergo conformational deformations (to generate or store ergons) at right locations and right time points (determined by gnons encoded in linear sequences of proteins); therefore, the all-important conformational strains (or conformons) of molecular machines cannot be described solely in terms of their energy content or the sequence information associated with the local structural deformations alone but must be described as hybrid physical entities containing both energy and information, i.e., as gnergons. This reminds us of the "fusion" of space and time to form "spacetime" or "events" in Einstein's special theory of relativity (Wald, 1977).

After having convinced myself that most living processes can be rationally accounted for on the molecular level in terms of gnergy (Ji, 1985a, 1990), I was naturally led to ask the question whether or not gnergy might play an equally fundamental role in nonliving processes. Although physicists rarely talk about information and concern themselves almost exclusively with energy considerations, I wondered whether information is also implicated in physics in some manner. A rather superficial survey of the physics literature, combined with discussions with my physicist colleagues over the past 8 years, indicated to me that there is no experimental nor theoretical reasons to exclude the possibility that *gnergy* rather than *energy* may be ultimately responsible for all natural phenomena, both living and nonliving. This led me to formulate a theoretical model of the universe based on the assumption that gnergy and not energy initiated the Big Bang (see below). This idea was presented at the Third International Seminar on the Living State held in Shillong, India, on December 13-19, 1986 in the form of a model of the universe called the "Shillongator." This name was adopted, in part, to express my appreciation for the far-sighted meeting held in Shillong organized by Professor R. K. Mishra and his colleagues. My attendance at the meeting, which was made possible by a travel grant from Shillong, was instrumental in formulating the model. Another reason for the name is to indicate the underlying assumption that the universe obeys the principle of self-organization just as all organisms do (see Section 1.4.7.).

1.11.1. The Big Bang and Gnergy

The emergence of Homo sapiens on this planet can be viewed as an inevitable consequence of the Big Bang that gave rise to our universe some 15 billion years ago (Parker, 1986). Since our body as well as other living organisms is composed of elements H, C, N, O, Fe, etc. whose syntheses in this universe depended critically on the initial conditions of the Big Bang (Davies, 1984), it is logical to conclude that the Big Bang was a necessary condition for the existence of living systems on this planet.

There are now convincing observational and theoretical reasons to believe that the presence of living organisms, including the sentient observers, in this universe is not an accident but rather was the consequence of the unique structure of this universe, as reflected in the laws of nature and the numerical values of the fundamental constants such as Newton's gravitational constant (G), the speed of light (c), Planck's constant (h), and the proton mass (m_p). Calculations show that small perturbations of the fundamental constants of nature away from their actual values lead to model universes that are devoid of C, N, O and Fe atoms that are essential for life. Such theoretical considerations based on astronomical data has firmly established the notion, known as the Anthropic Principle, that the structure and the dynamics of this universe and the evolution of living organisms in it are causally connected (Barrow and Tipler, 1986).

The first form of life, namely the first molecular systems that could self-replicate, is thought to have emerged in the biosphere of the earth about 2-3 billion years ago (Quastler, 1964). Although most current theories of the origin of life seem to assume that the emergence of the first self-replicating molecular systems was more or less a chance event, it appears more likely that the origin of life was an inevitable result of the initial conditions of the biosphere at the time of the origin of life. A similar idea was recently expressed by Chaisson (1987). Just as the current living cells self-replicate utilizing the genetic information stored in DNA, the first self-replicating systems of this planet might have utilized the "cosmological information" encoded in the form of the physical and chemical characteristics of the microenvironment of the biosphere prevalent at the time of the origin of life. This view seems to be in agreement with the origin of life theory advanced by A.G. Cairns-Smith (1985). It is possible that biological information did not originate out of no information but rather resulted as a consequence of the transduction of the pre-existing "cosmological information." That is, the cosmological information encoded in the initial conditions of this planet at the time of the origin of life might have been necessary and sufficient to cause living systems to evolve spontaneously; i.e., these

"Biocybernetics": A Machine Theory of Biology 155

initial conditions might have acted as a "cosmological DNA." The Shillongator is a model of the universe that incorporates the spontaneous evolution of living things as an integral component of the evolving cosmos viewed as a self-organizing system. The inclusion of living systems in cosmological models may be justified on two grounds: (1) Living systems, although minuscule in size and mass on the cosmological scale, is an intrinsic part of the universe, and (2) Without the sentient beings, *Homo sapiens*, no mental construct of the universe (as opposed to the physical universe) can exist.

Physicists describe most non-living structures and processes in terms of various forms of energy (thermal, mechanical, electrical, kinetic, potential, chemical, etc.) without employing the concept of information. This situation may have arisen either because "information" as defined by Shannon (Pierce, 1980) is not implicated in physics or because the role of information in non-living processes is expressed in terms other than "information." In a sense, the spins of electrons, and the colors and flavors of quarks (Davies, 1984; Glashow, 1975; Talbot, 1986; Segrè, 1980) may be viewed as carrying information, just as the molecular shapes of enzymes, DNA and RNA are said to carry genetic information. It is possible that, due to the extra-dimensionality of superstrings in contrast to point particles, superstrings contain some type of information as well (Green, 1984; Kaku and Trainer, 1987). The Higgs fields that gauge theorists are forced to "put in by hand" to endow mass to massless bosons in weak interactions (Adair, 1987) and the spacetime curvature of Einstein may be construed as carrying Shannon information.

The ultimate source of the information encoded in non-living systems (e.g., superstrings, quarks, atoms, molecules, rocks, clays, etc.) is postulated to be the Big Bang; that is, the Big Bang that occurred 15 billion years ago must have contained not only an immense amount of energy but also an enormous (likely even infinite) quantity of information. Such a possibility seems to gain some credibility when we take into account the conclusions of some cosmological theories (Guth and Steinhardt, 1985) that our universe may not be unique but represents only one of an infinite number of possible universes, some of which may exist in parallel with ours, although communicating with them is not allowed by the laws of physics. If we assume that there were W_o number of universes out of which our universe was selected at the time of the Big Bang, then the maximum information content of our universe would be estimated to be $\log_2 W_o$ bits. We may refer to such information "generated" at the time of the Big Bang as the "primeval information" or the Big Bang information, and the information generated since the Big Bang as the "post-Big-Bang information." As the temperature of the universe decreased after the Big Bang, the primeval information may have been "transduced" into, or "manifested" in, the post-Big Bang information. The "cosmological DNA" information postulated to have guided the origin of life (see above) can be viewed as a component of the post-Big Bang information.

The total information content of the earth probably remained more or less constant, until the first form of life originated about 12 billion years after the Big Bang. Subsequently, it increased explosively as the result of the multiplication of living cells and the production of human knowledge (Wojciechowski, 1987). In contrast, the energy density of the earth decreased as the temperature of the universe dropped rapidly after the Big Bang and remained constant after the earth began to circle around the sun (Figure 1.15).

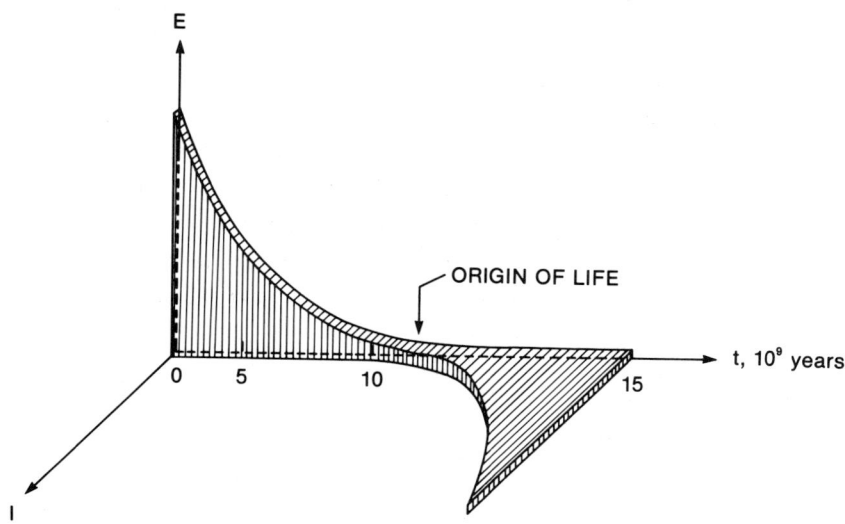

Figure 1.15. The evolution of the universe. According to the Big Bang theory, the temperature and the energy density of the universe has been decreasing since the Big Bang that occurred about 15 billion years ago. This is indicated in the E-t plane, where E = energy density, and t = time. Notice that the t-axis is approximately logarithmic. The sudden increase in the information density (defined as the amount of biological information divided by the volume of the biosphere) occurred with the emergence of the first self-replicating systems in the biosphere on the earth about 3 billion years ago. This information density (I) is plotted on the I-t plane orthogonal to the E-t plane. It is assumed that the information content of the universe as a whole at the time of the Big Bang was non-zero (see text); i.e., the decay curve on the E-t plane had a finite (or even infinite) thickness along the I-axis. This thickness probably varied with time between $t = 0$ and $t = 12 \times 10^9$ years when life originated but is depicted as constant in the figure for simplicity. The main point of this figure is that the cosmological evolution, including the evolution of living systems, cannot be fully described in the E-t space alone but only in the E-I-t space.

1.11.2. The Shillongator

On the basis of the above analysis, it may be concluded that the living processes on this planet are causally linked to the Big Bang via two threads—the information thread (i.e., the conversion of the primeval information and the "cosmological DNA" information into biological information) and the energy thread (i.e., a part of the primeval energy of the Big Bang remains in the sun, from which all of the free energy required to support life on this planet is derived). This expanded view of the biological evolution suggests a new model of the universe in which both the physics of the Big Bang and the biology of living systems play essential roles on an equal footing. The model incorporates the recent findings of the superstring theory that all the known fundamental particles in nature (fermions and bosons) can be derived from superstrings (Parker, 1986; Green, 1984; Kaku and Trainer, 1987). Superstrings are thought to be one-dimensional structures, 10^{-33}cm in length, that can undergo various vibrational and rotational motions, giving rise to different elementary particles.

Our model of the universe starts with superstrings (Figure 1.16) which are thought to have given rise to quarks and leptons by 10^{-4} seconds after the Big Bang (Parker, 1986) (step 1); quarks in turn formed subatomic particles (step 2), namely protons and neutrons, by approximately 20 seconds after the Big Bang (Parker, 1986). These subatomic particles further condensed and formed atoms of various elements in stars by 10^6 years after the Big Bang (Parker, 1986) (step 3). Most of these elements were made available on the earth to form numerous molecules before life could originate (step 4) at about 12 billion years after the Big Bang. The first self-replicating molecular systems appeared on this planet, most likely using the "cosmological DNA" as the necessary template (or information source) (step 5) and "frustrated" primordial RNA (see Appendix 1.C). Some unicellular organisms subsequently gave up their independence and cooperated to form multicellular organisms such as man (step 6). Multicellular organisms further organized spontaneously into various societies (step 7). The final step 8 does not imply a "physical" causality as implied by all the other arrows in the model but rather a "logical" causality, by which is meant the fact that the "discovery" or the "creation" of the notion of superstrings required the cooperative efforts of a whole society of scientists and was not the result of the effort of one or a few individuals, reminiscent of the termite society where a critical number of termites is known to be required before they spontaneously organize their activities to build their nest. Another unique aspect of step 8 is that it involves a movement in time from the present to the past, whereas all the other steps in the model implicate forward temporal movements, from

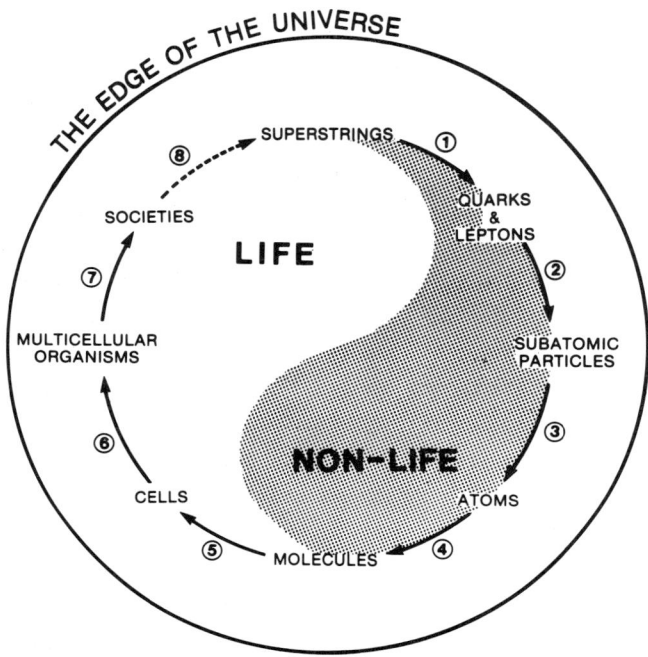

Figure 1.16. The Shillongator—a biological model of the universe. According to the Shillongator, our universe began with immense amounts of both energy and information, information (I) being defined according to Shannon's formula, $I = \log_2 (W_0/W)$, where W_0 = the number of all possible choices, and W = the number of choices actually realized (Pierce, 1980). When $W = 1$, I is maximum and equal to $\log_2 W_0$. Since according to some cosmological models, our universe is not unique but may represent one of almost infinite number of possible universes, W_0 is infinite and so is the information content of our universe (see text). The Shillongator model postulates that the initial energy and information existed as a hybrid entity called "gnergy" which may be contained in superstrings (Green, 1984; Kolb et al., 1985). The "primeval information" stored in superstrings is thought to be manifest as the post-Big Bang information as the temperature of the universe cooled, giving rise to quarks, leptons, subatomic particles, atoms, stars and galaxies. The molecules carrying the primordial information were already present on the surface of the earth that were required to form the first self-replicating molecular systems or cells (step 5). Cells further evolved to form multicellular organisms and societies. As the result of the cooperative activities of the whole society of *Homo sapiens*, the mental creation of knowledge constructs such as superstrings has been possible. The key postulate of the Shillongator is that the whole evolutionary process from superstrings to the society of *Homo sapiens* represents a self-organizing process, made possible by the "unfolding" or "explication" of the primeval information and driven by the cosmological entropy production attending the equilibration of the Big Bang energy in space and time. There are two kinds of symmetry breakings thought to be responsible for the evolution of first the non-living state of matter and then the living state—the "equilibrium symmetry breakings" (steps 1 through 4) and the "dissipative symmetry breakings" (steps 5 through 8), respectively. For more details, see text.

the past to the future.

The model of the universe depicted in Figure 1.16 appears unique in that it encompasses a much wider range of time, from the Big Bang to the present, than is usual in other cosmological models which are primarily concerned with the events that transpired during the very early periods of the universe (from 10^{-43} seconds to 10^6 years) (Parker, 1986) and thereby automatically exclude biology. To the best of my knowledge no other cosmological models contain both superstrings and human society within a single theoretical framework.

By naming the present cosmological model the Shillongator, I wish to emphasize the possibility that the universe may possess a self-organizing capability as do other well-known self-organizing systems such as the Belousov-Zhabotinskii reaction (Müller et al., 1985; Winfree, 1985), the Brusselator (Nicholis and Prigogine, 1977; Babloyantz, 1986), and living cells (Section 1.8). If this proves to be correct, it may be possible to conclude that there exists a new symmetry unifying all physical and biological processes that occur in our universe; this new symmetry may be identified with *the Principle of Self-Organization*, intensively studied in recent decades by I. Prigogine (1977, 1978, 1980), Haken (1983), Winfree (1980, 1985), and others. Jantsch (1980) appears to be the first to describe the universe according to this principle. The Gaia hypothesis of Lovelock (1979) may be regarded as an example of the application of the principle of self-organization to the biosphere.

Any physical system capable of self-organization must possess both the free energy and information sources, namely gnergy. These are identified for the Belousov-Zhabotinskii reaction, the living cell, and the universe in Table 1.21.

If the universe as a whole can be treated as a self-organizing system, we must be able to identify the energy and information sources that are both necessary and sufficient to cause its subsequent evolution, including the biological evolution. The requisite energy source is well known (Parker, 1986; Guth and Steinhardt, 1985), but the source of the needed information is far from clear. Based on the assumption that there exists a "symmetry" between the living and the nonliving, I postulate that our universe began with an immense amount of gnergy at the time of the Big Bang. Both energy and information together guided the subsequent cosmological evolution. This information is called the "primeval information" (see above). It is further postulated that both energy and information existed as inseparable components of higher dimensional, undifferentiated entities carrying gnergy, here identified with superstrings.

Table 1.21. The free energy (ergon) and Shannon information (gnon) sources of gnergy driving self-organization

Self-Organizing Systems	Source of	
	Free Energy (Ergon)	Shannon Information (Gnon)
1. Belousov-Zhabotinskii Reaction	Free energy-releasing chemical reactions	Initial conditions (chemical structures of reactants, reactant concentrations, temperature, pressure, reaction volume, shape of reaction vessel, etc.)
2. Living cell	Free energy-releasing chemical reactions	DNA, cytoplasm
3. Universe	Big-Bang energy	"Primeval information" (see Appendix 1.F)

According to the Shillongator model of the universe, then, the cosmological evolution (including the biological one) can be regarded as an inevitable manifestation of the primeval information, driven by the cosmological entropy production attending the equilibration of the Big Bang energy through space and time (the Second Law of Thermodynamics). That is, all self-organizing processes occurring in the universe may be viewed as the result of manifesting and "explicating" the primeval information (or better primordial gnergy) existing in an "implicate" form since the time of the Big Bang. This viewpoint appears to share common elements with the theory of implicate order proposed by Bohm (1980; Briggs and Peat, 1984). In addition, Bohm's emphasis on the "wholeness" of the universe is clearly evident in the Shillongator (Figure 1.15), which treats living systems, from single cells to the human society, as integral parts of the whole universe in which the Principle of Self-Organization is repeatedly manifest at different levels of cosmological structuration.

"Biocybernetics": A Machine Theory of Biology

All self-organizing systems may be decomposed into equilibrium structures and dissipative structures (Prigogine, 1977, 1978, 1980). For example, in the living cell viewed as the canonical self-organizing system, the phospholipid bilayers forming the plasma and subcellular membranes, certain cytoskeletal proteins and even intracellular water may be identified as equilibrium structures. In contrast, the membrane potentials (-50 to -80 mV, negative inside) resulting from asymmetric distributions of K^+, Na^+, protons, etc. provide excellent examples of dissipative structures on the microscopic level, since these membrane potentials would disappear as soon as the various ion pumps embedded in the cellular membrane stop dissipating free energy. In order for cells to carry out their biological functions, both equilibrium and dissipative structures must interact under the control of the genetic programs stored in DNA. In chemotaxis of human neutrophils, for example, the spatiotemporal alterations of the structure of the cell membrane (the equilibrium structure) and the intracellular Ca^{++} ion gradient (the dissipative structure) appear to cooperate to cause the movement of these cells toward ingestible particles (i.e., phagocytosis) (Saywer et al., 1985). As is well known, both equilibrium and dissipative structures exist in the non-living state as well. Living systems may emerge through unique combinations of selected sets of equilibrium and dissipative structures. That is, if there exist W_o number of all possible patterns of the interactions between equilibrium and dissipative structures in nature, only a small subset of these (W) can lead to living systems. The maximum (m) average (a) information content of a living (l) system (I_{mal}), then, can be estimated using Shannon's formula:

$$I_{mal} = \log_2 (W_o/W) \text{ bits} \tag{1.46}$$

The Second Law of Thermodynamics and the Principle of Deterministic Chaos (see Section 1.8.11) ensures the maximization of W_o; and the biological evolution minimizes W, thereby increasing biological specificity. Both of these changes tend to increase the maximum average information of living systems (I_{mal}).

Living systems can "live" because the biological information contained in them (i.e., I_{mal}) provides constraints on, or controls over, the spatiotemporal evolution of the spontaneous chemical reactions and physical processes occurring inside them; biological information is "manifest" in various symmetry-breaking chemical reactions and physical processes such as active transport, muscle contraction, and oxidative phosphorylation (Ji, 1974a,b, 1979, 1985). In fact, the dissipative structures found in the living cell can be regarded as the consequences of symmetry-breaking chemical reactions, including the vectorial chemical reactions of

P. Mitchell (Skulachev and Hinkle, 1981), catalyzed by free energy-driven enzymes such as Na^+/K^+ ATPase, actomyosin systems, and the inner mitochondrial membrane enzymes engaged in oxidative phosphorylation. These enzymes can be viewed as "symmetry-breaking molecules" or "molecular symmetry breakers." The mechanisms by which these molecular "machines" carry out various intracellular symmetry-breakings depend on conformons as the source of both energy and information (that is, gnergy) as discussed in Section 1.5.2.

The symmetry breakings that go on inside the living cell and in higher living structures such as humans and the human society (i.e., steps 5, 6, 7, and 8 in Figure 1.16) can be readily distinguished from the symmetry breakings that occurred in the non-living state of matter in the universe (i.e., steps, 1, 2, 3, and 4); the former requires free-energy dissipation by "symmetry-breaking molecules" working in cycles, whereas the latter results from unidirectional equilibration processes accompanying the cooling of the universe. On this basis, we can refer to the symmetry breakings responsible for the living state as the "dissipative symmetry breakings" ("dissipative" in the sense of I. Prigogine, 1977, 1978, 1980), and the symmetry breakings attending the unidirectional equilibration of the universe as the "equilibrium symmetry breakings." The latter appears to be closely related to the spontaneous symmetry breakings implicated in gauge field theories (Davies, 1984; Quigg, 1983; Pagels, 1982; Lopes, 1981). These two types of symmetry breakings can occur on two levels—the macroscopic and the microscopic levels. Consequently, we can identify four kinds of symmetry breakings in nature—the macroscopic equilibrium (ME) (e.g., formation of stars, galaxies, and ferromagnets), the macroscopic dissipative (MD) (e.g., formation of vortices in turbulent flows, formation of winds, hurricanes, and clouds), the microscopic equilibrium (me) (e.g., formation of quarks, nucleons, atoms, molecules, crystallization of salt solutions, freezing of liquids), and finally the microscopic dissipative (md) symmetry breakings (e.g., all intracellular energy-coupled biochemical and biophysical processes including active transport, muscle contraction, and biosyntheses). If we are correct in viewing the emergence of the four forces of nature (gravitational, weak, strong, and electromagnetic) as the result of "microscopic equilibrium" symmetry breakings (Note that all these forces emerged before the universe was about 20 seconds old (Parker, 1986), it seems logical to ask the question as to whether or not "microscopic dissipative" symmetry breakings will also have a unique force associated with them. That is, there may exist a new force that drives all the energy-coupled, goal realizing molecular work processes that occur inside the living cell. I postulate that this new force is identical with the *cell force* discussed in Section 1.8.9.

[After completing the section on the Shillongator around April, 1990, the author became aware of the theoretical writings of W. M. Elsasser (1961, 1975, 1987). Although Elsasser's theory of organisms does not directly deal with the origin of the universe, the incorporation of his theoretical results into biocybernetics has thrown new light on the nature of the origin of the universe. The formulation of the original version was couched in terms of the evolution of concrete material objects from superstrings to human societies. In contrast, the newer version described in Appendix 1.F is formulated in more abstract terms, placing emphasis on the concepts and principles underlying the evolution of the universe rather than on the nature of the concrete objects. Hence, we may refer to the original version of the Shillongator as the "concrete version" and the newer version as the "abstract version." Instead of revising the whole section to accommodate the new insights, I decided to leave the concrete version of the Shillongator intact and describe the abstract version as an appendix. It is hoped that a comparison of these two versions of the Shillongator will lead to a deeper understanding of the theoretical underpinnings of the biological model of the universe.]

1.12. APPLICATIONS

Biocybernetics is a unique physical theory in two respects.

(A) <u>Its multidisciplinarity</u>: It is built on the principles, concepts and analogies derived from practically every major branches of the natural sciences known up to the late twentieth century, namely (1) quantum mechanics, (2) statistical mechanics, (3) equilibrium and far-from-equilibrium (or irreversible) thermodynamics, (4) chemical kinetics, (5) chaodynamics (i.e., the study of nonlinear dynamical systems showing the phenomenon of deterministic chaos), (6) special and general relativity theories, (7) nuclear physics, (8) information (or communication) theory, (9) cybernetics, and last but not least (10) the theory of biological evolution (appropriately updated based on recent experimental and paleontological findings (Pollard, 1984; Ho and Saunders, 1984)). This multidisciplinary aspect of biocybernetics leads me to suspect that biocybernetics may fulfill the prediction made by the eminent paleontologist G. G. Simpson who wrote in 1964:

> ... Einstein and others have sought unification of scientific concepts in the form of principles of increasing generality. The goal is a connected body of theory that might ultimately be *completely* general in the sense of applying to *all* material phenomena. The goal is certainly a worthy one, and the search for it has been fruitful. Nevertheless, the tendency to think of it as *the* goal of science or *the* basis for unification of the sciences has been

unfortunate. It is essentially a search for a least common denominator in science. It necessarily and purposely omits much of the greatest part of science, hence can only falsify the nature of science and can hardly be the best basis for unifying the sciences. I suggest that both the characterization of science as a whole and the unification of the various sciences can be most meaningfully sought in quite the opposite direction, not through principles that apply to all phenomena but through phenomena to which all principles apply. . . . . . I have . . . .indicated what those latter phenomena are: they are the phenomena of life. . . . . Biology, then, is the science that stands at the center of all science, and it is here, in the field where all the principles of all the sciences are embodied, that science can truly become unified.

Biocybernetics as formulated in this chapter suggests the ways in which Simpson's program of unifying the sciences might be carried out, one of the most important approaches being the establishment of physical theories based on gnergy, the amalgamation of Gibbs free energy and Shannon information, that can be described using the fiber bundle theory (see Sections 1.1 and 1.8.8). It is probably because biocybernetics embodies such a wide variety of physical theories that it is so "omnipotent" in its ability to influence our thinking about many natural phenomena, including the phenomenon of humankind.

(B) <u>Its Omnipotency</u>: The domain of application of biocybernetics ranges from the molecular dynamics of enzymes (Section 1.5), to the cell (Section 1.8), to the human body (Section 1.9), to the structure of human society (Section 1.10), and to the nature of the universe (Section 1.11).

1.12.1. What is Life?

Erwin Schrödinger raised this question in 1946 in his influential little book with this title. For a recent debate concerning the merits or the lack thereof of this book, readers are referred to M. Perutz (1987) and I. Prigogine (Chapter 2 of this book). Almost 50 years after Schrödinger's book, there is still no consensus among scientists as to the nature of the basic physical principles and mechanisms that might be responsible for life—this despite the numerous fundamental discoveries undreamed of by Schrödinger, such as the DNA double helix, the universality of the genetic code and biochemistry in all cells, detailed 3-dimensional X-ray structures of hundreds of proteins and their dynamics as revealed by NMR and other spectroscopic techniques, and the current revolutions in molecular biology as applied to enzymology, developmental biology, pharmacology, immunology, cancer, and evolutionary biology.

The question of life and death can be discussed at various levels of biological organizations—(i) cells, (ii) tissues (e.g., cerebral cortex, skin, coronary artery, etc.), (iii) organs, and (iv) organisms; and each level is expected to possess unique properties. The following are some of the general statements derived from biocybernetics that may be useful in the scientific discussion of the phenomenon of life on the cellular level:

(1) All living systems are composed of one or more cells; i.e., the cell is the simplest living system in nature.

(2) The cell is an open thermodynamic system, exchanging energy and matter with its environment.

(3) The cell consists of two major classes of chemical species; h- or heavy-particles and l- or light-particles (see Section 1.8.7 for definitions). Although both h- and l-particles are ultimately composed of electrons, quarks, and gauge bosons, it is not necessary to break h- and l-particles down to these constituents to understand how living systems work. It is possible to view h- and l-particles as new particles or "chunks" on higher levels of organization (Hofstadter, 1980), having properties not derivable from their constituents. This viewpoint does not negate the fact that the fundamental particles are necessary for the existence of h- and l-particles; i.e., without electrons, quarks, and gauge bosons, h- and l-particles cannot exist, and without h- and l-particles, no life can exist.

(4) The cell can be viewed as a 6-dimensional structure in the sense that it takes a 6-dimensional mathematical space to completely specify the moment-to-moment state of the cell (i.e., the internal state, S^c, in Equation (1.19)). These 6 dimensions are x, y, z, t, c, and b, where the first four coordinates constitute the familiar spacetime of physics, the c coordinate represents the "$\underline{c}$hemical" space which is needed to specify the nature of the multifarious chemical transformations that take place inside the cell, and the b coordinate signifies the "$\underline{b}$iological" space needed to specify the fitness value of each chemical events occurring inside the cell. By the "fitness value" is meant to indicate the degree of the contribution of a given intracellular chemical process to the overall survivability or function of the cell. It is evident that there is a gradual increase in the dimensionality of the space required to describe relativity (4-dimensional), chemistry (5-dimensional) and biology (6-dimensional). In Section 1.8.2, we discussed 4-structures in physics and biology. In these discussions the 5th and 6th dimensions were "suppressed" in order to focus on the concept of "dynamic structures" in contrast to classical static structures.

(5) The 6-dimensional space required to describe the cell can be represented geometrically using the fiber bundle theory (Section 1.8.8.). We can identify this

space with the L-space described in Figure 1.11; the c and b spaces with the fiber space. To recognize the pioneering researches of M. Eigen (sequence space of RNA) and E. Mayr (theory of biological evolution) that contributed to the development of the concept of the L-space, we may alternatively call the c space the "Eigen space" and the b space the "Mayr space." We can summarize the relationship among the various spaces that constitute the L-space, the mathematical space of biology, as follows:

$$\text{L-Space} = \text{Spacetime} + \text{Eigen Space} + \text{Mayr Space} \quad (1.47)$$

(6) Just as the conservation of the isotopic spins of nucleons, when incorporated into the Yang-Mills gauge theory (Yang ,1977), leads to the prediction of the existence of the strong force, it seems possible that an analogical use of the Yang-Mills equation (see Table 1.16) allows one to predict that the conservation of the fitness value of biopolymers in the cell (i.e., the survivability of the cell) throughout biological evolution leads to the existence of a new force operating in the cell called the "cell force" (see Section 1.8.9).

(7) The strong force is what holds the nucleons together inside atomic nuclei, despite the electrostatic repulsion among protons; similarly, it is postulated that the cell force holds biopolymers together inside the cell in their spatiotemporally organized functional states, despite the randomizing influences of thermal motions.

(8) The existence of the cell force is synonymous with the adaptability of the cell—i.e., the cell force allows the cell (i) to recognize those input signals beneficial to its survival and those detrimental to it ("cognitive function"), and (ii) to effectuate those intracellular processes that enhance what is beneficial to the cell and suppress what is detrimental to it ("effector function"). Because of the "cognitive" and "effector" functions, the cell can exhibit "intelligent behaviors." The cognitive and effector functions together will be referred to as the "cellular intelligence (CI)." The following terms are inter-related:

$$\text{The Cell Force} = \text{Adaptation} \quad (1.48)$$

$$= \text{Fitness or Survivability} \quad (1.49)$$

$$= \text{Cognition} + \text{Effectuation} \quad (1.50)$$

$$= \text{Cellular Intelligence (CI)} \quad (1.51)$$

The multiplicity of related terms such as shown by Equations (1.48) through (1.51) seems to be forced upon us by the fact that the cell is too complex a physical system to be adequately described in terms of just one concept or one term per each of its fundamental properties.

(9) An interesting corollary to the cell force/gravitational force analogy discussed in Section 1.8.9 is that the chemical events occurring inside the cell cannot be completely accounted for on the basis of the rules of chemistry operating in test tubes, just as the motions of test particles in a freely falling elevator (general relativity) cannot be understood in terms of the rules of behavior of the same test particles studied in an elevator that is falling at a constant speed (special relativity) (see also Table 1.17). As pointed out earlier, this conclusion is consistent with the predictions made by Bohr (1933), Schrödinger (1946), Elsasser (1958, 1975, 1987), and Rothstein (see Chapter 7) that biology may not be completely explained on the basis of the laws of physics and chemistry alone, although the logical reasoning followed by these authors are quite different from mine. They reached this conclusion based on the complementarity principle derived from atomic physics or on the statistical mechanical considerations of genes, or on the principle of finite classes. In contrast, I have arrived at the above conclusion based on the postulated existence of a new force operating in the cell and its anticipated effects on intracellular chemical events in analogy to the effect of the gravitational force on the motions of material bodies in spacetime. A similar conclusion has been reached by Smith and Welch in Chapter 6 through a more rigorous formal reasoning than followed here. We may summarize these thoughts and assertions by one sentence: Life is meta-chemical. That is, chemistry alone cannot explain life.

(10) It is possible that one of the most fundamental rules governing the phenomenon of life at all levels is the principle of emergent properties (PEP) formulated in Section 1.4.4; i.e., life is an emergent property, and the cell force is an emergent property. This means that the living property of the cell cannot be completely accounted for in terms of the sum of the individual component chemical and physical processes that proceed in the cell. Biologists trying to describe cell life based on individual chemical reactions and physical processes going on inside the cell may be akin to chemists attempting to describe the "beauty" of a painting such as, say, the "Grand intérieur (Nice, 1919 or 1920)" by H. Matisse, based on the chemical composition of the pigments used by the painter. The subtle nature of the "beauty" that Matisse was striving to capture on his canvas is revealed in the incompletely erased traces of the semicircular chair and the window located on the bottom and the top portion of the painting, respectively. Organization is the essence of beauty, as it is of life. Organization is the property of the whole and disappears

when the whole is broken into component parts.

1.12.2. Mechanisms of Cell Death

(1) Isolated cells have been used extensively to study mechanisms of chemically induced cell death. The most commonly employed experimental criterion for cell death is the rupture of the cell membrane leading to the leakage of cytosolic enzymes such as lactate dehydrogenase (LDH) and the uptake of extracellular dyes such as trypan blue (see Orrenius, Chapter 9; de Groot and Knoll, Chapter 11, and Belinsky et al., 1984). When isolated cells (either freshly isolated and used within several hours, or cultured in appropriate media for several days before use) are treated with cytotoxic agents such as chemicals or visible light, various morphological changes can be observed to occur, including the formation of the so-called "blebs" on the cell membrane visible under light microscopy or disintegration of subcellular organelles revealed by electron microscopy. However, the genesis of these morphological changes do not necessarily signal cell death, since morphologically altered cells can maintain the cell membrane integrity for hours as revealed by their ability to exclude trypan blue (S. Ji, S. Finch, K. Jungermann and A. Stier, 1986, unpublished observation). In the absence of any functional criteria, therefore, it is impossible to determine whether the cellular morphological changes preceding cell death are signs of cell injury or cellular adaptation and protection. In other words, although it may be possible to specify the x, y, z, t and c coordinate values of an intracellular event preceding cell death, it may be difficult (or even impossible in most cases) to determine its b coordinate, namely its "fitness" value or its "semantics" (see Section 1.8.9). Therefore, it may be very difficult, or even impossible, to completely specify the moment-to-moment sequence of intracellular changes that can unambiguously link the exposure of the cell to a cytotoxic agent to the demise of the cell. This conclusion, which is amply supported by experimental observations reported in the literature over the past two or three decades, seems to agree well with the theoretical deductions discussed in Section 1.12.1; i.e., cell death cannot be completely described in terms of chemical and physical mechanisms alone, just as cell life cannot be so described.

(2) Although chemistry and physics may not be sufficient to account for life, the chemical and physical approaches are invaluable in biological research, because they enable biologists to identify those factors that can cause cell death; hence, by removing these factors from our environment, life can be protected from potential harms. In addition, the accumulation of individual pieces of experimental data on cell death will play an essential role in establishing our basic understanding of life on the

cellular level. The ultimate goal of cell death research is not so much to describe the phenomenon of cell death in concrete molecular details from the beginning of cell injury to its demise using specific chemical and physical mechanisms (even if one can do so), but to discover the general principles and rules of cell death, just as the mathematicians in the seventeenth century established algebra (i.e., the method of reasoning about numbers by employing letters instead of concrete numbers and signs to represent their relations (Millington and Millington, 1966)) by discovering the "axioms" of algebra, based on the patterns established in arithmetic, the art of counting and computing with concrete numbers. Arithmetics had to precede algebra; so it is natural to expect that "experimental" toxicology (or biology in general) will precede "theoretical" toxicology (or biology). But it is unfortunate that most toxicologists appear to believe that theorizing about toxicological problems is a waste of time and hence not worth funding nor training future toxicologists in and that the experimental approach is the only way to solving all the basic questions encountered in toxicology. The same criticism can be directed to biological research in general as currently practiced. It may be that the next century will witness the emergence of two clearly defined branches of biology: the "arithmetic" biology emphasizing the discovery of new facts, and the "algebraic" biology focused on discovering patterns and principles underlying life.

(3) Biocybernetics views the cell (the Bhopalator) as a self-reproducing molecular machine. Like all machines, the cell is an open system, operating most of the time far from equilibrium. The normal functioning of the cell depends on a well-defined set of inputs (see I^c in Equation (1.17)), such as reactants for exergonic reactions and chemicals for synthetic purposes, removal of outputs (i.e., O^c), including CO_2, H_2O, and heat, and the operation of the intracellular chemical and physical processes (i.e., S^c) according to the rules encoded in its genome, namely d^c and l^c in Equation (1.17). Therefore, the Bhopalator suggests that there are five major classes of mechanisms by which the cell can be killed:

(i) The I^c mechanisms: Cell deaths caused by an inadequate supply of input chemicals essential for cell life, such as O_2 for all aerobic cells, glucose for neurons, essential amino acids, essential fatty acids, inorganic ions, etc. (e.g., anoxic cell death).

(ii) The O^c mechanisms: Cell death caused by the accumulation of waste products of cell metabolism, including CO_2, H_2O, heat, urea, ketoacids, etc. (e.g., metabolic acidosis).

(iii) The S^c mechanisms: Cell deaths caused by the cell's inability to occupy appropriate internal states, namely sets of intracellular physicochemical fluxes or IDS. For IDS to be maintained, cells must contain all its structural and metabolic components within certain concentration ranges within the cell volume. The reason that cells die when the cell membranes rupture may be interpreted as due to this mechanism. Excessive phosphorylation or dephosphorylation of intracellular proteins may also be injurious to cells due to the S^c mechanism.

(iv) The d^c mechanisms: Cell deaths caused by the impaired "cognitive" function of the cell, namely the cell's inability to recognize input signals and undergo associated internal state (S^c) changes, either because of an uncoupling or an inhibition of the cognitive mechanisms of the cell (e.g., cellular atrophy due to the lack of a receptor, inhibition of receptor binding, or uncoupling an occupied receptor from the generation of an associated second messenger inside the cell, etc.).

(v) The l^c mechanisms: Cell deaths caused by the loss of the cell's ability to carry out its "effector" functions (e.g., the cell's inability to adapt to stress). Just as cells can be programmed to die (e.g., "apoptosis" (Wyllie, 1981)), perhaps "cells are programmed to live." It is possible that there are three kinds of genetic programs in the cell—(i) the *cell death program* (or the "apoptosis" program), (ii) the *cell life program* (or "immortality" program), and (iii) the *neutral program* (i.e., neutral with respect to cell life or death but may be essential for cellular differentiation). Cells may die via the l^c mechanism when the balance between the cell life process and the cell death processes favors the latter.

There are many examples from the literature that can be cited to support the five proposed classes of the cell death mechanisms; but this cannot be pursued here due to space limitation.

The most unexpected mechanism of cell death suggested by the Bhopalator is the l^c mechanism; i.e., cell death caused by the cell's inability to carry out a set of "effector" functions which constitute cellular intelligence (CI). The other necessary component of CI is the "cognitive" functions. In other words, the Bhopalator postulates that there exists a set of gene-directed cellular processes that are essential for its survival. I identify these gene-directed functions of the cell as those physicochemical processes responsible for readjusting intracellular metabolic fluxes in such a way as to enhance the survival of the cell under all kinds of stresses the cell might encounter, either chemical or physical. Such a set of functions will be called "cellular adaptive (or defense) functions (CAF)," the genetic program responsible for CAF may be called the "CAF genetic programs," and the associated genes may be called the *caf* genes. The well-known self-repair processes may be examples of CAF. Another capability of CAF may be to turn off, in times of stress, those unessential cellular processes to conserve ATP and other metabolites for the purpose of repairing cellular injuries and maintaining critical cellular processes. A corollary to this postulate is that cells capable of activating CAF in times of stress can survive longer than the cells wherein CAF mechanism is either inhibited or uncoupled.

1.12.3. Kinetics of Cell Death

(1) When the phenomenon of death is studied on the cellular level, it is found that not all the individuals of a given cell population die at the same time (within experimental error) after exposing the cell population to a cytotoxic agent. The temporal span within which cells die can vary from 3 to 10 hours (see Figure 1.A3, Appendix 1.D). Why is there so much spread in the survival times (or lifetimes) of individual cells? Cells in control populations also die in most experimental conditions, even without any exposure to toxic agents; exposure to toxic agents only hastens the rate of cell death.

For example, the visible light exposure of freshly isolated rat hepatocytes leads to a complete annihilation of cells within a time span of about 2 hours (i.e., the spread of the lifetimes of individual cells, designated S_{clt}, is 2 hours), whereas the spread of the lifetimes of the individual cells in the control population is about 5 hrs. That is, the light exposure reduced the value of the S_{clt} from 5 to 2 hours. Similar measurements using cultured hepatocytes (in collaboration with K. Jungermann) indicated that 5 mM acetaminophen caused the S_{clt} to decrease from 9 hours to 4.5 hours.

The mechanisms responsible for the reduction of S_{clt} by toxic agents are not known. One testable hypothesis is that a population of cells act as a living unit (i.e., an organism) and cells within it cooperate to prolong the lifetime of the population itself. We may then speak of the "population lifetime (plt)" in contrast to the "cell lifetime (clt)." When a certain subset of a cell population dies, chemical factors may be released from dying cells that act as signals for surviving cells to undergo "internal state transitions" in order to adapt to the environmental conditions that had led to the demise of some cells in the population. Such an "intelligent" mechanism would tend to spread out the lifetimes of cells in a population over a larger time span than in a cell population incapable of such a mechanism. Furthermore, such an adaptive capability of a cell population as a whole would require the presence of some mechanisms for intercellular communication. When such intercellular communication is blocked, cells may die within a narrower time span, predominantly determined by the properties of individual cells. This would explain not only the effect of visible light and 5 mM acetaminophen on decreasing the S_{clt} values but also the shift of the cell population lifetimes observed; the cell population lifetime was reduced from 6.8 hours to 4.2 hours by visible light exposure, and from 12-13 hours to 4.6 hours by 5 mM acetaminophen. For future reference, we may refer to this mechanism of the cytotoxic actions of chemicals and physical agents as the "intelligent cell population hypothesis." Since cells are already thought to be intelligent, should we be surprised if we find that populations of cells also show intelligent behaviors?

(2) Granted that we have provided a satisfactory explanation for the "hastening" of cell death by toxic agents by blocking intercellular communication. But how can we explain the fact that control cells also die? What are the mechanisms responsible for the death of cells not exposed to any toxic agents?

The two most likely reasons for the death of cells without being exposed to any toxic agents are (i) that the experimental procedures employed for isolating cells already caused the same set of cell injuries that can be caused by exogenous toxic agents, and (ii) the cells surviving the isolation procedures have not suffered the same kind of cell injuries inducible by exogenous toxic agents but have undergone changes in the internal states (i.e., "cell states" (Ji, 1987)) in such a way that the balance between the cell life-promoting processes and the cell death-promoting processes are tipped toward the latter. To make the second hypothesis viable, it is necessary for us to further speculate that there are three kinds of intracellular processes: (i) Those processes that contribute to cell life, (ii) those processes that contribute to cell death, and (iii) those that are neutral with regard to their impact on cell viability, at least during the experimental time period. The causes underlying these three kinds of intracellular processes may be classified as "external" or "internal," depending on whether a given process is caused by environmental factors (i.e., extrinsic to the cell) or by the activation or inhibition of genetic programs intrinsic to cellular genome, respectively (see Section 1.8.11). We may call genes responsible for the intrinsic intracellular processes essential for cell life the "cell life genes (CLG)" and those responsible for the intrinsic processes that promote cell death the "cell death genes (CDG)." The genes coding for apoptosis (Wyllie, 1980) may be equated with CDG, and the genes responsible for immortality of transformed cells may belong to the CLG family.

(3) Cytotoxic agents can cause cell deaths in two fundamental ways: (i) The CI-dependent mechanisms (as discussed in Section 1.12.2), and (ii) the PI-dependent mechanisms (as discussed in Section 1.12.3(1)). The symbols CI and PI indicate "cellular intelligence" and "the cell population intelligence," respectively. If a toxic agent kills cells by a CI- and PI-dependent mechanism, the slope of the cell death kinetics plot should increase and the cell population lifetime should decrease, as was found to be the case with visible light on freshly isolated hepatocytes and 5 mM acetaminophen on cultured hepatocytes (Appendix 1.D). If cytotoxic mechanisms cause cell death through PI-dependent mechanisms, the associated cell death kinetics plot would show an increased slope (i.e., a decreased spread of cell lifetimes)

without any changes in the cell population lifetime. Finally, if certain cytotoxicants kill cells via the CI-dependent mechanism without interfering with the PI-dependent mechanism, then one should observe a decrease in the cell population lifetime without any change in the slope of the cell death kinetics plot.

1.12.4. Molecular Mechanisms of Oxidative Phosphorylation

(1) The phenomenon of oxidative phosphorylation involves the coupling of the exergonic (free energy-releasing) redox reaction of respiratory substrates (SH_2) (Equation (1.52)) and the endergonic (free energy-consuming) synthesis of adenosine-5'-triphosphate (ATP) from adenosine-5'-diphosphate (ADP) and inorganic phosphate (P_i) (Equation (1.53)), catalyzed by appropriate enzymes located in the inner mitochondrial membrane. The magnitude of the Gibbs free energy transferred from the exergonic to the endergonic reactions is estimated to be 10-20 Kcal/mol of ATP synthesized (Slater, 1969):

$$SH_2 + 1/2\ O_2 \quad \rightarrow \quad S + H_2O \quad (1.52)$$

$$ADP + P_i + H^+ \quad \rightarrow \quad ATP + H_2O \quad (1.53)$$

(2) The widely accepted hypothesis concerning the coupling mechanism of oxidative phosphorylation is the chemosmotic hypothesis proposed by P. Mitchell in 1961 (Skulachev and Hinkle, 1981), for which he received a Nobel Prize for Chemistry in 1978. The chemosmotic hypothesis postulates (i) that the mitochondrial redox reaction generates protons in such a way that they are asymmetrically distributed across the inner membrane, thus storing a part of the redox free energy in the form of a transmembrane electrochemical gradient of H^+ (so-called the "proton motive force"), and (ii) that the transmembrane proton gradient (i.e., "osmotic" free energy) drives the synthesis of ATP ("chemical" free energy) by the spontaneous diffusion of protons across the inner membrane, thereby shifting the equilibrium shown in Equation (1.53) to the right (i.e., conversion of osmotic free energy to chemical free energy, hence the name "chemosmotic"). Almost all of the biochemistry textbooks now in print represent the chemosmotic "mechanism" of oxidative phosphorylation in simple and unencumbered diagrams, belying the uncertainties and mechanistic complexities surrounding one of the most basic processes in biology yet to be resolved. The simplicity of the form of the solution proposed by the chemosmotic hypothesis is so alluring that even some sophisticated biologists and physicists are persuaded to accept it as truth (but see Westeroff et al.,

1984 and Slater et al., 1985).

(3) However, beginning in the early 1980's, researchers in bioenergetics began to express their views that not all is well with the chemosmotic mechanism of energy coupling and that some of the newer experimental data require models not involving the bulk-phase proton gradient (as assumed by P. Mitchell) but instead "intramembrane" protons as suggested by R. J. P. Williams (1961) (Green and Ji, 1972; Ji, 1976, 1977, 1979; Westerhoff et al., 1984; Slater et al., 1985).

(4) Both Mitchell and his opponents agree that oxidative phosphorylation is the result of coupling two basic mitochondrial processes: (i) The "redox reaction-driven" uphill movement of protons (Equation (1.52)), and (ii) the "phosphorylation-driving" downhill movement of protons (Equation (1.53)). The major difference between the chemosmotic hypothesis and alternative models has to do with the question as to where precisely the high-energy intermediate protons are located—the bulk phase outside the mitochondrial inner membrane as proposed by Mitchell and his followers or the inner membrane phase as advocated by his opponents. The weight of the experimental evidence that now exists seems to me to favor the latter interpretation. However, when the system under consideration is as complex as mitochondria, it may be impossible to generate experimental data that are capable of providing a clear-cut choice between competing mechanistic models. Perhaps this is another example of the impossibility of explaining any fundamental biological phenomena in terms of the laws of physics and chemistry alone that Bohr (1933), Schrödinger (1946), and Elsasser (1958, 1975, 1987) spoke about and the results of this chapter seem to support.

(5) My rejection of the chemosmotic hypothesis as providing the fundamental explanation for the molecular mechanisms underlying oxidative phosphorylation is not based on any experimental observations (which I think will prove to be mostly neutral with regard to their ability to discriminate between the chemosmotic and non-chemosmotic hypotheses) but rather on the following theoretical grounds:

(i) I believe in the symmetry of biophysical principles in the sense that all fundamental bioenergetic processes in the cell (e.g., enzymic catalysis, active transport, muscle contraction, oxidative phosphorylation, photosynthesis) will obey common mechanistic principles. We may call this assumption the "symmetry principle of bioenergetic coupling mechanisms" or simply "symmetry principle of bioenergetics."

(ii) If the transmembrane proton gradient is an obligatory component in the molecular mechanism of energy coupling in oxidative phosphorylation, then the molecular mechanism of muscle contraction must not follow the same energy coupling mechanism as operating in mitochondria, because muscle contraction cannot be driven by any transmembrane proton gradient due to the lack of membranes in the actomyosin contractile system. To "conserve" the symmetry of coupling mechanisms in bioenergetics, I prefer to think that both oxidative phosphorylation and muscle contraction involve an identical "intermediate form of free energy," which may be designated as IFFE. Depending on the enzymic structures catalyzing energy coupling, IFFE may be used directly to drive muscle contraction or converted into a transmembrane proton gradient, which subsequently can drive phosphorylation of ADP if experimental conditions are right (see steps 2, 3 and 4 in Figure 1.3). Or more likely IFFE may directly drive ATP synthesis during oxidative phosphorylation without forming any transmembrane proton gradient. In 1976, I formulated a set of molecular mechanisms to explain various functions of mitochondria in which IFFE can be converted either into the transmembrane proton gradient or be directly utilized by the phosphorylation reaction of ADP (i.e. steps 3 and 4 in Figure 1.3). Apparently Slater et al. (1985) and Boyer (cited in Slater et al. 1985) were unaware of this proposal, when they recently published similar mechanisms of oxidative phosphorylation.

(iii) Neither Mitchell nor his opponents have suggested any enzymologically realistic molecular mechanisms, in my opinion, by which protons can be driven uphill coupled to respiration or ATP hydrolysis, regardless of the direction of the subsequent downhill proton movement, whether across the mitochondrial inner membrane or within it. If one can come up with a mechanism by which protons can be "pumped" uphill within the membrane, the same mechanism can surely be used to pump protons across the membrane. This may be viewed as the result of applying the "symmetry principles of bioenergetic coupling mechanisms." So the fundamental mechanistic question is not so much "*where* the protons go after they are pumped" but rather "*how* the protons are pumped in the first place."

(iv) To the best of my knowledge, the only enzymologically realistic molecular mechanisms proposed in the literature that can pump protons, either across or within the inner mitochondrial membrane coupled to exergonic enzymic reactions, is the one proposed by Green and myself in 1972 and further developed subsequently (Ji, 1974a, 1976, 1977 and 1979. The key concept underlying the proton pumping mechanism is that of the conformon extensively discussed in Section 1.5. That is, I claim that the molecular nature of IFFE is conformons. The ligand-operated molecular vending machine hypothesis of ligand-protein interactions described in Figure 1.13 can provide molecular mechanistic explanations as to how conformons can be utilized to pump protons and to drive ATP synthesis.

(v) It is perhaps not a coincidence that the same concept of conformons that provides satisfactory molecular mechanisms for oxidative phosphorylation can also be used to explain active transport in general (Ji, 1974a), muscle contraction (Ji, 1974a), enzymic catalysis (Ji, 1979b), cytochromes P-450 actions (see below), directed molecular motions during gene expression (Ji, 1990), and the origin of life itself (see Table 1.10 and Figure 1.C.1 in Appendix 1.C). This omnipotency of conformons probably derives from the fact that conformons possess both free energy (to do work) and genetic information (to control work).

1.12.5. Mechanisms of Gene Expression

(1) One of the most fundamental and novel properties of the living cell is that it is the smallest physical system in nature that can "read" genetic programs (or information) stored in DNA and "perform" molecular work processes specified in them. Therefore, the cell can be regarded as the "molecular computer" whose program is written in the nucleic acid language with an alphabet consisting of 4 nucleotides. The "molecular computer" has input and output devices mostly constructed out of proteins, and the central processing unit (CPU), also consisting of proteins and nucleic acids. The "molecular CPU" converts the genetic information written in the deoxyribonucleic acid language (i.e., genes) into the ribonucleic acid language (i.e., pre-mRNA and mRNA) and then into the protein language (i.e., polypeptides) with its alphabet consisting of 20 amino acids. This is the well-known "one-gene-one-polypeptide" or "multigene-one-polypeptide" paradigm of molecular biology (Leder, 1981; Milstein, 1986).

(2) According to biocybernetics and the cellular communication theory (CCT) (Section 1.8.11), however, these polypeptides are not the final form of gene expression but are further "translated" (catalyzed by themselves) into the dynamic structures, variously called the dissipative structures of Prigogine, IDS (intracellular dissipative structures), intracellular ion gradients (ICIG), intracellular chemical waves (ICCW), or the Prigoginian form of genetic information (Ji, 1985a, 1987, 1988). In other words, polypeptides "self-translate" the genetic information stored in their primary structures, namely the linear sequence of amino acids. This "self-translation" (or "autotranslation") of amino acid sequence information of polypeptides into IDS serves as the output device of the "molecular computer," the cell.

(3) One interesting consequence of applying the biocybernetics approach to gene expression is the suggestion that it may be necessary to modify the current "one-gene-to-one-polypeptide" or "multigene-to-one-polypeptide" paradigm to include IDS. The new paradigm may be referred to simply as the "DNA-to-polypeptide-to-IDS" hypothesis. Since biocybernetics views IDS as the immediate (or proximate) causes for all cell functions (Section 1.8.3; Ji, 1985a,c; Ji et al., 1986), it follows that the

new paradigm provides a complete molecular mechanistic sequence linking genome to cell functions, i.e., *genotype to phenotype*. Such a dynamic and holistic view of the mechanisms of gene expression afforded by biocybernetics should prove useful in integrating the massive experimental data now existing in the literature regarding the biochemistry and molecular biology of gene expression (Pardee, 1989).

(4) The suggested replacement of the term "one-gene-one-polypeptide" or "multigene-one-polypeptide paradigm" with the term "DNA-polypeptide-IDS paradigm" seems to eliminate several dilemmas and ambiguities in the present day molecular biology:

(i) The meaning of genes is not as simple as it used to be, as pointed out by Leder (1981). One important new perspective to come out of the recent advances in molecular immunology is the discovery that the immune system can generate millions and billions of proteins by shuffling a few hundred germ-line genes, thus invalidating the classical notion that every protein chain is encoded in a particular germ-line gene and that the total genome present in embryonic cells remains unchanged as somatic cells differentiate. By focusing our attention on the dynamics of the DNA molecule itself rather than on its linear sequences, we can avoid the inconsistencies in terminologies and concepts that ultimately result from "unjustifiably too detailed" pictures of genes, namely static nucleotide sequences and 3-D structures.

(ii) According to biocybernetics, DNA molecules (and their associated proteins) inside the cell are dynamic molecular machines that are driven by conformons (Section 1.5). The DNA machines can be divided into two major components: "*structural genes*," and "*spatiotemporal genes*." The former specifies the amino acid sequence of polypeptides as usual, and the latter specifies *which* structural genes are to be expressed *where*, *when*, and for *how long*. The "coding" regions (exons) of DNA are associated with structural genes and intervening sequences (introns) and noncoding regions of DNA most likely constitute the spatiotemporal genes. It should be emphasized that genetic information is not the monopoly of structural genes as is frequently assumed in molecular biology spatiotemporal genes (or regulatory sequences) contain just as much and as essential genetic information as structural genes. This conclusion follows immediately from applying the information theory of Shannon to the biology of the cell (Ji, 1988). In other words, we can view whole DNA molecules as containing genes that specify not only amino acid sequences but the ultimate goal of genetic information, namely cell functions. This is another reason for favoring the "DNA-polypeptide-IDS" paradigm over the currently used "one-gene-(or multigene)-one-polypeptide" paradigm.

(5) The Bhopalator model of the cell suggests a much more dynamic picture of the interaction between DNA and the intracellular biochemical processes catalyzed

by enzymes than the current literature suggests. The "DNA-polypeptide-IDS" interactions are postulated in the Bhopalator as the essence of the living cell and thought to be continuous in time and constant in action in the sense that every change in IDS can affect the DNA structure and its function, and likewise every change in the DNA structure can alter IDS. It is now well-known that the covalent modification of DNA (e.g., DNA methylation without changing gene sequence *per se*), which depends on the enzymic activities of the cytosol, can profoundly affect the pattern of gene expression that can be passed on from one generation of cells to another during development (Holliday, 1989). In the parlance of biocybernetics, DNA methylation is affected by IDS, and IDS is affected by DNA; in other words, the information flow between IDS and DNA mediated by polypeptides can be reversible! There is no physical principle that will be violated when information flows from IDS to DNA. Whether in fact such reversed information flow does occur or not depends solely on whether or not the biological evolution has favored it. Since IDS can be affected by the cell's environment (e.g., reception of growth factors or hormones) (see steps 19, 6, 7, and 8 in Figure 1.7, Section 1.5) and since IDS can affect the heritable pattern of DNA expression (see step 10 in Figure 1.7, Section 1.5; Holliday, 1989), it would follow that the "acquired" characteristics of one cell generation can be transmitted to the descendants, an example of the "lamarckian" process on the cellular level, the Lamarckism being defined as the theory that acquired characteristics may be transmitted to the descendants. The transmission of acquired characteristics from one cell generation to the next is well established in some experimental systems (Cullis, 1984; Schimke, 1980). If no Lamarckism occurs in humankind, this must be not because lamarckian processes are in violation of some fundamental physical or chemical principles, but because the biological evolution has selected such processes out through the evolution of specific mechanisms to prevent such processes from happening.

1.12.6. Cytochrome P-450 Isozymes Diversity: Biological Functions and Molecular Mechanisms of Their Actions

(1) Cytochrome P-450 isozymes are hemoproteins that absorb maximally near 450 nm when their heme iron is reduced and complexed with CO. Although the presence of a CO-binding pigment in liver microsomes (i.e., small sperical vesicles derived from the endoplasmic reticulum after disruption of cells by centrifugation) was first reported independently by Klingenberg and Garfinkel in 1958, it was Omura and Sato who characterized the basic properties of this hemoprotein and named it "cytochrome P-450" in 1964. The major catalytic property of cytochrome

P-450 is to "split" an oxygen molecule and insert one of the two oxygen atoms so generated into the C-X bond of lipophilic organic compounds to produce C-O-X, where X is most frequently H, but can also be N or S. The other oxygen atom combines with electrons originating from NADPH and two protons from medium to form water. These enzymes are also called "mixed-function oxidases," "monooxygenases," or even "hydroxylases."

(2) Generally accepted notions regarding the possible role of cytochrome P-450 isozymes in biology are (i) the rendering of hydrophobic xenobiotics hydrophilic by hydroxylation so as to promote their excretion, (ii) hydroxylation of endogenous substrates in the biosynthesis of steroids, and (iii) other endogenous metabolic functions.

(3) As indicated in Table 1.10, footnote 4, there are now suspected to be about 200 different cytochrome P-450 isozymes, widely distributed in living systems, from mammals to yeasts (Alvares, 1981). Some workers estimate that these isozymes as a group may be able to metabolize up to 200,000 environmental chemicals; i.e., one isozyme may be able to metabolize up to 10^3 different substrates on average!

(4) Despite the extremely wide substrate specificity, there are examples of purified cytochrome P-450 isozymes capable of exhibiting high "stereospecificity" or "regioselectivity;" e.g., different forms of rat cytochromes P-450 monooxygenate testosterone at the 7α-, 16α-, and 6ß-positions. So the intriguing question is, "How can cytochrome P-450 metabolize so many different substrates, with occasional retention of regioselectivity?"

(5) The conformon theory of enzyme catalysis summarized in Table 1.10 suggests a possible model of cytochrome P-450 that can not only account for all of the observed properties of cytochrome P-450 isozymes, but also provide realistic mechanisms for carrying out two seemingly irreconcilable catalytic functions: wide substrate specificity involving thousands of substrates without losing regioselectivity toward some compounds. The hypothesis consists of the following key elements:

(i) Classical enzymes carry out two functions: (a) specific recognition of substrates (i.e., "specificity") mediated by the Volkenstein-Jencks conformons (see Table 1.10), and (b) enhancement of chemical reaction rates (i.e., "catalysis") mediated by the Franck-Condon conformons. These two functions are most likely carried out by the receptor domain and the effector domain of cytochrome P-450 (see Figure 1.13).

(ii) Unlike classical enzymes wherein the Volkenstein-Jencks conformons and the Franck-Condon conformons are permanently coupled, cytochrome P-450 systems may exist in two states: (a) the *coupled state* where the Volkenstein-Jencks conformons and the Franck-Condon conformons are "functionally" coupled so that both specificity and catalysis occur together, and (b) the *uncoupled state* where only the Franck-Condon conformons are active without interacting with the Volkenstein-Jencks conformons so that catalysis without specificity can occur.

(iii) In the uncoupled state of cytochrome P-450 systems (i.e., cytochrome P-450 isozymes with or without accessory components such as NADPH-cytochrome c reductase, b_5, or other factors yet to be discovered), nonspecific catalysis is most likely effectuated through the production of the superoxide anion free radical which may be so situated in the heme cavity that it can interact with xenobiotics either within or outside the heme pocket. It is possible that certain amount of selectivity for xenobiotics can be observed with uncoupled cytochrome P-450 isozymes due to the chance complementarity of the topological characteristics of the internal or external surface of the heme pocket and the molecular shapes of xenobiotics.

(iv) The site within cytochrome P-450 where substrates bind to generate the Volkenstein-Jencks conformons will be called the "substrate recognition site," "substrate binding site" or simply the receptor (or R site), while the site where catalysis actually proceeds (i.e., where the activated oxygen atom is inserted into the C-X bond of a substrate during regioselective monooxygenation reaction, or where the superoxide anion-mediated electronic rearrangements of substrates occur, presumably driven by the Franck-Condon conformons) will be called the "catalytic site" or simply the effector (or E site). In the coupled state of cytochromes P-450, substrate binding may be a prerequisite for catalysis; whereas, in the uncoupled state of cytochromes P-450, catalysis can proceed without substrate binding. One possible mechanism to accomplish such a result would be to assume that in the coupled state the catalytic step must be driven by the free energy generated from substrate binding (see the conformon transfer step in Figure 1.13), while in the uncoupled state the oxygen molecule bound to the heme iron is already "activated" without any input of free energy from the substrate binding site.

An alternative description of the above mechanism can be given using the conformon terminology: (a) In the coupled state of cytochromes P-450, the microenvironment of the E site is such that the activation of oxygen molecules cannot occur unless the Volkenstein-Jencks conformons generated from the substrate binding event is transferred from the R site to the E site, where it is converted into the Franck-Condon conformons, which then provide the requisite free energy to lower the activation free energy barrier for monooxygenation. (b) In the uncoupled state, the structural link between the R site and the E site is broken so that the E site can now activate oxygen molecules into superoxide anion free radicals without any free energy input from the substrate binding event. (c) It is possible that, in the coupled state of cytochromes P-450, the unoccupied R site exerts an inhibitory action on the

E site (e.g., the heme pocket may be in a closed state and hence O_2 cannot get in) so that the latter cannot activate oxygen molecules; however, when substrates bind to the R site, the resulting Volkenstein-Jencks conformons are transferred to the E site thereby activating it (e.g., by opening up heme cavity so that O_2 can now readily bind to the heme moiety), leading to a spontaneous reduction of oxygen to form the superoxide anion free radical. (d) In other words, the E site can exist in at least two states, E^T (from "tight") and E^R (from "relaxed"), with E^T inefficient toward oxygen activation and E^R efficient toward oxygen activation. There are two ways to induce the E^T to E^R transition, one through stereospecific binding of substrates (mostly endogenous but some xenobiotics resembling endogenous substrate by chance) and the other by nonspecific interactions of cytochromes P-450 with xenobiotics.

$$(R - E^T) \;+\; s \;\rightarrow\; (R_s - E^R) \tag{1.54}$$

$$(R - E^T) \;+\; x \;\rightarrow\; (R \ldots E^R)(x) \tag{1.55}$$

where (R-E) represents a cytochrome P-450 system with the substrate binding site R and the catalytic site E tightly coupled (see the dash), the superscripts T and R indicate the conformational state of the E site, s is the stereospecific substrate for cytochrome P-450, R_s indicates the specific binding of s to the R sites, x is nonspecific xenobiotic substrate, $(R \ldots E^R)$ indicates the "dissolution" of the structural coupling between the R and E sites within a cytochrome P-450 molecule so that E goes into its R state which is now directly accessible to x without mediation through E, and (x) reflects nonspecific interactions between cytochrome P-450 and xenobiotics (e.g., perturbation of the endoplasmic reticulum membrane by x, leading to conformational changes of cytochromes P-450). I believe that most, if not all, of the experimental data on cytochrome P-450 found in the current literature can be accommodated by Equations (1.54) and (1.55). Furthermore, these equations are capable of generating testable predictions.

1.12.7. Mechanisms of Xenobiotic Induction of Cytochromes P-450: The "Xenogene" Hypothesis

Various xenobiotics are known to induce cytochrome P-450 isozymes in the liver and other organs (Alvares, 1981; Atchison and Adesnik, 1983). However, the precise mechanisms responsible for this induction is not yet known. The discussion on the possible structure and function of cytochromes P-450 presented above suggest the following possible mechanisms for the induction of cytochromes P-450 by xenobiotics:

(1) Cells are endowed with a set of genes dedicated to *recognizing* and *removing* potentially harmful xenobiotics to which the ancestors of the cells in question were never exposed. Such xenobiotics will be called "evolutionarily unfamiliar" or "evolutionarily novel" compounds. "Food" or "nutrients" may be defined as evolutionarily unfamiliar xenobiotics that cells have "found" useful for their survival and for which cells therefore have "evolved" the necessary enzymes to utilize them, through biological evolution.

(2) There are three distinct mechanisms for removing xenobiotics from the cell's interior: (i) active transport of parent xenobiotics, (ii) metabolic conversion of lipophilic xenobiotics into hydrophilic products which diffuse out of the cell spontaneously, and (iii) metabolic conversion of lipophilic xenobiotics into hydrophilic products which are then actively transported out of the cell. For each of the three kinds of xenobiotic-removing mechanisms, the cell needs one or more enzyme systems, and the totality of the genes responsible for the production of such xenobiotic-removing enzymes is called the "xenogenes" (Ji, 1990). The genes coding for cytochromes P-450 may constitute a subset of xenogenes and can be represented as xgc, from $\underline{x}$enogenes for $\underline{c}$ytochromes P-450.

(3) The xenogenes coding for cytochromes P-450 (i.e., xgc) can be divided into the structural genes to be designated xgc^S and the spatiotemporal genes designated xgc^{ST} (see Section 1.12.5). The structural genes of cytochromes P-450 in turn divide into the genes specifying the amino acid sequence of the substrate binding site (or the receptor) and those determining the amino acid sequence of the catalytic site (the effector) of cytochromes P-450. We may designate these genes as $xgc^S(R)$ and $xgc^S(E)$, where the R and E within parentheses indicate the substrate binding (or receptor) site and the catalytic (or effector) site, respectively.

(4) The number of the $xgc^S(R)$ genes will be designated as n(R) and that of the $xgc^S(E)$ as n(E). In analogy to the V and C genes encoding the variable and constant regions of antibodies, it is postulated that a complete cytochrome P-450 structural gene is assembled by combining one of the $xgc^S(R)$ genes with one of the $xgc^S(E)$ genes through some mechanisms postulated to be similar to those of gene recombination and conversion that occur during lymphocyte differentiation (Leder, 1981; Milstein, 1986). Just as the antibody diversity resides mainly in the V gene family, so it is postulated that the diversity of cytochromes P-450 is mainly due to the multiplicity of the $xgc^S(R)$ genes. Based on the assumption that the number of cytochrome P-450 isozymes estimated at present is 200, it is concluded that n(R) is at least 200, and n(E) is likely to be a much smaller number.

(5) The postulated mechanism of xenobiotic induction of cytochrome P-450

isozymes consists of at least 9 distinct steps as described in detail in Appendix 1.E.

(6) The only additional feature to be added here is the postulated mechanisms that will enable the cell to "express" the $xgc^S(R)$ genes encoding the cytochrome P-450 isozyme best fit to metabolize specific xenobiotics; i.e., the cytochrome P-450 isozyme whose substrate binding site is *complementary* to the shape of xenobiotic XH_2. The postulated mechanism consists of the following steps:

(i) XH· formed in step 5 in Figure 1.A.4 (Appendix 1.E) covalently bind to a specialized protein to be called the "xenobiotic transducer protein (TP)";

$$XH· \; + \; TP \to TP\text{-}XH \qquad (1.56)$$

where TP-XH is postulated to act analogously to a hapten-modified carrier protein in the immune response. TP is thought to consists of two segments—the XH_2-binding domain (i.e., the receptor domain in Figure 1.13) and the DNA-binding domain (or the effector domain).

(ii) TP has the unusual property that, when XH· covalently binds to the xenobiotics-binding domain, the DNA-binding domain undergoes a conformational change in such a way that its 3-dimensional shape reflects some essential characteristics of the 3-dimensional shape of the TP-binding surface of XH_2. Such a molecular feat is within the capacity of the ligand-protein interaction mechanisms depicted in Figure 1.13.

(iii) The DNA-binding domain of TP-XH now "searches" for the right binding sites on the spatiotemporal gene (i.e., xgc^{ST}) that "selects" and "activates" the correct member of the $xgc^S(R)$ genes that encodes a specific cytochrome P-450 isozyme whose receptor domain structure is complementary to the molecular shape of the TP-binding region of XH_2. Through steps (ii) and (iii), TP in effect "transduces" the molecular shape information of XH_2 to the conformational shape information of TP so that the appropriate spatiotemporal gene encoding cytochromes P-450 can be activated by XH. This can be viewed as a variation of the conformon-messenger transduction hypothesis discussed in Section 1.8.11. The mechanism by which $xgc^{ST}(R)$ selects and expresses the correct $xgc^S(R)$ genes for XH_2 is unknown.

1.12.8. The Biocybernetics Model of the Biological Evolution

The Darwinian theory of biological evolution contains two main goals: (i) The establishment of the *fact of evolution*, and (ii) To propose *natural selection* as the primary mechanism of evolution (Gould, 1982). Darwin succeeded in accomplishing the first but not the second goal. Recent protobiological (e.g., spontaneous formation of peptides from amino acids), molecular biological (e.g., dynamic instabilities of genes with respect to recombinations, multiplicity of mutations of DNA with no discernable phenotypic consequences, etc.), and paleobiological data (e.g., punctuated equilibria) all combined to generate a formidable movement against the original suggestion of Darwin, later elaborated by his followers (Mayr, 1988), that organisms evolved through the mechanisms of natural selection of the fittest phenotypes produced by random mutations of genes (Depew and Weber, 1985; Ho and Saunders, 1984). One of the most important new elements introduced into the current evolutionary debate seems to be the idea of self-organization (Prigogine, 1977, 1978, 1980; Prigogine and Allen, 1982), a key theoretical building block of biocybernetics described in this chapter.

The concept of evolution is probably the single most important principle of nature that originated in biology, comparable to the principles of thermodynamics and quantum mechanics. Recent theoretical developments in particle physics and cosmology indicate that evolution is not limited to biology but is fundamental in nonliving world as well; i.e., it is not just the biological world that has evolved, but the whole universe has been evolving. The universality of evolution is now a well-established fact. It is my basic assumption that the study of biological evolution will facilitate the understanding of the cosmological evolution and *vice versa*. It was partly based on this assumption that I have proposed that our universe was "programmed" to evolve and produce sentient beings who can uncover the origin of itself (see Appendix 1.F). If this idea is correct, it would seem logical to conclude that the origin of life on this planet 2-3 billion years ago was not a random and rare event, as some cosmological models assume, but rather a programmed and "predetermined" event, whose realization was aided by the primordial physical and chemical environments unfolding in this part of the universe at that particular epoch of the history of the universe. The nonrandom synthesis of polypeptides from mixtures of amino acids discovered by S. Fox (1984) may be viewed as a historical remnant of the cosmological programs stored in the gnergy tetrahedron of the Shillongator (Appendix 1.F).

If the biological evolution and the cosmological evolution are not separate and independent of each other as is now commonly thought but are instead intimately "coupled" as indicated in the formulation of the Shillongator (see Section 1.11 and Appendix 1.F), then it is possible (i) that the complete understanding of one kind of evolution will critically depend upon our knowledge about the other, and (ii) that physics and chemistry on the one hand and biology on the other are not as separate and autonomous in nature as they might have appeared in the world of fragmented sciences of the last several centuries but represent different unfoldings of the same primeval substance of our universe.

1.12.9. Application of the Law of Requisite Variety to Human Evolution

To apply the Law of Requisite Variety (Appendix 1.A) to human evolution, we have to make the following assumptions: (i) *Homo sapiens* can be treated as a machine in certain essential aspects obeying the Law of Requisite Variety. (ii) The environment dominates *Homo sapiens* (i.e., the environment changes *Homo sapiens* but not *vice versa* at least up until the Industrial Revolution). (iii) The variety of human behaviors (i.e., bodily motions) is directly proportional to the variety of the neuronal connections in the brain which in turn is proportional to the number of brain cells and hence to the brain size. (iv) A significant portion of the brain mass is dedicated to maintaining the survival of the cells in the body through various autonomous processes, including circulation, digestion, reproduction, and immune defenses against harmful microorganisms.

The following are some of the observational data that were mentioned in the lecture given by R.E. Leakey at Rutgers on October 16, 1989 and for which biocybernetics appears to provide reasonable explanations:

(1) The Brain Size/Body Size Ratio: This ratio of *Homo sapiens* is known to be the largest among all primates (Gould, 1977) and to have almost doubled about 2-3 million years ago. The increase in the brain size/body size ratio must have increased the variety of the neuronal connections in the brain not dedicated to the maintenance of the body and hence "free" to increase the variety of voluntary human behaviors. This would have led to a greater variety of environmental niches populated by *Homo sapiens* and to a greater probability of survival of *Homo sapiens* when large-scale changes in environmental conditions annihilated most of the living organisms, including some Homo sapiens and closely related primates.

(2) The Greater Variety of Species found in East Africa relative to that found in West Africa: This observation was attributed by R.E. Leakey to the greater variety of environmental conditions available in East Africa relative to West Africa. His interpretation appears to be consistent with the prediction derivable from the Law of Requisite Variety applied to the group of animals in East Africa; that is, the greater the variety of the environment, the greater the variety of the internal states of surviving individuals and hence the greater variety of animal species. This follows from equation (1.A1) in Appendix 1.A; if we equate V_O with a small and fixed number of outcomes of the interactions between an organism and its environment that are compatible with reproduction, then V_M must increase as V_E increases.

(3) The Bipedal Apes: Dr. Leakey emphasized the fundamental importance of the transition of human ancestors from a four-legged state to the bipedal state. This view can be strongly supported by the result of applying the Law of Requisite

Variety, since the quadripedal-to-bipedal transition can be regarded as a form of increasing the variety of the bodily motions of our human ancestors (i.e., the degree of freedom of bodily motions increased dramatically). There must have been a corresponding increase in the brain size/body size ratio. Because of the increased variety of the bodily motions now available, the bodily environment must have increased its variety as well, thus leading to a sort of positive feed back on the growth of the brain size/body size ratio.

(4) <u>The Hairless Body Surface of Homo sapiens</u>: Dr. Leakey suggested that the hairless body surface of *Homo sapiens* might have served as an efficient system of dissipating bodily heat, thus allowing *Homo sapiens* to make long distance travels. I find this interpretation unconvincing, because there must be a lot of other hair-covered animals who do more long distance traveling than humans do. A more logical explanation of this fact, suggested by the Law of Requisite Variety, is that the hairless body surface acted as a more sensitive receptor for environmental input stimuli than hair-covered body surface, thus increasing the variety of input signals to the brain which in turn led to an increase in the variety of the neuronal connections and associated increase in the brain size/body size ratio. The hairless skin might act as the sole signal receptor for *Homo sapiens* during the early months and years of rearing, before other channels of communication become available such as the vision and speech. This explanation would gain strong support if it can be shown that tactile stimulations of the skin of experimental animals in their rearing stages lead to increased neuronal connections in the brain.

(5) <u>The Long Rearing Periods of Homo sapiens</u>: It is my understanding that the infant Homo sapiens spends much longer rearing periods under the complete dependence on parents than other primates. This would have two consequences: (i) To maintain the homeostasis of the physiology of the young in the face of unfavorable environmental conditions, the variety of the behaviors of the caring parent must increase and hence its brain size/body size ratio, and (ii) During the long incubation time, the young may receive a large variety of tactile stimuli through the hairless skin in proportion to the length of the rearing period, leading to a greater variety of the neuronal connections and of behavioral potentials of the young brain.

1.12.10. Biocybernetics Model of the Inflammatory Response: the "Londonator"

As indicated in Section 1.9, the average adult human body is composed of 10^{12}-10^{14} cells. In order for these cells to survive as a unit, they must coordinate their activities according to a set of rules and through elaborate intercellular communications. The normally functioning body accomplishes a set of physiological

"Biocybernetics": A Machine Theory of Biology

goals, one of which is self defense against a host of potentially harmful events, including infection by microorganisms and physical injuries.

The endogenous self-defense mechanisms that have evolved in the human body can be classified into the "systemic" and "local" defense mechanisms: The purpose of these defense mechanisms is (i) to recognize and be activated by pathogenic agents, (ii) to localize the effects of pathogenic agents at the site of injury, (iii) to remove pathogenic agents (broadly defined as any agents causing abnormalities in the body), and (iv) to repair any tissue damages caused by pathogenic agents and restore the normal state of the body. Clearly the systemic self-defense mechanisms will be activated when pathogenic agents affect the whole body at once as would be the case with the invasion of the body through systemic circulation by pathogenic microorganisms or whole body irradiation. Local defense mechanisms will be activated when specific loci in the body are injured by pathogenic agents, be they microorganisms, toxic chemicals, or physical injuries.

Both systemic and local defense mechanisms can be treated as spatiotemporally organized chemical reaction-diffusion systems composed of elements belonging to different levels of biological organizations, ranging from metabolic pathways to cells to tissues and to organs. We can treat these defense mechanisms as examples of biological machines and represent them algebraically:

$$M^{SD} = (I^{SD}, S^{SD}, O^{SD}, d^{SD}, l^{SD}) \tag{1.57}$$

$$M^{LD} = (I^{LD}, S^{LD}, O^{LD}, d^{LD}, l^{LD}) \tag{1.58}$$

where the superscripts SD and LD refer to "systemic defense" mechanisms and "local defense" mechanisms, respectively. I is the set of all pathogenic agents that can trigger SD or LD, S is the set of all the internal states in which the SD or LD machines can exist (i.e., the kinds of metabolic pathways activated, the kind of cells involved, etc.), O is the physiological consequences of the activation of SD or LD, d is the mechanism by which I triggers M into action and l is the mechanism by which M produces O (see Equation (1.8) in Section 1.4).

A. <u>The Inflammatory Response</u>: (1) The inflammatory response may be one of the most important, if not the only, local self-defense mechanisms operating in the body. It is triggered whenever local tissues are injured by pathogenic agents. The signs of the inflammatory response are (i) *rubor* (redness due to vasodilation), (ii) *tumor* (swelling due to the increased vascular permeability toward blood proteins), *dolor* (pain due to the release of mediators that stimulate pain receptors), *calor* (heat production due to increased local blood flow secondary to vasodilation), and *functio laesa* (lost function due to the abnormal physiological state of the affected tissue). These signs of inflammation were recognized over 2,000 years ago (Boyd and Sheldon, 1979) and can be viewed as the elements of the set O^{LD} in Equation (1.58).

(2) Probably the most important breakthrough in our understanding of the molecular mechanisms responsible for the inflammatory response came when J. R. Vane and his associates discovered in the early 1970's in London that aspirin, an effective nonsteroidal antiinflammatory agent, inhibited prostaglandin synthetase, thus implicating the arachidonic acid cascade in the inflammatory response (Vane, 1978). Although the molecular, metabolic and cellular mechanisms underlying the inflammatory response are much more complex and involve more than just the prostaglandin cascade (Boyd and Sheldon, 1979), because of the importance of the discovery of Vane and his group in advancing our knowledge of the inflammatory mechanisms in general, it may be justified to call the biological machine responsible for the inflammatory response (i.e., Equation (1.58)) the "Londonator."

(3) One of the most disheartening aspects of the scientific study of the inflammatory response is the immense complexity of the phenomenon, implicating just about every control system operating in our body, namely N (e.g., neurogenic mechanisms of vasoconstriction around inflammatory sites), C (e.g., endothelial cellular responses leading to vascular permeability changes), E (e.g., increased release of glucocorticoids systemically (Munck et al., 1984; Munck and Guyre, 1989)), and I (e.g., migration of granulocytes and monocytes from the blood compartment into tissues to perform phagocytosis and initiate tissue repair processes).

(4) It is possible that the myriads of processes triggered by local tissue injuries can be classified according to two major criteria: (a) On the basis of the control systems involved (i.e., N, C, E or I), and (b) on the basis of the size of the body parts affected (i.e., L, local, or S, systemic). According to this scheme, then, we can recognize maximally eight different classes of processes triggered by local tissue injuries—N^S, N^L, C^S, C^L, E^S, E^L, I^S, and I^L. Space does not allow me to cite specific examples here of the inflammatory processes reported in the current literature that can be interpreted as belonging to each of these eight classes. The experimental data on inflammation published in the literature is almost beyond comprehension by one person. If the above classification scheme, which is based on the Piscatawaytor model of the human body (Figure 1.14), provides a theoretical framework to organize the vast amount of observational data on inflammation into a coherent body of knowledge, then the Piscatawaytor indeed will have accomplished its role as the "Periodic Table of Medicine" as predicted in Section 1.9.3.

(5) The Londonator can be regarded as the first clear-cut example of what may be called the "tissue machine" consisting of perhaps 10^3-10^6 cells, intermediate

in size between individual cells and organs. Also, the tissue machine can be viewed as the smallest biological machine whose function depends on all of the four control systems, namely N, C, E and I. The true complexity of the Londonator is revealed by the fact that the prostaglandin cascade, which is only a part of the metabolic and cellular processes involved in the inflammatory response, contains 30-40 metabolites derived from arachidonic acid. These metabolites have relatively short lifetimes, in the range of 1 to 10 minutes thus limiting the geometric domains in which they act. It is possible that there are numerous other arachidonic acid metabolites which have even shorter lifetimes that have not yet been detected but play equally important roles in inflammation. Therefore it seems quite reasonable to conclude that the number of different internal states available to the Londonator is quite huge and perhaps even comparable to the diversity of antibodies (10^6-10^9).

(6) When presented with the current literature on any small aspect of the inflammatory response, we cannot help but ask the question, "Why are the inflammatory processes so complicated?" Most investigators, including myself until recently, tend to think that the complexity of the inflammatory response is one of those problems in biology, for which there is no satisfactory explanation.

One major advantage of applying the machine concept to inflammation is that we can now utilize the Law of Requisite Variety described in Appendix 1.A to explain the *Why?* question raised above. This law in effect says that in order for a machine to accomplish a narrowly defined set of goals (i.e., small V_O in Equation 1.A1, Appendix 1.A, the essence of "homeostasis") under a wide variety of environmental conditions which determine the nature of input signals (i.e., large V_E), the variety of the internal states (i.e., V_M) must be comparably large. To apply this law to inflammation, we must identify the following equalities:

(i)   Variety of I^{LD} in Equation (1.58)
       = V_E in Equation (1.A1)
       = Variety of species, strains, and individuals of animals involved, sex, age, nutritional states, body loci, time of day, etc. that are associated with tissue injuries

(ii)   Variety of S^{LD} in Equation (1.58)
       = V_M in Equation (1.A1)
       = Variety of the patterns of the metabolic and cellular processes that are activated through the N, C, E and I control systems

(iii)   Variety of O^{LD} in Equation (1.58)
       = V_O in Equation (1.A1)
       = Variety of the consequences of the inflammatory response, namely (a) injury confinement by the N, C, E and I mechanisms, (b)

removal of pathogenic agents by the N, C, E and I mechanisms, and (c) repair of tissue damage and controlled cellular proliferation under the control of the N, C, E and I mechanisms

If V_O is much smaller than V_E as seems likely when (i) is compared with (iii), then according to Equation (1.A1), V_M must be a large number, as we already estimated above—i.e., 10^6 to 10^9, in analogy to the antibody diversity in immunology.

B. <u>The Systemic Stress Response</u>: (1) One of the predictions made by the Piscatawaytor is that no part of the human body is an island. A corollary to this prediction is that, when the body is injured locally, the whole body will respond in such a way as to isolate the injurious event, remove injurious agents and repair the damages done, by activating whatever self-defense "programs" or "plans," whether local or systemic, that are available within the body.

(2) One important example of the systemic responses that are triggered by inflammation (i.e., local tissue injuries) may be the increase in the blood concentration of glucocorticoids. According to Munck, Guyre and Holbrook (1984) and Munck and Guyre (1989), the physiological role of stress-induced increases in glucocorticoid levels is to protect the body not against the source of stress itself but against the normal defense reactions that may be overactivated by stress. We may call this notion the "Munck-Guyre-Holbrook hypothesis" of glucocorticoid actions. Their hypothesis seems to account for a vast amount of the experimental data on glucocorticoid physiology that have accumulated during the past 5 decades.

(3) The Munck-Guyre-Holbrook hypothesis of glucocorticoid physiology is consistent with the Londonator model of the inflammatory response, if we assume that stress-induced increases in glucocorticoid levels represents an important component of the E^S mechanism (defined above) accompanying inflammation. This mechanism contributes to confining tissue injuries at the site of original tissue damage, thereby preventing the inflammatory response from spreading throughout the whole body. In addition to E^S, other systemic defense processes are likely to be activated, namely N^S, C^S and I^S, with different response times. Without such preventive measures, the Londonators everywhere in the body would be eventually activated with the original pathogenic agent as the initiator and the damaged tissues acting as the sources of the secondary signals that trigger neighboring tissues into further inflammatory responses. In other words, in the normal physiological state, the Londonators in every part of the body are ready to be activated; however, once a locus in the body is injured and the associated Londonator is activated, this may send signals to the remaining Londonators in the body to "desensitize" them through the N, C, E and I mechanisms. This hypothesis would predict that there will be four different classes

of the intercellular messengers mediating this postulated systemic "desensitization" process of the Londonators, one each for the N, C, E and I control systems. The Munck-Guyre-Holbrook hypothesis of glucocorticoid actions seems to address the E component of this systemic desensitization process.

(4) The system of cells in our body that implements the systemic self-defense mechanisms (in contrast to local self-defense mechanisms) may be viewed as a self-organizing chemical reaction-diffusion system and a machine. Such a group of cells will be referred to as the "Hanoverator," in recognition of the imaginative theoretical and experimental investigations carried out by Munck and his colleagues in Hanover, New Hampshire, laying the foundation for understanding the role of glucocorticoids in the self-defense mechanisms of the body (Munck et al., 1984; Munck and Guyre, 1989). Just as the Londonator is a theoretical model for the inflammatory response, the local self-defense mechanism, the Hanoverator is the self-defense mechanism on the systemic level, in which glucocorticoids, a component of the E control system, play a prominent role. As indicated already, the Hanoverator is expected to be regulated not only by E but also by N, C and I control systems. The detailed structural organization and the mechanism of the operation of the Hanoverator are yet to be worked out, but my current hypothesis is that the Hanoverator is composed of the entire set of Londonators in an organism plus one or more groups of cells, probably located in the brain, that regulate and coordinate the activities of individual Londonators, just as regulatory genes control the turnings on and off of right structural genes in the nucleus of the cell.

1.12.11. Frustrated Self-Defense Mechanisms (FSDM) Hypothesis of Disease Development: Role of Uncontrolled Free Radicals

The current biomedical literature is replete with publications dealing with the possible involvement of free radicals (i.e., molecular species with unpaired electrons in valence orbitals, such as $HO^\cdot$ and $^\cdot CCl_3$) in all kinds of diseases, including cancer (Pryor, 1976; Cerutti, 1985; Halliwell and Gutteridge, 1985). Since it is now well established that cells possess numerous self-defense mechanisms against excessive intracellular production of chemically reactive molecular species including free radicals (e.g., GSH, vitamins C and E, superoxide dismutase, etc.), intracellular free radicals need not be always hazardous to cells. Only when these self-defense mechanisms are frustrated or deranged, would free radicals accumulate and do damage to cells and tissues. Therefore, it seems logical to distinguish two kinds of free radicals: "controlled" and "uncontrolled" and to anticipate that, under certain situations, cells may actually capitalize "controlled" free radicals to accomplish some useful cellular functions (see below).

It is unfortunate that most authors tend to assume without proof that free radicals are always "bad" and cause cell and tissue damages. It is important to recognize that the mere demonstration of correlations between free radicals and diseased states does not necessarily prove any causal relations, since at least some free radicals may be the consequences of diseases rather than the causes for diseases. It is possible that, in most if not all cases, the experimental detection of high levels of free radicals in living systems reflects the normal oxidative metabolic processes gone awry, or a breakdown of self-defense mechanisms, or both. Of course, once free radicals accumulate beyond critical levels, they can cause cellular and tissue damages secondarily. Cerutti (1985) calls such a state of affairs the "prooxidant state." The question still remains whether the prooxidant state of Cerutti represents the cause or the consequence of diseased states. In other words, uncontrolled free radicals can be either the cause or the consequence of the frustrated self-defense mechanisms, or both.

To analyze the potentially complex role that free radicals can play in the etiology of various diseases, it seems to me mandatory to have a theoretical model of disease development that takes into account the best available set of empirical knowledge in biology and medicine and relevant theoretical principles.

The frustrated self-defense mechanisms (FSDM) hypothesis of disease development was formulated on the basis of the laws and principles of biocybernetics discussed in this chapter. This hypothesis is schematically presented in Figure 1.17. In the formulation of the FSDM hypothesis, the concept of machine (with the attendant notions of "inputs," "outputs," and "internal states") played a crucial role in that this concept provided a theoretical framework to bring together a large number of diverse elements into a coherent mechanistic scheme. In this sense, the scheme shown in Figure 1.17 can be regarded as the "machine theory of disease development," a concrete application of the principles of biocybernetics to biomedical sciences.

The key postulates contained in the FSDM hypothesis are explained below:

(i) All diseases of the human body (and other mammals) result from "frustrating" (i.e., inhibiting, uncoupling, or otherwise deranging the normal course of the progression of) one or more steps in self-defense mechanisms. In other words, diseases result if and only if the body is stressed (chemically, microbially, physiologically, physically, or psychologically) beyond the capacity of the body to withstand the imposed stresses.

"Biocybernetics": A Machine Theory of Biology

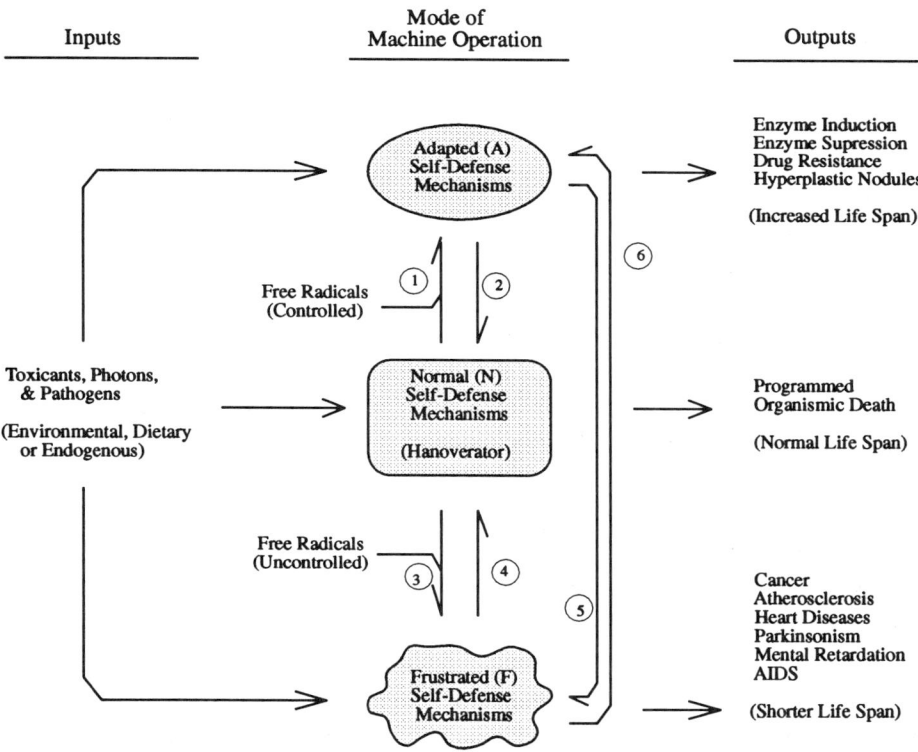

Figure 1.17. A schematic representation of the frustrated self-defense mechanisms (FSDM) hypothesis of disease development. The totality of the biochemical, cellular, and physiological processes designed to protect the body against pathogenic agents (see Inputs) is viewed as a self-organizing chemical reaction-diffusion system called the Hanoverator. Various diseases (see Outputs) result when the Hanoverator operates in the "frustrated" or F mode.

(ii) The body possesses two kinds of self-defense mechanisms (SDM)—the local SDM (e.g., the inflammatory response) and the systemic SDM (e.g., increased plasma glucocorticoids under stress) called the Londonator and the Hanoverator, respectively (see Section 1.12.10). Both are self-organizing chemical reaction-diffusion systems and examples of biological machines. The Hanoverator is composed of Londonators (10^6 - 10^9 of them?) and regulatory units (most likely located in the brain) that control and coordinate Londonators in the interest of protecting the whole body.

(iii) The Hanoverator can operate in three distinct modes: the normal (N), adapted (A), and frustrated (F) modes.

(iv) The mode of operation of the Hanoverator is determined by the steady-state level of uncontrolled free radicals (UFR) in the specialized cells (likely located in the brain, particularly in the hypothalamus) that regulate the activities of the Hanoverator. High levels of UFR favor the transition of the Hanoverator from the N or the A to the F mode (see Steps 3 and 5 in Figure 1.17).

(v) Diseases develop if and only if the Hanoverator is in the F mode (see Outputs in Figure 1.17). The specific nature of diseases is determined by the characteristics of the steps of the self-defense mechanisms that are actually frustrated by a given pathogenic agent (i.e., toxicants, radiations, pathogenic microorganisms from environment, diet, or endogenous sources such as gut-derived endotoxins; see Inputs in Figure 1.17).

(vi) When the body is exposed to low levels of pathogenic agents, the free radical levels in the critical cells in the brain are thought to be under the control of the appropriate self-defense mechanisms and the Hanoverator is stimulated to undergo a state transition from the N to the A mode (see Step 1). When the Hanoverator is in the A mode, the body shows signs of adaptation including enzyme (e.g., cytochrome P-450) induction or suppression, drug resistance, and hyperplastic nodules of the liver (Farber, 1987, 1990; Farber and Sarma, 1987). When the stress is removed, the Hanoverator can revert back to the N mode (Step 2).

(vii) It is possible that the A mode of the Hanoverator can be forced to undergo a transition to the F mode (see Step 5), but this may take much higher levels of free radicals than required for the N to F mode (i.e., Step 3). The F mode of the Hanoverator may be induced to undergo a transition to either the N (Step 4) or the A (Step 6) mode.

(viii) The mode of operation of the Hanoverator is postulated to play a fundamental role in determining the life span of individual mammals. In general, the higher the free radical levels in the critical regulatory cells in the brain (which will favor the A to N to F mode transitions in Figure 1.17), the shorter the life span of individuals (see the different life span outcomes indicated in the parentheses in Figure 1.17). More on this in Section 1.12.13.

(ix) To increase the fidelity of the performance of the self-defense mechanisms, the body utilizes a set of n independent self-defense mechanisms (e.g., DNA repair, anti-oxidant defenses, immune surveillance, autoregulation of microcirculation, stress responses, etc.) against a given disease entity, D. We may designate the number of distinct self-defense mechanisms engaged in protecting the body against a disease entity D as n(D). The literature data on cancer indicate that n may vary from 2 to 7 (see Section 1.12.12).

To the best of my knowledge, the scheme in Figure 1.17 is at present the only one of its kind in the biomedical literature that provides a theoretically coherent mechanism to integrate the following set of the concepts, processes and phenomena that are widely discussed in the current literature in environmental toxicology, aging, cancer, and nutrition (Ames et al., 1985)—(i) *free radicals*, (ii) *aging*, (iii) *cancer*, (iv) *adaptation*, (v) *environmental* and dietary sources of free radicals, and (vi) *self-defense mechanisms*. What is unique about the scheme is the fact that it is based on the principles of biocybernetics that have proved applicable to numerous other branches of biology and medicine as demonstrated in this chapter. It is hoped that the proposed theoretical scheme in Figure 1.17 will be helpful not only to basic researchers in the various disciplines cited above but also to workers in risk assessment and risk management who are faced with complex and important environmental health problems, for the solution of which basic biomedical sciences have so far contributed little.

1.12.12. The "Frustrated Self-Defense Mechanisms (FSDM)" Hypothesis of Chemical Carcinogenesis

The inflammatory response is one of the most important self-defense mechanisms that our body possesses as mentioned earlier. The FSDM hypothesis of disease development as applied to chemical carcinogenesis predicts that at least some of the cancers induced by low dose exposure to chemicals over long periods of time may result from the frustrated inflammatory response.

Low-dose exposures to certain chemicals may lead to specific cancers in specific areas of the body of specific animals of specific species, strains, sex, age and nutritional states, if such cancers result from the "frustrated" inflammatory response. Such a high specificity of cancers caused by low-dose exposures to carcinogens can be predicted logically from the large variety of the internal states available to the Londonator (see Section 1.12.10). In order for pathogenic agents to cause cancers via the "frustrated" inflammatory response pathway, they must elicit two related events: (i) The activation of the inflammatory response (i.e., to activate the appropriate Londonator), and (ii) the "frustration" of the normal course of progression of the inflammatory response through the inhibition of the mechanisms that normally stop cellular proliferation after tissue repair is completed.

It is generally agreed that chemical carcinogenesis proceeds through three distinct stages: (i) Initiation, (ii) promotion and (iii) progression (Farber, 1987; Farber and Sarma, 1987; Schulte-Hermann, 1987). One of the major research goals in the field of carcinogenesis is to unravel the molecular, cellular and physiological mechanisms responsible for these three stages of carcinogenesis.

The numerous thought-provoking publications on carcinogenesis by E. Farber and his colleagues during the past two or three decades (Farber, 1982, 1987, 1990; Farber and Sarma, 1987) seem to agree with the basic tenets of the FSDM hypothesis of chemical carcinogenesis. The following are some of the specific examples of the application of the FSDM hypothesis to the questions raised by E. Farber:

(1) Why is the time-span between the initiation step and the appearance of tumor so long, often taking 1/3 or 2/3 of the normal lifespan?

One simple answer suggested by the FSDM hypothesis is that the time required for cancer to develop after initiation is determined primarily by the efficiency with which the self-defense mechanisms (at the biochemical, cellular and physiological levels) of the organisms involved are working under the given environment. In other words, the length of the latent period (LP) of carcinogenesis (measured in years), the efficiency of the self-defense mechanisms (ESD) involved (measured as a fraction between 0 and 1), and the normal life span (NLS) of the animal under investigation (measured in years) can be simply related by defining ESD as follows:

$$(ESD) = (LP)/(NLS) \quad (1.59)$$

Or $\quad (LP) = (ESD)(NLS) \quad (1.60)$

According to Equation (1.60), when ESD = 0, LP is also zero, meaning that the latent period is now solely determined by the kinetics of the uncontrolled cellular proliferation measured under the condition where no self-defense mechanisms operate. On the other hand, when ESD = 1 (perfect defense), then the latent period would be identical with the normal lifespan of the animal; that is, no carcinogenic effect. What is also clear is that LP will vary widely from the very nearly zero fraction of NLS to the whole fraction, depending upon the experimental conditions employed. The

"Biocybernetics": A Machine Theory of Biology

value of 1/3 to 2/3 for the ratio (LP)/(NLS) cited by Farber (1987) may simply reflect the nature of the animals, carcinogen doses, and promotion conditions employed.

(2) What is the mechanistic relationship between initiation and promotion?

Using the rodent liver as the experimental model, Farber and others have shown that initiation with hepatocarcinogens followed by promoting conditions (e.g., phenobarbital, partial hepatectomy) caused profound phenotypic changes of liver cells characterized by alterations of intracellular enzyme distributions (e.g., decrease in cytochrome P-450, increases in phase II enzymes, etc.) and the appearance of the ability of initiated cells to resist chemical cytotoxicity (Farber, 1982, 1990). Farber postulates that these phenotypic changes result from the expression of the evolutionarily developed adaptive functions of hepatocytes (in agreement with step 1 of Figure 1.17). Farber further noted that the majority of the so-called "resistant" cells could remodel or redifferentiate back to normal cells when the chemical environment is normalized (step 2 in Figure 1.17), while only a very small minority of them (or even single cells) can develop into full tumors (in agreement with step 5).

The FSDM hypothesis of chemical carcinogenesis supports the adaptive interpretation of the promotion phenomenon proposed by Farber and further suggests a possible mechanism for linking the initiation step and the promotion step. During the initiation step, a minority of hepatocytes might have undergone DNA modifications affecting the locus where the "adaptive genetic programs" (an example of self-defense mechanism genes, either structural (protoocogenes?) or regulatory are stored so that, during the promotion step, these cells experience "frustrations" in executing their adaptive cellular programs and consequently exhibit uncontrolled cellular proliferations.

(3) What is "Progression?"

During the lecture given at the Annual Spring Meeting of the Mid-Atlantic Chapter of the Society of Toxicology held at Princeton, N.J., on May 10, 1990, E. Farber stated in effect that, due to its complexity, practically nothing is known about the mechanisms of cancer progression. Again the FSDM hypothesis of chemical carcinogenesis is consistent with Farber's assessment of the potential complexity of the mechanisms underlying tumor progression and anticipate the great difficulty in acquiring precise knowledge about the mechanistic details of this phenomenon. Unlike Farber's hypothesis of chemical carcinogenesis, which is primarily limited to the initiation and promotion steps, however, the FSDM hypothesis appears to provide a logically coherent theoretical framework that is broad enough not only to accommodate the existing experimental data on cancer (including progression) but also order them into a meaningful system of knowledge, through the use of the eight categories of both systemic (S) and local (L) self-defense responses regulated by the N, C, E, and I control systems of the body (see Section 1.12.10).

(4) Why does the cancer incidence rate obey a power law?

There are numerous reports in the cancer literature indicating that the cancer incidence rate increases with age obeying to a power law:

$$R_C = at^b \tag{1.61}$$

where R_C is the age-specific cancer incidence rate (i.e., the fraction of the population of individuals within a given age interval who die of cancer, divided by the age interval), t is the age, and a and b are constants. From the double logarithmic plots of the cancer incidence rate against age using the cancer data of oesophagus, stomach, rectum, pancreas, prostate, and skin in males in England and Wales, Cook et al. (1969) obtained a series of straight lines with slopes (i.e., b in Equation (1.61)) that ranged from 2 to 7 among different cancers. Others have reported similar findings (Peto et al., 1975; Ames et al., 1985).

One simple explanation for the power-law dependency of Equation (1.61) can be provided by assuming that more than one self-defense mechanisms must be frustrated before cancer can develop. This would be the case if multiple independent self-defense mechanisms have evolved to protect the body against a given type of disease entity. For convenience, we may refer to this notion as the "multiple-channels" hypothesis of bio-communications or cellular communications, since diseases can be viewed as the consequences of deranging communications among cells in biological systems (see Section 1.8.11).

Let us designate the probability of frustrating the ith self-defense mechanisms at time t as $p_i(t)$. Then the joint probability, $P_{FSDM}(t, n))$, of frustrating n self-defense mechanisms simultaneously at time t can be written as

$$P_{FSDM}(t, n) = \prod_{i=1}^{n} p_i(t) \tag{1.62}$$

If all $p_i(t)$ are equal, Equation (1.62) reduces to:

$$P_{FSDM}(t, n) = (p(t))^n \tag{1.63}$$

If we assume that the probability of frustrating a given self-defense mechanism increases (or the efficiency of a given self-defence mechanism decreases) with time,

"Biocybernetics": A Machine Theory of Biology

presumably due to the linearly decreasing vitality of longevity-determinant cells (LDC) as postulated in Section 1.12.13.(5) below, we can write

$$p(t) = ft \qquad (1.64)$$

where f is a proportionality constant. Combining Equations (1.63) and (1.64) leads to

$$P_{FSDM}(t, n) = gt^n \qquad (1.65)$$

where $g = f^n$. Therefore, Equation (1.65) indicates that the joint probability of frustrating n self-defense mechanisms simultaneously by age t and thereby inducing cancer at age t is dependent on the age raised to the nth power. As already pointed out, according to the data of Cooks et al. (1969), n ranges from 2 to 7.

(5) Why does carcinogenesis occur in stages?

Equation (1.65) also provides a rational explanation for the well established phenomenon that chemically induced cancers occur in multiple stages, recognized as *initiation, promotion*, and *progression*. These multiple stages may simply reflect the fact that the set of n self-defense mechanisms that has to be frustrated simultaneously before cancer can be induced get frustrated not all at once but one (or at most two) at a time.

If we divide the total span of the time required for a chemical to induce a cancer in mammals into 3 periods for convenience, and designate the individual frustrated self-defense mechanisms involved in a cancer as $F_1, F_2, \ldots, F_7$ (the number is arbitrary here), then the pathological symptoms of each period will be determined by the distribution of these frustrated self-defense mechanism over the 3 periods. A typical example may look like this:

$$\text{Period 1} = F_1 + F_3 + F_7 \qquad (1.66)$$

$$\text{Period 2} = F_2 + F_5 \qquad (1.67)$$

$$\text{Period 3} = F_4 + F_6 \qquad (1.68)$$

By convention, these periods are called the initiation step, the promotion step, and the progression step, respectively. According to this view, the ultimate goal of cancer research would be to define and characterize the molecular, cellular, and physiological processes underlying each of these frustrated self-defense mechanisms.

1.12.13. A Molecular Theory of Aging

In the previous two sub-sections, the importance of self-defense mechanisms (SDM) viewed as machines was demonstrated in understanding the etiology of diseases in general and cancer in particular. The possible role of SDM in determining the longevity of human and other mammals is clearly revealed in Figure 1.17 by the correlation existing between the efficiency of SDM (which is thought to decrease successively as the Hanoverator makes transitions from A to N to F) and the life span of individuals (indicated with the parentheses in the Output column). The purpose of this sub-section is to describe a coherent molecular theory of aging that provides fundamental explanations for aging processes, ranging from the molecular level to the whole organismic level.

The empirical observations suggesting that the average life spans of different species of animals are genetically determined include the following: (i) There exists a good power-law relationship between the life span of different species of animals and their brain or body weight (Kohn, 1971) (Since the brain size is most likely genetically determined, so must the life span), and (ii) There is an excellent linear correlation between the logarithms of the superoxide dismutase (SOD) activities (normalized for specific metabolic rates) and the maximal life span potentials (MLSP) of primate and rodent brains and livers (Cutler, 1984).

There may be other evidence supporting the notion of genetically programmed longevity, but these two observations are sufficient for me to postulate that the average life span of animals is genetically determined. Therefore, it seems reasonable for us to invoke the concept of "programmed organismic death" in analogy to the well-known phenomenon of "programmed cell death" or apoptosis (Wyllie, 1981). In Figure 1.17, I have identified the self-defense mechanism (i.e., the Hanoverator) in the normal mode as the cellular system ultimately responsible for executing the "programmed organismic death." How might the Hanoverator regulate the longevity of individuals? I propose the following set of postulates to provide realistic mechanisms for accomplishing "programmed organismic death."

(1) <u>The Brain Contains Longevity-Determinant Cells</u>: The maximum life span potentials (MLSP) (i.e., the maximum life span observed within a given species of animals) are determined by a group of cells located in the brain, most likely in the hypothalamus in agreement with Meites et al. (1984). These will be called the "longevity-determinant cells (LDC)."

(2) <u>The Vitality of LDC is Programmed to Decrease Linearly with Time</u>: The vitality (i.e., functional capacity) of LDC is postulated to be genetically programmed to decrease linearly with time, beginning with birth.

"Biocybernetics": A Machine Theory of Biology

(3) <u>Longevity-Determinant Cells Control the Efficiency of Self-Defense Mechanisms</u>: It is postulated that LDC have complete control over the efficiency of SDM in such a way that SDM operate with linearly decreasing efficiencies.

(4) <u>The Longevity of Individual Organisms is Determined by the Intersection between the Line Reflecting the Linearly Decreasing LDC Vitality and the Background Level of Uncontrolled Free Radicals in LDC</u>: This postulate is schematically represented in Figure 1.18. The x-coordinate of the intersection between the linearly decreasing line and the horizontal life labeled $Threshold_1$ determines the longevity. The value of the x-coordinate, which determines the longevity, can be reduced in three ways: (i) By raising the height of the horizontal lines (see $Threshold_2$), (ii) by increasing the slope of the LDC vitality line (not shown), or (iii) by both of these mechanisms.

Because of the deterministically chaotic behaviors intrinsic to all gene-directed phenotypes (Section 1.8.11), it is logical to anticipate that the x coordinates of the intersections between the LDC vitality lines and the free radical threshold will be randomly distributed with respect to age. If we assume that this distribution is Gaussian, then a "square-shaped" % survival curve can be calculated as shown in Figure 1.19, in good agreement with the actual human mortality data in the literature (Lamb, 1977).

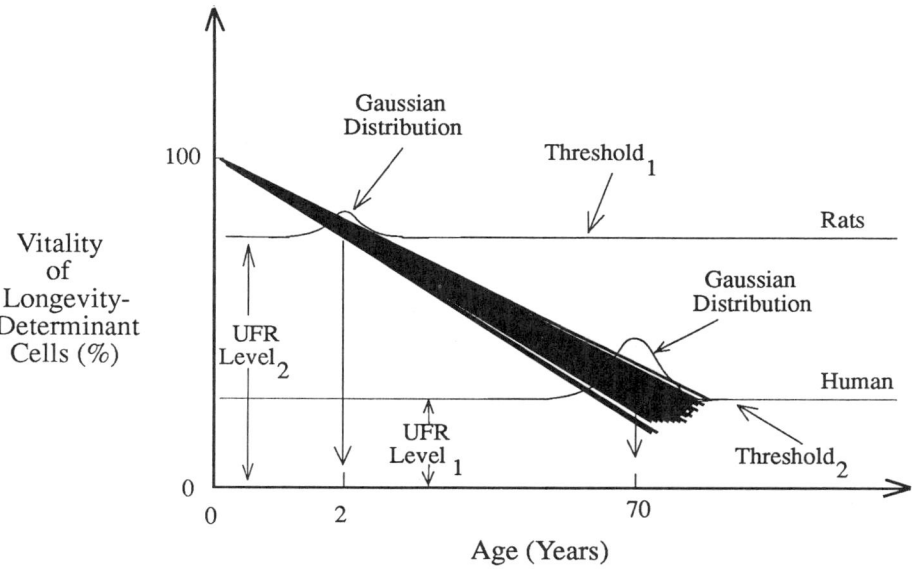

Figure 1.18. The postulated linear decrease in the vitality of the longevity-determinant cells (LDC) in the mammalian brain. The heights of $Threshold_1$ and $Threshold_2$ are thought to be determined by the background levels of uncontrolled free radicals (UFR).

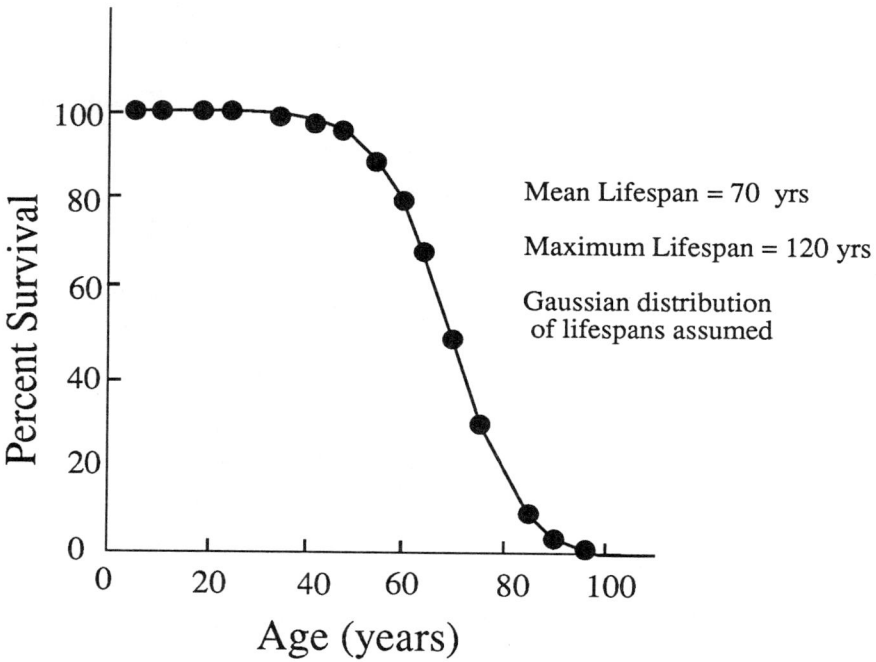

Figure 1.19. The percent survival curve calculated on the basis of the assumption that the life spans of individuals in a population are distributed normally, obeying the relation $f(t) = (1/\sigma \, (2\pi)^{-1/2}) e^{-(t-m)/2\sigma}$, where f(t) is the number of individuals with age t, m is the mean life span of the population, and σ is the standard deviation of the life span distribution.

(5) <u>The Vitality of Longevity-Determinant Cells (LDC) is Determined by the Conformational Strains Stored in DNA called Gedda Conformons</u>: The conformon hypothesis of molecular machines described in Section 1.5.2 provides a possible molecular mechanism for programming the functional capacity of LDC with respect to time. Gedda conformons are defined as the conformational strains embedded in DNA that encode the temporal information (or execute the temporal programs) of the

"Biocybernetics": A Machine Theory of Biology

cellular genome (Ji, 1985). The term "temporal information" simply indicates the genetic instructions that determine both the time at which a given gene is to be turned on and the time when the gene expression is to be turned off, the necessity of which in human genetics was first clearly recognized by Gedda (Gedda and Brenci, 1978). To make the present postulate more specific and thereby experimentally testable, the following sub-postulates are further detailed:

(i) The biological functions of longevity-determinant cells are driven by the Gedda conformons in their DNA. That is, without the Gedda conformons in DNA, LDC lose their functions.

(ii) The maximum number of the Gedda conformons that can be generated in the DNA of LDC is fixed. We will designate this number with the symbol G_0.

(iii) There are two ways of depleting DNA of Gedda conformons—(a) by driving LDC functions, namely the control of self-defense mechanisms of the body, and (b) by interaction with free radicals generated either endogenously from metabolism or exogenously due to various cellular stresses. The latter mechanism is schematically depicted in Figure 1.20.

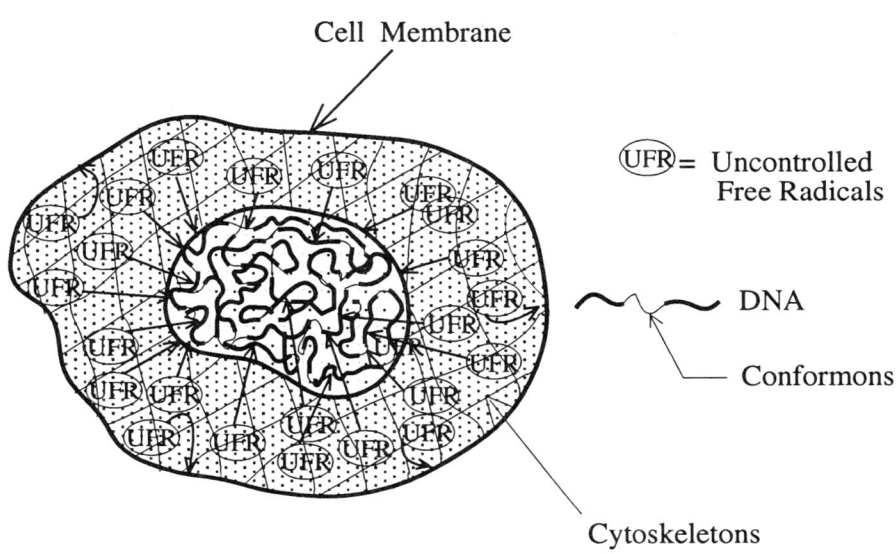

Figure 1.20. A schematic representation of the postulated dissipation of Gedda conformons stored in the DNA of longevity-determinant cells and uncontrolled free radicals (UFR) generated inside the cell. Gedda conformons are symbolized as stretched and thinned out segments of the DNA double helix strands.

(iv) The kinetic equation for the dissipation of Gedda conformons can be derived on the basis of Reaction (1.69):

$$G^* + F \xrightarrow{k} G\text{-}F + \text{Heat} \qquad (1.69)$$

where G^* is the DNA segment storing active Gedda conformons, F is free radicals, G-F is the covalently modified DNA segment that previously stored Gedda conformons, and k is the rate constant. Activated Gedda conformons (G^*) are thought to be generated from the "potential" Gedda conformons (G) by introducing conformational strains into DNA through as-yet-unidentified DNA- and ATP-dependent enzymes.

In other words, it is assumed in Reaction (1.69) that free radicals can attack only the "activated" Gedda conformons and the potential Gedda conformons are unaffected. These assumptions lead to the following kinetic expression;

$$-d(G)/dt = k(G^*)(F) \qquad (1.70)$$

where (G) is the total amount of the potential Gedda conformons remaining in DNA at time t, (G^*) the amount of the activated Gedda conformons present in DNA at time t, and (F) is the concentration of free radicals in the LDC.

If we assume that, at any time t, only a constant number, a, of Gedda conformons are activated in DNA and that the intracellular free radical concentration is kept constant at a value of b (due to the operation of appropriate self-defense mechanisms), then Equation (1.70) can be integrated to obtain

$$G = G_0 - abt \qquad (1.71)$$

which in effect states that the amount of the potential Gedda conformons remaining in DNA at t decreases linearly with time. This, I propose, is the molecular basis for the Postulate (2) made above and represented in Figure 1.18 and for explaining the well-known phenomena that numerous physiological functions of the human body decline linearly with time after maturity (Strehler, 1977).

(v) Constant b in Equation (1.71) is a function of the specific metabolic rate (SMR) (i.e., the rate of oxygen consumption divided by tissue or cell weight) of LDC,

$$b = c(SMR) \qquad (1.72)$$

where c is a constant or a mathematical function. Combining (1.71) and (1.72) leads to

$$G = G_0 - ac(SMR)t \tag{1.73}$$

If we assume that organisms die when the G value reaches a certain minimum threshold (see Figure 1.18), to be designated as G_d, the subscript d signifying death, then the maximal life span potentials (MLSP) of a species would be determined by the relation

$$G_d = G_0 - ac(SMR)(MLSP) \tag{1.74}$$

or $\quad (MLSP) = (G_0 - G_d)/ac(SMR) \tag{1.75}$

or $\quad \log (MLSP) = K - \log (SMR) \tag{1.76}$

where $K = \log ((G_0 - G_d)/ac) = $ a constant.

In other words, Equation (1.76) predicts a linear correlation with a slope of -1.00 between the logarithm of MLSP and the logarithm of SMR. This prediction was tested with the data read off from the best fitting curve in the plot of MLSP against SMR of some 50 mammalian species published by Cutler (1984). The results are plotted in Figure 1.21, which is in excellent agreement with the theoretical prediction made by Equation (1.76).

1.12.14. "Is The Mind A Computer? (And, If So, What Kind of Computer Is It?)"

Jerry Fodor of the Department of Philosophy at Rutgers gave a university-wide lecture under the above title in New Brunswick on December 4, 1989. One of his main conclusions is that rationality, the ability for the human brain to think rationally (rationality), may be one of the very few properties exhibited by the human brain that can be studied scientifically, using the Turing machine as a model. Other studies about the human brain such as psychology of intelligence, the psychology of creativity, the psychology of mental pathology, etc., although humanly of great interest, do not "seem to go very deep." The following quotation from his transcribed lecture (a copy of which was kindly sent to me by Dr. Fodor) summarizes his viewpoint:

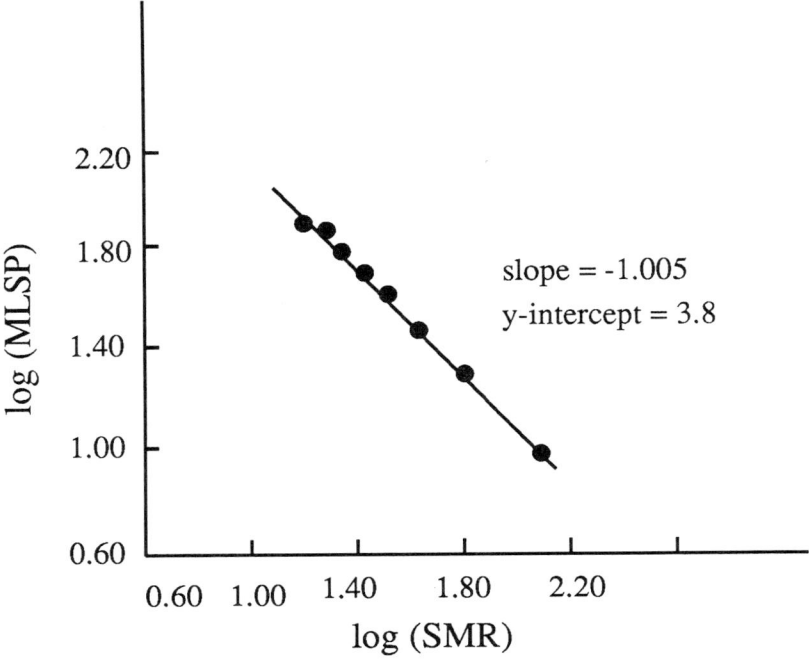

Figure 1.21. A double logarithmic plot of the maximal life span potentials (MLSP) against specific metabolic rates (SMR) of mammals. The 8 data points were read off from the best fitting curves of the MLSP vs. SMR plot published by Cutler (1984).

... nobody would bother to study the moon or the weather, except that they happen to be *our* moon and *our* weather, so we are humanly curious about them. But to expect a deep scientific theory of the Moon would be like expecting a deep scientific theory of Wisconsin. The moon and Wisconsin are just historical accidents. Happening to happen on the Moon or happening to happen in Wisconsin are, no doubt, bona fide properties that some events exhibit. But they aren't the sort of properties that give you much explanatory leverage. . . . My point is that there's every reason to believe that our brains are historical accidents too; they are just an example of what you get when you spend several millions of years piling one adventitious evolutionary adaptation on top of another. If so, then brains are just another one of those middle-sized objects that are interesting to us . . . ., but that are quite likely to prove inappropriate objects for serious scientific study. . . . The exception is Turing's proposal for, in effect, making serious science out of notions like rationality and truth. . . . and Turing has given us some reason to suppose that these properties may provide a scientific domain. The humanly plausible intuition that the function of minds is, in large part, to be rational is vindicated by a science that views mental processes as in large part computational.

The Turing machine is an abstract logical machine that models the computing capability of a general-purpose computer. It was introduced by Alan Turing in 1936 (Hopcroft and Ullman, 1979; Hopcroft, 1984; Hodges, 1983). It has three structural components: (i) A finite control, (ii) an input tape divided into squares or cells, and (iii) a tape head that can scan one cell of the tape at a time. In one move, the Turing machine executes the following set of motions, depending on the symbols scanned by the tape head and the state of the finite control:

(a) Changing the state of the finite control,

(b) Printing a symbol on the tape cell scanned, replacing what was written there, and

(c) Moving its tape head left or right by one cell.

Formally, a Turing machine (M^T) can be denoted as a 8-tuple:

$$M^T = (I^T, S^T, O^T, d^T, L, B, S_0, F) \tag{1.77}$$

where the superscript T refers to the Turing machine, I, S, O and d have already been defined in Section 1.4 and Figure 1.1, L is a finite set of allowable tape symbols, B is the blank space on the tape (an element of L), S_0 is the start state (an element of S^T), and F is a set of final states of the machine (a subset of S^T).

The Turing machine with the above characteristics is necessary and sufficient to model the computing power of the digital computer as we know it today (Hopcroft and Ullman, 1979).

To view the human brain as a machine is consonant with the general approach adopted in biocybernetics, the "machine" theory of biology (see Table 1.5). Although I fully agree with Fodor's conclusions (i) that the human brain shares a common property with artificial computers, i.e., *rationality*, and therefore (ii) that the Turing machine will provide a valuable paradigm for cognitive science, I feel that the whole scientific approach to brain research (including cognitive science) can be broadened and enriched by taking into account not only the fundamental similarities in behaviors between computers and the human brain but also their fundamental differences. The following are some of my thoughts based on "biocybernetics principles" that may be of some interest to experts in cognitive and computer sciences as they discuss various complex problems encountered in scientific research on the human brain:

(1) Computers are deterministic machines (otherwise they will not sell), while the human brain is a "deterministically chaotic machine" as its component, namely cells, are known to be (Section 1.8). Therefore, the human brain should exhibit both *deterministic* behaviors as well as *chaotic* behaviors— "deterministic" in the sense that human brain behaviors are globally constrained by the human genome, and "chaotic" in the sense that the behaviors of individual brains, although constrained by their genetic programs, are still beyond precise predictions. In other words, the human brain behaves like a "strange attractor" in chaodynamics (Tsonis and Tsonis, 1989).

(2) The Turing machine is an image of the human brain cast on the plane called "rationality." To illustrate this point, we may use the beautiful wooden blocks that decorate the cover of the book by D. Hofstadter (1980), entitled *Gödel, Escher, Bach: An Eternal Golden Braid*. The two pieces of wooden cubes hanging in air are so ingeniously carved as to cast three different sets of shadows on three mutually perpendicular planes, depending on the direction of illumination—B on the horizontal plane, EG on the right vertical plane, and GE on the left vertical plane. For convenience, I suggest that we refer to these images as the "Hofstadter images" and the wooden blocks the "Hofstadter cubes." The Hofstadter images show how an identical object can give rise to two or more seemingly "irreconcilable" images, depending on how one views that object—a clear geometric rendition of Bohr's Complementarity Principle and J. A. Wheelers' principle of observer participancy ("The Universe Looks at Itself," a Rutgers University Distinguished lecture delivered in New Brunswick on November 5, 1990). Using the Hofstadter cubes as an analogy, I suggest that the human brain has many seemingly irreconcilable images on "lower-dimensional" planes of observations. That is, the principle of causality breaks down between these images of the human brains in the sense that one image cannot be "explained" by other images. The Turing machine is one of these Hofstadter images of the human brain, but the human brain is "more" than the Turing machine. Even on the same plane of "rationality," depending on the mode of observations, the human brain may cast "rationality" images that do not overlap with the Turing images. In other words, the human brain may exhibit "rational" behaviors not "capturable" by the Turing machine metaphor.

(3) Any objective observations that can be made about the human brain behaviors should be given equal "opportunity" to contribute to our understanding of the nature of the human brain. It seems premature to select one or other methods of observations as intrinsically more "informative" than others. Such judgment may be

difficult to make in the absence of a complete knowledge about the object under investigation. Using the Hofstadter cubes again, someone might be tempted to claim that B is the "true" shape of the Hofstadter cubes because the vertical illumination is superior to the horizontal illumination.

(4) There is the question of whether or not "rationality" of the human mind that gave rise to digital computers should be viewed as the property of individual human brains, or whether it should be thought of as one of the "emergent" properties of the society of Homo sapiens, and individual brains are simply potential (not necessarily actual) "Turing machines" capable of performing the "rationality program" that has evolved because it gave a survival advantage to the human society as a whole. This "rationality program" that I am speaking of may be similar to the nest-building "genetic program" of termites whose execution is triggered only when the number of termites in a group exceeds a certain threshold (Lüscher, 1961). Such a "group" mechanism must have evolved because the termite nest is too big and too complicated in structure for one termite to handle, and hence it does not make "sense" to store such a "genetic program" within individual termites even if it is possible to do so. It makes better "sense" to store the nest-building program within a group of termites; i.e., the nest-building "genetic program" is the property of the whole termite society and not that of individual termites—an astounding "rational" act on the part of nature. Similarly, it seems possible that what we call "rationality" is actually a form of a genetic program that belongs to the human society as a whole and not to individual human brains. The human brain can be "taught" or "trained" to behave like a "Turing machine," as I am sure it can be taught to behave totally irrationally: Rationality and irrationality are different images of the human mind on the same plane of logical thoughts. If the human mind can be claimed to be rational; so can it be claimed to be irrational with an equal force. What is rational may not be the human brain per se but what the human society as a whole has produced as a group; the algebra of machines and electronics technology that is synonymous with the digital computer may not belong to individual human brains, just as the nest-building "rationality" of termite society does not belong to individual termite brains.

1.13. CONCLUDING REMARKS

When D. E. Green and I proposed the concept of conformons in 1972 in an attempt to account for oxidative phosphorylation in mitochondria in molecular terms, I did not realize how far the conformon concept would eventually propel me to go—far beyond mitochondriology and into the universal domain of information-free energy interactions underlying living processes. The writing of this chapter has provided me with an invaluable opportunity to organize my thoughts into a coherent physical theory, here called *biocybernetics*, that appears to possess the ability to account for a wide variety of phenomena in nature involving information-free energy interactions.

As evident from the nature of the various biological and biomedical problems addressed and the concrete and testable solutions suggested in this chapter, biocybernetics appears to be a physical theory that may have few antecedents in the history of science. One of the strongest indications that imbues me with confidence about the basic correctness of biocybernetics is the universality of conformons in fundamental biological processes evident in Table 1.10. It would be highly improbable for any theory to provide coherent explanations for such a wide range of biological phenomena, from enzymic catalysis to gene expression to the origin of life, had there been some basic flaws in the structure of the theory.

It is possible that conformons constitute the first member of a large family of fundamental physical entities in nature that possess both free energy and Shannon information, which I called "gnergons". Another example of gnergons is intracellular dissipative structures (IDS) that have been postulated to act as the direct or *efficient* causes for all cellular functions. One of the most unexpected consequences of biocybernetics is the possibility that the domain of validity of gnergons may not be limited to biology but extend to the origin and the evolution of our universe itself. If this conjecture proves to be correct in the future, we may witness the narrowing of the gaps that now separate physics from biology on the one hand and natural sciences from human sciences on the other through biocybernetics, the science of information and free energy in living systems.

ACKNOWLEGMENT

I have benefited greatly from stimulating discussions with my mentors, D. E. Green and B. Chance, and my various colleagues, postdocs, and students. In more recent years, I have been aided by frequent consultations with G. R. Welch, D. Kondepudi, A. C. Scott, M. Y. Han, S. D. Ray, R. Snyder, M. Iba, R. Guy, C. Gardner, D. Laskin, J. Laskin and F. Kauffman, for whose help and support I express my heartfelt thanks. My attendance at the International Seminars on the Living State-II and -III held in Bhopal (1983) and Shillong (1986), at the invitation of Professor R. K. Mishra, served as major stimuli for my theoretical work, which have led to the formulation of biocybernetics described in this chapter.

For any creative work, time is an essential ingredient. I have been very fortunate during the past nine years at Rutgers-The State University of New Jersey, for the College of Pharmacy, under the able leadership of Dean John L. Colaizzi, has provided me with an ideal environment in which to teach and carry out my theoretical research. Special thanks go to Cathy Raymore, Bernadine Chmielowicz, Vivian Gallino, and Judith Funari, the administrative staff of our Department, for their able and cheerful assistance in my teaching and research activities. Both my theoretical and experimental investigations since 1970 have been supported, directly or indirectly, by NIH (National Institutes of Health), NIAAA (National Institute of Alcohol Abuse and Alcoholism), Xerox Corporation, Max Planck Institute of Systems Physiology in Dortmund (Prof. M. Kessler), and the Department of Pharmacology and Toxicology (Dr. R. Snyder), College of Pharmacy, Rutgers University, for which I express my deep gratitude.

REFERENCES

Adair, R. K. (1987). The Great Design: Particles, Field, and Creation. Oxford University Press, New York, pp. 49-66.

Aebi, U. and Engel, J. (eds.) (1989). Cytoskeletal and Extracellular Proteins. Springer-Verlag, Berlin.

Alberts, B., Bray, D., Lewis, J., Raff, M., Roberts, K. and Watson, J. D. (1983). Molecular Biology of the Cell. Garland Publishing, Inc, New York.

Aleksandrov, A. D., Kolmogorov, A. N. and Lavrent'ev, M. A. (eds.) (1984). The Mathematics: Its Content, Methods, and Meaning. The M.I.T. Press, Cambridge, MA. Volume 3, p. 146.

Alvares, A. P. (1981). Cytochrome P-450s: Research Highlights of the Last Two Decades. Drug Metabol. Rev. 12(2): 431-436.

Allen, R. D. (1987). The Microtubule as an Intracellular Engine. Sci. Am. 256(2): 42-49.

Ames, B. N., Saul, R. L., Schwiers, E., Adelman, R. and Cathcart, R. (1985). Oxidative DNA Damage as Related to Cancer and Aging. In: Molecular Biology of Aging: Gene Stability and Gene Expression (R. S. Sohal, L. S. Birhnbaum and R. G. Cutter, Eds.). Raven, New York, pp. 137-144.

Anderson, P. W. (1983). Suggested Model for Prebiotic Evolution: The Use of Chaos. Proc. Nat. Acad. Sci. (U.S.) 80: 3386-3390.

Anderson, P. W. (1987). Computer Modeling of Prebiotic Evolution: General Theoretical Ideas on the Origin of Biological Information. Comments Mol. Cell. Biophys. 4: 99-108.

Anderson, P. W. and Stein, D. L. (1984). Broken Symmetry, Emergent Properties, Dissipative Structures, Life: Are they Related? In: Basic Notions of Condensed Matter Physics (Anderson, P. W., ed.) The Benjamin/Cummings Publishing Co., Menlo Park, CA, pp. 263-285.

Artymiuk, P. J., Blake, C. C. F., Grace, D. E. P., Oatley, S. J., Phillips, D. C. and Sternberg, M. J. E. (1979). Crystallographic Studies of the Dynamic Properties of Lysozyme. Nature 280: 563-568.

Ashby, W. R. (1964). An Introduction to Cybernetics. Methuen & Co., Ltd., London.

Atchison, M. and Adesnik, M. (1983). A Cytochrome P-450 Multigene Family. Characterization of a Gene Activated by Phenobarbital Administration. J. Biol. Chem. 258: 11285-11295.

Audrey, L. (1989). Chaos in Cancer. Medical Hypotheses 28: 143-144.

Babloyantz, A. (1986). Molecules, Dynamics, and Life: An Introduction to Self-Organization of Matter. John Wiley & Sons, New York.

Barnsley, M. F., Massopust, P., Strickland, H. and Sloan, A. D. (1987). Fractal Modeling of Biological

Structures. <u>Ann. N.Y. Acad. Sci.</u> <u>504</u>: 179-194.
Barrow, J. D. and Tipler, F. J. (1986). <u>The Anthropic Cosmological Principle</u>. Oxford University Press, New York.
Belinsky, S. A., Popp, J. A., Kauffman, F. C. and Thurman, R. G. (1984). Trypan Blue Uptake as a New Method to Study Zonal Hepatotoxicity in the Perfused Liver. <u>J. Pharmacol. Exp. Ther.</u> <u>230</u>: 755-760.
Bernstein, H. J. and Phillips, A. V. (1981). Fiber Bundles and Quantum Theory. <u>Sci. Am.</u> <u>245</u>(1): 122-137.
Bohm, D. (1980). <u>Wholeness and the Implicate Order</u>. Routledge & Kegan Paul, London.
Bohr, N. (1933). Light and Life. <u>Nature</u> <u>131</u>: 421-423.
Bohr, N. (1937). Biology and Atomic Physics. Address at the Physical and Biological Congress in Memory of Luigi Galvani, Bologna. In: <u>The Philosophical Writings of Niels Bohr, Volume II</u>. Ox Bow Press, Woodbridge, Conn. 1987, pp. 13-22.
Bohr, N. (1958). Quantum Physics and Philosophy - Causality and Complementarity. In: <u>Philosophy in the Mid-Century</u> (Klibansky, R., ed.). La Nuova Italia Editrice, Florence.
Boiteux, A., Hess, B. and Sel'kov, E. E. (1980). Creative Functions of Instability and Oscillations in Metabolic Systems. <u>Current Top. Cellular Reg.</u> <u>17</u>: 171-203.
Boyd, W. and Sheldon, H. (1979). <u>Introduction to the Study of Disease</u>, Eighth Edition. Lea & Febiger, Philadelphia. Chapter 5.
Boyer, P. D. (1977). Coupling Mechanisms in Capture, Transmission, and Use of Energy. <u>Ann. Rev. Biochem.</u> <u>46</u>: 955-966.
Briggs, J. P. and Peat, F. D. (1984). <u>Looking Glass Universe</u>. Cornerstone Library, New York. Chapter 2.
Cairns-Smith, A. G. (1982). <u>Genetic Takeover and the Mineral Origins of Life</u>. Cambridge University Press, London.
Cairns-Smith, A. G. (1985). The First Organisms. <u>Sci. Am.</u> <u>252</u>(6): 90-101.
Campbell, J. (1982). <u>Grammatical Man: Information, Entropy, Language and Life</u>. Simon & Schuster, Inc., New York, pp. 266-273.
Careri, G. (1984). <u>Order and Disorder in Matter</u>. The Benjamin/Cummings Publishing Co., Inc., Menlo Park, CA.
Careri, G. and Wyman, J. (1984). Soliton-assisted unidirectional circulation in a biochemical cycle. <u>Proc. Nat. Acad. Sci.</u> <u>81</u>: 4386-4388.
Careri, G. and Wyman, J. (1985). Unidirectional circulation in a prebiotic photochemical cycle. <u>Proc. Nat. Acad. Sci.</u> <u>82</u>: 4115-4116.
Cerutti, P. A. (1985). Prooxidant States and Tumor Promotion. <u>Science</u> <u>227</u>: 375-381.
Chaisson, E. (1987). <u>The Life Era: Cosmic Selection and Conscious Evolution</u>. W.W. Norton & Co., New York.
Chance, B., Sies, H. and Boveris, A. (1979). <u>Physiol Rev.</u> <u>59</u>: 527.
Codd, E. F. (1968). Cellular Automata. <u>Academic Press, Inc.</u>, Orlando.
Connor, J. A., Wadman, W. J., Hockberger, P. E. and Wong, R. K. S. (1988). Sustained Dendritic Gradients of Ca^{2+} Induced by Excitatory Amino Acids in CA1 Hippocampa 1 Neurons. <u>Science</u> <u>240</u>: 649-653.
Cook, P. J., Doll, R. and Fellingham, S. A. (1969). A Mathematical Model for the Age Distribution of Cancer in Man. <u>Int. J. Cancer</u> <u>4</u>: 93-112.
Coon, M. J. (1990). "Peroxides and Cytochrome P-450", a seminar given for the Joint Graduate Program of Toxicology at Rutgers University and R.W. Johnson Medical School, UMDNJ, on March 14, 1990.
Crick, F. H. C., Brenner, S., Klug, A. and Pieczenik, G. (1976). A Speculation on the Origin of Protein

Synthesis. Origin of Life 7: 389-397.
Cruzeiro, L., Halding, J., Christiansen, P. L., Skovgaard, O. and Scott, A. C. (1988). Temperature effects on the Davydov Soliton. Physical Rev. A. 37(3): 880-887.
Cullis, C. A. (1984). Environmentally induced DNA changes. In: Evolutionary Theory: Paths into the Future (J. W. Pollard, ed.). John Wiley & Sons, Inc., New York, pp. 203-216.
Cutler, R. G. (1984). Free Radicals and Aging. In: Molecular Basis of Aging (A. K. Roy and B. Chatterjee, eds.). Academic Press, Inc., New York, pp. 263-354.
Cvitanović, P. (1984). Universality in Chaos. Adam Hilger Ltd., Bristol.
Davies, P. (1982). The Edge of Infinity. Simon & Schuster, New York, p. 182.
Davies, P. (1984). Superforce: The Search for a Grand Unified Theory of Nature. Simon & Schuster, New York.
Davies, P. C. W. (1986). The Forces of Nature. Second Edition. Cambridge University Press, Cambridge.
Davydov, A. S. (1973). The Theory of Contraction of Proteins Under Their Excitation. J. theor. Biol. 38: 559-569.
Davydov, A. S. (1981). The Role of Solitons in the Energy and Electron Transfer in One-Dimensional Molecular Systems. Physica 3D: 1-22.
Delbrück, M. (1976). How Aristotle Discovered DNA. In: Physics and Our World: A Symposium in Honor of Victor F. Weisskopf (Huang, K., ed.). American Institute of Physics, New York, pp. 123-130.
Delbrück, M. (1986). Mind From Matter? Blackwell Scientific Publications, Inc., Palo Alto, CA., pp. 51-63.
Depew, D. J. and Weber, B. H. (eds.) (1985). Evolution at a Crossroads: The New Biology and the New Philosophy of Science. The M.I.T. Press, Cambridge, Mass.
D'Espagnat, B. (1989). Reality and the Physicist: Knowledge, Duration and the Quantum World. Cambridge University Press, Cambridge, pp. 7 and 119-134.
Ehrig, H. (1974). Universal Theory of Automata. B.G. Teubner, Stuttgart.
Eigen, M. (1986). The Physics of Molecular Evolution. Chemica Scripta 26B: 13-26.
Eigen, M. (1987). New Concepts for Dealing with the Evolution of Nucleic Acids. Cold Spring Harbor Symp. Quant. Biol. LII: 307-320.
Eigen, M., Winkler-Oswatitsch, R. and Dress, A. (1988). Statistical geometry in sequence space: A method of quantitative comparative reference analysis. Proc. Nat. Acad. Sci. 85: 5913-5917.
Eigen, M., Lindemann, B. F., Tietze, M., Winkler-Oswatitsch, R., Dress, A. and von Haeseler, A. (1989). How Old is the Genetic Code? Statistical Geometry of tRNA Provides an Answer. Science 244: 673-679.
Eliel, E. L., Allinger, N. L., Angyal, S. J. and Morrison, G. A. (1966). Conformational Analysis. Interscience Publishers, New York, p. 1.
Elsasser, W. M. (1958). The Physical Foundation of Biology. Pergamom Press, London.
Elsasser, W. M. (1961). Quanta and the Concept of Organismic Law. J. theor. Biol. 1: 27-58.
Elsasser, W. M. (1975). The Chief Abstractions of Biology. North-Holland Publishing Co., Amsterdam.
Elsasser, W. M. (1987). Reflections on a Theory of Organisms. Editions Orbis Publishing, Felighsburg, Quebec, Canada.
Englander, S. W., Kallenbach, N. R., Heeger, A. J., Krumhansl, J. A. and Litwin, S. (1980). Nature of the Open State in Long Polynucleotide Double Helices: Possibility of Soliton Excitations. Proc. Nat. Acad. Sci. 77: 7222-7226.
Epstein, E. R. (1987). Patterns in Space and Time Generated by Chemistry. Chemical & Engineering News, March 30, pp. 24-36.
Farber, E. (1982). Chemicals, Evolution, and Cancer Development. Am. J. Pathol. 108: 270-275.

Farber, E. (1987). Cancer Development and Its Natural History. <u>Cancer</u> <u>62</u>: 1676-1679.
Farber, E. (1990). Clonal Adaptation during Carcinogenesis. <u>Biochem. Pharmacol.</u> (in press).
Farber, E. and Sarma, D. S. R. (1987). Hepatocarcinogenesis: A Dynamic Cellular Perspective. <u>Lab. Invest.</u> <u>56</u>: 4-22.
Farmer, D., Toffoli, T. and Wolfram, S. eds. (1984). <u>Cellular Automata</u> North-Holland Physics Publishing, Amsterdam.
Ferris, J. P. (1984). The Chemistry of Lùfe's Origin. <u>Chem. Engin. News August, 27</u>: 22-35.
Field, R. J. (1985). Chemical Organization in Time and Space. <u>American Sci.</u> <u>73</u>: 142-150.
Finch, S. A. E. and Stier, A. (1988). A Perfusion Chamber for High-Resolution Light Microscopy of Cultured Cells. <u>J. Microscopy</u> <u>151</u>(1): 71-75.
Fox, S. W. (1984). Proteinoid Experiments and Evolutionary Theory. In: <u>Beyond Neo-Darwinism</u> (M. W. Ho and P. T. Saunders, eds.). Academic Press, London, pp. 15-60.
Frauenfelder, H. (1987). Function and Dynamics of Myoglobin. <u>Ann. N.Y. Acad. Sci.</u> <u>504</u>: 151-167.
Frost, A. A. and Pearson, R. G. (1961). <u>Kinetics and Mechanism</u>, Second Ed. John Wiley & Sons, Inc., New York.
Gagliardi, R. M. (1988). <u>Introduction to Communications Engineering</u>. John Wiley & Sons, New York.
Gavish, B. (1986). Molecular Dynamics and the Transient Strain Model of Enzyme Catalysis. In: <u>The Fluctuating Enzyme</u> (Welch, G. R., ed.). John Wiley & Sons, New York, pp. 263-339.
Gedda, L. and Brenci, G. (1978). <u>Chronogenetics: The Inheritance of Biological Time</u>. Charles C. Thomas Publishers, Springfield, Ill.
George, F. H. (1977). <u>The Foundations of Cybernetics</u>. Gordon and Breach Science Publishers, London.
Glansdorff, P. and Prigogine, I. (1971). <u>Thermodynamic Theory of Structure, Stability and Fluctuations</u>. (Wiley-Interscience, London).
Glashow, S. L. (1975). Quarks with Color and Flavor. <u>Sci. Am.</u> <u>233</u>: 38-50.
Glass, L. and Mackey, M. C. (1979). Pathological Conditions Resulting from lustabilities in Physiological Control Systems. <u>Ann. N.Y. Acad. Sci.</u> <u>316</u>: 214-235.
Goldstein, H. (1980). <u>Classical Mechanics</u>. Second Edition. Addison-Wesley Publishing Co., Reading, MA.
Gould, S. J. (1977). <u>Ever Since Darwin: Reflections in Natural History</u>. W.W. Norton & Co., New York, pp. 179-185.
Gould, S. J. (1982). Darwinism and the Expansion of Evolutionary Theory. <u>Science</u> 216: 380-387.
Green, D. E. (1974). A Framework of Principles for the Unification of Bioenergetics. <u>Ann. N.Y. Acad. Sci.</u> <u>227</u>: 6-45.
Green, D. E. and Ji, S. (1972). The Electromechanochemical Model of Mitochondrial Structure and Function. In: <u>The Molecular Basis of Electron Transport</u> (Schultz, J. and Cameron, B. F., eds.). Academic Press, New York, pp. 1-44.
Green, M. B. (1985). Unification of Forces and Particles in Superstring Theories. <u>Nature</u> 314: 409-414.
Gribbin, J. (1983). <u>Spacewarps</u>. Dell Publishing Co., Inc., New York.
Grivell, L. A. (1983). Mitochondrial DNA. <u>Sci. Am.</u> 248 (3): 78-89.
Gurd, F. R. N. and Rothgeb, T. M. (1979). Motions in Protein. <u>Adv. Prot. Chem.</u> <u>33</u>: 73-165.
Guth, A. H. and Steinhardt, P. J. (1985). The Inflationary Universe. <u>Sci. Am.</u> <u>250</u>(5): 116-128.
Haken, H. (1983). <u>Synergetics</u>. Springer-Verlag, Berlin. Chapters 7 through 11.
Halliwell, B. and Gutteridge, J. M. (1985). <u>Free Radicals in Biology and Medicine</u>. Clarendon Press, Oxford.
Hawking, S. W. (1988). <u>A Brief History of Time</u>. Bantam Books, Toronto.
Heichman, K. A. and Johnson, R. C. (1990). The Hin Invertasome: Protein-Mediated Joining of Distant Recombination Sites at the Enhancer. <u>Science</u> 249: 511-517.
Hernández-Cruz, A., Sala, F. and Adams, P. R. (1990). Subcellular Calcium Transients Visualized by

Confocal Microscopy in a Voltage-Clamped Vertebrate Neuron. Science 247: 858-862.
Hess, B. (1975). Energy Utilization for Control. Ciba Foundation Symposium 31 (new series). Elsevier, Amsterdam, pp. 369-392.
Hess, B. (1983). Non-Equilibrium Dynamics of Biochemical Processes. Hoppe-Seyler's Z. Physiol. Chem. 364: 1-20.
Hess, B., Boiteux, A., Busse, H. G. and Gerisch, G. (1975). Spatiotemporal Organization in Chemical and Cellular Systems. In: Membranes, Dissipative Structures and Evolution (Nicholis, G. and Lefever, R., eds.). John Wiley & Sons, New York, pp. 137-168.
Hess, B., Goldbeter, A. and Lefever, R. (1978). Temporal, Spatial, and Functional Order in Regulated Biochemical and Cellular Systems. Adv. Chem. Phys. 38: 363-413.
Ho, M. W. and Saunders, P. T. (eds.) (1984). Beyond Neo-Darwinism. Academic Press, Inc. London.
Hodges, A. (1983). Alan Turing: The Enigma. Simon and Schuster, New York.
Hofstadter, D. (1980). Gödel, Escher, Bach: An Eternal Golden Braid. Vintage Books, New York.
Holcombe, W. M. L. (1982). Algebraic Automata Theory. Cambridge University Press, U.K.
Holliday, R. (1989). A Different Kind of Inheritance. Sci. Am. 260(6): 60-73.
Hopcroft, J. W. (1984). Turing Machines. Sci. Am. May, 1984: 86-98.
Hopcroft, J. E. and Ullman, J. D. (1979). Introduction to Automata Theory, Languages, and Computations. Addison-Wesley Publishing Co., Reading, Mass.
Hunt, V. D. (1989). Superconductivity Sourcebook. John Wiley and Sons, New York.
Hyman, J. M., McLaughlin, D. W. and Scott, A. C. (1981). On Davydov's Alpha-Helix Solitons. Physica 3D: 23-44.
Jain, M. K., Berg, R. A. and Tandon, G. P. (1990). Mechanical Stress and Cellular Metabolism in Living Soft Tissue Composites. Biomaterials 11: 465-472.
Jantsch, E. (1980). The Self-Organizing Universe. Pergamon Press, Oxford.
Jencks, W. P. (1975). Binding Energy, Specificity, and Enzymic Catalysis: The Circe Effect. Adv. Enzymol. 43: 219-410.
Ji, S. (1974a). A General Theory of ATP Synthesis and Utilization. Ann. N.Y. Acad. Sci. 227: 211-226.
Ji, S. (1974b). Energy and Negentropy in Enzymic Catalysis. Ann. N.Y. Acad. Sci. 227: 419-437.
Ji, S. (1976). A Model of Oxidative Phosphorylation that Accommodates the Chemical Intermediate, Chemosmotic, Localized Proton and Conformational Hypotheses. J. theor. Biol. 59: 319-330.
Ji, S. (1977). A Possible Molecular Mechanism of Free Energy Transfer in Oxidative Phosphorylation. J. theor. Biol. 68: 607-612.
Ji, S. (1979). The Principles of Ligand-Protein Interactions and their Application to the Mechanism of Oxidative Phosphorylation. In: Structure and Function of Biomembranes (Yagi, K., ed.). Japan Scientific Societies Press, Tokyo, pp. 25-37.
Ji, S. (1985a). The Bhopalator - A Molecular Model of the Living Cell Based on the Concepts of Conformons and Dissipative Structures. J. theor. Biol. 116: 399-426.
Ji, S. (1985b). Conformons and Solitons: New Concepts in Bioenergetics. In: The Living State - II (Mishra, R. K., ed.). World Scientific Publishing Co., Singapore, pp. 563-573.
Ji, S. (1985c). The Bhopalator: A Molecular Model of the Living Cell. Asian J. Exp. Sci. 1: 1-33.
Ji, S. (1987). A General Theory of Chemical Cytotoxicity Based on a Molecular Model of the Living Cell, the Bhopalator. Arch. Toxicol. 60: 95-102.
Ji, S. (1988). Watson-Crick and Prigoginian Forms of Genetic Information. J. theor. Biol. 130: 239-245.
Ji, S. (1990). The Bhopalator—A Molecular Model of the Living Cell: New Developments. In: Molecular and Biological Physics of Living Systems (Mishra, R. K., ed.). Kluwer Academic Publishers, The Netherlands, pp. 187-214.
Ji, S. and Finette, S. (1985). Codons, Conformons and Dissipative Structures: A Molecular Theory of the

Living State. In: <u>The Living State - II</u> (R. K. Mishra, ed.). World Scientific Publishing Co., Ltd. Singapore, pp. 546-561.
Ji, S., Höper, J., Acker, H. and Kessler, M. (1978). The Effects of Low O_2 Supply in the Respiratory Activity, Reduced Pyridine Nucleotide Fluorescence, K^+ Efflux and the Surface PO_2 and PCO_2 of the Isolated, Perfused Rat Liver. <u>Adv. Exp. Med. Biol.</u> <u>94</u>: 545-552.
Ji, S., Ray, S. D. and Esterline, R. (1986). Intracellular Dissipative Structures (IDS) as Ultimate Targets of Chemical Cytotoxicity. <u>Adv. Exp. Med. Biol.</u> <u>197</u>: 871-889.
Jürgens, H., Peitgeu, H. -O. and Saupe, D. (1990). The Language of Fractals. <u>Sci. Am.</u> <u>263</u>(2): 60-67.
Kaku, M. and Trainer, J. (1987). <u>Beyond Einstein</u>. Bantam Books, Toronto.
Kalckar, H. M. (1969). <u>Biological Phosphorylations: Development of Concepts</u>. Prentice-Hall, Inc., Englewood Cliffs, N.J.
Karplus, M. (1987). Molecular Dynamics Simulations of Proteins. <u>Physics Today</u>, October, 68-72.
Kemeny, G. and Gorklany, I. M. (1974). Quantum Mechanical Model for Conformons. <u>J. theor. Biol.</u> <u>48</u>: 23-38.
Kessler, M., Höper, J. and Krumme, B. A. (1976). Monitoring of Tissue Perfusion and Cellular Function. <u>Anesthesiol.</u> <u>45</u>(2): 184-197.
Kimura, M. (1979). The Neutral Theory of Molecular Evolution. <u>Sci. Am.</u> <u>241</u>: 98-126.
Klonowski, W. and Klonowska, M. T. (1982). Biophysical Time on Macromolecular Level. <u>Biomathematics</u> <u>80</u>: 28-33.
Kohn, R. R. (1971). <u>Principles of Mammalian Aging</u>. Prentice-Hall, Inc., Englewood Cliffs, N.J., p. 140.
Kolb, E. W., Seckel, D. and Turner, M. S. (1985). The Shadow World of Superstring Theories. <u>Nature</u> <u>314</u>: 415-419.
Kondepudi, D. K. and Nelson, G. W. (1985). Weak Neutral Currents and the Origin of Biomolecular Chirality. <u>Nature</u> <u>314</u>(6010): 438-441.
Kubo, R. (1966). The Fluctuation-Dissipation Theorem. <u>Rep. Prog. Phys.</u> <u>29</u>: 255-284.
Kuhn, T. S. (1978). <u>Black-Body Theory and the Quantum Discontinuity</u>. Clarendon Press, Oxford.
Kuhn, H. (1983). Self-organization in Molecular Aggregates and Origin of Life. In: <u>Darwin Today</u>, VIII (Geissler, E., ed.). Akademie-Verlag, Berlin.
Lamb, M. J. (1977). <u>Biology of Aging</u>. John Wiley and Sons, New York, p. 35.
Lasser, A. (1983). The Mononuclear Phagocytic System: A Review. <u>Human Pathology</u> <u>14</u>(2): 108-126.
Lear, J. (1988). <u>Aristotle: The Desire to Understand</u>. Cambridge University Press, Cambridge.
Leder, P. (1981). The Genetics of Anitbody Diversity. <u>Sci. Am.</u>: 102-116.
Levi-Civita, T. (1977). <u>The Absolute Differential Calculus</u>. Dover Publications, Inc. New York.
Lewin, B. (1987). <u>Genes</u>. Third Edition. John Wiley & Sons, New York.
Lopes, J. L. (1981). <u>Gauge Field Theories: An Introduction</u>. Pergamon Press, Oxford.
Lovelock, J. E. (1979). <u>GAIA: A New Look at Life on Earth</u>. Oxford University Press, Oxford.
Lowenstein, J. M. (1985). Molecular Approaches to the Identification of Species. <u>American Scientist</u> <u>73</u>: 541-547.
Lumry, R. (1974). Conformational Mechanisms for Free Energy Transduction in Protein Systems: Old Ideas and New Facts. <u>Ann. N.Y. Acad. Sci.</u> <u>227</u>: 46-73.
Lumry, R. and Gregory, R. B. (1986). Free-Energy Management in Protein Reactions: Concepts, Complications, and Compensation. In: <u>The Fluctuating Enzyme</u> (Welch, G. R., ed.). John Wiley & Sons, New York, pp. 1-190.
Lüscher, M. (1961). Air Conditioned Termite Nests. <u>Sci. Am.</u> <u>205</u>(1): 138-145.
Mandelkow, E., Mandelkow, E. -M., Hotani, H., Hess, B. and Müller, S. C. (1989). Spatial Patterns from Oscillating Microtubules. <u>Science</u> <u>246</u>: 1291-1293.
Markus, M. and Hess, B. (1984). Transitions Between Oscillatory Modes in a Glycolytic Model System.

Proc. Nat. Acad. Sci. (U.S.) 81: 4394-4398.
May, R. M. (1976). Simple Mathematical Models With Very Complicated Dynamics. Nature 261: 459-467.
Mayr, E. (1988). Toward A New Philosophy of Biology. The Belknap Press of Harvard University Press, Cambridge, MA.
McClare, C. W. F. (1971). Chemical Machines, Maxwell's Demon and Living Organisms. J. theor. Biol. 30: 1-34.
McClare, C. W. F. (1972). A Quantum Mechanical Muscle Model. Nature 240: 88-90.
McNamee, J. E. (1990). Introduction to Fractals in Biomedical Research. Ann. Biomed. Eng. 18: 109-110.
McSwigger, J. A. and Cech, T. R. (1989). Stereochemistry of RNA Cleavage by the Tetrahymena Ribozyme and Evidence that the Chemical Step Is Not Rate-Limiting. Science 244: 679-683.
Medina, M. A. and Núñez de Castro, I. (1988). Evidence of an Intracellular Dissipative Structure. Z. Naturforsch. 43C: 793-794.
Meites, J., Hylka, V. W. and Sonntag, W. E. (1984). Cellular-Molecular Versus Neuroendocrine Concepts of Aging: A Need for Integration. In: Molecular Basis of Aging (Roy, A. K. and Chatterjee, B. eds.). Academic Press, Inc., New York, pp. 187-207.
Millington, T. A. and Millington, W. (1966). Dictionary of Mathematics. Harper & Row, New York.
Milstein, C. (1986). From Antibody Structure to Immunological Diversification of Immune Response. Science 231: 1261-1268.
Misner, C. W., Thorne, K. S. and Wheeler, J. A. (1973). Gravitation. New York: W.H. Freeman and Company.
Monod, J. (1971). Chance and Necessity. Vintage Books, New York, P. 105.
Mollenauer, L. F. and Stolen, R. H. (1982). Solitons in Optical Fibers. Laser Focus Magazine, April, pp. 193-198.
Montagu, A. (1980). Sociobiology Examined. Oxford University Press, Oxford.
Moore, W. J. (1963). Physical Chemistry, Third Ed., Prentice-Hall, Inc. Englewood Cliffs, N.J. pp. 213-228
Moriyasu, K. (1985). An Elementary Primer for Gauge Theory. World Scientific Publishing Co., Singapore.
Müller, S. C., Plesser, T. and Hess, B. (1985). The Structure of the Core of the Spiral Wave in the Belousov-Zhabotinskii Reaction. Science 230: 661-663.
Munk, A. and Guyre, P. M. (1989). Glucocorticoid Physiology and Homeostasis in Relation to Anti-Inflammatory Actions. In: Anti-Inflammatory Steroid Action: Basic and Clinical Aspects. Academic Press, Inc., New York, pp. 30-47.
Munck, A., Guyre, P. M. and Holbrook, N. J. (1984). Physiological Functions of Glucocorticoids in Stress and Their Relation to Pharmacological Actions. Endocrine Rev. 5(1): 25-44.
Murdock, D. (1989). Niels Bohr's Philosophy of Physics. Cambridge University Press, New York.
Murray, A. W. and Kirschner, M. W. (1989). Dominoes and Clocks: The Union of Two Views of the Cell Cycle. Science 246: 614-621.
Needham, G. (1988). 19th Century Realist Art. Harper & Row, Publishers, New York.
Nicolis, G. and Prigogine, I. (1977). Self-Organization in Nonequilibrium Systems: From Dissipative Structures to Order through Fluctuations. John Wiley & Sons, New York. Chapter 7.
Nicholls, D. G. (1982). Bioenergetics: An Introduction to the Chemiosmotic Theory. Academic Press, Inc., London.
Novikoff, A. B. and Holtzman, E. (1976). Cells and Organelles, Second Edition. Holt, Rinehardt and Winston, New York.
Olsen, M., Smith, H. and Scott, A. C. (1984). Solitons in a Wave Tank. Am. J. Phys. 52(9): 826-830.
Pagels, H. R. (1982). The Cosmic Code: The Quantum Physics as the Language of Nature. Simon and Schuster, New York. Chapters 3, 4 and 5.
Pardee, A. B. (1989). G_1 Events and Regulation of Cell Proliferation. Science 246: 603-608.

Parker, B. (1986). Einstein's Dream: The Search for a Unified Theory of the Universe. Plenum Press, New York.
Pauling, L. and Wilson, E. B. (1963). Introduction to Quantum Mechanics. Dover Publications, Inc., New York.
Perutz, M. F. (1987). Physics and the Riddle of Life. Nature 326: 555-558.
Peto, R., Roe, F. J. C., Lee, P. N., Levy, L. and Clark, J. (1975). Cancer and Ageing in Mice and Men. Br. J. Cancer 32: 411-426.
Pierce, J. R. (1980). Introduction to Information Theory. Second, Revised Edition. Dover Publications, Inc., New York.
Plomin, R. (1990). The Role of Inheritance in Behavior. Science 248: 183-188.
Polkinghorne, J. C. (1985). The Quantum World. Princeton University Press, Princeton, N.J.
Pollard, J. W. (ed.) (1984). Evolutionary Theory: Paths into the Future. John Wiley & Sons, New York.
Polozov, R. V. and Yakushevich, L. V. (1988). Nonlinear Waves in DNA and Regulation of Transcription. J. theor. Biol. 130: 423-430.
Pool, P. (1967). Impressionism. Thames and Hudson, Ltd., London.
Prigogine, I. (1961). Introduction to Thermodynamics of Irreversible Processes. Second Edition. Interscience Publishers, New York.
Prigogine, I. (1977). Dissipative Structures and Biological Order. Adv. Biol. Med. Phys. 16: 99-113.
Prigogine, I. (1978). Time, Structure and Fluctuations. Science 201: 777-785.
Prigogine, I. (1980). From Being to Becoming: Time and Complexity in the Physical Sciences. San Francisco: W.H. Freeman and Co.
Prigogine, I. and Allen, P. M. (1982). The Challenge of Complexity. In: Self-Organization and Dissipative Structures (Schieve, W. C. and Allen, P. M., eds.), University of Texas Press, Austin, pp. 3-39.
Prigogine, I., Nicolis, G. and Babloyantz, A. (1972). Thermodynamics of Evolution. Physics Today, December: 38-44.
Prigogine, I. and Stengers, I. (1984). Order Out of Chaos: Man's Dialogue with Nature. Toronto: Bantam Books.
Pryor, W. A. (ed.) (1976). Free Radicals in Biology. Volume 1. Academic Press, New York.
Quastler, H. (1964). The Emergence of Biological Organization. Yale University Press, New Haven.
Quigg, C. (1983). Gauge Theories of the Strong, Weak, and Electromagnetic Interactions. The Benjamin/Cummings Publishing Corp., Menlo Park, CA.
Raisbeck, G. (1964). Information Theory: An Introduction for Scientists and Engineers. Cambridge, Mass: The M.I.T. Press.
Rebbi, C. (1979). Solitons. Sci. Am. 240(2): 92-114.
Reynolds, W. L. and Lumry, R. (1966). Mechanisms of Electron Transfer. New York: Ronald Press.
Ridley, B. K. (1984). Time, Space and Things. Second Edition, Cambridge University Press, Cambridge, England.
Ross, J., Müller, S. C. and Vidal, C. (1988). Chemical Waves. Science 240: 460-465.
Rothstein, J. (1971). Informational Generalization of Entropy in Physics. In: Quantum Theory and Beyond (Bastin, T., ed.). Cambridge University Press, Cambridge, pp. 291-305.
Sawyer, D. W., Sullivan, J. A. and Mandell, G. L. (1985). Intracellular Free Calcium Localization in Neutrophils During Phagocytosis. Science 230: 663-666.
Schieve, W. C. and Allen, P. M., Eds. (1982). Self-Organization and Dissipative Structures: Applications in the Physical and Social Sciences. University of Texas Press, Austin.
Schimke, R. T. (1980). Gene Amplification and Drug Resistance. Sci. Am. 243(5): 60-69.
Schrödinger, E. (1946). What Is Life? The Physical Aspect of the Living Cell. Cambridge University Press, Cambridge.

Schulte-Hermann, R. (1987). Initiation and Promotion in Hepatocarcinogenesis. Arch. Toxicol. 60: 179-181.
Schwinger, J. (1986). Einstein's Legacy: The Unity of Space and Time. Scientific American Library, Scientific American Books, Inc., New York.
Scott, A. C. (1982). Dynamics of Davydov Solitons. Physical Rev. A. 26(1): 578-595.
Scott, A. C. (1985a). Solitons in Biological Molecules. Comments Mol. Cell. Biophys. 3(1): 15-37.
Scott, A. C. (1985b). Biological Solitons. In: Dynamical Problems in Soliton Systems (Takeno, S., ed.). Springer-Verlag, Berlin. pp. 224-235.
Searle, J. (1984). Minds, Brains and Science. Harvard University Press, Cambridge, Mass.
Segrè, E. (1980). From X-Rays to Quarks. University of California, Berkeley.
Shaitan, K. V. and Rubin, A. B. (1982). Equation of Motion of the Conformon and Primitive Molecular Machines for Election (Ion) Transport in Biological Systems. Mol. Biol. (USSR) 16(5): 1004-1018.
Shaw, R. (1981). Strange Attractors, Chaotic Behavior, and Information. Flow. Z. Naturforsch. 369: 80-112.
Simpson, G. G. (1964). This View of Life. Harcourt, Brace & World, Inc., New York.
Skulachev, V. P. and Hinkle, P. C. (1981). Chemiosmotic Proton Circuits in Biological Membranes. Addison-Wesley Publishing Company, Inc., London.
Slater, E. C. (1969). The Oxidative Phosphorylation Potential. In: The Energy Level and Metabolic Control in Mitochondria (S. Papa, J. M. Tager, E. Quagliariello and E. C. Slater, eds.). Adriatica Editrice, Bari, Italy, pp. 225-259.
Slater, E. C. (1977). Mechanism of Oxidative Phosphorylation. Ann. Rev. Biochem. 46: 1015-1026.
Slater, E. C., Benden, J. A. and Herweiger, M. A. (1985). A Hypothesis for the Mechanism of Respiratory-Chain Phosphorylation not Involving the Electrochemical Gradient of Protons as Obligatory Intermediate. Biochim. Biophys. Acta 811: 217-231.
Sobell, H. M. (1985). Actinomycin and DNA Transcription. Proc. Nat. Acad. Sci. (U.S.) 82: 5328-5331.
Sobell, H. M., Banerjee, A., Lozansky, E. D., Zhou, G. P. and Chou, K. C. (1982). The Role of Low Frequency (Acoustic) Phonons in Determining the Premelting and Melting Behaviors of DNA. Comm. Int. J. Mol. Sci. (Wuhan, China) 2(4): 113-128.
Strehler, B. L. (1977). Time, Cells, and Aging. Academic Press, Inc., New York, p. 112.
't Hooft, G. (1980). Gauge Theories of the Forces Between Elementary Particles. Sci. Am. June 104-138.
Talbot, M. (1986). Beyond the Quantum. Macmillan Publishing Company, New York.
Tank, D. W., Sugimori, M., Connor, J. A. and Llinás, R. R. (1988). Spatially Resolved Calcium Dynamics of Mammalian Purkinje Cells in Cerebellar Slice. Science 242: 773-777.
Taylor, E. F. and Wheeler, J. A. (1966). Spacetime Physics. W.H. Freeman and Co., San Francisco.
Thurman, R. and Kauffman, F. C. (1980). Factors Regulating Drug Metabolism in Intact Hepatocytes. Pharmacol. Reviews 31(4): 229-251.
Tribus, M. and McIrvine, E. C. (1971). Energy and Information. Sci. Am. 225: 179-188.
Tsonis, P. A. and Tsonis, A. A. (1989). Chaos: Principles and Implications in Biology. CABIOS 5(1): 27-32.
Tzagoloff, A. (1982). Mitochondria. Plenum Press, New York.
Ueda, T., Matsumoto, K., Akitaya, T. and Kobatake, Y. (1986). Spatial and Temporal Organization of Intracellular Adenine Nucleotides and Cyclic Nucleotides in Relation to Rhythmic Motility in Physarium Plasmodium. Exp. Cell Res. 162: 486-494.
Vane, J. R. (1979). Prostaglandins, Pain and Aspirin. In: Pain & Prostaglandins: New Clinical Perspectives (J. R. Vane, G. Weissmann, and R. B. Zurier, eds.). Burroughs Wellcome Co., Research Triangle Park, N.C.
Volkenstein, M. V. (1972). The Conformon. J. theor. Biol. 34: 193-195.
Volkenstein, M. V. (1986). Electronic-Conformational Interaction in Biopolymers. In: The Fluctuating

Enzyme (Welch, G. R., ed.). John Wiley and Sons, New York, pp. 403-419.
Vol'kenshtein, M. V., Dogonadze, R. R., Madumarov, A. K., Urushadze, Z. D., and Kharkats, Y. I. (1972). The Theory of Enzyme Catalysis. Molek. Biologiya 6: 431-439, (Translated from Russian).
Wald, R. M. (1977). Space, Time and Gravity. The University of Chicago Press, Chicago, p. 3.
Waldrop, M. M. (1983). The New Inflationary Universe. Science 219: 375-377.
Wang, J. C. (1985). DNA Topoisomerases. Ann. Rev. Biochem. 54: 665-697.
Watson, J. D., Hopkins, N. H., Roberts, J. W., Steitz, J. A. and Weiner, A. M. (1987). Molecular Biology of the Gene. The Benjamin/Cummings Publishing Co., Menlo Park, CA.
Welch, G. R. (1977). On the Role of Organized Multienzyme Systems in Cellular Metabolism: A General Synthesis. Prog. Biophys. Mol. Biol. 32: 103-191.
Welch, G. R. and Keleti, T. (1981). On the "Cytosociology" of Enzyme Action in vivo: A Novel Thermodynamic Correlate of Biological Evolution. J. theor. Biol. 93: 701-735.
Welch, G. R. and Kell, D. B. (1986). Not Just Catalysts - Molecular Machines in Bioenergetics. In: The Fluctuating Enzyme (Welch, G. R., ed.). John Wiley & Sons, New York, pp. 452-492.
West, B. J. (1990). Physiology in Fractal Dimensions: Error Tolerance. Ann. Biomed. Engin. 18: 135-149.
West, B. J. and Goldberger, A. L. (1987). Physiology in Fractal Dimension. American Scientist. 75: 354-365.
Westerhoff, H. V., Melandri, B. A., Venturoli, G., Azzone, G. F. and Kell, D. B. (1984). Mosaic Proton Coupling Hypothesis for Free Energy Transduction. FEBS Lett. 165: 1-5.
Wheeler, J. A. (1990). The Universe Looks at Itself. A Rutgers University Distinguished Lecture delivered in New Brunswick, NJ, on November 5, 1990.
Wicker, J. S. (1987). Entropy and Information: Suggestions for Common Language. Phil Sci. 54: 176-193.
Wiener, N. (1948). Cybernetics: Or Control and Communication in the Animal and the Machine. The M.I.T. Press, Cambridge, MA.
Wier, W. G., Cannell, M. B., Berlin, J. R., Marban, E. and Lederer, W. J. (1987). Cellular and Subcellular Heterogeneity of $[Ca^{2+}]_i$ in Single Heart Cells Revealed by Fura-2. Science 235: 325-328.
Williams, R. J. P. (1961). Possible Functions of Chains of Catalysts. J. theor. Biol. 1: 1-17.
Wilson, A. C., Cann, R. L., Carr, S. M., George, M., Gyllensten, U. B., Helm-Bychowski, K. M., Higuchi, R. G., Palumbi, S. R., Prager, E. M., Sage, R. D. and Stoneking, M. (1985). Mitochondrial DNA and Two Perspectives on Evolutionary Genetics. Biol. J. Linn. Soc. 26: 375-400.
Wilson, E. O. (1982). Sociobiology: The New Synthesis. The Belknap Press of Harvard University Press, Cambridge, MA.
Winfree, A. T. (1980). The Geometry of Biological Time. Springer-Verlag, Berlin.
Winfree, A. T. (1985). Organizing Centers for Chemical Waves in Two and Three Dimensions. In: Oscillations and Travelling Waves in Chemical Systems (Field, J. and Burger, M., eds.). John Wiley and Sons, New York.
Wojciechowski, J. A. (1987). Computing Systems in an Ecology of Knowledge Perspective. Future Computing Systems 1(3): 1-19.
Wyllie, A. H. (1981). Cell Death: A new Classification Separating Apoptosis from Necrosis. In: Cell Death in Biology and Pathology (I. D. Bowen and R. A. Lockshin, eds.). Chapman and Hall, London, pp. 9-34.
Yang, C. N. (1977). Magnetic Monopoles, Fiber Bundles, and Gauge Fields. Ann. N.Y. Acad. Sci. 294: 86-97.
Zee, A. (1986). Fearful Symmetry. MacMillan Publishing Co., New York.
Zgierski, M. Z. (1975). Mechanisms of Conformon Creation in α-Helical Structures. J. theor. Biol. 55: 95-106.

APPENDICES

APPENDIX 1.A

The Law of Requisite Variety

"When a system (a machine) is influenced by its environment in a dominating manner (i.e., the environment can affect the machine but the machine cannot influence its environment to any significant extent), the only way for the machine to reduce the degree of the influence from its environment is to increase the variety of its internal states." As is often said, "Only variety can destroy variety."

This law can be expressed quantitatively as follows; if we designate the variety of the environment (i.e., the number of different environmental conditions) with the symbol V_E, and the variety of the internal states of the machine with V_M, then the variety of the outcomes (V_O) of the interactions between the machine and its environment cannot be less than V_E/V_M; i.e.,

$$V_O \geq V_E/V_M \qquad (1.A1)$$

Clearly, this equation indicates that the only way to decrease V_O under the condition of constant V_E is to increase V_M. For a more detailed discussion, see W.R. Ashby (1964).

Examples of the Application of the Law of Requisite Variety: (1) Without a thermostatic system in a home, the variety of the indoor temperature would be identical to the variety of the outdoor temperature. To reduce the variety of the indoor temperature so that it remains within a narrow temperature range despite widely fluctuating outdoor temperatures, the home-heating system (including the thermostat) must be able to exist in a number of different internal states (i.e., the thermostat settings). (2) In order for a computer to carry out the same program (i.e., accomplish the same outcome of performing correct execution of a given program) under a variety of input conditions, the computer must possess a large variety of internal states (or configurations) in which it can exist. That is, the more complex the internal connections of a computer, the greater its capacity to compute. Another equivalent statement would be that simple computers cannot carry out complex computations, which we know to be true.

APPENDIX 1.B

Nucleic Acid and Protein Communication Channels

As shown in Figure 1.A1 below, it is postulated that there exists a mechanistic analogy between the coding of amino acids via triplets of nucleotides by biological evolution and the coding of rate constants of enzymes via a fixed number of amino acids, spatio-temporally organized within an enzymic active site, at the time of a catalytic act.

The number, X, of the amino acid residues that constitute a conformon can be estimated as follows. The maximum amount of the free energy contained in one conformon can be assumed to be comparable to that of the free energy of hydrolysis of ATP, namely 16 Kcal/mol = 2.66×10^{-20} cal = 1.11×10^{-19} joules (Slater, 1969). Since the minimum amount of energy required to transmit one bit of information is known to be 3.0×10^{-21} joules/bit at 37 °C (Pierce, 1980), the maximum amount of the information that can be transmitted via one conformon is

$$(1.11 \times 10^{-19} \text{ joules/conformon})/(3.0 \times 10^{21} \text{ joules/bit}) = 37 \text{ bits/conformon} \tag{1.A2}$$

Furthermore, since one conformon is formed by selecting x amino acids from a pool of 20 amino acids (Figure 1.A1), we can estimate the maximum amount of information that can be transmitted by one conformon, using the simplified version of Shannon's formula (Raisebeck, 1964; Pierce, 1980):

$$I = \log_2 20^x = x \log_2 20 = 4.33 \, x \quad \text{bits/conformon} \tag{1.A3}$$

By equating Equations (1.A2) and (1.A3), we obtain x = 8.5, or approximately 9, a number consistent with the knowledge that it requires at least 3 contact points between an enzyme and a substrate for optical isomers to be discriminated at the active site.

It appears reasonable to postulate that the living cell embodies two kinds of molecular communication systems—(1) the nucleotide (N) communication system whose symbols are made up of nucleotide triplets encoding 20 amino acids, and the protein (P) communication system utilizing a set of 3 - 9 amino acids to code for rate constants of enzymic reactions (Figure 1.A1). Because of the fact that the information retrieval from nucleic acids in the modern-day living cell depends on the catalytic functions of enzymes (Alberts et al., 1983), it may be concluded that the rate

"Biocybernetics": A Machine Theory of Biology

of information retrieval from the N system (driven by conformons embedded in DNA, namely DNA conformons) must be equal to the rate of information flow through the P system (driven by conformons embedded in proteins, namely protein conformons). If we now assume that the rates of information flow through the N and P systems can be estimated using Shannon's channel capacity equation (Pierce, 1980), taking into account the fact that the band width (i.e., the degree of freedom of choosing symbols from the source of information) is 3 out of 4 for the N system and 9 out of 20 for the P system, we can generate the following pair of equations;

$$C_N = W_N \log_2 (1 + p_N/n_N) \tag{1.A4}$$

$$C_P = W_P \log_2 (1 + p_P/n_P) \tag{1.A5}$$

where the subscripts N and P stand for the nucleotide and protein communication systems, respectively, C is the channel capacity specifying the rate of information transmission in bits/second or bits/symbol from an information source, W is the band width, p is the power (i.e., the rate of energy dissipation) of the signal being transmitted, and n stands for the power of noise.

Figure 1.A1. Postulated analogy between codons and conformons.

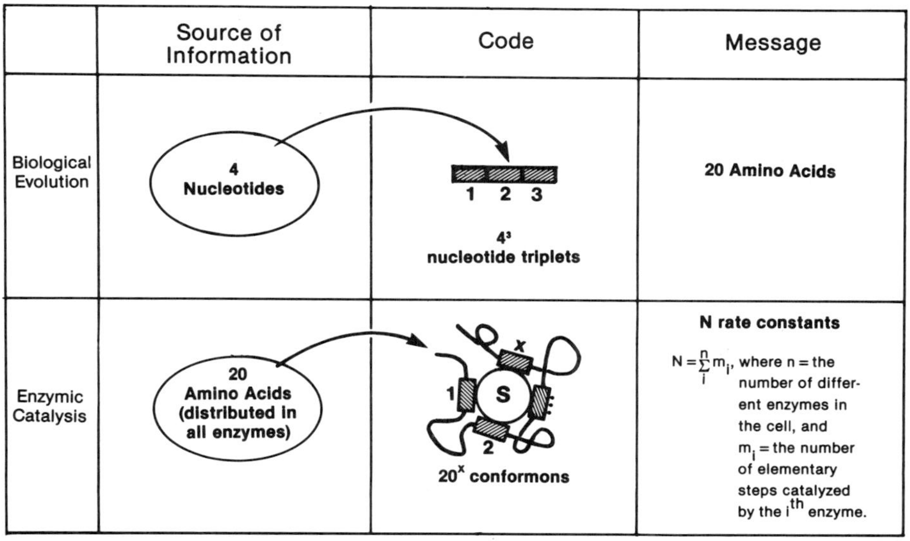

	Source of Information	Code	Message
Biological Evolution	4 Nucleotides	1 2 3 4^3 nucleotide triplets	20 Amino Acids
Enzymic Catalysis	20 Amino Acids (distributed in all enzymes)	20^x conformons	**N rate constants** $N = \sum_{i}^{n} m_i$, where n = the number of different enzymes in the cell, and m_i = the number of elementary steps catalyzed by the ith enzyme.

If we now assume that $W_N = \log_2 4^3 = 6$ bits/symbol, and $W_P = \log_2 20^9 = 37$ bits/symbol and that $C_N = C_P$, then we can combine Equations (1.A4) and (1.A5) to obtain the following expression relating the rates of energy dissipation of the N and P systems;

$$p_N/n_N = (1 + p_P/n_P)^{6.17} - 1 \qquad (1.A6)$$

Since we can estimate that n_N and n_P are in the order of RT = 0.62 Kcal/mol at 37°C, we find that $p_N = 4.8 \times 10^8$ Kcal/mol when $p_P = 16$ Kcal/mol, the free energy of the hydrolysis of ATP. Therefore, the ratio, p_N/p_P, is 3×10^7. This indicates that the operation of the nucleotide communication system requires a much greater free energy dissipation than the protein communication system, by a factor of about 10^7.

APPENDIX 1.C

The Conformon-Based Model of the Origin of Life—the "Princetonator"

The model is shown schematically in Figure 1.A2 and consists of the following key components;

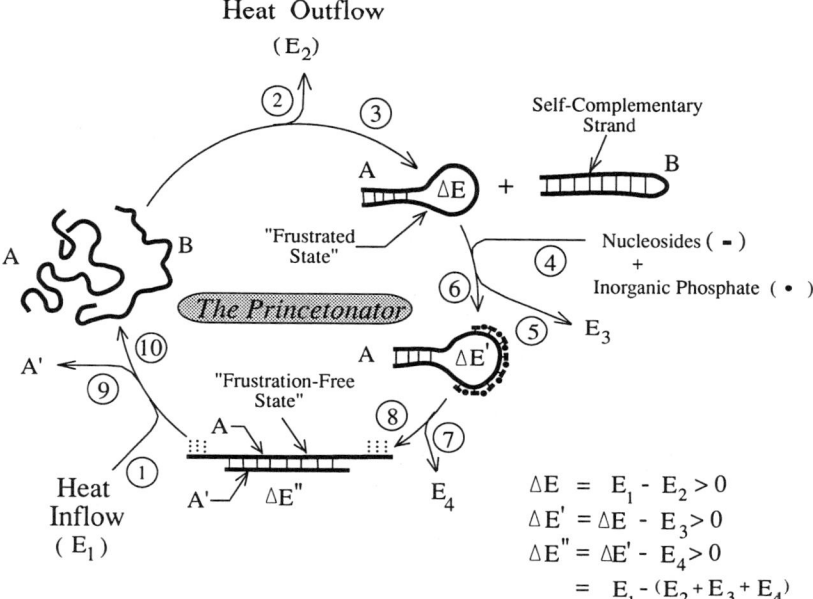

Figure 1.A2. A model of the origin of biological information—the Princetonator.

(1) The primordial soup contains two kinds of biopolymers designated as A and B and is subjected to the "thermal cycle," i.e. the periodic changes in temperature due to the daily rotation of the earth around its axis. This leads to a flux of solar energy through the earth (see steps 1 and 2) and a storage of the balance of the input (E_1) and output free energies (E_2) in the biosphere.

(2) A and B "crystalize" when the temperature of the earth surface drops at night (see step 3). Due to the sequence differences, more "frustrations" are formed in A than in B, each frustration storing "conformational strain" free energy of ΔE.

(3) Nucleosides (symbolized as -) and inorganic phosphate (symbolized as dots) bind to the exposed surface of the frustrated segments of A, driven by a partial dissipation of free energy (E_3) into heat (see steps 4, 5 and 6).

(4) The free energy, $\Delta E'$, drives the polymerization of the bound nucleotides and inorganic phosphate, thus replicating a sizable segment of A and generating heat, E_4 (see steps 7 and 8).

(5) As the temperature of the biosphere rises again with the approach of the daytime, the double stranded biopolymers separate due to thermal fluctuations, thus regenerating the original template, A.

(6) The partial replicates of A now combine to form complete replicates, through the repetition of steps (1) through (5).

Clearly, the Princetonator provides a theoretically feasible molecular mechanism for biopolymer self-replication in the primordial soup that is driven solely by the solar radiation. The free energy and the "genetic" information necessary and sufficient for the self-replication is supplied by the frustrations embedded in "crystallized" biopolymers, and such frustrations will be referred to as the "Anderson conformon" to indicate the close theoretical connection between the concept of frustrations that Anderson (1983) introduced into biophysics and that of conformons (Green and Ji, 1972) utilized to explain enzymic catalysis and gene expression (Ji, 1974b, 1979, 1985, 1990).

APPENDIX 1.D

Cell Death Kinetics

The kinetics of light-induced cell death was studied using a "cell perifusion chamber" recently developed in the laboratory of A. Stier in Göttingen (Finch et al., 1988). Hepatocytes were isolated from rats according to routine procedure and were allowed to adhere onto a coverslip which was coated with either fetal calf serum or purified collagens overnight at 37° C. Cells readily attached themselves to the coverslip upon incubation for 15 minutes in the perifusion medium consisting of NaCl

(80 mM), KCl (40 mM), HEPES (50 mM), CaCl (1 mM), $MgSO_4$ (2 mM), Na_2HPO_4 (1 mM), pH 7.4, equilibrated with 95% O_2, and 5% CO_2 at 37° C. The medium was pumped through the chamber (20 mm x 8 mm x 0.1 mm in dimension = 16 µl) at a flow rate of 10 µl/min using a precision pump (B. Braun Melsungen, Type 871022, Göttingen, W. Germany). To visualize dead cells, trypan blue was added in the perifusion medium at a concentration of 0.1% by weight. The perifusion chamber was observed with an inverted Zeiss microscope at a magnification of 160. A field of 100 - 130 hepatocytes was selected and individual cells were identified with numbers. The time at which dead cells were stained with trypan blue was recorded over a period of 4 to 12 hours. Relative to the experimental period, the time required for trypan blue to stain the nucleus of dead cells was short—less than one minute. The temperature of the perifusion chamber was kept constant at 36° - 37° C with a Zeiss hot-air blower which was automatically controlled with a thermocouple sensor. The result presented below is a part of a series of the cell-death experiments carried out by the author in A. Stier's laboratory in the summer of 1986 in collaboration with Drs. S. Finch and K. Jungermann.

Typical results of cell death kinetics are shown in Figure 1.A3. The control cell population was illuminated with visible light (from a Zeiss microscope illumination power set at 5 V) only during the short time period needed to count dead cells (less than 3 - 5 minutes) and otherwise kept in the dark. In contrast, the test cell population was exposed to the same light throughout the experiment. The cells that died before a given time point were counted and plotted on the y axis against time. The control experiment commenced at 12:40 in the afternoon, and the test experiment at 1:00 PM. The light-exposed cells began to die within 2 hours with 50% of the cells dead in 4.0 hours. The control cells began to die within about 4 hours, and 50% of the control cells died in about 6.5 hours. One of the major effects of light exposure, therefore, is the shortening of both the lagtime (from 4.0 to 2.0 hours) and the time at which 50% of cells dies. We may call the time required for a toxic agent to kill 50% of a population of organisms the "lethal time 50 (LT_{50})" in analogy to LD_{50}, the dose of toxic agents that is required to kill 50% of a population of organisms. Toxic agents are expected to decrease both LD_{50} and LT_{50} in general. The molecular mechanisms underlying the light-induced cell death are unknown at present. It is possible that light-induced generation of free radicals (via photodynamic actions of endogenous aromatic compounds) are responsible at least in part in deranging cellular control of metabolic processes, which leads to cell death.

Another interesting effect of light exposure is the increase in the rate of cell death, as indicated by the rise in the slope of the steepest portion of the curves shown in Figure 1.A3. We may refer to such a plot as the "cell-death kinetics plot." The slope of a cell-death kinetics plot also reflects the spread of the lifetimes of individual cells; the steeper the slope, the narrower the spread of individual cell lifetimes; the smaller the slope, the greater the spread of the lifetimes of individual cells. Therefore, we can speak of the "spread of cell lifetimes" as an important parameter of a cell population that is affected by toxic agents. We may conveniently designate this parameter with the symbol, S_{clt}. Experimentally, we may define S_{clt} as the difference in the x coordinate values of the two points generated by the intersections between the line passing through the most points on the cell-death kinetics plot and the x axis, on the one hand, and the horizontal line whose y coordinate is the total number of cells in the population under investigation, on the other. From Figure 1.A3, the S_{clt} = 2.0 hr for the light-exposed cell population, and S_{clt} = 4 - 5 hr for the control cell population. In experiments with cultured hepatocytes exposed to acetaminophen without continuous light exposure, it was found that S_{clt} decreased from 9.0 hours to 4.5 hours by 5 mM acetaminophen (S.Ji, S. Finch, A. Stier and K. Jungermann, 1986, unpublished observation). Therefore, in both types of experiments, the toxic agents caused the S_{clt} values to decrease by a factor of about 2.

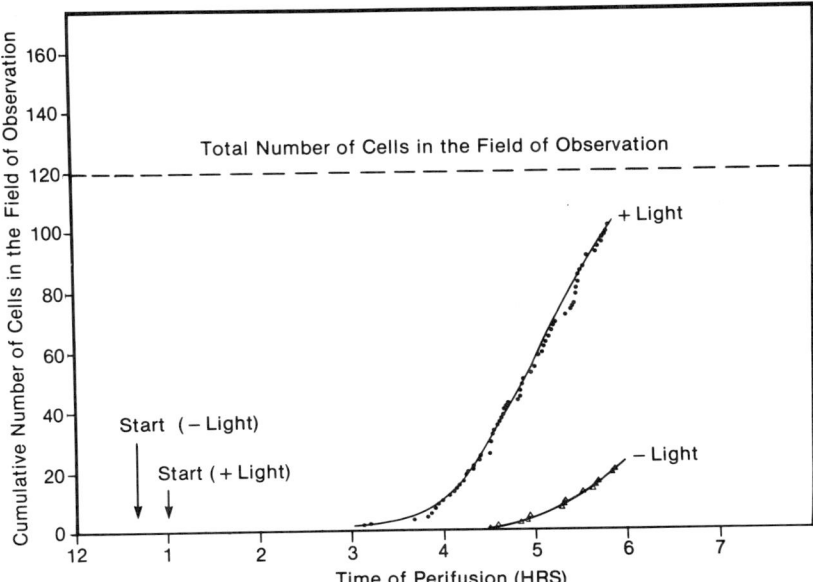

Figure 1.A3. A cell death kinetics plot. See text for detailed account.

APPENDIX 1.E

The "Xenogene" Hypothesis

The purpose of this hypothesis is to provide a testable mechanism responsible for xenobiotic-induced synthesis of cytochromes P-450 in the liver cell. The following are the key assumptions constituting the proposed mechanism:

(1) Cells have evolved to utilize the free radical chemistry (as compared to what may be called the "paired-electron" chemistry) to get rid of lipophilic xenobiotics that invade the intracellular space.

(2) The free radical chemistry is the only way for cells to metabolize evolutionarily unfamiliar chemicals (i.e., new chemicals that have not been around long enough for cells to have adapted to), because free radicals can initiate covalent rearrangements of numerous compounds "at a distance," i.e., without enzymes coming into close contact with them. In normal enzyme catalysis, the tight binding of a substrate by an enzyme is a prerequisite for the subsequent catalysis, presumably because a part of the binding free energy conserved in conformational strains of the enzyme-substrate complex (i.e., the Volkenstein-Jencks conformons in Table 1.10) is necessary to lower the activation free energy barrier for catalysis. This is the so-called Circe effect (Jencks, 1975).

(3) The price that the cell must pay for employing the free radical-mediated metabolic mechanisms is the loss of control over the behavior of the free radical intermediates formed, leading to potential interference of intracellular metabolism, often leading to cell injury and cancer (Ji, 1990).

(4) There are nine key steps in the proposed mechanism (see Figure 1.A4);

(i) Lipophilic xenobiotics, XH_2, that penetrates the cell membrane is entrapped by the endoplasmic reticulum due to the liphophilicity of XH_2 (see step 1 in Figure 1.A4).

(ii) XH_2 causes conformational perturbation and "uncouples" the R and E sites in cytochromes P-450 (step 2).

(iii) The conformationally perturbed and "uncoupled" cytochromes P-450 produce superoxide anion, $O_2^{-\cdot}$ which may diffuse over a varying distance from the center of the heme depending on the nature of the microenvironment (step 3).

(iv) Conversion of O_2^- into $HO\cdot$ (step 4).

(v) Partially enzymatic or nonenzymatic interaction between $HO\cdot$ and XH_2 to produce $XH\cdot$ (step 5).

(vi) Partial conversion of $XH\cdot$ to oxidized products for excretion (step 6).

(vii) Covalent attachment of $XH\cdot$ to a hypothetical binding site on the DNA or DNA/protein complex that contains the genes encoding cytochrome P-450

"Biocybernetics": A Machine Theory of Biology

isozymes. The set of genes encoding the enzymes catalyzing various intracellular processes that are needed for removing xenobiotics from the intracellular milieu (e.g. phase I and phase II enzymes, cell membrane transporters, etc.) are called the "xenogenes" and the sites on the xenogenes that "recognize" xenobiotics will be called the "X-recognition site" (step 7).

(viii) X-induced synthesis of m-RNA for enzymes including cytochromes P-450 that have specific affirmities for XH_2 (step 9).

(ix) Synthesis of enzymes directed by the m-RNA (step 9).

(x) It is assumed that the cell possess the mechanisms by which it expresses only those genes within its genome that are best *fit* to remove XH_2. In the absence of any such genes, the cell removes xenobiotics nonspecifically, utilizing steps 1 through 6 only. Also, during the early phase of xenobiotic invasion, the cell depends on the nonspecific mechanism of removing xenobiotics, until the cell adapts, namely replaces the nonspecific mechanisms with more specific mechanisms available within the repertoire of the "xenogenes" (see Section 1.12.7).

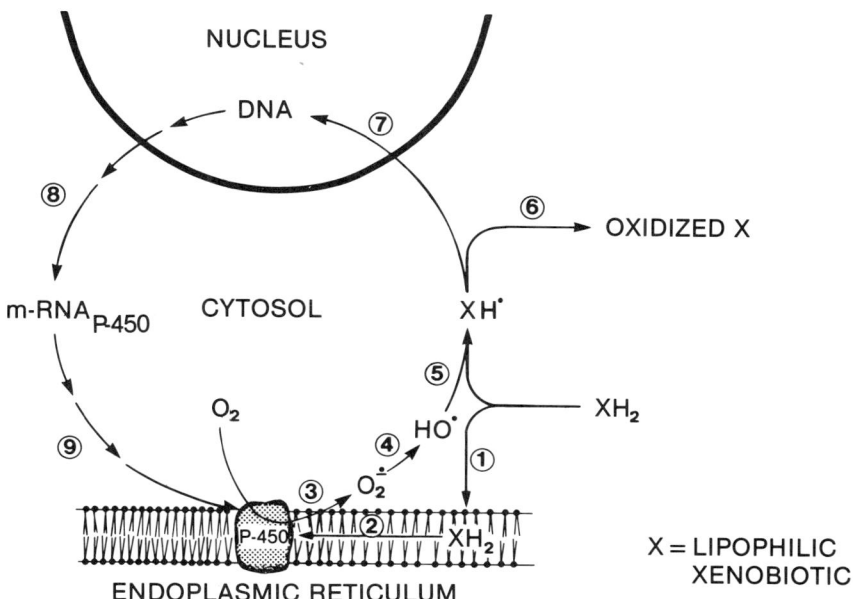

Figure 1.A4. Hypothetical mechanism of xenobiotics—induced synthesis of cytochromes P-450. See text for details.

APPENDIX 1.F

The Shillongator—the Abstract Version

(1) <u>The Bohr-Elsasser Theorem</u>: The major new insight that has resulted from my recent encounter with the theoretical publications of W. M. Elsasser (1961, 1975, 1987) is that the laws of physics and chemistry, although necessary, may be insufficient to account for living systems and processes completely. Elsasser based his conclusion on the argument that the laws of physics and chemistry depend, for their mathematical derivations, on the assumption that the size of the classes (also called sets in mathematics or ensembles in physics) of objects under consideration is infinite; i.e., these laws are strictly applicable to *infinite classes*. In comparison, the size of the classes of living organisms studied in biology is invariably finite and very small, and therefore the usual mathematical equations which converge only at infinities break down when applied to *finite classes*. It was N. Bohr (1933) who first pointed out that the laws of physics and chemistry may be insufficient to completely account for the phenomenon of life, just as the laws of the Newtonian mechanics were insufficient to explain the structure of the hydrogen atom. As is well known, only with the development of quantum mechanics was the stability of the hydrogen atom completely accounted for. Bohr therefore suggested that it might be necessary for biologists to accept life as a given in analogy to physicists who accept the quantum of action as an irreducible fact of nature. It is my opinion that the major significance of the theoretical results of Elsasser is to have provided a logical foundation for the validity of Bohr's idea about the nature of life. For this reason, it may be legitimate to refer to the important conclusion reached by Bohr and Elsasser as the "Bohr-Elsasser Incompleteness Theorem of Physicochemical Explanations of Life" or simply as the "Bohr-Elsasser theorem."

(2) The formulation of the Bohr-Elsasser theorem reinforced my conviction that biology implicates a set of fundamental laws and principles not derivable from the existing laws of physics and chemistry and therefore indirectly supports the biology-derived concepts such as gnergy and the cell force (Section 1.8.9.). Before reading the publications of Elsasser, I had assumed, like many others, that the biological properties somehow emerged from the nonliving state of matter spontaneously around 2-3 billion years ago—i.e., the history of the biological properties need not be assumed to go back more than 2-3 billion years. Although I cannot logically exclude this possibility as yet, Elsasser's writings and Bohm's theory of the implicate order (Bohm, 1980) influenced my thinking in the direction of seriously entertaining the alternative possibility that the biological properties of the

universe may trace their origin all the way back to the beginning of the universe and that the emergence of life on this planet is the consequence of "unfolding" and manifesting the properties of the primeval substance of the universe, just as the age-dependent properties of embryos can be viewed as the manifestations of the key developmental programs stored in fertilized eggs. This line of thinking eventually led me to postulate that the universe originated from gnergy, the primeval substance thought to be composed of a complementary (in the sense of Bohr (1933)) union of four essential entities, namely *energy*, *matter*, *life*, and *information*. In analogy to the wave-particle duality of light, I propose to use the term "energy-matter-life-information tetrahedrality of gnergy" to indicate the notion that gnergy is neither energy nor life nor matter nor information but can manifest such properties or entities under right set of physical conditions. Geometrically, we can represent the tetrahedrality as a tetrahedron. The tetrahedron with the four apeces occupied by energy, matter, life and information will be referred to as the "gnergy tetrahedron."

(3) Before presenting the detailed description of the new version of the Shillongator, it may be useful to list the major postulates underlying it. These postulates are best formulated within the context of the Aristotelian epistemology, according to which full understanding of any object in nature is possible only when four different causes are clearly defined—the material, efficient, formal, and final causes (Barrow and Tipler, 1986).

(i) <u>The Material Cause</u>: The universe evolved out of the gnergy tetrahedron, the postulated primeval substance composed of a complementary union of four different entities—energy, matter, life, and information. What is novel here is the introduction of the concept of "tetrahedrality" as a logical extension of the "duality" of light in respect to its wave and particle characteristics. Although some of the ancient and modern models of the universe (Barrow and Tipler, 1986) assume that the universe is composed of one or more of the same set of entities postulated here, what distinguishes the Shillongator from these models is the idea that the primeval substance of the universe is neither energy nor matter nor life nor information but a complementary union of all of these.

(ii) <u>The Efficient Cause</u>: Different properties of the gnergy tetrahedron, including the emergence of life on this planet, are thought to be manifested as the result of a series of spatiotemporal symmetry breakings undergone by the aging and cooling universe. As was discussed in Section 1.11.2, life is postulated to be associated with "dissipative symmetry breakings" in addition to "equilibrium symmetry breakings" commonly occurring in nonliving processes such as various phase transitions induced by cooling. The concept of dissipative symmetry breakings is an extension of "dissipative structures" thoroughly investigated by Prigogine and his schools in Brussels and Austin during the past three decades (Prigogine and Stengers, 1984) and is consistent with the idea championed by Prigogine that irreversible processes serve as the source of order in the universe.

(iii) <u>The Formal Cause</u>: It is postulated that our universe is a self-organizing physicochemical system, obeying the Principle of Deterministic Chaos—i.e., this universe is a *deterministically chaotic machine ("chaomachine")* in the sense that the global behaviors of this universe are programmed in the the gnergy tetrahedron, while its detailed dynamical behaviors are unpredictable. Numerous ancient and not-so-ancient philosophers and cosmologists already thought about the universe as a machine or a clockwork world (Barrow and Tipler, 1986). The Shillongator is not a simple repetition of these models but rather represents a hybrid of two opposing cosmological views—the *mechanical and hence deterministic* and the *chaotic and nondeterministic* universes. The gross features of the universe such as the formation of galaxies, stars and planets and the emergence of life on this planet are thought to be "programmed" into the gnergy tetrahedron but the details of local events in the universe are thought to occur more or less randomly, albeit within the constraints imposed by the gnergy tetrahedron. In the parlance of chaodynamics (Shaw, 1981; Tsonis and Tsonis, 1989) and Prigogine's irreversible thermodynamics, our universe is a strange attractor driven by irreversible processes. Or in the language of biocybernetics, *our universe is a self-organizing chaomachine.*

(iv) <u>The Final Cause</u>: This universe is postulated to have a purpose—a purpose of Knowing Itself, through the agency of *Homo sapiens* destined to uncover the origin of the universe. In other words, we are living in a *Universum sapiens*, the self-knowing universe; i.e., the purpose of the self-organizing universe of ours is self-knowing. This aspect of the Shillongator is highly reminiscent of the Aristotelian view that the natural world is intelligent and functions according to some deliberate design (Barrow and Tipler, 1986). What is common to both the Aristotelian universe and the Shillongator seems to be the underlying assumption that there exists a set of invariant physical principles that operates in both the universe as a whole and the world of living organisms; i.e., the living and the nonliving are symmetric with respect to the application of these laws. It is interesting to note that these two cosmological models have been motivated primarily by the experimental observations of living organisms. In a sense, biocybernetics which can be viewed as a grand synthesis of the biological and physical sciences of the late twentieth century seems to confirm the conclusion drawn by Aristotle over two thousand years ago that the universe is *biological, intelligent* and *teleological*, just as modern physics has substantiated the reality of the atom conjectured to exist through pure thought by Democritus also more than two millennia ago.

(4) The new version of the Shillongator incorporating the four causes delineated above is schematically represented in Figure 1.A5.

The following comments highlight the key features of the abstract version of the Shillongator.

(i) One of the most unusual aspect of the Shillongator is the assumption that the universe began as the gnergy tetrahedron defined in (3)(i) above. Unlike the modern cosmological models that are based on quantum mechanics and hence forbidden by the Heisenberg uncertainty principle from speculating about the state of the universe earlier than 10^{-43} sec after the Big Bang, the Shillongator specifies the state of the universe even before the Big Bang singularity. In other words, the epistemological underpinning of the Shillongator is that the human mind is capable of perceiving reality in two ways—through the mathematical reasoning and the non-mathematical logical reasoning. By the very nature of the method of reasoning employed, all mathematical models of the universe, both quantum gravitational and relativistic, are incapable of specifying the state of the universe before $t = 0$. In contrast, the Shillongator which is a *logical* cosmological model based primarily on biological observations is free to characterize the possible state of the universe even before the Big Bang singularity, without being branded "metaphysical." In other words, the Shillongator is "meta-mathematical" and "meta-quantum mechanical" but not necessarily metaphysical.

(ii) The Shillongator postulates that all of the physical and biological phenomena observed in the universe are the manifestations of the intrinsic properties enfolded in the gnergy tetrahedron. The unfolding of the intrinsic properties of the primeval substance is postulated to have taken place through six discrete transitions called "breaks" as the universe aged and cooled—(a) the "Einstein-Bohm-Wheeler" break postulated to have occurred at $t = 0$, an essential link between the gnergy tetrahedron and the world described by quantum mechanics and relativity theories, (b) the "Hawking-Penrose-Guth" break, giving rise to the cosmological events at or near the earliest times (10^{-43} sec) after the Big Bang singularity that are allowed for by the Heisenberg uncertainty principle, (c) the "Planck-Heisenberg" break, connecting the microworld (m) of superstrings, quarks, leptons and gauge bosons and the macroworld (M) of galaxies, stars, and planets and their associated heavy elements that began to form about 10^6 years after the Big Bang, (d) the "Bohr-Elsasser" break that marked the beginning of the reification of life (L) on this planet about 2-3 billion years ago, (e) the "Wilson" break, ushering in the era of human social activities and organizations about 2-3 million years ago leading to human civilizations (C), and finally (f) what I call the "*Homo abstractus*" break, marking the postulated emergence of the "new" species or subspecies of humankind during the past 2-3 millennia that is characterized by not only the mental capacity to carry out *abstract thinking* (as reflected in the development of mathematics, languages, sciences, arts, etc.) but also the *instinctive yearning* for new knowledge about themselves and their surroundings, including the origin of the universe.

(5) The justifications for using the names of the various scientists in connection with the breaks in the Shillongator are briefly indicated below:

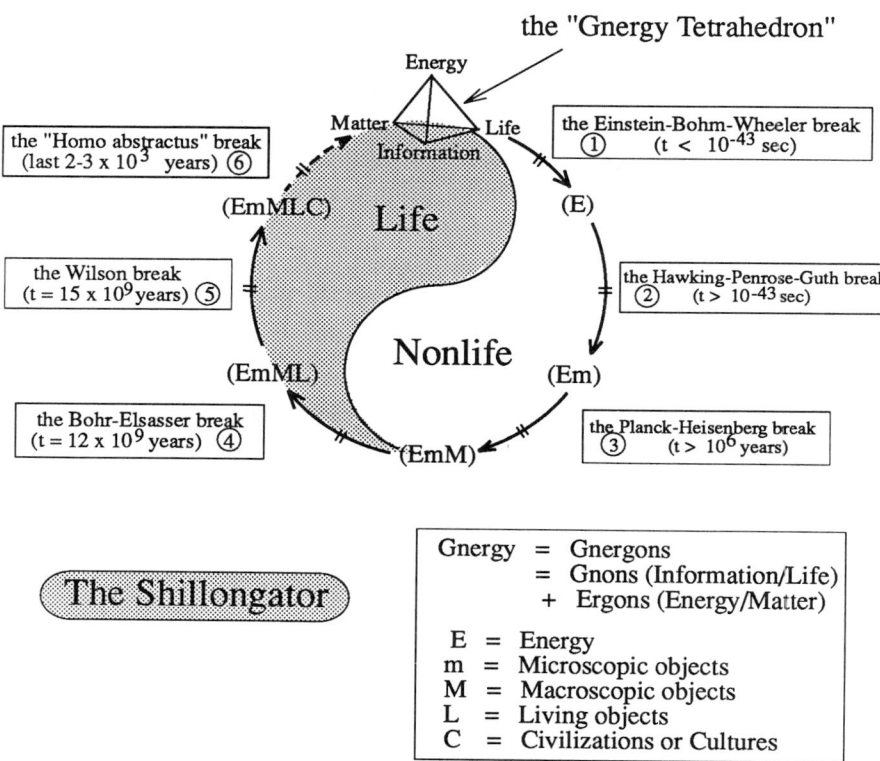

Figure 1.A5. The abstract version of the Shillongator, a biological model of the universe, first proposed at the International Seminar on the Living State - III, held in Shillong, India, on December 13-19, 1986.

"Biocybernetics": A Machine Theory of Biology 235

(i) <u>Einstein</u>—In contrast to Bohr, Einstein believed in "mathematical" reality revealed through pure thought. The gnergy tetrahedron is thought to be an example of the reality revealed through pure thought. Unlike Einstein's mathematical reality, however, the gnergy tetrahedron is best viewed as what D'Espagnat (1989) refers to as "logico-empirical" reality perceived through pure thought aided by empirical observations.

(ii) <u>Bohm</u>—The concepts of "implicate order," "enfolding" and "unfolding" of order, and "holomovement" that Bohm has enunciated over the years as a quantum physicist are reflected in the Shillongator in the form of the gnergy tetrahedron "enfolding" energy, matter, information and life and "unfolding" these entities as the universe undergoes a series of "phase transitions" or "breaks."

(iii) <u>Wheeler</u>—The idea of "pre-geometry" of Wheeler may correspond to the gnons of gnergy thought to be composed of a complementary union of gnons (life/information) and ergons (energy/matter). According to Wheeler, the Big Bang singularity is the threshold where space and time are transcended but the physical world, in some sense, survives (Davies, 1982). The Shillongator identifies the pre-Big Bang world with the gnergy tetrahedron. In addition, the Shillongator shares many similarities with Wheeler's "participatory universe" or "self-synthesizing" universe (Wheeler, 1990).

(iv) <u>Hawking and Penrose</u>—For their pioneering contributions to our knowledge on the nature of the Big Bang singularity and the earliest cosmological events thereafter, permitted by the Heisenberg uncertainty principle.

(v) <u>Guth</u>—For his inflationary universe model which resolves numerous problems associated with the standard model of the Big Bang by inserting the inflationary step in the evolutionary history of the universe (Waldrop, 1983).

(vi) <u>Planck and Heisenberg</u>—For their discovery of the discontinuity between the micro- and macroworlds.

(vi) <u>Bohr</u>—For having suggested for the first time the possible discontinuity between the living and the nonliving worlds.

(vii) <u>Elsasser</u>—For having developed a logical and epistemological base for upholding Bohr's idea of the insufficiency of the laws of physics and chemistry for completely explaining life.

(viii) <u>Wilson, E. O.</u>—For revealing the possible genetic and evolutionary underpinnings of human social behaviors thought to be essential for developing advanced civilizations and the creation of information.

(ix) "<u>Homo abstractus</u>"—We have been referred to as *Homo sapiens*, the "knowing" primate. "Knowing" is not uniquely human, since recent biological evidence suggests that practically all organisms have varying degrees of knowing capabilities; otherwise they cannot survive. According to biocybernetics, the simplest living organism, the cell, is intelligent in the sense that the cell knows what is beneficial and what is detrimental to its survival and can activate those cellular programs best fit to promote its survival; the cellular intelligence and the cell force are both essential for the conservation of life (see Table 1.16).

It seems legitimate to think about the possibility that the evolutionary process of humankind did not stop 2-3 million years ago at the stage of the "knowing primate" but has been in progress, especially during the past 2-3 thousand years, to a new stage characterized by the ability of the human brain to carry out abstract thinking, leading to the emergence of the current "information society." Thus, we may speak of two species or subspecies of humankind, *Homo sapiens* that originated 2-3 million years ago, and *Homo abstractus* that emerged only since about 2-3 thousand years ago. The human evolution postulated here is not one of morphology (the external space), the usual criterion of evolution, but rather that of the mental sphere (the internal space). One of the crowning achievements of Homo abstractus may be its comprehension of the origin of the universe and hence of himself.

(6) The Shillongator as depicted in Figure 1.A5 represents not only an unusually broad-based synthesis of natural sciences, physics, chemistry, biology and cosmology, but also possesses the capability to bridge the gap between natural sciences and metaphysics, the latter through the Homo abstractus break (see the dotted arrow). This break is unusual in two ways—(i) it indicates the final cause of the universe, namely the self-knowledge, and (ii) it provides a rational explanation for the Anthropic Principle (the idea that the structure of this universe is so constructed as to allow life and sentient beings to evolve in it (Barrow and Tipler, 1986)), since the self-knowing universe needs Homo abstractus to achieve self-knowledge, and since the Anthropic Principle is a necessary condition for the existence of Homo abstractus in this universe. In other words, the Shillongator postulates that the Anthropic Principle is necessary and sufficient for this universe to know itself, a metaphysical conclusion.

(7) Einstein (Zee, 1986) is quoted to have said,

" . . . . I want to know how God created this world.
 . . . . . I want to know His thoughts. . . . . "

If it weren't for Einstein, such questions would have never been regarded rational and would have been relegated permanently to the domain of religion. I believe that it was his genius to have glimpsed the possible logical answers to the question about the state of the universe even before creation. We may define the *Einsteinian questions* as those questions concerning the "what?," "How?" or "Why?" of the universe that border science and metaphysics. Clearly, the four causes of the Shillongator as delineated in (3) above supplies a possible set of scientific answers to the Einsteinian questions.

(8) One of the unforeseen insights provided by the Shillongator is the possibility that the ultimate answers to all of the major problems investigated by various fields of contemporary science (e.g., the quantum reality question, role of observers in cosmology and quantum physics, mechanisms of origin of life, mechanisms of biological evolution, mechanisms of cell life and death, etc.) may not be attained independently and locally as is commonly believed but may come about all simultaneously, as the result of the discovery of the purpose and the fundamental principles of the operation of the universe. The latter possibility would be logically anticipated if all the observable phenomena of the universe, both living and nonliving, owe their existence to the "unfolding" of the "cosmological programs" stored in the primeval substance of the universe, the gnergy tetrahedron, as postulated by the Shillongator.

Chapter 2

SCHRÖDINGER AND THE RIDDLE OF LIFE

Ilya Prigogine

It is a privilege to participate in this conference on "Molecular Theories of Cell Life and Death." It happens that this conference is being held in the very year of the centenary of Erwin Schrödinger's birth. This is why I think it appropriate to devote my lecture to some of the problems that Schrödinger discussed in his book "What is Life?" published for the first time in 1945 (Schrödinger).

This book had a strange fate. It was certainly very successful, as it contributed to make people aware of the fascinating problems of modern biology. On the other hand, it has been, and still is, sharply criticized by many scientists such as M. F. Perutz, who wrote in a recent paper (Perutz, 1987): "*To my disappointment, a close study of his book and of the related literature has shown me that what was true...was not original, and most of what was original was known not be true even when it was written.*" Perutz goes as far as writing that "*The apparent contradiction between life and the statistical laws of physics can be resolved by invoking a science largely ignored by Schrödinger. That science is chemistry.*"

I consider this judgement as unjust. In this short note, I therefore shall present some comments on Schrödinger's contribution, and specially on the following statements:

—Living matter feeds on "negative entropy." (Chapter VI, § 57)
—Living matter escapes approach to equilibrium. (Chapter VI, § 56)
—The supports of genetic information may be considered as aperiodic solids.

(Chapter V, § 45)

—We should seek for a physical explanation for the coherence displayed by biological processes. (Chapter VII, § 65-66)

Let us briefly consider these statements in succession. The second law of thermodynamics had, starting from its very first formulation by Clausius in 1865, a very broad scope, as it asserted what is possible for every "*isolated*" physical systems (which exchange neither matter nor energy with their environment). For these systems, the second law implies the existence of a function, the entropy S, which increases monotonously until it reaches a maximum value for the state of thermodynamic equilibrium ($dS/dt \geq 0$).

This formulation has since been extended to "*open*" systems (which exchange matter and energy with their environment). For such systems, one has to distinguish two terms which contribute to the evolution of S, namely $d_e S$, the entropy flowing across the boundaries of the system, and $d_i S$, the internal entropy production, this latter being necessarily positive or zero.

$$dS = d_e S + d_i S \geq 0$$

In the frame of this formulation, we see that Schrödinger's statements have to be understood as expressing that a living organism is in a "steady state." Then, the entropy flow compensates the (positive) internal entropy production; this implies the existence of a "negative entropy flow" ($d_e S + d_i S \leq 0$). This may lead to misunderstandings when one considers "*closed*" systems; we may indeed take the usual definition of entropy evolution for closed systems (which exchange only energy, and not matter, with their environment), $d_e S = dQ/T$, where dQ is the heat received, and T the temperature. The existence of steady state implies that $dQ < 0$, which means that there are heat flows from the living organism to its environment, and alternatively that the internal temperature of the living system is higher than that of the environment. Such a simplified mode is correct; however, it is not very relevant for our understanding of biological processes.

To understand Schrödinger's statements, we thus have to turn to the real living systems, which are *open* systems. Here the entropy flow $d_e S$ has a much more general meaning, as it includes a flow of free energy carried by matter.

The explicit form of entropy flow for open systems is now presented in many textbooks. Therefore, I shall not write it down in full detail here. Let us stress that it contains both a convection and a diffusion term; the convection term contains a flow of free energy, which to express it briefly, "feeds" the living system (Glansdorff and Prigogine, 1971; De Groot and Mazur, 1962).

Let us add that today we understand better the meaning of entropy production. Already in very simple cases, we see that production of order is coupled with production of entropy. Let us consider a simple example: thermodiffusion. A two-component system, containing hydrogen and nitrogen, is submitted to an external heat constraint, maintaining a thermal gradient. As the result, one of the components, say hydrogen, accumulates in one of the compartments, while the other, say nitrogen, accumulates in the other compartment. The entropy production (always positive or zero) is the sum of two terms, heat flow (positive) and diffusion (negative). We thus have two coupled processes: heat flow produces entropy, while diffusion goes against the concentration gradient—if the diffusion could be isolated from heat conduction, it would give a negative entropy production. The global effect is thermodiffusion, involving both order and disorder production.

The dual role of entropy, leading to both "order" and "disorder," is essential in the production of biomolecules, which occurs always as coupled processes, in which the entropy production, related to the destruction of some metabolites, is used to produce the biologically relevant molecules, such as proteins and nucleic acids.

Schrödinger's statement that living matter escapes approach to equilibrium is rather obvious today; it means that the genetic material acts as an external constraint, maintaining the living system in far-from-equilibrium conditions.

The problem of non-linear dissipative processes has made enormous progress since Schrödinger's time. A number of biological processes have been analyzed from this point of view (Nicolis and Prigogine, 1988). I do not think it is necessary to go here into details; the important point is that this "constructive" role of non-equilibrium has now been recognized in many fields of modern physics, including material science, chemical explosions, catalysis, and so on... (Walgraet, 1987; Vidal and Pacault, 1984).

Schrödinger's statement that the supports of genetic information lies in "aperiodic solids" is specially intriguing. Indeed, we know now that biological molecules contain sequences of "messages" or "texts" whose material support are one-dimensional aperiodic solids containing "information." Now, this information is closely related to the aperiodic character emphasized by Schrödinger. I have already mentioned the role of non-equilibrium in the production of macroscopic structures, of "order." From this point of view, chemistry plays a very specific role. Indeed, one could say that it may "encapsulate" irreversible time into matter. In this way, irreversible processes may be made more permanent and transmitted over longer periods of time. This is of special importance for us, as we should be able to describe in these terms a world where the very existence of biological system implies some recording of irreversible processes into matter. This is in contrast with

Schrödinger and the Riddle of Life 241

hydrodynamics. Take for example the Bénard instability; once one stops the heat flow, the Bénard patterns vanish. On the contrary, chemical molecules produced under non-equilibrium conditions keep some memory of the deviations from equilibrium which existed at the moment of their production.

We may illustrate these statements on a very simple example; let us take two monomers X and Y, which can transform each into the other or produce a polymer. For example, every time one of them presents a concentration higher than some critical level, it becomes part of the polymer chain. Now, what happens if we produce this polymer under near-to-equilibrium conditions? There the concentrations are violently fluctuating, according to the Poisson law, and the result would be a disordered chain without any coherence, such as XXYXYYYXYXYYXXXYX... One would obtain a highly disordered structure. What happens if we produce the polymer from a system whose behaviour corresponds to a limit cycle? Obviously, the chain would be somewhat like XYXYXYXYXY... Here, we see already that nonequilibrium has been "encapsulated" into the structure of the polymer. This being said, a periodic structure like XYXYXYXYXY... is not a very hopeful candidate for generating biomolecules. This is why I find very interesting the work done by Nicolis and G. and J. Rao (Nicolis and Subba Rao, 1987). They have coupled the polymer-producing reaction with a chaotic reaction (details are not important here), and produce a chain which, in a sense, is "between" the ordered and the disordered ones. In this example, the succession of X and Y may be described through a Markov chain of higher order. Now, such structures have some information "content." One of the statements of information theory is that in order to measure the information content of a message, one has to specify how long the program able to generate this message has to be. Obviously, the program associated with the chain XYXYXY... is very short. However, if the chain contains long-range correlations, one would need a much longer program to generate it. This example shows how chemistry may, at least in principle, be "translated" into information; it is still a "paper example," but it will hopefully be followed by real-world experiments in a not too distant future.

I come now to the last point, which was discussed in Chapter VII of Schrödinger's book. Schrödinger was indeed quite impressed by the precision of biological processes. He compares biological systems to clocks; but we know now that there are not only mechanical clocks, but also chemical clocks, again based on nonequilibrium processes. At the time of Schrödinger's work, the only long-range coherent processes known in many-body systems were low-temperature processes associated with superfluidity or superconductivity. This is why the natural analogy for this time was with low-temperature physics. This is no longer necessary today, as we now know that nonequilibrium physics and chemistry display numerous examples of long-range coherence.

These few examples show the lasting interest of Schrödinger's beautiful book. This book has certainly been one of the sources of my own interest for nonequilibrium processes. For me, the fact that nonequilibrium may be a source of order remains still a quite surprising fact.

We are still far from a detailed explanation of the origins of life; notwithstanding, we begin to see the type of science which is necessary. Certainly, as emphasized by Perutz, this science will encompass chemistry; but not only chemistry: chemistry combined with deterministic chaos. What we need is a better understanding of the mechanisms which lead from the law of physics and chemistry to "information." Traditional information theory was too vague, to put it in a somewhat clear-cut way, because it is not deeply enough rooted in physics and chemistry, while traditional physics and chemistry were too narrow. We need to build a closer bridge between these two fields, which would allow us to find a more global interpretation of natural laws.

REFERENCES

De Groot, S. R. and Mazur, P. (1962). Non-Equilibrium Thermodynamics. North Holland, Amsterdam.

Glansdorff, P. and Prigogine, I. (1971). Thermodynamics Theory of Structure, Stability and Fluctuations. John Wiley & Sons, Ltd, New York.

Nicolis, G. and Prigogine, I. (1988). Exploring Complexity Freeman (to appear) From Chemical to Biological Organization; (Markus, M., Müller, S. C., and Nicolis, G., eds). Springer Berlin, Springer.

Nicolis, G. and Subba Rao, J. (1987). Generation of Spatially Asymmetric, Information-rich Structures in far from Equilibrium Systems. In: Coherence and Chaos in Dynamical Systems, Manchester University Press, Manchester, (to appear).

Perutz, M. F. (1987). Nature 326: 555-559.

Schrödinger, A. (1945). What is Life? Cambridge University Press, Cambridge.

Vidal, C. and Pacault, A. (1984). Nonequilibrium Dynamics in Chemical Systems. Springer, Berlin.

Walgreat, D. (ed.) (1987). Patterns, Defects and Microstructures in Nonequilibrium Systems, Dordrecht, Nijhoff; Vidal, C., and Pacault, A., 1984, Nonequilibrium Dynamics in Chemical Systems, Springer, Berlin.

Chapter 3

A NOTE ON CHIRAL SYMMETRY BREAKING AND THE ORIGIN OF BIOMOLECULAR CHIRALITY

Dilip K. Kondepudi

ABSTRACT

The origin of biomolecular chirality is a problem that interest physicists, chemists and the biologists. Developments in the theory of self-organization in nonequilibrium systems have provided new insights into the mechanisms that might have led to the present homochirality of life. Since the discovery of parity violation in weak interactions, the possibility of a link between the handedness of the weak force and that of biomolecules has been of great interest. At first glance, it might appear that the effect of the weak force on chemical processes would be too weak to have a noticeable influence. Here we note that, when we consider the large volumes (such as the oceans) and times in which the chemical evolution might have taken place, the effects of the handedness of weak interactions could be considerable.

3.1. INTRODUCTION

Since Pasteur discovered the homochiral nature of biomolecules, the processes that could have led to such a state of broken symmetry have been of interest to chemists and physicists alike. Modern advances in biochemistry have given us a very precise definition of this homochirality: the proteins are made of L-amino acids and the DNA is made of D-sugars. Though one does find polypeptides that contain D-amino acids in the biosphere (Bently, 1969) these instances are very rare. At what stage during the evolution of life did this homochirality arise? What process resulted in this homochirality? Is the particular homochirality of L-amino acids and D-sugars a matter of chance or is it a consequence of the asymmetry between the 'left' and the 'right' that we discovered in the realm of fundamental interactions between elementary particles? These are questions that seem formidable at the present time but we hope to eventually answer them.

The developments in our understanding of far-from-equilibrium chemical systems that have occurred in the last two decades, particularly in the theoretical studies of spontaneous symmetry breaking, give us a very useful and general framework on which to consider this problem. One interesting outcome of this approach is an understanding that it is possible for the very small chiral asymmetry of weak interactions (i.e., in the interaction between the electron and the nucleons, or that which can be found in β-decay) to have an influence on symmetry breaking chemical processes during biomolecular evolution. This influence is in the form of making it overwhelmingly probable for one enantiomer to dominate rather than the other. Thus, the dominance of L-amino acids becomes linked to the particular handedness of weak interactions and is not a matter of chance. The effects of parity violating interactions on chemical reaction rates are indeed very small; for weak-neutral-current interactions, estimates show it to be only about one part in 10^{17}. Nevertheless, when evolutionary time scales and large reaction volumes (such as the oceans) are considered, there exists a natural mechanism in which the emergent biomolecular chirality is determined by the chiral asymmetry in the parity violating weak interactions. I shall briefly describe this process in this short note.

The existence of such a process is by no means a proof that it had occurred. It is only a possibility that we must consider in trying to solve the grand puzzle of the origin of biomolecular chirality. We can not rule out this possibility on the grounds that parity violation in weak interactions has "too small" an effect on chemical reactions.

3.2. CHIRAL SYMMETRY BREAKING IN NON-EQUILIBRIUM CHEMICAL SYSTEMS

When a chemical system containing chiral molecules is in thermodynamic equilibrium the concentration of the L- and the D-enantiomers will be equal. This follows from the principle of detailed balance which is valid for a system in thermodynamic equilibrium. According to this principle, every transformation, in

A Note On Chiral Symmetry Breaking

particular chemical transformation, is balanced by its exact opposite. Under conditions that maintain the system far from thermodynamic equilibrium, however, if appropriate reaction mechanism are present, the system can make a transition to a state in which the concentrations of the two enantiomers are not equal. Even though the reaction kinetics are identical for the L- and the D-enantiomers (i.e., the reaction kinetics are chirally symmetric) the system can "break" this symmetry by making a transition to a state with unequal enantiomer concentrations. Generally, such symmetry breaking is possible when there is some kind of chiral autocatalysis, as was noted by Frank (1953) and several others (Decker, 1979; Walker, 1979).

This phenomenon can be illustrated through a simple model reaction scheme.

$$S + T \underset{k_{-1}}{\overset{k_1}{\rightleftarrows}} X_{L,D} \quad\quad (a)$$

$$S + T + X_{L,D} \underset{k_{-2}}{\overset{k_2}{\rightleftarrows}} 2X_{L,D} \quad\quad (b)$$

$$X_L \underset{k_R}{\overset{k_R}{\rightleftarrows}} X_D \quad\quad (c)$$

$$X_L + X_D \overset{k_3}{\longrightarrow} P \quad\quad (d)$$

In this scheme, S and T are non-chiral reactants that react chemically to form a chiral molecule X as an L-enantiomer, X_L, or as a D-enantiomer, X_D, with equal probability. Reaction (b) is the autocatalytic step in which the enantiomers of X catalyze the production of their own kind. In reaction (d), the enantiomers mutually destroy each other irreversibly to form a product P. Reaction (c) is the racemization reaction. This model, and extensions of it have been studied extensively (Kondepudi and Nelson, 1984; 1985) to show how symmetry breaking occurs in it. Here I shall summarize the main points.

First, we must consider the system under conditions away from thermodynamic equilibrium. This is done by assuming an input of the reactants S and T and removal of the product P. Thus we assume that the concentration of S and T are maintained at a fixed level through this input. Then the concentrations of X_L and X_D are determined by the concentrations of S and T. On analyzing the steady states of this system, one finds that there is a threshold for the variable $\lambda = [S][T]$, where [S] and [T] are the concentrations of S and T respectively, above which the system makes a transition to a state in which $[X_L] \neq [X_D]$; below this threshold, the concentrations of the two enantiomers are equal.

Let us denote the threshold value of λ by λ_c and define an asymmetry parameter $\alpha = [X_L] - [X_D]$. The above mentioned breaking of symmetry is described by the equation.

$$\frac{d\alpha}{dt} = A\alpha^3 + B(\lambda - \lambda_c)\alpha \tag{3.1}$$

In fact, using group theoretic methods (Sattinger, 1979) it can be shown (Kondepudi and Nelson, 1984) that, for any system that breaks chiral symmetry, one can always derive an equation of the above form by suitably defining the variables λ and α. This general equation, it must be noted, is valid only in the vicinity of the threshold λ_c. For the considerations that follow, the behavior of the system in this region is all we need. A and B are constants that depend on the particular kinetic scheme under consideration. General expressions for calculating these constants can be found in Kondepudi and Nelson (1984).

An analysis of equation (3.1) (with the inclusion of the unavoidable random fluctuations) tell us that if λ is made to increase from a value below λ_c to a value above λ_c, the system will randomly make a transition to a state in which $[X_L] > [X_D]$ or a state in which $[X_L] < [X_D]$. Both states will be reached with equal probability.

One of the most interesting aspects of this system is its sensitivity to very small chiral interaction that favors the formation of one of the enantiomers, say X_L. We consider the sensitivity in the presence of random fluctuations of root-mean-square value $\sqrt{\epsilon}$. In an Arrhenius reaction mechanism, if we assume that due to the chiral interaction the activation-energy barrier for X_L is reduced by an amount ΔE (compared to that of X_D) then it can be shown that the above equation is modified to:

$$\frac{d\alpha}{dt} = -A\alpha^3 + B(\lambda - \lambda_c)\alpha + Cg + \sqrt{\epsilon F(t)} \tag{3.2}$$

in which $g = \Delta E/kT$ (k is the Boltzmann constant and T is the temperature) and $F(t)$ is the function representing the random fluctuations. C is a constant that depends on the reaction kinetics. The presence of the chiral interaction will make it more probable for the system to make a transition to state in which $\alpha > 0$, that is $[X_L] > (X_D]$. The question is: how much more is probable is it? In trying to assess this effect of the chiral interaction which favors the formation of X_L, one might think that if Cg is less than $\sqrt{\epsilon}$ then the effect will be small and that the system will reach

A Note On Chiral Symmetry Breaking

the two states, $\alpha > 0$ and $\alpha < 0$, with almost equal probability. It is somewhat of a surprise to find that this is not true. Even if Cg is less than $\sqrt{\epsilon}$, the system can be greatly affected by the chiral interaction and make a transition to the state $\alpha > 0$ with overwhelming probability (Kondepudi and Nelson, 1985; Kondepudi et al., 1986). The sensitivity depends on the rate at which λ moves from a value below λ_c to a value above λ_c. If this rate is γ, then the probability P_+ of transition to a state in which $[X_L] > [X_D]$ is given by (Kondepudi and Nelson, 1985; Kondepudi et al., 1986):

$$P_+ = \frac{1}{\sqrt{2\pi}} \int_{-\infty}^{N} Exp(\frac{-x^2}{2}) dx \quad \text{in which} \quad N = \frac{Cg}{\sqrt{\epsilon/2}} (\frac{\pi}{B\gamma})^{\frac{1}{4}} \tag{3.3}$$

This phenomenon is quite general and is valid for any system that breaks a two-fold symmetry. The validity of equation (3.3) has been well tested by both analog electronic and digital numerical simulation of equation (3.2). With this result we are now in a position to assess the possibility of a link between parity violation in weak interactions and biomolecular chirality.

3.3. SELECTION OF BIOMOLECULAR CHIRALITY BY EXTREMELY SMALL PARITY NON-CONSERVING INTERACTIONS

Ever since the discovery of parity violation in β-decay, there have been many suggestions for a possible link between it and the biomolecular chirality (Thieman, 1981; Walker, 1979). There was—and still is—much skepticism in the possibility of weak interactions influencing chemical processes because the magnitude of the influence is very small. For instance, the factor g in the above equations at best (Hegstrom, 1985) is of the order 10^{-11}. If the recently discovered weak-neutral-current interaction between the electron and the nucleon is taken into consideration, g is only of the order 10^{-17}. Even so, in light of the above mentioned sensitivity, we reconsidered the problem (Kondepudi and Nelson, 1985). We found that on bio-evolutionary time scales of 10^5 to 10^6 years, if the reaction is assumed to occur in large volumes such as the oceans, even a value of $g \approx 10^{-17}$ can have a strong influence as a selector of the enantiomer that will dominate.

The scenario in which this happens is the following. In the primordial oceans some nonchiral reactants, such as S and T in the above model, are accumulating because they are being formed in the atmosphere and descending into the oceans. Their oceanic concentrations are slowly increasing. In the oceans, they react to form a chiral molecule X which is chirally autocatalytic. We assume that, due to the autocatalysis and other favorable reactions, the system is capable of breaking chiral symmetry when the concentrations of the nonchiral molecules reach a certain threshold value. As the concentrations of the nonchiral precursors increase and go past the threshold value, the system makes a transition to a state of broken symmetry. In this process, due to weak interactions, the reaction rate of one of the enantiomers is larger by one part in 10^{17}. The concentrations of the reactants are in the 10^{-3}M range. The reaction takes place such that in a time of about 100 years oceanic mixing maintains homogeneity over at least a few miles. The entire process takes about 15000 years. Then using the above model to calculate the constants B and C and substituting these values in formula (3.3) we find that P_+ is 0.98. Weak interaction effects are not insignificant!

In this context it is interesting to note that numerical evaluation of the effects of weak neutral currents on the ground states of amino acids showed that it is the biologically dominant L-enantiomer that has lower energy (Mason and Tranter, 1984; Tranter, 1985). This implies that at equilibrium the concentration of L-enantiomer will be larger by a factor of about 10^{-17}. This does not imply, however, that the reaction rate for the formation of the L-enantiomer is necessarily larger than that of the D-enantiomer.

Hence, among all the possible explanations for the observed biomolecular chirality, we cannot rule out the possibility that weak interactions were responsible for the dominance of L-amino acids in the biosphere. Their influence is not negligibly small on a bio-evolutionary scale. Of course, any other *systematic* chiral influence can be equally influential in the selection of chirality; but the ones suggested so far are rather contrived and/or depend on very particular geographical conditions.

REFERENCES

Bently, C. H. (1969). Molecular Asymmetry in Biology, Academic Press, Vol. 1 New York.

Decker, P. (1979). The Origin of Molecular Asymmetry Through the Amplification of "Stochastic Information" (NOISE) in Bioids, Open Systems Which can Exist in Several Steady States. J. Mol. Evol. 4: 49-65.

Frank, F. C. (1953). On Spontaneous Asymmetric Synthesis. Biochim. Biophys. Acta 11: 459-463.

Hegstrom, R. A. (1985). Weak Neutral Current and β Radiolysis Effects on the Origin of Biomolecular Chirality. Nature, 315: 749-750.

Hegstrom, R. A. and Kondepudi, D. K. (1990). The Handedness of the Universe. Sci. Am. 262(1): 108-115.

Kondepudi, D. K. and Nelson, G. W. (1984). Chiral Symmetry Breaking States and Their Sensitivity in Nonequilibrium Chemical Systems. Physica 124A: 465-496.

Kondepudi, D. K. and Nelson, G. W. (1985). Weak Neutral Currents and the Origin of Biomolecular Chirality. Nature 314: 438-441.

Kondepudi, D. K., Moss, F. and McClintock, P. V. E. (1986). Observation of Symmetry Breaking, State Selection and Sensitivity in a Noisy Electronic System. Physica 21D: 296-306.

Kondepudi, D. K., Prigogine, I. and Nelson, G. W. (1985). Sensitivity of Branch Selection in Nonequilibrium Systems. Physics Lett. 11A: 29-32.

Mason, S. F. and Tranter, G. E. (1984). The Parity-Violating Energy Difference Between Enantiomeric Molecules. Mol. Phys. 53: 109-111.

Sattinger, D. H. (1979). Group Theoretic Methods in Bifurcation Theory Lecture Notes in Mathematics, No. 762, Springer-Verlag, Berlin.

Thieman, W. (ed.) (1981). Origin of Life 11 (this volume devoted to the origin of biomolecular chirality).

Tranter, G. E. (1985). The Parity Violating Energy Difference Between the Enantiomers of α-Amino Acids. Mol. Phys. 56: 825-838.

Walker, D. C. (ed.) (1979). Origins of Optical Activity in Nature. Elsevier, New York.

Chapter 4

MATHEMATICAL MODELS OF CELL BIOCHEMISTRY

W. Mike L. Holcombe

ABSTRACT

The metabolic organization of a cell is modelled at several levels by different types of machine. Examples of these models are given and their interrelationship discussed. A general mathematical model of a machine is proposed to provide a unified framework for this analysis. The organizational complexity of a dynamic system such as a cell is examined and a measure of this complexity is suggested as an indicator of cell death.

4.1. INTRODUCTION

At the atomic level and at the genetic level it is possible to envisage the construction of a rigorous mathematical framework in which to describe biochemical activity. At higher levels, however, the situation is more problematical and this paper addresses the problem of trying to construct a precise mathematical language for the analysis of the biochemical organization of a cell. The benefits of a successful foundation are many, but principally it will allow us to use mathematical tools for the analysis of these systems and computers for their simulation and modelling. With these aims in view, we develop a series of models to describe different aspects of biochemical organization in cells and a mathematical framework which unifies these

Mathematical Models of Cell Biochemistry 251

models. We also suggest ways of defining a precise specification language, which could provide a means of defining and implementing organizational models on a powerful parallel computer. The applications of such a tool are many and varied and would present us with opportunities for major advances in understanding the theory of complex systems and the way they behave when perturbed. Applications in biotechnology and related areas would also be possible.

4.2. WINDOWS ON THE CELL

4.2.1. The Metabolic Machine

The first model is concerned with simple metabolic pathways. We use the concept of a *finite state machine* and so we need to consider the definition of a finite set, Q, of *states*; a finite set, Σ, of *inputs*; and a finite set, Φ, of *outputs*. Initially the inputs and outputs will simply be regarded as "abstract symbols," but they do have a concrete molecular interpretation here also (Holcombe, 1982, 1985; Ji, 1985; Krohn et al., 1967).

To explain how these choices are made we will consider an elementary pathway—the Citric Acid Cycle, or Krebs cycle. (This has been chosen because of the Sheffield-Krebs connection.) In this model we ignore enzymes and concentrate only on the substrates and the coenzymes. The substrates are regarded as *states* and the coenzymes as *inputs*. The *outputs* are either reduced coenzymes or the Co-A from acetyl Co-A. The reduced machine is obtained by "merging" states which participate in reactions without the involvement of coenzymes. The next-state function is defined by the lower diagram in Figure 4.1 as is the output function. Here $1 \xrightarrow{(\alpha, \alpha)} 4$ means: if the machine is in state 4.1 and receives the input a, the next state is state 4 and the output is a.

Other simple metabolic systems can be modelled with this type of machine. There is a substantial algebraic theory for these machines and some of the consequences of this theory may have interesting interpretations within the biochemical arena. In our interpretation the metabolic machine is processing data from the set Σ, of coenzymes (or cofactors) and the resultant output data is then utilized elsewhere in the cellular system. The way in which the metabolic machine interacts with the rest of the system is the next issue we discuss.

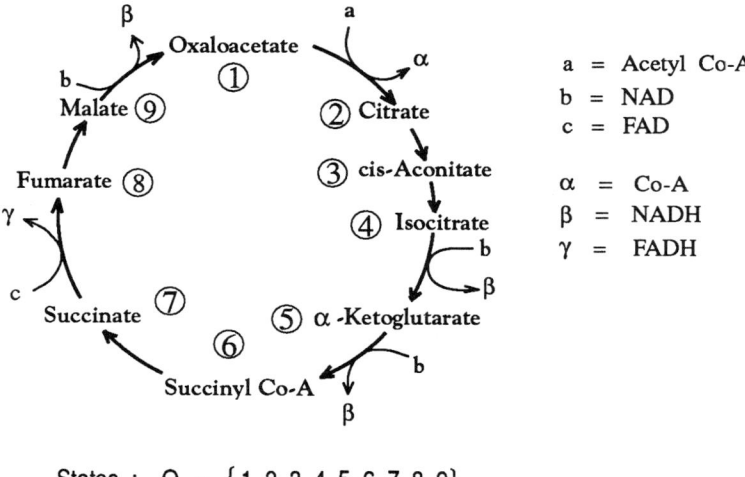

States : $Q = \{1, 2, 3, 4, 5, 6, 7, 8, 9\}$

Inputs : $\Sigma = \{a, b, c\}$

Outputs : $\Phi = \{\alpha, \beta, \gamma\}$

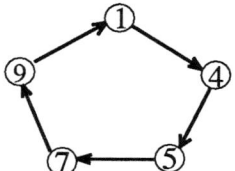

Figure 4.1. The Krebs cycle as a metabolic machine.

Prior to becoming available as input data to the machine described above, the cofactors will originate from other metabolic systems and will have undergone processing or encoding themselves. For example, acetyl Co-A is produced from pyruvate originating in aerobic organisms, from the glycolysis pathway which ultimately has as its origin glucose and an environmental input. The transformation of the other cofactors, e.g., NADH back to NAD, is a result of pathways which ultimately generate ATP and is thus an important source of energy. Diagrammatically we can represent the metabolic machine and its input processing precursors as illustrated in Figure 4.2.

The modelling of the input data *encoding* can be achieved again by the use of a finite state machine, the only real difference is in the way this machine is connected to the Krebs machine. Acetyl Co-A appears as the last state in the glycolysis sequence and this state is used as an input to the Krebs machine, we cannot connect the machines in *series* because of this feature.

Mathematical Models of Cell Biochemistry

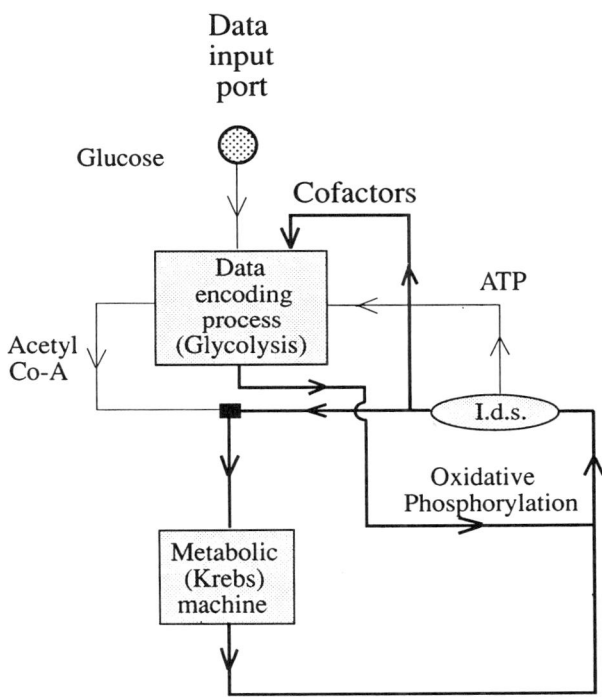

Figure 4.2. The metabolic machine and its link with input data encoding and i.d.s (intracellular dissipative structure) processing.

The role of the IDS (intracellular dissipative structure; see Section 1.8.5, Chapter 1) in the organization of the energy of the cellular metabolisms has yet to be fully defined. At the moment we will regard the IDS' simply as sources of energy that control the activity of various types of machines in the system.

From this picture of a metabolic machine and its immediate environment there are two directions in which we may travel. One direction, downwards, takes us to the level of the individual state transitions, we call this the *conformational level*. The other direction, takes us up to a higher level, the place where the enzymatic control of the metabolic machines is to be found. We call this the *enzyme control level*.

4.2.2. The Conformational Level

In the transition from state q_1 to state q_2 in a metabolic machine, there will usually be involved some enzyme $e = e[q_1, q_2]$ (see Figure 4.3). In many cases, especially in a reduced machine, a coenzyme, σ, is also part of the reaction

$$q_1 + e + \sigma \rightarrow q_2 + e + \Phi \tag{4.1}$$

where Φ is the "formal" output. The significant part of the reaction at the conformational level is concerned with the way in which different conformational states of the enzyme feature. By conformational states, we mean the 3-dimensional arrangement of atoms in a molecule that can be achieved without breaking or forming covalent bonds.

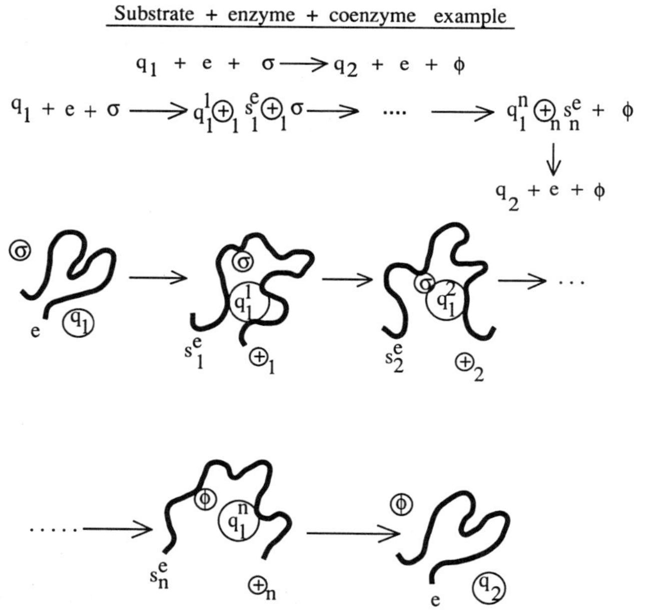

Substrate + enzyme + coenzyme example

$$q_1 + e + \sigma \longrightarrow q_2 + e + \phi$$

$$q_1 + e + \sigma \longrightarrow q_1^1 \oplus s_1^e \oplus \sigma \longrightarrow \cdots \longrightarrow q_1^n \oplus s_n^e + \phi$$
$$\downarrow$$
$$q_2 + e + \phi$$

The states are (q_1, σ), (q_1^1, σ), ..., (q_1^n, ϕ) and the inputs are $e, s_1^e, \ldots, s_n^e$. The conformational changes "drive" the machine.

Figure 4.3. A conformational machine

Mathematical Models of Cell Biochemistry

We can define a sequence $S_1^e, S_2^e, \ldots, S_n^e$ of conformational states of the enzyme e, and the way in which substrate binding, activation, deactivation and product release occur is described in terms of this sequence. Thus $q_1 + e + \sigma \to q_2 + e + \Phi$ can be written in more detail as:

$$q_1 \oplus_1 S_1^e \oplus_1 \sigma \to q_1^2 \oplus_2 S_2^e \sigma \to \ldots \to q_1^n \oplus_n S_n^e \oplus_n \Phi$$

$$\to q_2 + S_1^e + \Phi \tag{4.2}$$

where $\oplus_i$ indicates the binding at stage i and q_1^i is a configuration of substrates. At the end of the reaction, the enzyme takes up its initial "stable" conformation ready for the next operation of the reaction.

The description of this process is given by the sequence $S_1^e, \ldots, S_n^e$ of the conformations of e together with the definition of the "binding sums" $\oplus_1, \ldots, \oplus_n$. We can call the sequence

$$< S_1^e, q_1^1, \oplus_1 >, \ldots < S_n^e, q_1^n, \oplus_n > \tag{4.3}$$

a *conformational sequence*, since it describes the way in which the enzyme-substrate complex behaves. If no coenzyme is involved we simply delete σ from the reaction (4.2), and if other substrates are involved then (4.2) will be adjusted accordingly.

The description of $\oplus_i$ must be based on the geometry of S_i^e and q_1^i and their bonding relationship with σ. Although we haven't described this formally as a machine, it is clear that this could be done; it would be a particularly simple form of a finite state machine.

4.2.3. The Enzyme Control Net

Without the appropriate enzymes available to catalyze the various chemical steps of a metabolism, the reactions cannot take place. The way in which the availability of enzymes controls the processes is modelled by an enzyme control net (see below). We consider two special classes of objects:

places—often the sites of enzyme synthesis, and thus controlled genetically, say, at a ribosome. Other examples of places are receptor molecules in the plasma membrane, where environmental influence stimulates the production of proteins; and (ii) transitions—usually sites of metabolic machine activity. Essentially the *places* are the source of enzymes produced by enzyme machines, which require two types of input, substrates (amino acids, second messengers etc.) and *energy* in the form of ATP. The genetic description, in terms of nucleotide sequences determines the form of the polypeptide chain. Suppose that the coding sequence is a b c# where # denotes an end marker, and each of a, b, and c, is a codon generating the peptide P_1, P_2, and P_3 from amino acid A_1, A_2, and A_3 (see Figure 4.4). The enzyme machine can then be defined as a finite state machine with

States   : a, b, c
Inputs   : A_1, A_2, A_3
Outputs : P_1, P_2, P_3

and operates as shown in Figure 4.4.

a, b, c, ...   = Codons
A, B, C, ...   = Amino acids
P_1, P_2, P_3, ... = Peptides
= Endmarker

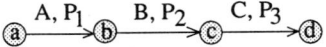

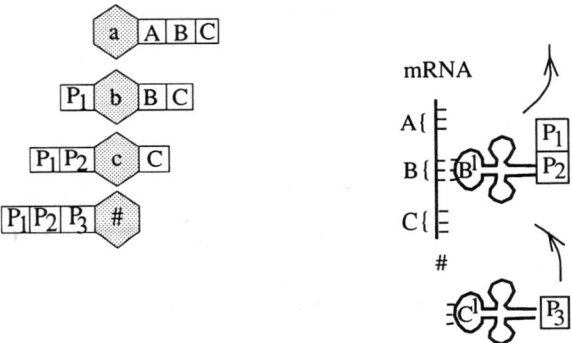

A^1 is the anticodon of A, etc.

Figure 4.4. An enzyme machine

Mathematical Models of Cell Biochemistry 257

Thus we regard the codon sequence as defining a set of machine states, rather than as an input sequence for a machine. Our inputs are the building blocks of proteins. The transitions and places are connected together in a complex intercommunicating system.

The machine described by the enzyme control net, processes tokens through the system (Figure 4.5). Before a transition can take place (or fire) it must have received the appropriate number of suitable tokens from the places that "feed" into it. For example, transition t_1 can only fire if tokens A and B have been donated by the places P_1 and P_2. When t_1 has fired tokens E and F are sent to places P_3 and P_4. Now place P_3 cannot donate toke G to transition t_3 until token D arrives at P_3. Then token G can be sent to t_3 and t_3. Thus the machines represented by the transitions t_1, t_2, t_3, .. are controlled by the availability of tokens donated by the places. We regard the transitions as metabolic machines, the places as enzyme machines and the tokens as coenzymes. A transition is *enabled* if all the necessary enzymes required for its operation are available. Then, whenever data (i.e. coenzymes) and energy (perhaps in the form of ATP) are available, the metabolic machine can operate and data processed. The eventual degradation of any enzymes will result in the machine failing to operate and this may be corrected by the enzyme control net sending suitable tokens to the transition site.

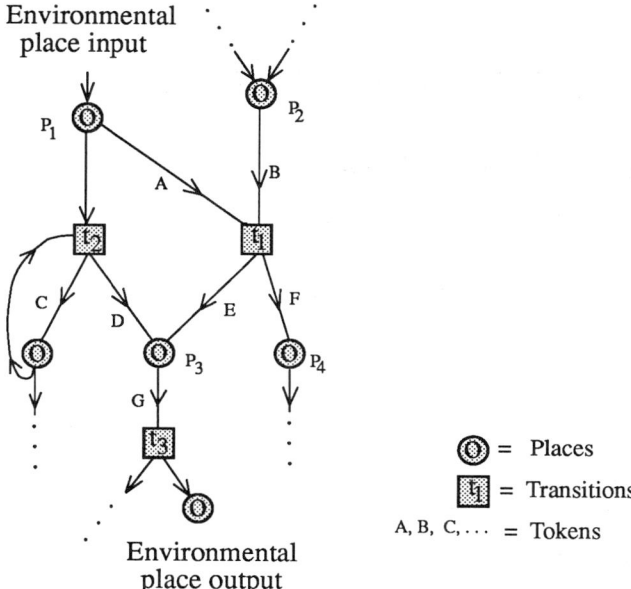

Figure 4.5. An enzyme control net

The production of enzymes at the places is achieved by enzyme machines which will require the sort of data for processing (say amino acids) and energy for driving the processing (e.g., ATP) as explained above. Similar structures to the enzyme control net, such as Petri nets and data flowgraphs, are currently being used to describe and analyze parallel computing processes.

4.2.4. Bringing It All Together

The final model consists of a network of metabolic machines embedded in an enzyme control net and powered by an energy source (see Figure 4.6). As we have mentioned before, it is possible to "zoom" in at various elements of this structure, for example, the metabolic machine and look at it at several levels, including the lower level of the conformational machine. Notice that both the network of metabolic machines and the enzyme control net have interfaces with the environment of the complete cellular system.

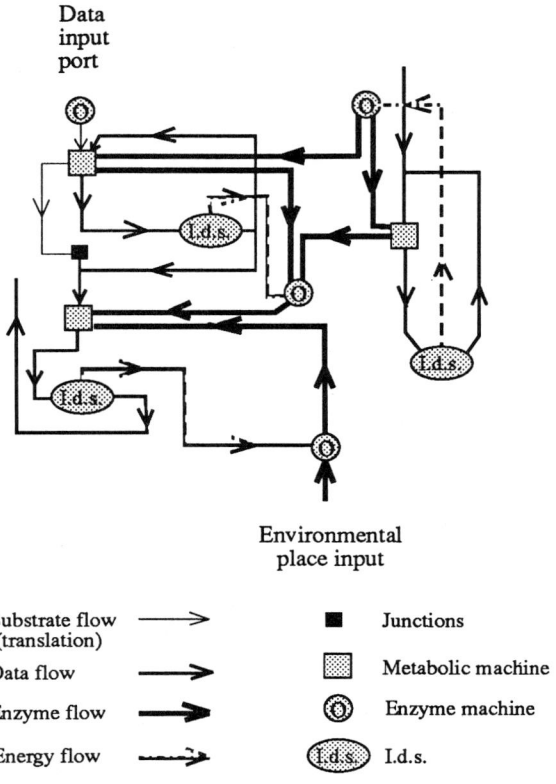

Figure 4.6. The complete system

Mathematical Models of Cell Biochemistry

4.3. A GENERAL ABSTRACT MACHINE

Each individual machine, whether metabolic, enzyme or whatever is performing a processing role. In some cases, such as glycolysis, the role is one of encoding input data into a form suitable for processing by subsequent machines. The enzyme machines, in general, are processing chemicals (obtained, for example, by absorption from the environment using another type of encoding process) and producing some output, which itself may be decoded somewhere. All of these machines, including the more complex parallel control and energy systems (such as the enzyme control net) can be viewed in this way. We introduce a suitably defined *data type*, X, which may have to include control and energy variable information.

The initial input data is encoded by some relation or function into the machine data type. The machine, which has a finite state control unit then processes these data by successively applying functions or relations to it. The finite state control unit ensures that we can distinguish between successful computations and unsuccessful ones. At the end of a successful computation the resulting processed data is then decoded into some suitable output data type. This idea is captured by the definition of an X-machine due to Eilenberg (1974).

It represents the most general approach to computing yet considered, in an algebraically more useful framework than the traditional idea of a Turing Machine. (The position of the recently postulated quantum computer is not, at the moment, quite clear.)

Thus an input data element $y \in Y$ is first of all encoded by α to form a subset $\alpha(y) \subseteq X$ (see Figure 4.7). We now look for a successful path from q_0 to q^1 in the finite state control. This path;

$$q_0 \xrightarrow{\phi_1} q_1 \xrightarrow{\phi_2} q_2 \xrightarrow{\phi_3} \ldots \xrightarrow{\phi_n} q^1 \qquad (4.4)$$

will have a label $\phi_1 \phi_2 \ldots \phi_n$ which defines a relation on X since each ϕ_i is a relation on X. (There may be several paths from q_0 to q^1 if the machine is non-deterministic.) We now apply $\phi_1 \phi_2 \ldots \phi_n$ to the subset $\alpha(y)$ to produce

$$(\phi_1 o \phi_2 o \ldots o \phi_n)(\alpha(y)) \qquad (4.5)$$

another subset of X. Now we apply the decoding relationship $\omega: X \quad Z$ to obtain

$$\omega((\phi_1 o \phi_2 o \ldots o \phi_n)(\alpha(y))) \qquad (4.6)$$

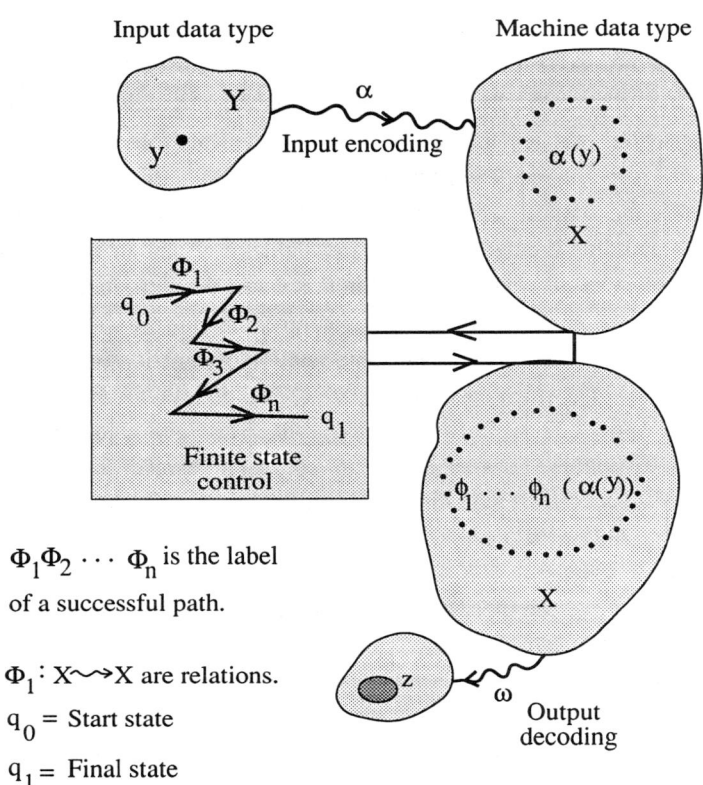

Figure 4.7. An X-Machine

our final output data set from Z.

Using this model of computation it is possible to precisely describe all of our previous notions of a machine and thus to formulate a general framework for the discussion of biochemical organization.

Furthermore, in other research, I am trying to construct a *formal specification language* using X-machines. This will, eventually, form the "front end" of a systems development tool which would be run on a highly parallel computer. A system would be *specified* (or defined) in the language and the machine would enable testing of the specification to be carried out and the automatic generation of correct code so that the system can be simulated. This goal is still some way off but eventually we hope to

Mathematical Models of Cell Biochemistry

simulate and examine the organization of complex systems such as we have been discussing.

4.4. THE TOPOLOGY OF DEATH

Our final contribution is a discussion on the way in which we might measure the organizational complexity of a system and use it to differentiate between a living cell and a dead cell, and between a growing cell and a dying cell.

If we regard the cell as a (large) number of processors connected together in a complex and highly parallel way we can examine the way in which information is flowing through the system. This information may be data (e.g., coenzymes, amino acids, etc.) or control information (e.g., enzymes) or energy information (e.g., ATP). The success of the organism or cell, depends on the flow of information being sufficient to keep the processors operating when they need to operate.

Let us specify N processors, $P_1, P_2,...,P_N$. Each processor P_1 will be connected (as far as information sharing goes) with certain other processors.

We define an NxN matrix A as shown in Figure 4.8 (some of weights might be negative to account for *inhibitory* channels).

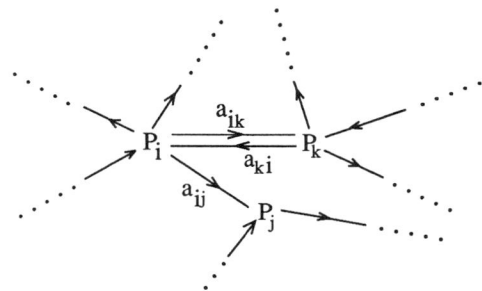

- Adjacency matrix $A(t) = (a_{ij}(t))$ is an N x N matrix of real numbers.
- $a_{ij}(t)$ gives the <u>weight</u> of the connection $P_i \longrightarrow P_j$ at time t.
- $a_{ij}(t) = 0$ means there is no connection $P_i \longrightarrow P_j$ at time t.
- $a_{ii}(t) = 0$ by convention, for all t.
- Readiness vectors: - both 1 x N vectors.
- $\tau(t) = (...,i,...,0,...)$, to describe transmission readiness, has 1 in position i if P_i is ready to transmit along a channel and 0 otherwise.
- $R(t) = (...,0,...,1,...,0)$ is a similar vector describing receiving readiness.

Figure 4.8. A processor network with information channels

The adjacency matrix A(t) is likely to change with time as some channels are altered by events. At a given time t, and given a knowledge of A(t), and the two readiness vectors, T(t) and R(t), we can analyze the amount of successful communication that can go on in the system. We envisage that the matrix A(t) will change only slowly with time, but T(t) and R(t) can change suddenly. The amount of successful communication can be calculated thus:

T(t)·A(t) provides us with information about the target processors that need to be communicated with and the "urgency" of that need from the weighing factors. The scalar product T(t)·A(t)*R(t) produces a number x(t) which indicates the success of the communication. As the cell ages and communication becomes more difficult, we find x(t) declining with $\frac{d}{dt}$(x(t)) negative. Some threshold value must define the certainty of death. During the living phase of the cell x(t) can be altering as the cell attempts to repair communication channels that might have become damaged, perhaps due to toxins, etc. Hence we can describe the living and dying phases of the cell using the function x(t).

$$x(t) = (T(t) \cdot A(t)) * R(t)$$

where * is the scalar product.

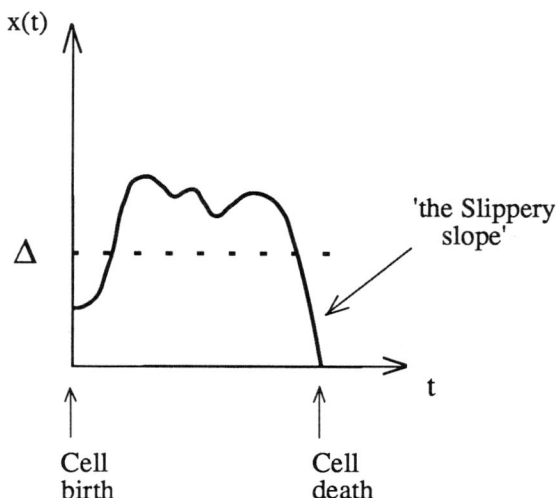

Δ is the death threshold.

If x(t) is $<\Delta$ and dx/dt < 0, then death is inevitable.

Figure 4.9. The measure of dynamic organization complexity

REFERENCES

Eilenberg, S. (1974). <u>Automata, Languages and Machines</u>, Vol A. New York, Academic Press.
Holcombe, W. M. L. (1982). <u>Algebraic Automata Theory</u>. Cambridge University Press, UK.
Holcombe, W. M. L. (1985). <u>Towards a Formal Description of Intracellular Biochemical Organization.</u> Dept. Research Report CS-MH-1, University of Sheffield.
Ji, S. (1985). The Bhopalator: A Molecular Model of the Living Cell Based on the Concepts of Conformons and Dissipative Structures. <u>J. theor. Biol.</u> <u>116</u>: 399-426.
Krohn, K., Langer, R. and Rhodes, J. L. (1967). Algebraic Principles for the Analysis of a Biochemical System. <u>J. Comp. and Syst. Sci.</u> <u>1</u>: 119-36.

Chapter 5

DAVYDOV'S SOLITON

Alwyn C. Scott

ABSTRACT

Davydov's theory for storage and transport of biological energy in protein is described and related to recent infrared absorption measurements in crystalline acetanilide. Some aspects of the quantum theory are considered in detail.

5.1. INTRODUCTION

In living organisms a fundamental mechanism for the transfer of energy into functional proteins or enzymes is the hydrolysis of adenosine triphosphate (ATP) into adenosine diphosphate (ADP) according to the reaction

$$ATP^{-4} + H_2O \rightarrow ADP^{-3} + HPO_4^{-2} + H_3O^+ \tag{5.1}$$

Under normal physiological conditions about 10 kcal/mol or 0.422 eV of free energy is released by this reaction (Fox, 1982) leading to several interesting questions: How is this free energy transferred into protein? How is it stored there? How does it move inside a protein? How is it transformed into useful work?

To answer questions of this sort a theory was proposed by Davydov (1973) which focused attention on the self-trapping of molecular vibrational energy in the amide-I (or CO stretch) vibration of the peptide unit (CONH), a basic structural element of all proteins. According to this theory, it was proposed that the localization of amide-I vibrational energy would alter the surrounding structure (primarily the hydrogen bonding) and that this local alteration would, in turn, lower the amide-I energy enough to prevent its dispersion.

At about the same time as the original paper by Davydov, Careri (1973) published some unexpected spectral measurements in the amide-I region of crystalline acetanilide ($CH_3CONHC_6H_5$), or ACN. As the temperature was lowered from room temperature, he observed an anomalous amide-I band (at 1650 cm^{-1}) growing up on the red side of the normal amide-I band (at 1665 cm^{-1}). This 1650 cm^{-1} band was called anomalous because it could not be explained with accepted concepts of molecular spectroscopy (e.g., Fermi resonance, Davydov splitting, etc.). At first Careri suspected some unusual one-dimensional phase transformation might provide an explanation, but no such evidence was found after several years of experimental work. Recently a self-trapping theory was proposed which is closely related to that of Davydov and explains the salient experimental facts (Careri et al. 1983, 1984; Eilbeck et al., 1984).

The present situation, therefore, is that the 1650 cm^{-1} band in ACN seems to provide direct experimental evidence for a self-trapped state of molecular vibrational energy. The "red shift" of 15 cm^{-1} from the normal band can be considered as the binding energy of a Davydov-like soliton, and this interpretation leads to quantitative predictions of biological significance.

This chapter is organized into three phases. The first is a review of Davydov's soliton theory and the experimental observations in crystalline acetanilide. The second phase is a detailed comparison of various attempts to provide a quantum mechanical explanation for self-trapping of molecular vibrations. Finally some questions of biological significance are briefly considered.

5.2   DAVYDOV'S SOLITON THEORY

This section is intended to provide a brief summary of Davydov's soliton theory for the convenience of the reader. Such a summary is helpful to appreciate the differences between the theory of self-trapping proposed for proteins and the theory proposed recently to explain experimental measurements on crystalline acetanilide. It is also necessary in order to see how the quantum theory developed by Davydov as a basis for self-trapping is related to other quantum analyses. Several surveys of this work are available for further reference (Davydov 1979a, 1982a).

Careful inspection of the α-helix structure of protein reveals three channels situated approximately in the longitudinal direction with the sequence

etc. H-N-C=O---H-NC=O---H-N-C=O---H-N-C=O etc.

where the dashed lines represent hydrogen bonds. For a detailed analysis it is necessary to consider the interaction of all three channels, but one is sufficient to lay out the basic ideas.

A single channel is governed by the energy operator

$$\hat{H} = \hat{H}_{CO} + \hat{H}_{ph} + \hat{H}_{int} \tag{5.2}$$

Taking the components of $\hat{H}$ in order, $\hat{H}_{CO}$ is an energy operator for the CO stretch (amide-I) vibration including the effects of nearest neighbor dipole-dipole interactions. Thus

$$\hat{H}_{CO} = \sum_n E_o \hat{b}_n^\dagger \hat{b}_n - J\, \hat{b}_{n+1}^\dagger \hat{b}_n + \hat{b}_n^\dagger \hat{b}_{n+1} \tag{5.3}$$

where E_o is the fundamental energy of the amide-I vibration, $-J$ is the nearest neighbor dipole-dipole interaction energy, and $\hat{b}_n^\dagger (\hat{b}_n)$ are boson creation (annihilation) operators for amide-I quanta on the nth molecule.

$\hat{H}_{ph}$ is the energy operator for longitudinal (acoustic) sound waves. Thus

$$\hat{H}_{ph} = \frac{1}{2}\sum_n [M^{-1}\hat{p}_n^2 + W(\hat{u}_n - \hat{u}_{n+1})^2] \tag{5.4}$$

where M is the mass of a molecule, W is the spring constant of a hydrogen bond, $\hat{p}_n$ is a longitudinal momentum operator for the nth molecule, and $\hat{u}_n$ is the corresponding longitudinal position operator.

$$\hat{H}_{int} = \chi_a \sum_n (\hat{u}_n - \hat{u}_{n-1}) \hat{b}_n^\dagger \hat{b}_n \tag{5.5}$$

where χ_a is the derivative of amide-I vibrational energy with respect to the length (R) of the adjacent hydrogen bond. Thus

$$\chi_a = dE_o/dR. \tag{5.6}$$

Davydov minimizes the average value of $\hat{H}$ with respect to the wave function

$$|\psi\rangle = \sum_n a_n(t) \exp[\hat{\sigma}(t)] \hat{b}_n^\dagger |0\rangle \tag{5.7}$$

where

$$\hat{\sigma} \equiv -\frac{i}{\hbar} \sum_n [\hat{\beta}_a(t)\hat{P}_n - \pi_n(t)\hat{u}_n] \quad (5.8)$$

A straightforward calculation shows that

$$\beta_n(t) = <\psi|\hat{u}_n|\psi> \quad (5.9)$$

and

$$\pi_n(t) = <\psi|\hat{P}_n|\psi>. \quad (5.10)$$

The wavefunction in (5.7) will by called Davydov's ansatz throughout this chapter. One of the aims here is to study the range of validity of this ansatz.

Assuming that Davydov's ansatz approximates the true wave function, (5.9) and (5.10) show that β_n and π_n are the average values of the position and momentum operators, respectively. Furthermore, a_n is the probability amplitude for finding a quantum of amide-I vibrational energy on the nth molecule. The normalization condition $<\psi|\psi> = 1$ implies that

$$\sum_n |a_n|^2 = 1. \quad (5.11)$$

Thus Davydov's ansatz describes the dynamics of a single quantum of amide-I vibrational energy.

Minimization of $<\psi|\hat{H}|\psi>$ with respect to a_n, β_n and π_n leads to the differential-difference equations

$$(i\hbar \frac{d}{dt} - E_o) a_a + J(a_{n+1}) - \chi_a(\beta_n - \beta_{n+1})a_n = 0 \quad (5.12a)$$

$$M \frac{d^2\beta_n}{dt^2} - W(\beta_{n+1} - 2\beta_n + \beta_{n+1}) = \chi_a[|a_{n+1}|^2 - |a_n|^2] \quad (5.12b)$$

Extensive numerical and theoretical analysis of (5.12) yields the following results (Scott 1982, 1983, 1984; MacNeil and Scott 1984): (i) it is reasonable to expect soliton formation at the level of energy released by ATP hydrolysis (5.1), and (ii) such a soliton travels rather slowly with respect to the speed of longitudinal sound waves. This suggests neglecting the kinetic energy of longitudinal sound by assuming $\beta_n = 0$, whereupon

$$\beta_n - \beta_{n+1} = -\frac{X_a}{W} |a_n|^2 \qquad (5.13)$$

and, in this "adiabatic approximation," (5.12) becomes

$$(i\hbar \frac{d}{dt} - E_o) a_n + J(a_{n+1}) + \gamma_a |a_n|^2 a_n = 0 \qquad (5.14)$$

where

$$\gamma_a \equiv X_a^2/W, \qquad (5.15)$$

Davydov has emphasized that a solitary wave solution of (5.12) cannot be created directly by absorption of a photon because of an unfavorable Franck-Condon factor. This is because the necessary intermolecular displacement in (5.12b) cannot occur in a time that is short enough for photon absorption. The Franck-Condon factor will be discussed in detail in the following section.

5.3. SELF-TRAPPING IN CRYSTALLINE ACETANILIDE

Just as in the α-helix, careful inspection of the crystal structure of acetanilide reveals channels situated in the b-direction with the sequence

etc. H-N-C=O---H-N-C=O---H-N-C=O---H-N-C=O etc.

Recent infrared absorption measurements on microcrystals of acetanilide show an unexpected band at 1650 cm^{-1} which rises with decreasing temperature to become the dominant spectral feature below 100 °K. When this band was discovered, Careri suspected it to be caused by a subtle phase change along b-direction of the crystal (Careri, 1973), but careful studies over a period of several years failed to reveal any such evidence. The lack of a viable alternative eventually led to the suggestion that the 1650 cm^{-1} band might be caused by direct absorption of an infrared photon into a self-trapped state similar to that proposed by Davydov. The qualifier "similar" is important because, as was noted above, the Franck-Condon factor is unfavorable for direct photon absorption by a self-trapped solution of (5.12).

Davydov's Soliton

The corresponding theory proceeds, as in the previous section, by defining the energy operator

$$\hat{H} = \hat{H}_{CO} + \hat{H}_{ph} + \hat{H}_{int} \tag{5.16}$$

where $\hat{H}_{CO}$ is again given by (5.3) but with (Eilbeck et al., 1984).

$$J = 3.96 \text{ cm}^{-1}. \tag{5.17}$$

In the present analysis, however, self-trapping is assumed to be caused by interaction with an optical phonon rather than an acoustic phonon. Thus

$$\hat{H}_{ph} = \tfrac{1}{2}\sum_n [m^{-1}\hat{P}_n^2 + w\hat{q}_n^2] \tag{5.18}$$

and

$$\hat{H}_{int} = \chi_o \sum \hat{q}_n \hat{b}_n^\dagger \hat{b}_n. \tag{5.19}$$

Minimization of $<\psi|\hat{H}|\psi>$ with respect to the parameters of the Davydov ansatz wavefunction (5.7), where

$$\hat{\sigma} \equiv -\frac{1}{\hbar}\sum_n [q_n(t)\hat{P}_n - P_n(t)\hat{q}], \tag{5.20}$$

leads to the dynamic equations

$$(i\hbar \frac{d}{dt} - E_o)a_n + J(a_{n+1} + a_{n-1}) - \chi_o q_n a_n = 0 \text{ and} \tag{5.21a}$$

$$m\frac{d^2 q_n}{dt^2} - w q_n = \chi_o |a_n|^2. \tag{5.21b}$$

As before

$$q_n(t) = <\psi|\hat{q}_n|\psi>. \tag{5.22}$$

The adiabatic approximation ($q_n = 0$) reduces (5.21) to

$$(i\hbar \frac{d}{dt} - E_o) a_n + J(a_{n+1} + a_{n-1}) - \gamma_o |q_n|^2 a_n = 0 \qquad (5.23)$$

where

$$\gamma_o \equiv \chi_o^2 a/w. \qquad (5.24)$$

A detailed numerical study of a system of equations similar to (5.23) but representing one hundred molecules of ACN in two coupled channels has recently been carried out by Eilbeck et al. (1984). This work shows that the red shift from the normal amide-I band at 1665 cm^{-1} to the 1650 cm^{-1} band is best fit by choosing

$$\gamma_o = 44.7 \text{ cm}^{-1}. \qquad (5.25)$$

We turn next to an estimate of the Franck-Condon factor for direct photon absorption by a self-trapped state of (5.21). Before absorption $|a_n|^2 = 0$, and after absorption $|a_n|^2 = 0$ over a localized region such that (5.11) is satisfied. Thus the ground state wavefunction of (5.21b) must shift from

$$\phi_o = (\frac{w}{\pi \hbar \omega})^{1/2} \exp(-q_n^2 \frac{w}{2\hbar\omega}) \qquad (5.26)$$

before absorption to

$$\tilde{\phi}_o = (\frac{w}{\pi \hbar \omega})^{1/2} \exp[-q_n + \gamma_o |a_n|^2)^2 \frac{w}{2\hbar\omega}] \qquad (5.27)$$

after absorption, where

$$\omega = (w/m)^{1/2} \qquad (5.28)$$

is the frequency of the optical mode that is mediating the self-trapping. The transition probability for soliton absorption is therefore reduced by the Franck-Condon factor

$$[\int \phi_o \tilde{\phi}_o^* \, dq_n]^2 \geq \exp(-\gamma_o/2\hbar\omega), \qquad (5.29)$$

Davydov's Soliton

which is close to unity for

$$\gamma_o \ll \hbar\omega. \tag{5.30}$$

The frequency (ω) of the optical mode can be determined from the temperature dependence of the 1650 cm^{-1} line. Such temperature dependence is expected, because the probability of (5.21b) being in its ground state, and therefore able to participate in self trapping, is $[1-\exp(-\hbar\omega/kT)]$. Thus as the temperature is raised, the low temperature factor given in (5.29) should be reduced by the additional factor $[1-\exp(-\hbar\omega/kT)]^2$. A least square fit to intensity data is obtained for $\hbar\omega = 131$ cm^{-1}. Together with (5.25) this implies $\exp(-\gamma_o/2\hbar\omega) = 0.84$.

Further evidence tending to favor a self-trapping explanation for the 1650 cm^{-1} band is the recent observation of the overtone series shown in Table 5.1 (Scott et al., 1985). Since the overtones $N \geq 2$ are self-trapped states involving more than one quantum of the amide-I vibration, it is interesting to consider states that avoid the constraint of (5.11).

Table 5.1. Overtone Series for the ACN Soliton

N	$\nu(N)$
1	1650 cm^{-1}
2	3250
3	4803
4	6304

5.4. THE QUANTUM THEORY OF SELF-TRAPPING

In this section we approach the problem from a classical perspective. Starting with the classical amide-I coordinates, P_n and Q_n, for which the Hamiltonian is $\Sigma_n[P_n^2 + Q_n^2]$, it is convenient to define the complex mode amplitudes

$$A_n \equiv \omega_o^{1/2}(P_n + iQ_n). \tag{5.31}$$

In terms of these complex mode amplitudes (including dipole-dipole interactions)

$$H_{CO} = \sum_n [\frac{E_o}{\hbar}|A_n|^2 - J(A^*_{n+1}A_n + A^*_n A_{n+1})] \tag{5.32}$$

where

$$\omega_o = E_o/\hbar \tag{5.33}$$

is the classical oscillation frequency of an amide-I vibration. (From here on we will assume $\hbar = 1$ and measure energy and frequency in the same units.)

With a classical interaction energy

$$H_{int} = \chi \sum_n q_n |A_n|^2, \tag{5.34}$$

where q_n is the coordinate of some low-frequency phonon with adiabatic energy

$$H_{ph} = \frac{1}{2} w \sum_n q_n^2 \tag{5.35}$$

one arrives at the total classical Hamiltonian

$$H = H_{CO} + H_{ph} + H_{int}. \tag{5.36}$$

Minimizing (5.36) with respect to the q_n requires

$$q_n = -\frac{\chi}{w}|A_n|^2, \tag{5.37}$$

whereupon (4.6) can be reduced to

$$H = \sum_n E_o|A_n|^2 - J(A^*_{n+1}A_n + A^*_n A_{n+1}) - \frac{1}{2}\gamma|A_n|^4 \tag{5.38}$$

where

$$\gamma \equiv \chi^2/w. \tag{5.39}$$

The corresponding dynamical equation for A_n is

$$(i\frac{d}{dt} - E_o)A_n + J(A_{n+1} + A_{n-1}) + \gamma|A_n|^2 A_n = 0. \tag{5.40}$$

Davydov's Soliton

In additional to the energy, H, another constant of the motion along solutions of (5.40) is the number

$$N = \sum_n |A_n|^2. \tag{5.41}$$

To this point the discussion of the present section has been entirely classical. We now consider quantization in four special cases: (i) $J \ll \gamma$, (ii) $\gamma \ll J$, (iii) semiclassical quantization, and (iv) the Davydov ansatz. In each case it will be of particular interest to calculate an overtone series corresponding to that presented in Table 5.1 for crystalline acetanilide.

5.4.1. The Case $J \ll \gamma$

In this case we neglect the dipole-dipole interaction terms in (5.38) and (5.40), and write the energy

$$H = \sum_n h_n \tag{5.42}$$

where

$$h_n = E_o |A_n|^2 - \frac{1}{2}\gamma |A_n|^4. \tag{5.43}$$

Under quantization, the terms in (5.42) become operators

$$h_n \to \hat{h}_n \tag{5.44}$$

through replacement of the complex mode amplitudes by creation and annihilation operators for bosons. Thus

$$A_n \to \hat{b}_n, \tag{5.45a}$$

$$A_n^* \to \hat{b}_n^\dagger. \tag{5.45b}$$

Since the ordering of these operators is not determined by (5.43), we take the averages

$$|A_n|^2 \to \frac{1}{2}(\hat{b}^\dagger \hat{b} + \hat{b}\hat{b}^\dagger) \text{ and} \tag{5.46}$$

$$|A_n|^4 \to \tfrac{1}{6}(\hat{b}^\dagger \hat{b}^\dagger \hat{b}\hat{b} + \hat{b}^\dagger \hat{b}\hat{b}^\dagger \hat{b} + \hat{b}^\dagger \hat{b}\hat{b}\hat{b}^\dagger + \hat{b}\hat{b}^\dagger \hat{b}\hat{b}^\dagger + \hat{b}\hat{b}\hat{b}^\dagger \hat{b}^\dagger + \hat{b}\hat{b}^\dagger \hat{b}^\dagger \hat{b}). \tag{5.47}$$

where the subscript have been dropped for typographical convenience. Noting that $\hat{b}^\dagger$ and $\hat{b}$ have the properties $\hat{b}^\dagger|N\rangle = \sqrt{N+1}\,|N+1\rangle$ and $\hat{b}|N\rangle = \sqrt{N}\,|N-1\rangle$ (where $|N\rangle$ is an harmonic oscillator eigenstate), it is straight-forward to show that

$$\hat{h} = (E_o - \tfrac{1}{2}\gamma)(\hat{b}^\dagger \hat{b} + \tfrac{1}{2}) - \tfrac{1}{2}\gamma \hat{b}^\dagger \hat{b}\hat{b}^\dagger \hat{b}. \tag{5.48}$$

Thus

$$\hat{h}|N\rangle = E(N)|N\rangle \tag{5.49}$$

where

$$E(N) = (E_o - \tfrac{1}{2}\gamma)(N + \tfrac{1}{2}) - \tfrac{1}{2}\gamma N^2. \tag{5.50}$$

In summary, eigenvectors of the operators defined through (5.43), (5.44), (5.46) and (5.47) are identical to those of an harmonic oscillator, but the corresponding eigenvalues are given (5.50).

The form of (5.50) is significant. It can be written

$$E(N) = E^C + E^L + E^{NL} \tag{5.51}$$

where E^C is the ground state ($N = 0$) energy, $E^L \propto N$ and

$$E^{NL} = -\tfrac{1}{2}\gamma N^2. \tag{5.52}$$

This "nonlinear" contribution is directly measured from the overtone series in Table 5.1.

5.4.2. The Case $\gamma \ll J$

In this case the classical equation (5.40) reduces to the nonlinear Schrödinger

(NLS) equation of soliton theory. To see how this goes, assume the repeat distance between molecules of d and replace the discrete variable n by a continuous variable, $x = n$, which measures distance in units of d. Then (5.40) takes the form

$$(i\frac{\partial}{\partial t} - E_o + 2J) A + J \frac{\partial^2 A}{\partial x^2} + \gamma |A|^2 A = 0. \tag{5.53}$$

Quantization of this equation was originally performed using the Bethe ansatz method and recently it has been shown that such solutions can be efficiently constructed from a quantum version of inverse scattering theory (Sklyanin and Faddeev, 1978; Thacker and Wilkinson, 1979).

Under quantization, the functions A and A* are replaced by annihilation and creation operators for boson fields, $\hat{\phi}$ and $\hat{\phi}^\dagger$. At equal times these have the commutation relations $[\hat{\phi}(x), \hat{\phi}*(y)] = [\hat{\phi}^\dagger(x), \hat{\phi}^\dagger(y)] = 0$ and $[\hat{\phi}(x), \hat{\phi}^\dagger(y)] = \sigma(x - y)$. In terms of the previous discussion it is evident that $\hat{\phi}(x)$ is equivalent (under scaling) to $\hat{b}_n$ in the continuous limit $n = x$. In effecting this limit two procedures are customary: (i) neglect consideration of the ground state energy which is unbounded in the limit, and (ii) "normal order" all operator expressions, i.e., move all creation operators to the left.

Since $\hat{b}\hat{b}^\dagger = \hat{b}^\dagger\hat{b} + 1$, normal ordering of (5.48) and neglect of the ground state energy imply

$$\hat{h}_n = (E_o - \gamma)\hat{b}_n^\dagger\hat{b}_n - \frac{1}{2}\gamma\hat{b}_n^\dagger\hat{b}_n^\dagger\hat{b}_n\hat{b}_n. \tag{5.54}$$

Thus to put (5.53) in standard form for quantum analysis, let

$$A = \phi \exp[-i(E_o - 2J - \gamma)t] \tag{5.55}$$

and scale time as $t \to t/J$. Then (5.53) becomes

$$i\phi_t + \phi_{xx} + \frac{\gamma}{J}|\phi|^2\phi = 0 \tag{5.56}$$

where a subscript notation is used for the partial derivatives. Under quantization $\phi \to \hat{\phi}$ and (5.56) becomes the operator equation

$$i\hat{\phi}_t + \hat{\phi}_{xx} + \frac{\gamma}{J}\hat{\phi}^\dagger\hat{\phi}\hat{\phi} = 0 \tag{5.57}$$

with energy operation

$$\hat{H} = \int dx \, \hat{\phi}_x^\dagger \hat{\phi}_x - \frac{1}{2} \frac{\gamma}{J} \int dx \, \hat{\phi}^\dagger \hat{\phi}^\dagger \hat{\phi} \hat{\phi}, \tag{5.58}$$

number operator

$$N = \int dx \, \hat{\phi}^\dagger \hat{\phi} \tag{5.59}$$

and momentum operator

$$P = -i \int dx \, \hat{\phi}^\dagger \hat{\phi}_x. \tag{5.60}$$

The quantum inverse scattering method provides exact wavefunctions, $|\psi\rangle$, that diagonalize $\hat{N}$, $\hat{P}$ and $\hat{H}$ as follows:

$$\hat{N}|\psi\rangle = N|\psi\rangle \tag{5.61}$$

where

$$N = \text{integer} \geq 0 \text{ and} \tag{5.62}$$

$$\hat{P}|\psi\rangle = Np|\psi\rangle \tag{5.63}$$

where p is a real number, and

$$\hat{H}|\psi\rangle = Np^2 + \frac{\gamma^2}{48J^2}(N - N^3)|\psi\rangle. \tag{5.64}$$

Furthermore in the limit $\gamma/J \to 0$

$$|\psi\rangle \to \int dx \, e^{ipx} \hat{\phi}|0\rangle. \tag{5.65}$$

Equations (5.62) and (5.64) imply an overtone series

$$E(N) = E^L + E^{NL} \tag{5.66}$$

where $E^L \propto N$ and

$$E^{NL} = -\frac{\gamma^2}{48J^2} N^3. \tag{5.67}$$

5.4.3. Semiclassical Quantization

In the parameter range $\gamma \approx J$, it is possible to impose elementary quantum conditions on stationary solutions of (5.40). Writing such a solution in the form

$$A_n = \left(\frac{J}{\gamma}\right)^{1/2} \alpha_n \exp\left[-i\left(\frac{E_o}{J} + \omega\right)t\right] \tag{5.68}$$

reduces (5.40) to the standard form

$$\omega \alpha_n + \alpha_{n+1} + \alpha_{n-1} + \alpha_n^3 = 0. \tag{5.69}$$

Using a shooting method (Scott and MacNeil 1983) it is possible to find a family of numerical solutions for (5.69) with the following properties: (i) $\alpha_n = \alpha_{-n}$, (ii) for $n \geq 0$, $\alpha_n > \alpha_{n+1}$, and (iii) $\lim_{n \to \infty} \alpha_n = 0$. From such a solution the conserved quantities H and N defined in (5.38) and (5.41) are readily calculated as

$$N = \frac{J}{\gamma} \sum_n \alpha^2 \tag{5.70}$$

and

$$H(N) = E_o - J \frac{\sum_n \alpha_n \alpha_{n-1}}{\sum_n \alpha_n^2} N - \frac{1}{2}\gamma N^2 \frac{\sum_n \alpha_n^4}{\left(\sum_n \alpha_n^2\right)^2} \tag{5.71}$$

Semiclassical quantization is effected by noting that stationary solutions are of the form

$$A_n(t) = A_{n0} \exp[-i\theta(t)] \tag{5.72}$$

where $\dot{\theta} = dH(N)/dN$. Thus N and θ are conjugate variables and the quantum condition

$$\oint N \, d\theta = 2\pi(\text{integer}) \tag{5.73}$$

together with the definition of N (5.41) imply

$$N = \text{integer} \geq 0. \tag{5.74}$$

Equation (5.71) has the form

$$E(N) = E^L + E^{NL} \tag{5.75}$$

where

$$E^{NL} = -\frac{1}{2}\gamma N^2 \frac{\sum_n \alpha_n^4}{(\sum_n \alpha_n^2)^2} \tag{5.76}$$

In the limit $J \ll \gamma$, $\alpha_n \ll \alpha_0$ for $|n| \geq 1$ so (5.76) evidently reduces to (5.52). In the limit $\gamma \ll J$ it is straightforward to show that (5.76) reduces to (5.67). Thus (5.76) is expected to provide an accurate calculation of E^{NL} over the entire parameter range.

It is now possible to consider how data from the overtone series for the 1650 cm^{-1} band in acetanilide compare with these calculations. From (5.17) and (5.25).

$$\gamma_0/J = 11.3. \tag{5.77}$$

This lies in the range of (5.76) for which

$$E(N) \doteq E_0(N) - \frac{1}{2}\gamma N^2, \tag{5.78}$$

so the line at 1650 cm^{-1} implies

$$E_0 = 1672.3 \text{ cm}^{-1}. \tag{5.79}$$

From the measured values of overtone frequency, $v(N)$ in Table 5.1, the nonlinear contributions to the overtone spectrum can be calculated as

$$E^{NL} = v(N) - NE_0. \tag{5.80}$$

In Table 5.2 we compare these calculations with those computed from (5.78).

Table 5.2. Nonlinear Terms in Acetanilide Overtone Series

N	$-E^{NL}$ (cm^{-1})	$-\frac{1}{2}\gamma N^2$ (cm^{-1})
1	22	22
2	95	89
3	214	201
4	385	357

5.4.4. Davydov's Ansatz

We are now in a position to evaluate Davydov's ansatz. In the context of an adiabatic approximation, the wavefunction introduced in (5.7) takes the form

$$|\psi\rangle = \sum_n a_n(t) \hat{b}_n^\dagger |0\rangle, \qquad (5.81)$$

where the $a_n(t)$ are solutions of (5.14). The form of the Davydov ansatz has the following properties:

(i) In the limit $J \ll \gamma$, it reduces to the first eigenfunction $|1\rangle$, in (5.49).

(ii) In the limit $\gamma \ll J$, it reduces to the asymptotic form of Bethe's ansatz in (5.65).

(iii) Between these two limits, Davydov's ansatz gives energies that agree with semiclassical calculations.

Thus one concludes that Davydov's ansatz is a useful approximation to the exact wavefunction over the entire parameter range $0 \leq \gamma/J < \infty$ with the constraint (5.11) which implies $N = 1$.

5.5. BIOLOGICAL SIGNIFICANCE OF SELF-TRAPPING

Measurements on crystalline acetanilide (ACN) confirm Davydov's theory of self-trapped states (solitons) in hydrogen-bonded polypeptide chains. Furthermore, Table 5.1 shows that the "N = 2" state in ACN can absorb almost all (95%) of the free energy released in hydrolysis of adenosine triphosphate (ATP). It is reasonable to suppose that a corresponding state can form on the hydrogen bonded polypeptide chains of α-helix.

Over a decade ago McClare argued that the free energy released in ATP hydrolysis should transfer resonantly into a protein in order to avoid thermal degradation (McClare, 1972a,b, 1974). To store and transport this energy he posited an "excimer" state in protein which would be closely related to the amide-A band of α-helix at 3240 cm^{-1}. McClare's excimer is qualitatively similar to the "conformon" of Green and Ji (1972) and Ji (1974, 1979, 1985, 1987; Section 1.5.2, Chapter 1) and the basic properties of both are provided by a Davydov soliton in the "N = 2" state (Davydov 1973, 1974, 1977, 1979b, 1982b). In the past, such suggestions have been rejected or ignored by the biochemical community because a localized region of free energy within a protein was believed to be physically impossible. Since this view is no longer tenable, the early proposals of Davydov, McClare, and Green and Ji must be reevaluated. A recent paper by Careri and Wyman (1984), suggesting a soliton mechanism for cyclic enzyme activity, provides a first step in this direction.

REFERENCES

Careri, G. (1973). Search for Cooperative Phenomena in Hydrogen-Bonded Amide Structures. In Cooperation Phenomena (H. Haken and M. Wagner, eds.) Springer, Berlin, pp. 391-394.

Careri, G., Buontempo, U., Carta, R., Gratton, E. and Scott, A. C. (1983). Infrared Absorption in Acetanilide by Solitons. Phys. Rev. Lett. 51: 304-307.

Careri, G., Buontempo, U., Galluzzi, F., Scott, A. C., Gratton, E. and Shyamsunder, E. (1984). Spectroscopic Evidence for Davydov-like Solitons in Acetanilide. Phys. Rev. B30: 4689-4702.

Careri, G. and Wyman, J. (1984). Soliton-Assisted Unidirectional Circulation in a Biochemical Cycle. Proc. Natl. Acad. Sci. (USA) 81: 4386-4388.

Davydov, A. S. (1973). The Theory of Contraction of Proteins Under Their Excitation. J. theor. Biol. 38: 559-569.

Davydov, A. S. (1974). Quantum Theory of Muscular Contraction. Biofizika 19: 670-676.

Davydov, A. S. (1977). Solitons and Energy Transfer Along Protein Molecules. J. theor. Biol. 66: 379-387.

Davydov, A. S. (1979a). Solitons, Bioenergetics, and the Mechanism of Muscle Contraction. Int. J. Quant. Chem. 16: 5-17.

Davydov, A. S. (1979b). Solitons in Molecular Systems. Phys. Ser. 20: 387-394.

Davydov, A. S. (1982a). Biology and Quantum Mechanics, Pergamon, New York.

Davydov, A. S. (1982b). Solitons in Quasi-One Dimensional Structures. Soc. Phys. Usp. 25: 898-918 [Usp. Fiz. Nauk 138: 603-643.

Eilbeck, J. C., Lomdahal, P. S. and Scott, A. C. (1984). Soliton Structure in Crystalline Acetanilide. Phys. Rev. B30: 4703-4712.

Fox, R. F. (1982). Biological Energy Transduction Wiley, New York p. 216.

Green, D. E. and Ji, S. (1972). The Electromechanochemical Model of Mitochondrial Structure and Function. In: The Molecule Basis of Electron Transport (J. Schultz and B. F. Cameron, eds.) Academic Press New York, pp. 1-43.

Ji, S. (1974). A General Theory of ATP Synthesis and Utilization. Ann. N.Y. Acad. Sci. 227: 419-437.

Ji, S. (1979). The Principles of Ligand-Protein Interaction and their Applications to the Mechanism of Oxidative Phosphorylation. In Structure and Function of Biomembranes (Yagi, K., ed.). Japan Scientific Societies Press, Tokyo, pp. 25-37.

Ji, S. (1985). Conformons and Solitons: New Concepts in Bioenergetics. In: The Living State - II (R. K. Mishra, ed.), World Scientific Publishing Co., Singapore, pp. 563-573.

Ji, S. (1987). A General Theory of Chemical Cytotoxicity based on a Molecular Model of the Living Cell, the Bhopalator. Arch. Toxicol. 60: 95-102.

MacNeil, L. and Scott, A. C. (1984). Launching a Davydov Soliton: II. Numerical Studies. Phys. Ser. 29: 284-287.

McClare, C. W. F. (1972a). A Quantum Mechanical Muscle Model. Nature 240: 88-90.

McClare, C. W. F. (1972b). A "Molecular Energy" Muscle Model. J. theor. Biol. 35: 569-595.

McClare, C. W. F. (1974). Resonance in Bioenergetics. Ann. N.Y. Acad. Sci. 227: 74-97.

Scott, A. C. (1982). Dynamics of Davydov Solitons. Phys. Rev. A26: 578-595.

Scott, A. C. (1983). Erratum: Dynamics of Davydov Solitons. Phys. Rev. A27: 2767.

Scott, A. C. (1984). Launching a Davydov Soliton: I. Soliton Analysis. Phys. Rev. 29: 279-283.

Scott, A. C. (1985). Solitons in Biological Molecules. Comments Mol. Cell. Biophys. 3(1): 15-37.

Scott, A. C., Gratton, E., Shyamsunder, E. and Careri, E. (1985). The IR Overtone Spectrum of the Vibrational Soliton in Crystalline Acetanilide. Phys. Rev. B32, 5551-5553.

Scott, A. C. and MacNeil, L. (1983). Binding Energy Versus Nonlinearity for a "Small" Stationary Soliton. Physics Lett. 98A(3): 87-88.

Skylanin, E. K. and Faddeev, L. D. (1978). Quantum Mechanical Approach to Completely Integrable Field Theory Models. Sov. Phys. Dokl. 23: 902-904 [Dok]. Acad. Hauk SSSR 243, 1430-1433].

Thacker, H. B. and Wilkinson, D. (1979). Inverse Scattering Transform as an Operator Method in Quantum Field Theory. Phys. Rev. D19: 3360-3665.

"At the very outset of our studies we are faced by one of the most difficult problems with which the biologist has to deal, namely the structure, chemical and physical, and the elementary properties of protoplasm."

Bayliss, 1920

Chapter 6

CYTOSOCIOLOGY: A FIELD-THEORETIC VIEW OF CELL METABOLISM

Harry A. Smith and G. Rickey Welch

ABSTRACT

This paper proposes a unique approach to describing the interconnectiveness of cellular metabolism. We apply basic concepts of "curved" or "connected" geometrics describing various natural processes (general relativity, gauge field theories) to a holistic description of cellular processes. The first approach entails a geometric view of cellular metabolism in which metabolic processes are treated as "geodesic" paths on a curved "metabolic manifold" described by a "metabolic metric." The second approach describes how the concept of gauge fields may be used to illustrate the unique properties of microenvironments within the living cell, with particular attention given to how fundamental metabolic properties such as free energy changes and kinetic parameters of enzymes can be described as being "gauged" according to the specific environment in question. Our results suggest that matter in biological systems be viewed as distinct states which are "alive" when existing in a self-propagating, internally-negentropic region of space-time "curved" by global (boundary) free-energy dissipation.

6.1. INTRODUCTION

One of the major outstanding problems of the biological sciences is the

Cytosociology: A Field-Theoretic View of Cell Metabolism

precise manner in which the spatiotemporal order of the living state gives rise to those unique processes deemed characteristic of life. That such a problem remains extant is interesting, in itself, considering the exemplary recognition biologists have given to the organization of living beings. Yet after nearly two centuries of "modern" biological thought we are only now perceiving the direction that an understanding of this problem will take. As evidenced by the eclectic nature of these proceedings, the organization of life and death will be comprehended only by a merging of a variety of disciplines, mostly biological, but not excluding the mathematical, physical and social realms.

Motivated by the above epistemological perspective we discuss the nature of living matter from a dynamic, sociological and process-oriented view. In order to accomplish this goal, we provide a conceptual framework which takes the unique biological aspects of cellular dynamics and describes them in terms of *fields*—a concept which has provided modern physics with a powerful tool for understanding the dynamical interactions of elementary particles. The field concept is particularly useful for studying the "sociological" aspects of cellular metabolism since it provides a means of discussing long-range interactions without such distasteful notions as "action at a distance." Moreover, field dynamics naturally lends itself to a geometric view in which symmetries and broken symmetries (other key elements of the spatio-temporal order of the living state) are especially well described.

The relatively young science of biology is not without its own field constructs. In the past, most consideration thereof has dealt with morphogenetic gradients in tissue development. In the last 10-15 years, the application of certain physico-chemical principles to biological phenomena has been paving the way for the establishment of a hierarchical field structure for biological space-time—analogous to, and consonant with, that in physics.

Here, we will elaborate the issue rather specifically at the level of the *metabolic infrastructure* of the living cell, for the following reasons: i) the cell is the smallest functional unit of life, and ii) the material events involved in cellular processes are directly amenable to basic physicochemical principles. Mathematical similarities and analogies between the biological processes and various theoretical constructs in physics suggest a manner by which epiphenomena of the living state may ultimately be drawn into a unity (or "implicate order," Bohm, 1980) with the current holistic views of the universe.

We present here two different, but not unrelated, field theories of the living state. The first entails a geometric view of cellular metabolism in which metabolic processes are treated as "geodesic" paths on a curved "metabolic manifold" described by a "metabolic metric." The second approach describes how the concept of gauge fields may be used to illustrate the unique properties of microenvironments within the living cell, with particular attention to how fundamental metabolic properties such a free energy changes and kinetic parameters of enzymes can be described as being "gauged" according to the specific environment in question.

6.2. CYTOSOCIOLOGY OF CELL METABOLISM

In recent decades, electron microscopy has revealed a richly diverse particulate infrastructure in living cells. This structure encompasses the extensive membrane reticulation as well as the hyaloplasmic space. The latter is further found to be laced with a dense array of cytoskeletal elements and an interlocking microtrabecular lattice (Porter, 1984; Clegg, 1984). It is clear that cellular dynamics must, therefore, take into account this high degree of "surface-phenomena" if it is to be reflective of the activity of the living state. A question that naturally arises is the extent to which the catalytic processes of the cell are associated with, or take advantage of, this organization.

Accumulating evidence indicates that the majority (if not all) of the enzymes of intermediary metabolism operate *in vivo* in association with particulate structures. A casual examination of the literature indicates extensive organization in virtually all major metabolic pathways (Clegg, 1984; Friedrich, 1984; Porter, 1984; Sitte, 1980; Srere, 1981, 1987; Welch, 1977a, 1985, 1987), involving, in part, formation of protein-protein complexes and/or adsorption to cytological substructures. The "concentrations" of proteins in association with cytomembranes and organelles indicate high, crystal-like densities (Sitte, 1980; Srere, 1981). Moreover, there is a remarkable homology in the surface area-to-volume ratio for *all* membranous cytological structures. These considerations have led Sitte (1980) to propose that all cytomembraneous elements have evolved to function as "protein collectors" in the operation of cell metabolism

Thus, cell biology presents us with a rather simple, biphasic view of cellular infrastructure: i) a solid phase, encompassing extensive membrane surfaces and the fibrous lattice-work; and ii) a soluble, aqueous phase which may, in itself, have considerable structure (Clegg, 1984). From the evidence presented above, we must focus on the solid phase as the primary site of intermediary metabolism, with the soluble phase functioning largely in subservient roles such as thermal buffering, distribution of substrates, and maintenance of proper ionic and "regulatory" environments. It is the *solid state* nature of metabolic organization which affords the cell its enormous degree of control over matter and energy flux. The enhancements of enzymic activity and control related to such organization are, by now, well known

(Welch, 1977b), and the concept of *channeling* of substrates through metabolic pathways has been a subject of continuing interest (Friedrich, 1984; Srere, 1981; Welch 1977b; Welch and Clegg, 1987). It is apparent that organization of metabolic pathways leads to more efficient "communication" and "cooperation" among cellular elements and processes, thus implying that we must take a more group-dynamical approach towards elucidating fundamental metabolic processes. Our rationale for applying the concepts of field theory to metabolic dynamics is, therefore, well grounded in the experimental framework which has elucidated the organization of metabolism.

Within these "organizational modes" of metabolism and their associated microenvironments, the familiar concepts of scalar chemical reactions, uniform concentrations, homogeneity, and isotropy must be abandoned in favor of their dialectical opposites. One cannot help but suppose that with these radical shifts in our understanding of the basic physical chemistry of cellular organization will come glimmers of insight as to the manner in which this organization contributes to those unique properties we singularly associate with the "living state." Indeed, such shifts in paradigms by which thermodynamics has been applied to the living, organized state have led the Brussels school (Nicolis and Prigogine, 1977; Prigogine, 1980) to an elegant formulation of non-linear, non-equilibrium thermodynamics in which the formation of highly organized dissipative *global* structures occur within a natural framework. The lessons we have learned from the monumental efforts of the Brussel's school are clear; organization cannot be described using simple linear superpositions of interacting states. Rather, we must break free of the bounds of linear thought (so entrenched in classical biochemistry) and seek out non-linear models in our quest to bring biological organization into the realm of the tractable. We suggest that the field theoretic view described herein offers such a model of *local* cellular dynamics. As such, this view supplements and augments the global models of spatiotemporal organization which have been so successful in drawing one of the fundamental questions of biology under the umbrella of modern physical thought.

6.3. CYTOSOCIOLOGY OF ENZYME ACTION IN VIVO

Traditionally, we have defined "enzymes" simply as "proteinaceous catalysts." Now, we must add more *biological* flavor, asserting that enzymes catalyze specific reactions at rates *and under appropriate conditions*, commensurate with the vitality of the cell—with the idea of locational specificity as part-and-parcel of the defining character.

In these localized microenvironments, traditional "macroscopic" descriptions of metabolic processes, employing the *standard differential equations* for reaction/diffusion dynamics, simply break down. In particular, the idea of a "bulk concentration" will not apply, in most instances, in the cellular microenvironments. Here, concentrations of enzymes and their respective substrates are, in many cases, of the same order of magnitude. It is highly plausible, that there are "molecular channels" in organized multienzyme systems *in vivo*, wherein each individual enzyme is subject to *a local, "quantized" substrate concentration*. In such localized regimes metabolic processes are executed in a *vectorial* manner, in some cases coupled to nonequilibrium energy sources.

The significance of enzyme organization *in vivo* has been discussed from a sociological perspective by Keleti (1975) and Welch (1977a,b). More recently, these authors (Welch and Keleti, 1981) have proposed a model of enzyme function and evolution based on a progressive "socialization" of transition-state dynamics within the enzyme molecule. These concepts were seminal in developing the more elaborate scheme of a geometrized cell metabolism described below.

It appears that enzyme active sites have evolved a complementarity not to the reactant or product species, but rather to the transition-state intermediate between the two (Fersht, 1985). It is this complementarity which is the root of the catalytic power of enzymes. This aspect of enzyme function is entirely in consonance with our "process"-oriented view of cellular dynamics, particularly when we recognize that the transition state is, by definition, a molecule (or molecular aggregate) *in chemical flux*. The efficacy of an enzyme in establishing precisely the right conditions necessary to realize such catalytic power (as well as control thereof) is now known to be influenced by a rich montage of cellular (and in many hormonal cases, extracellular) factors. Any evaluation of enzyme function *in vivo* must, therefore, consider these influences as more than contaminating nuisances to a "proper" study of the isolated enzyme. Rather, they should be viewed as the *most proper* conditions by which an enzyme's natural role may be appreciated. We are thus led to view the enzyme's activity as being more of a functional continuum than an isolated discrete process. Enzymes represent, in this light, an organizational conduit facilitating the transport of a matter-energy flux in a "preferred" direction commensurate with the cell's needs at any moment in time. The question which naturally arises from such considerations is whether there is a clearly defined adaptive value of this "socialization" of enzyme activity which can be related to the evolution of enzymic systems in the biological world.

Welch and Keleti (1981) dissected the rate constants of enzyme action into a composite form which depicts the "sociological" influences brought to bear on the processes represented by these constants. By separating the standard free-energy change of the transition state into an "intrinsic" part and a part which represents contributions from extrinsic factors, one can immediately write the rate constant as a product of factors, one of which involves the exponential product of the energy changes of the "sociological" interactions of the enzyme. A variety of such interactions bear on the rate constants owing to the interaction of the enzyme with its environment (Welch, 1984). Such interactions seem to have developed progressively throughout the "fine-tuning" of metabolism and were apparently accompanied and potentiated by three trends in protein evolution (Keleti, 1984). First, protein structures had to become large enough to accommodate specific transitions (Koshland, 1976). Second, surface groups came to play a key role in determining the specific position of the proteins with the social infrastructure of the cell (i.e. association with membranous structures, cytoskeletal elements, nuclear scaffolds etc. Anderson, 1976; Srere, 1981). Third, specific groups on the surface of the proteins came to be associated with the transmission of some of the "sociological" influences to the active site. We begin to formulate a picture of enzyme dynamics which encompasses communication and interaction between the enzyme's *external* environment (the cellular matrix) and its *internal* environment (the active site). What then of the rest of the protein molecule? Might we not suggest that it serves as a "homeostatic buffer" maintaining the active site at a "constancy" commensurate with catalytic activity? Indeed, this view is entirely congruent with our experimental knowledge of the dynamical role of the noncatalytic regions of the molecule.

As mentioned above, evolution has favored the strong binding of the transition state with weak binding of substrates (Fersht, 1985). This notion can be formulated more precisely as an *extremal principle* where variations in the binding energy of the transition state tend to be negative, and variations for substrate binding tend to be positive. It appears that during early stages of cellular evolution, the development of sociological interactions provided for the maximization of catalytic power via such tendencies. Importantly, these factors lead to a unification and optimization of the overall catalytic power for *sequences* of enzymes. Keleti (1975, 1978) and Welch (1977a,b) have suggested that the organization of these pathways contributed to increased efficiency, as some factors increase the "valleys" while others reduce the "peaks" of the overall free-energy profile.

One is led to ask whether there is a similar evolutionary "extremal principle" applicable to the "cytosociological" form of the transition-state activation energies. Welch and Keleti (1981) proposed such a principle for the energy barrier $\Delta G_T^{\ddagger}$ corresponding the parameter k_{cat}/K_m. This is an apparent second order rate constant, which has been a focal point in the evolution of enzymes kinetics (Fersht, 1985).

Thus for the total free energy change $\Delta G_T^{\ddagger}$, we have

$$\Delta G_T^{\ddagger} - \Delta G_T^{\ddagger *} + \sum_{i=1}^{n} \Delta G_i^{\ddagger} \qquad (6.1)$$

where $\Delta G_T^{\ddagger *}$ is the intrinsic value and $\Delta G_i^{\ddagger}$ the extrinsic ("social") factors. We ought to be able to write according to our extremal principle

$$\delta_e (\Delta G_T^{\ddagger *}) < 0 \qquad (6.2)$$

as the condition for favorable evolutionary changes in the *intrinsic* free-energy transition-state barrier for an individual enzyme (Welch and Keleti, 1981). Yet we *cannot* write a similar condition for the summation term of equation (6.1) representing the extrinsic "cytosociological" factors. The essential reason is that when the enzyme is embedded in its natural setting, some of the extrinsic factors are of a *regulatory* nature—which, under suitable metabolic conditions, produce rate lowering influences on $\Delta G_T^{\ddagger}$. This may be especially evident in enzyme clusters where the multienzyme systems have evolved to a state where the *overall* catalytic process is the focus of evolutionary optimization at the expense of optimizing the rate of enhancement of each individual species. In fact, some individual activity may actually be lowered from the free (non-interacting) state as a means of maximizing the overall metabolic flux through the assembly. It is likely that evolution of enzyme activity first dealt with the intrinsic factors affecting $\Delta G_T^{\ddagger}$, with later emphasis (at higher levels) on the extrinsic factors. It is proper, then, to view evolution as optimizing the organization, specialization, and regulatory interactions of organized enzyme systems. Evolution appears to optimize the whole organism which we take to imply the resultant of the totality of enzyme reactions occurring in an interactive network of metabolic society. Welch and Keleti (1981) summarized these ideas in a formulation which offers a metrical view of cellular organization. Following their argument, we write the following weighted sum (resultant), encompassing all of the enzymatic components of the organism's metabolism:

$$\Gamma^{\ddagger} - \sum_{\substack{(r \neq s)}}^{n} \beta_{rs} \gamma_s^{\ddagger} \qquad (6.3)$$

Cytosociology: A Field-Theoretic View of Cell Metabolism

where β_{rs} are the cytosociological factors which weigh the influence of $\gamma_s^{\ddagger}(\sum_{i=1}^{n_s} \Delta G_i^{\ddagger})$ for the rth reaction. In general, the β_{rs} will be complex functions of one or more of the n contributions to $\gamma_s^{\ddagger}$. Hence, the β_{rs} are also the subject of evolutionary change.

In a more mathematical form, the sociological coupling involved in equation (6.3) may be specified by defining $\beta_{rs} = \partial \gamma^{\ddagger} / \partial \gamma^{\ddagger}$ (where $r \neq s, = 1$), including the important constraint that β_{rs} is, itself, a function of the coordinates (see below for the reason for this added constraint). We can illustrate the effect of β_{rs} by the following transformation

$$\overline{g}^{\ddagger} = \begin{pmatrix} g_1^{\ddagger} \\ \vdots \\ g_m^{\ddagger} \end{pmatrix} - \beta_{rs} \begin{pmatrix} \gamma_1^{\ddagger} \\ \vdots \\ \gamma_m^{\ddagger} \end{pmatrix} \tag{6.4}$$

that is, β_{rs} is a matrix which operating on $\overline{\gamma}^{\ddagger}$ produces a new vector $\overline{g}^{\ddagger}$ whose components are

$$g_r^{\ddagger} = \sum_{s=1}^{m} \beta_{rs} \gamma_s^{\ddagger} (r = 1, 2, \ldots, m) \tag{6.5}$$

In this light, $\Gamma^{\ddagger}$ is the resultant of the component forms $g_{\gamma}^{\ddagger}$ and β_{rs} represents a "metabolic metric" defining the cytosociological interactions of metabolism. Where such interactions occur, β_{rs} takes on a rich structure incorporating all of the sensual intimacies of enzyme dynamics described above. Many of these interactions are easily disrupted by experimental extraction techniques and transform the "metric" to an *in vitro* form δ_{rs}, the Kronecker delta, defined by

$$\delta_{rs} = \begin{cases} 1 \text{ if } r = s \\ 0 \text{ if } r \neq s \end{cases} \tag{6.6}$$

It is clear that in order to effectively understand the enzyme in its natural environment, one can rely very little upon the traditional scheme of "grind and purify" (the "delta" approach) which is sure to provide misleading information about any metabolic system relying upon complex and subtle interactions for its *biological* function. Rather we should seek out those technics which maintain as close as possible the *natural group-dynamical state* of the metabolic system (the "beta" approach). As an example of the *beta* approach we may mention a recent series of observations made by Berry et al. (1987) indicating linear couplings of various forces and flows in intact hepatocytes. These observations suggest a high degree of organization and coordination occurring in a large number of metabolic systems.

If $\Gamma^{\ddagger}$ is restricted to the kinetic processes in the cell, then it can be interpreted as a "potential function" containing all the information for the transition-state barrier of all the kinetic processes of intermediary metabolism (Welch and Keleti, 1981). The biological extremal principle then takes on the following form:

$$\delta_e(\Gamma^{\ddagger}) > 0 \qquad (6.7)$$

which "depicts the highest and most continuing form of enzyme evolution" (Welch and Keleti, 1981). The purpose of this work is to place such notions on a firmer theoretic footing and to make explicit the field structure inherent in this view of metabolism.

6.4. ENZYMES: FIELD-EFFECT ELEMENTS

In order that any field picture of cellular dynamics have physical meaning, we must first answer the question: are there any localized entities in the cell which actually seem to manifest a field-type phenomena? We submit that the answer is yes and is to be found in the dynamics of enzymatic activity. The major emphasis of enzymology has been on the active center of the enzyme molecule. The transition-state free-energy change ($\Delta G^{\ddagger}$) has been used as a "window" into the workings of this reaction site, decomposing catalytic processes into entropic ($\Delta S^{\ddagger}$) and enthalpic ($\Delta H^{\ddagger}$) contributions. The customary sources of catalytic power can via $\Delta S^{\ddagger}$ or $\Delta H^{\ddagger}$ be formulated in the guise of *transition-state stabilization*. From the enthalpic ($\Delta H^{\ddagger}$) side, a number of energetic factors lead to decreases in $\Delta G^{\ddagger}$. Generally, these fall into one of the three categories: (i) catalysis by rack mechanisms (whereby the bound, ground-state substrate is strained or distorted), (ii) general acid-base catalysis (involving proton transfer to or from the transition state), and (iii) electrophilic/nucleophilic catalysis (usually involving covalent stabilization of the transition state). Underlying these various catalytic modes is the importance of *local electric fields*, in the microenvironments of enzyme active centers (Fersht, 1985).

Considering the relatively small volume occupied by the actual reaction site, one might ask: *Why are enzymes so big?* The protein as a whole is apparently designed to provide a specific solvent medium for a given chemical reaction

(Somogyi, Welch and Damjanovich 1984; Welch, Somogyi and Damjanovich, 1982) wherein the combined chemical and protein subsystems engage in a fluid and variable exchange of free energy (Lumry and Biltonen, 1969; Lumry and Gregory, 1986), facilitating the entrance of the bound chemical system into its transition state. Accordingly, we are led to picture the protein matrix as an intermediary, a "deterministic" mediator, between a localized chemical reaction coordinate and the ambient medium. A crucial point here is that, while the local active-site configuration defines the physicochemical nature of the substrate-product transition state(s), the protein molecule determines the rate at which this state(s) is reached (Ji, 1974, 1979, these proceedings). From a cytosociological perspective we have also noted that the protein matrix serves as an internal environment buffering the effects of a chaotic environment on an active center constantly poised in a state of readiness commensurate with cellular needs.

If the enzyme is viewed as a macromolecular free-energy transducer, an immediate question concerns which kinds of useful energy are transduced. It is quite clear that mechanical (i.e., kinetic/potential) forms are involved (Lumry and Biltonen, 1969; Ji, 1974, 1979; Welch, Somogyi, and Damjanovich, 1982). Considering the predominance of activated protonic/electronic states during catalytic events at the active center, it is natural to ponder a role of the protein molecular as a protical/electrical transducer as well. There is an increasing indication, experimentally and theoretically, that such transduction does indeed occur in individual proteins and that it may be involved in catalytic processes (Somogyi, Welch, and Damjanovich, 1984; Welch and Berry, 1983). Interest in this type of energy transduction is heightened further when we relate the cytological juxtaposition of organized enzyme systems and sites of protonic/electronic sources (Somogyi, Welch, and Damjanovich, 1984; Welch, 1984; Welch and Berry, 1983). It is quite plausible that, in many cases, the enzyme molecule connects the active center not only mechanically to the surroundings, but, also protically/electrically.

Protein structure contains a number of elements which suggest modes of energy transduction. Obvious possibilities are regions of local secondary structure, viz., α-helix and β-structure. These have been implicated, by many workers (reviewed by Welch and Berry, 1983, 1985) in generating local electric fields, protonation/deprotonation events, proton semiconduction, vibrational excitation, *inter alia*. Also *hydrogen-bond networks*—within single proteins and among conjoined proteins—are of great interest as possible conduits for protochemical processes (i.e., chemical processes wherein protons play a crucial role—editor's note) in enzyme action.

Long-range, mobile protonic states are finding increasing relevance to many cellular processes, stemming from the original pioneering suggestion as to their role in electron-transfer phosphorylation (Mitchell, 1979). This kind of energy continuum is emerging as a unifying theme in cell metabolism. Its importance demands that we begin looking at enzyme structure-function with new perspectives.

Superimposed on these designs may be long-range electronic semiconduction states, which may be important in some enzymes, especially in organized states *in vivo* (Welch and Berry, 1985). Various modes of electronic (and electron-phonon) coupling between enzyme and substrate have been offered as theoretical possibilities (Green and Ji, 1972; Ji, 1974, 1979, these proceedings; Conrad, 1979; Volkstein, 1981). And, long-range electric-field effects have been discussed by a number of workers (Fröhlich and Kremer, 1983; Welch and Berry, 1983; Westerhoff, Tsong, Chock, Chen and Astumian, 1986).

Naturally, the question arises as to the interaction between excited conformational (e.g., vibrational) states and excited protonic/electronic states in the enzyme molecule. This consideration is particularly important when one notes that virtually all enzymatic mechanisms involve some kind of mechanical force acting through electrostatic interactions generated across chemical bonds in the substrate.

There have emerged various theoretical models, which propose that spatiotemporal ordering of the fluctuational behavior of the protein molecule serves an integral role in enzyme catalyses. A notable principle, from all of the models, is the unity of the enzyme molecule and the ambient medium. We find that biological evolution has apparently come to grips with the random field in the medium by designing a macromolecular structure that is capable of correlating statistically that random energy source in an anisotropic fashion. External and internal hydrogen bonding plays an integral role in the dynamic tertiary (and quaternary) structure of globular proteins. Following tenets of certain of the dynamic enzyme models (Somogyi, Welch, and Damjanovich, 1984), one sees that gaps ("faults," "defects") in the internal bonding arrangement might elevate locally the free energy of the system, electrostatically and mechanically, as well as protonically. Catalytic functions would, then, depend on precise internal fault states (Lumry and Gregory, 1986). Such faults (or defects) arise, in a protein dissolved in aqueous solution, in conjunction with binding/relaxation of bound water, fluctuating proton-transfer processes and charge-density fluctuations at the surfaces, etc. [Inside the protein, these faults can migrate, for example, by proton-hopping (Nagle, and Tristram-Nagle, 1983). The energy required for a single "hop" is about kT (i.e., thermal energy).] Thus, isolated in bulk solution, the internal defect pattern of a given protein is subject to a random generator from the solute-solvent system. The catalytic turnover numbers for isolated enzymes *in vitro* reflect this random field.

Consideration of the roles of mobile protons in enzyme structure, function, and evolution has led to the postulation that externally derived high-energy protons may function in the modulation of enzymatic events in organized states *in vivo* (reviewed by Welch and Berry, 1983, 1985; Welch and Kell, 1986). When "linked up," via hydrogen-bond networks, to the energy continuum of a proton electrochemical potential difference (e.g., at a membrane interface), the conformation-dynamical aspects of enzyme action may be governed by a proton flow. Moreover, the localized proton flow might serve a stoichiometric role as a source/sink of protons for *protochemical* events at active centers, as well as functioning in substrate translocation.

Hydrogen bond structures in proteins may also couple to other energy continua in organized states. Another likely source would be the electric fields at the surfaces of cytological particulates, noted above. A mode of operation here has been expounded by Fröhlich (1980, 1986), who conjectured that coherent excitations of proteins should play an important role in their biological activity. Over the years theoretical calculations have supported this claim, and some experimental evidence has been found (Fröhlich and Kremer, 1983). Specifically, this notion maintains that the activities of localized enzymes couple to electric fields (and other energy sources) via excitation of metastable, highly polar states in the proteins. The modality involves longitudinal electric dipolar oscillations of hydrogen bond units, which are modeled as active *phonons* (i.e., quasi-particle lattice vibrations). Such systems can store external energy in specific modes (and subsequently do work with it), via a phonon condensation phenomenon analogous to the condensation of a Bose-Einstein gas, if the energy is pumped in above a critical rate.

More recently, Westerhoff and coworkers (Westerhoff and Kamp, 1987; Westerhoff, Tsong, Chock, Chen, and Astumian, 1986) have developed a model for membrane-transport proteins driven by an electric field. This involves coupling of the field to dipoles (e.g., α-helices) in the protein structure. Model calculations have shown that such a system can store and utilize field energy for work processes—under a variety of realistic conditions. Finally, we note that possible relevance of mobile electronic states in protein dynamics. Interaction of electronic and nuclear degrees of freedom is well-known in solid-state physics. An example is electron-phonon coupling in crystal lattice dynamics. The roles of such modes in enzyme action have been discussed by others (Green and Ji, 1972; Ji, 1974, 1979; Conrad, 1979; Fröhlich, and Kremer, 1983; Nagle and Tristram-Nagle, 1983; Volkenstein, 1972, 1981; Welch, 1986; Welch and Berry, 1983, 1985). The idea of long-range electronic semiconduction was proposed long ago by Szent-Györgyi (1941) as an integrative principle in cellular processes. Generally, this notion has not been taken seriously (at least for large supramolecular distances) because of the observed nature of the conduction band in isolated proteins, perceived problems in insulating the flow, etc.—though one should note that in organized states *in vivo* such problems may be obviated.

6.5. A PRIMER OF FIELD THEORY

In the following section, we shall develop two field constructs of metabolic organization and enzyme function. In order to make these sections self-contained, we explain some of the terms and programs used in what follows. The reader well-versed in these matters may omit this section with no difficulty in following our subsequent arguments. For the novice, however, the following will provide a useful, albeit condensed, introduction to basic field theory as we employ it.

We think of a *field* as being a mathematical function ascribing certain unique properties to points in space-time. For example, the *electron* field function may be considered as defining in any region of space-time the following properties: (1) *Mass*—a measure of how our selected region of space-time "curves" the space-time near it, (2) *Electric Charge*—a measure of how this region of space couples to or experiences the electromagnetic force, (3) *Spin*—a measure of how the field function transforms under a specific transformation of space-time components (Lorentzian transformation) as well as a measure of how the region of space interacts with a magnetic field, (4) Other *Internal Charges* such as lepton number, baryon number, color, flavor, etc., as well as (5) *Kinetic Properties* such as momentum. When all these properties are specified in the function, we can reasonably say we have described the particle field in question. By knowing the field structure we also know a great deal of its isolated properties and can deduce, for example, its equations of motion by an extremal principle (*vide infra*). If we wished to pass beyond the classical limit and describe a *quantized* field, we would expand the field function as a series of exponentials multiplied by Fourier Coefficients. For example a massless scalar field is expanded as:

$$\phi(x,t) = \int \frac{d^3k}{[(2\pi^3)ZW_k]^{y_z}} \{a(k)e^{i(k\cdot x - w_k t)} + a^+(k)e^{-i(k\cdot x - W_k t)}\} \qquad (6.8)$$

Where $a(k)$ and $a^+(k)$ are the coefficients of expansion, and the exponential term represents the wave character of the field function. We next consider these coefficients as being *operators* (in the quantum mechanical sense) and apply canonical conservation relations to them as:

$$[a(k), a^+(k')] = \delta^3(k-k')$$
$$[a(k), a(k')] = [a^+(k), a^+(k')] = 0 \qquad (6.9)$$

where [A, B] is the commutator AB - BA, k and k' are two different momentum vectors, $\delta^3(k-k')$ is the Dirac delta function and we have used rationalized units in which Planck's constant and the speed of light are set equal to one. This procedure establishes relationships among the operators such that the quantum nature of the fields is realized within the formalism of the theory. For example one can write the quantum mechanical Hamiltonian (in the absence of potentials) for the field defined above as follows: $H_o = \int d^3k w_k a^+(k)a(k)$, this function defines the total energy of the field. The operators a(k) and $a^+(k)$ are interpreted as destruction and creation operators, respectively. In practice $a^+(k)$ is used to create a one particle state from the vacuum state:

$$|k> \approx a^+(k)|0> \qquad (6.10)$$

where a(k) annihilates a one particle state to the vacuum state.

$$|0> \approx a(k)|k> \qquad (6.11)$$

In order to describe interactions between fields, we invoke the Lagrangian formalism which has served field theories so well. The Lagrangian of a particle given as **L = KE - PE** (kinetic energy minus potential energy). For fields, we must take into account the fact that they extend throughout space. We thus deal more appropriately with the Lagrangian *density* ($\mathcal{L}$) from which the total Lagrangian is obtained by integration over all space. We will for purposes of simplicity ignore the differences between the two. In general, the Lagrangian density is a function of the free field function Φ, and its time and space derivatives which we collect together as $\delta_\mu (\Phi)$ ($\mu = 0, 1, 2, 3$ and includes spatial and temporal components) thus we have $\mathcal{L}(\Phi, \delta_\mu \Phi)$. From the Lagrangian, it is possible to deduce the equations of motion of the field by Hamilton's principle which states that the evolution of the system is such as to minimize $\mathcal{L}(\Phi, \delta_\mu \Phi)$—the *principle of least action*.

We may schematize this idea as follows:

$$[\delta \mathcal{L}(\phi, \partial \mu \phi) = 0] \Rightarrow D\phi$$

So far, we have only described isolated fields. In order to include interaction between one or more fields (i.e., the "social" activities of the fields), we adopt the following approach. Suppose that two fields Φ_1 and Φ_2, are to interact. We then write the total Lagrangian as a sum of three parts

$$\mathcal{L} = \mathcal{L}_0(\phi_1) + \mathcal{L}_0(\phi_2) + \mathcal{L}_{int}(\phi_1\phi_2) \qquad (6.12)$$

where $\mathcal{L}_0$ represents the Lagrangian for the isolated fields and $\mathcal{L}_{int}$ is a term describing the interactions of the field. *It must be emphasized that it is the precise mathematical form of $\mathcal{L}_{int}$ which embodies the physical content of the interactions.*

There are well defined rules for constructing a physically meaningful Lagrangian containing interactions. Briefly, the function must be a scalar quantity and hence relativistically invariant. In practice this means that all the various field components (scalars, vectors, tensors, spinors, etc.) must be arranged so that each term in the Lagrangian is, itself, a scalar quantity i.e., each covariant component must be "multiplied" by its corresponding contravariant component $(X_\mu X^\mu)$. This principle is especially important in the interaction term where fields differing in tensorial rank appear. The following represents acceptable interaction terms. $\Phi_1 \Phi_2$ (scalar-scalar), $A^\mu A_\mu$ (vector-vector), $A^\mu \delta_\mu$ (vector derivative) whereas ΦA^μ and $\Phi \delta_\mu$ are not since they wind up "looking" like a vector and not a scalar. It will be seen that this principle will be of immense importance in describing the proper forms of the fields themselves, as well as the result of symmetry operations on the fields. A summary of the logical structure of the Lagrangian formalism is schematized in Figure 6.1. A number of excellent monographs and texts on quantum field theories are available (Agarwal, 1977; Bjorken and Drell, 1965; Itzykson and Zuber, 1980; Mandl and Shaw, 1984; Ramond, 1981; Ryder, 1985) and the interested reader is directed to them for further information regarding the points developed in this section. More specific topics are developed as needed in the following sections.

6.6. BIOLOGICAL SPACE-TIME

Physics has taught us that the mathematical characterization of the real world requires abandonment of the simple four-dimensional world of human perception, in favor of abstract multidimensional spaces. So must ultimately be the case for biological phenomena. Prigogine (1980) notes that "biological space" is atypical of the Euclidean geometrical form used in classical physics. In living systems we have an organized space "in which every event proceeds at a moment and in a region that make it possible for the process to be coordinated as a whole. This space is

Cytosociology: A Field-Theoretic View of Cell Metabolism

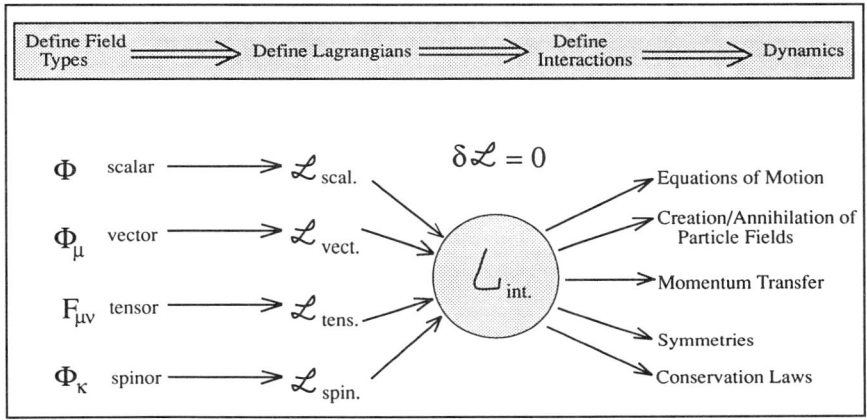

Figure 6.1. Logic of Lagrangian Field Theory.

functional, not geometrical.... In this space the events are processes localized in space and time and not merely trajectories." A distinguishing feature of a "metabolic field" lies in the existence of an underlying *material substratum*, which directs the spatiotemporal course (trajectory) of molecular events therein. Not only are the enzymes contained in these organized systems simply the catalysts of specific chemical transformation, but also they (together with conjoined cytomatrix proteins and/or lipoprotein arrays) form the very substratum that is the "scene of the action."

The development of a true molecular theory of the living state can be fostered by the use of analogical thinking and common motifs from the parental sciences of physics and chemistry. Conventional biochemistry, despite its great wealth of accumulated knowledge, is of itself inadequate for the task at hand. One point which the subject of metabolic organization has shown us is, that we cannot "reduce" the intracellular reaction-diffusion system to a simple linear superposition of individual "particle states" in ordinary space-time. It is rather apparent that metabolic space-time has an intrinsic "connectedness"—a substratum. There are two physical constructs, general relativity and gauge field theory, which have revolutionized our understanding of "curved" (or "connected") geometries in natural processes. We suggest that such constructs will prove useful in the description of biological processes as well.

6.7. CELL METABOLISM: AN AFFINE GEOMETRY

The theory of general relativity characterizes gravitational interaction between material objects as a curvature in the local geometry of space-time. Accordingly, the moon follows an orbit around the earth because that represents the "shortest distance between two points" (i.e., a geodesic path) in the regional space-time. Let us see how such a paradigm may apply to intermediary metabolism.

Let us take a metabolizing system composed of N components (metabolites). We suppose that metabolic space-time is Riemannian, such that a "line element," ds, can be defined as follows:

$$ds^2 = g_{\mu\nu}(x)dx^\mu dx_\nu \qquad (6.13)$$

where superscript and subscript Greek indices denote "contravariant" and "covariant" forms, respectively. In the Einstein notation, when an index is repeated in a mathematical term (e.g., Equation 6.13), it signifies summation on the index. The x-variables might denote, in three dimensions, the spatial positions and/or velocities of the N metabolites. By analogy to classical mechanics, we could proceed by a "phase space" representation. Since each metabolite has three space coordinates and three velocity coordinates, one would have $\mu, \nu = 1,2,...6N$ as well as $x^0 =$ time. The matrix $g_{\mu\nu}(x)$ (with appropriate scaling factors) is the *metric tensor* (Misner, Thorne, and Wheeler, 1973), serving as a measurement device which enables one to calculate the scalar magnitude of vectors and the scalar product of two vectors. In our case, the "vectors" refer to diffusional mass transport or to oriented chemical-reaction flux. If the metabolizing system were completely homogeneous in space, (i.e., not "connected") then $g_{\mu\nu}(x)$ reduces to the *identity matrix* of Euclidean space.

By analogy to relativity theory, we further suppose that the metabolite "particles" follow *geodesic paths* in the living cell. The differential equation of the curve having an extremal length is called the "geodesic equation" (Misner, Thorne and Wheeler, 1973). To obtain this equation, one must find the conditions which lead to stationarity for the following integral:

$$I = \int L ds \qquad (6.14)$$

This involves a solution to the well-known relation in variational calculus:

$$\delta I = \delta \int L ds = 0 \qquad (6.15)$$

where L is a Lagrangian function. From textbooks on differential geometry, one finds that the standard geodesic equation is as follows:

$$\frac{d^2 x^\beta}{ds^2} + \Gamma^\beta_{\mu\nu} \frac{dx^\mu}{ds} \frac{dx^\nu}{ds} = 0 \tag{6.16}$$

or, when an external field is present,

$$\frac{d^2 x^\beta}{ds^2} + \Gamma^\beta_{\mu\nu} \frac{dx^\mu}{ds} \cdot \frac{dx^\nu}{ds} = W^\beta \tag{6.17}$$

The entity $\Gamma^\beta_{\mu\nu}$ is called a "Christoffel symbol." More generally, in Riemannian geometry it is known as the *affine connection*, or connection coefficient. It is so called, because it *connects* the components of a vector at one point with its components at a nearby point—allowing for the possibility that, when something "flows" through space, not only does it change by simple transport but also by virtue of a change in our measurement basis (i.e., by virtue of space-time curvature).

Mathematically, $\Gamma^\beta_{\mu\nu}$ is defined by the following relation (see Misner, Thorne, and Wheeler, 1973);

$$\Gamma^\beta_{\mu\nu} = \frac{1}{2} g^{\beta\lambda} \left(\frac{\partial g_{\lambda\mu}}{\partial x^\nu} + \frac{\partial g_{\lambda\nu}}{\partial x^\lambda} - \frac{\partial g_{\mu\nu}}{\partial x^\lambda} \right) \tag{6.18}$$

An important ramification of the affine connection of a Riemannian space-time is seen in the notion of *differentiation* used in calculus. In a curved space-time the derivative (e.g., the gradient) of a function takes new meaning. For example, the *covariant derivative* D_α of the (contravariant vector) V^μ is defined as

$$D_\alpha V^\mu = \frac{2V^\mu}{2X^\alpha} + \Gamma^\mu_{\alpha\lambda} V^\lambda \tag{6.19}$$

The first part of this expression is just the usual form of a directional derivative, while the second term corrects the flow due to the spatial change in "basis" vectors used in measuring the quantity (Misner, Thorne, and Wheeler, 1973).

Clearly, for simple Euclidean space, we have

$$\frac{d^2x^\beta}{ds^2} = 0 \qquad (6.20)$$

or, with an external field,

$$\frac{d^2x^\beta}{ds^2} = W^\beta \qquad (6.21)$$

In general relativity, Equation (6.20) says that particles follow simple, unperturbed rectilinear motion. In our metabolic system, it says that each reaction-diffusion process follows its own, individual course—in a homogeneous bulk-phase system and not connected (other than by simple chemical stoichiometry) to any other process. Or in the presence of an external field (W^β), each component process couples independently in a homogenous manner.

In order to build a "metabolic field theory" analogous to that in general relativity, the specification of a geodesic condition and the associated metric (as in Equations 6.13-6.18) is only half the picture. There must also be some functional form by which the flow of matter and energy is tied to the geometry of the field structure—analogous to the field equations of gravity. Relating to Section 6.3, we immediately realize that the key element in describing metabolic infrastructure is the metric tensor $g_{\mu\nu}(\underline{x})$. In relativity theory the geometrical elements $g_{\mu\nu}$ play the role of *potentials* (analogous to gravitational potential energy in Newtonian theory), which can be used to construct the gravitational field equations. On one side of such equations is an expression for the curvature of space-time (as dictated by the $g_{\mu\nu}$), and on the other side is a mass-energy term. In the parlance of physics, one describes the meaning of these equations by saying that "the presence of matter tells space-time how to curve, while (symmetrically) the curvature of space-time tells matter how to move" (Misner, Thorne and Wheeler, 1973). This gives matter a new, "systemic" meaning. In relativistic field theory, we are presented with the view that mass is to be regarded as concentrated potential energy that moves on through space. Accordingly, matter merges with the field, becomes an off-spring of the field, and sees its essence (*viz.*, inertial mass) vary with the contained energy (as influenced by field interactions).

Welch and Keleti (1981) explored the analogy with the field quality of "matter-in-flux" in cell metabolism, as it relates to enzyme organization *in vivo* (see also Welch and Kell, 1986). Following guidelines from physics, we would suggest that in biology matter itself should not be the only object of analysis. Rather, it should include the "self-activity" of matter. Using basic tenets of thermodynamics, it was proposed (Welch and Keleti, 1981) that the *chemical potential* (the partial molar free energy due to each and every mass component) might represent the "self-activity" (or energetic manifestation) of matter in the living system. A chemical (or electrochemical) potential gradient, either in real space or along a "reaction coordinate," is the very driving force for all biological material transformations. This reasoning leads to the important realization that, in a teleonomic sense, it is the *creation* and *utilization* of requisite (electro)chemical potentials of mass components which actually sets apart matter in the "living state" from that in inanimate forms. In essence, the inner conditions imposed on matter by the biological system define the chemical potentials of mass components, the "self-activity," the "energetic manifestation," of that matter. Also, the same kind of mass component might have a different "self-activity" in different parts of the same biological system (or in different systems), depending on the imposed condition associated with the particular process related to that component. Hence, the "energetic manifestation" of matter in biology takes on a distinctly *relativistic* guise—as compared with that same matter in its "inert" (or "passive"), inanimate forms.

The functional form of the metabolic metric $\beta_{rs} = \partial \gamma_r^{\ddagger} / \partial \gamma_s^{\ddagger}$ suggests that it can be interpreted as a measure of how one set of free energy changes (r) are "coupled" to another set(s). Thus if the metabolizing system were completely homogeneous in space (i.e., not connected) then β_{rs} would be the identity matrix as $g_{\mu\nu}$ does in flat Euclidean space. We note here that although we have chosen our "coordinates" to be the potentials $\gamma_i^{\ddagger}$, related to transition-state energies, we may be perfectly general and consider further the possibility of a broader connectedness. We may define our coordinate basis as consisting of numerous parameters of the metabolizing system, some of which may not be directly related to the transition-state energy but rather to the overall metabolic character of the cell.

Although we have discussed connectedness in terms of the β_{rs} components themselves, this view is somewhat misleading. The true value of defining and using β_{rs} as we have lies in whether or not β_{rs} is a function of the coordinates. In analogy to general relativity, when $\beta_{rs} = \beta_{rs}(x^{\mu})$, then we have the important result that any "affine connection" $\Gamma_{\mu\nu}^{\beta}$ constructed for the metabolizing system in terms of the metric β_{rs} does not vanish (since $\dfrac{\partial \beta_{rs}}{\partial x^\mu} \neq 0$) indicating that not only are the metabolic components connected, but that the nature of this connectedness is position dependent. Hence, the subtle cytosociological interactions of the cell are highly dependent on the microdomains in which those interactions occur—a result whose prediction depends on a position-dependent metric.

Pursuing the gravitational analogy to its fullest, it would seem that any formalism for the "metabolic field equations" must relate the metabolic metric to the distribution of the metabolites—with the *free energy* of such components a crucial factor. In suggesting a didactic course here, we would adopt a direct analogy to the "principle of geometrization of physics," which may be stated as follows (McVittie, 1949):

> A distribution of matter and radiation in any region of space-time and the Riemannian geometry appropriate to that region have the same qualitative and quantitative properties.

This principle is what guided Einstein in choosing the appropriate mathematical relationship between the properties (e.g., density, momentum, energy) of matter *and* the values of the metric coefficients. Specifically, his field equations of gravity relate the components $g_{\mu\nu}$ (on one side of the equation) to the components of the *energy-momentum tensor* $D_{\mu\nu}$ of matter (on the other side of the equation); this is possible because the tensorial forms on both sides of the equation are *of the same rank*. Knowing the nature of the energy-momentum tensor, one then solves the equations for the potentials $g_{\mu\nu}$. (As a disclaimer, we note that our results herein are not so lofty as those of Einstein!)

Here, we give some qualitative indications as to a possible reification of the "geometrization" principle for cell metabolism. A graphical view of these concepts is offered in Figure 6.2. First (and perhaps foremost), we need a tensor of rank = 2, to match the tensorial rank of $g_{\mu\nu}$. Moreover, this tensor must, somehow, relate to the flow of (electro)chemical potential. Let us suppose that the (electro)chemical potentials (ϕ_i) of substrates in the living cell define a scalar field, $\Phi(\underline{x})$. Now, we return to the Lagrangian formalism. The dynamics of the field $\Phi(\underline{x})$ is determined by the Lagrangian density $\mathcal{L}(\phi, \nabla_\mu \phi)$ where, generally "∇_μ" denotes a space-time gradient) through extremization of an action integral (similar to Equation 15). This extremization leads (via the Euler-Lagrange relation) to the desired field equations. From basic physics (Morse and Feshbach, 1953), we suspect that our field equation

Cytosociology: A Field-Theoretic View of Cell Metabolism

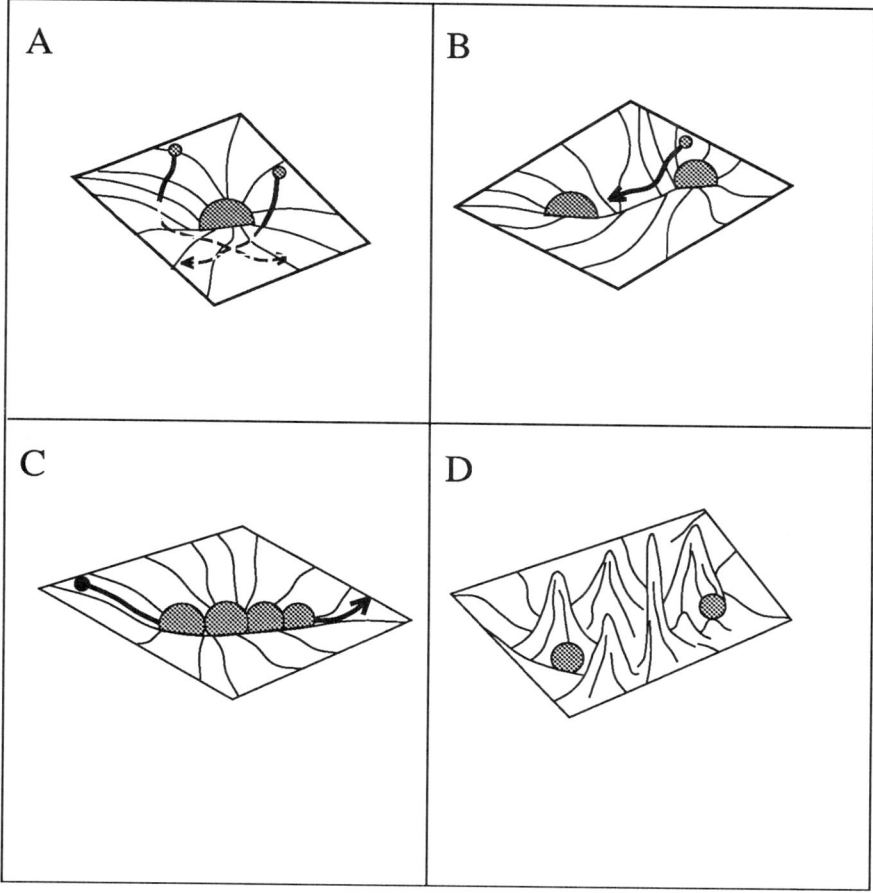

Figure 6.2. Curvature of Metabolic Space.

Capsule A. How the presence of a perturbing potential may alter the course of a substrate through two pathways.

Capsule B. Selective directing of a metabolite from one perturbing site to another.

Capsule C. Channeling

Capsule D. A more realistic curved space.

will contain the second derivatives of ϕ (x_μ). And, we anticipate that our scalar field will be affected by a *source function*, $Q(\underline{x}_\mu)$ (itself a scalar field), which in the limit of "flat" (Euclidean) space follows the *Poisson equation*:

$$\nabla^2 \phi \sim Q \tag{6.22}$$

where here "∇^2" is just the spatial Laplacian operator. (In relativity theory (Misner, Thorne, and Wheeler, 1973), one finds that Einstein's gravitational field equations reduce to the form of Equation 6.22 in the Newtonian limit).

The source term, Q, obviously must relate to spatial locations where the chemical potential of a given substance is produced or consumed (or otherwise concentrated). Espousing the organizational view of cell metabolism, we might use something similar to a function called the "kinetic power" K_Γ, proposed by Keleti and Welch (1984). It is defined as $k_{cat} [E]_T/K_m$ where k_{cat}, K_m and $[E]_T$ denote the catalytic turnover constant, the Michaelis constant, and the total enzyme density, respectively, for a given enzyme reaction. As detailed elsewhere (Keleti and Welch, 1984), K_Γ contains, explicitly or implicitly, all factors (e.g., enzyme concentration, substrate/product diffusion, substrate/product binding, catalytic turnover) which bear upon a given metabolic transformation *in situ*. Considering the structural heterogeneity of the metabolic infrastructure, we assign a scalar field $K_\Gamma(x)$ to the cytoplasm. Then, dimensionality requirements would suggest something akin to the following;

$$\nabla^2 \phi \sim K_\Gamma^2 \tag{6.23}$$

From particle physics (Ramond, 1981; Ryder, 1985), we find that the most general form of the Lagrangian density for a scalar field $\phi(\underline{x})$ is as follows:

$$\mathcal{L} \sim \frac{1}{2}(\nabla_\mu \phi)^2 - V(\phi) \tag{6.24}$$

where now "∇_μ" represents the gradient in space and time, and V is a scalar function. The first part is a "kinetic" term and the second a "potential" term. This Lagrangian leads to the following kind of field equation:

$$\nabla^2 \phi - \frac{\partial^2 \phi}{\partial x^2} + \frac{\partial^2 \phi}{\partial y^2} + \frac{\partial^2 \phi}{\partial z^2} - \frac{1}{c^2}\frac{\partial^2 \phi}{\partial t^2} \sim -\frac{2V}{2\phi} \tag{6.25}$$

(where c is an appropriate dimensionality constant). In "curved" Riemannian spacetime Equations (6.24) and (6.25) become (with Greek indices μ, ν spanning space *and* time coordinates);

$$\mathcal{L} = \frac{1}{2}\sqrt{-g}\,[g^{\mu\nu}\nabla_\mu\phi\cdot\nabla_\nu\phi - V(\phi)] \tag{6.26}$$

and

$$\frac{1}{\sqrt{-g}}\nabla_\mu(\sqrt{-g}\cdot g^{\mu\nu}\nabla_\mu\phi) = -\frac{2V}{2\phi} \tag{6.27}$$

where g is the determinant of $g_{\mu\nu}$ (Ramond, 1981; Ryder, 1985). Thus, we are led to a "metabolic field" Lagrangian in a form something like the following:

$$\mathcal{L} \sim \frac{1}{2}\sqrt{-g}\,(g^{\mu\nu}\nabla_\mu\phi\nabla_\nu\phi - K_\Gamma^2\phi) \tag{6.28}$$

Using the chemical potential in our field formalism implies the validity of such thermodynamic constructs as Gibbs free energy. Basically, we are employing the "local equilibrium assumption," which is well-known in irreversible thermodynamics. The ultimate applicability thereof in cellular microenvironments is a subject of some concern (Welch, 1985). We eschew the issue here and assume that the chemical potential is definable.

We are now in a position (at least qualitatively) to unite our foregoing Lagrangian formalism with the geodesic paths for particle motion. Referring to Equations (6.13) through (6.15), one can parametrize the geodesic curve according to entities other than pure distance, $\underline{s}$. In fact, one might choose any of a number of so-called "affine parameters" say, σ, which depend on the path in such a way as to leave the form of the geodesic equation (Equation 6.16) the same. That is,

$$\frac{d^2x^\beta}{d\sigma^2} + \Gamma_{\mu\nu}\frac{dx^\mu}{d\sigma}\cdot\frac{dx^\nu}{d\sigma} = 0 \tag{6.29}$$

Following a "dynamic" Lagrangian method for geodesic determination (Misner, Thorne, and Wheeler, 1973), it is readily found that the geodesic of Equation (6.29) is an affinely parametrized curve which extermizes the integral

$$I = \frac{1}{2}\int g_{\mu\nu}\frac{dx^\mu dx^\nu}{d\sigma\, d\sigma}d\sigma \qquad (6.30)$$

in the sense of δI, where

$$L = \frac{1}{2}g_{\mu\nu}\frac{dx^\mu}{d\sigma}\frac{dx^\nu}{d\sigma} \qquad (6.31)$$

The affine parameters (σ) must relate to each other (and to s) linearly. This suggests a possible route to parametrize a (should we say "biodesic?") path according to such factors as free-energy dissipation (or conservation). Such thermodynamic parametrization of the path leads to the realization that spatial coordinates and time enter into Equations (6.29) through (6.31) on a completely equal (or symmetric) footing. That is, free-energy dissipation gauges not only the spatial form of the trajectory but also leads to the definition of a *thermodynamically proper time* for the flow process (Richardson and Rosen, 1979; Van Rysselberghe, 1963).

The mathematical form of Equation (6.31) is strikingly similar to that of the *dissipation function*, ψ, of irreversible thermodynamics.

$$\psi = \sum_{i,j} J_i \cdot F_i \qquad (6.32)$$

where J_i and F_i are the conjugate flows and forces, respectively. In more explicit form, one has

$$\psi = -\sum_{i,j} L_{i,j} \nabla u_i \cdot \nabla u_j \qquad (6.33)$$

where u_i is the "potential energy" of species i (and therefore, ∇u_i a driving force on its flow) and L_{ij} a phenomenological coupling coefficient. The resemblance of Equations (6.31) and (6.33) (and its significance in the geometrization of thermodynamics) has been stressed by a number of workers (Peusner, 1986; Richardson and Rosen, 1979).

Then, the next question concerns, *Which dissipative process* should we use in Equation (6.33), as regards our "metabolic Field?" We suggest a singular role of *diffusion*. The association is obvious for actual material-transport phenomena, whereby u_i is just the (electro)chemical potential and L_{ij} a type of phenomenological (Onsager) diffusion coefficient. Enzyme-based chemical reactions (including the catalytic turnover *per se*) can also be cast in a diffusion form. In the usual thermodynamic scheme (Prigogine, 1967), the dissipation function ψ_c, for chemical reactions is given by

$$\psi_c = -A \cdot v \tag{6.34}$$

where A is the chemical affinity and v the reaction velocity. The affinity is defined as

$$A = -\left(\frac{\partial G}{\partial \xi}\right)_{T,P} \tag{6.35}$$

where G is Gibbs free energy, ξ the "extent of reaction," and T and P have several meanings.

By virtue of the canonical relation between free energy and chemical potential, one might generalize this definition of A for organized ("vectorialized") enzyme systems, such that it has the form "$\nabla_\xi \phi$" (where the subscript denotes the reaction coordinate along some spatial direction). Assuming the chemical reaction to be in the linear thermodynamic regime, the velocity is given in the form (see Prigogine, 1967):

$$V = L_c \cdot \nabla_\xi \phi \tag{6.36}$$

(This assumption is rather restrictive; we employ it here purely for illustrative purposes—see Welch, 1985). In this design L_c takes the form of a "chemical" diffusion coefficient. To see the relevance of this approach, refer to the enzyme "kinetic power," K_Γ, defined above. When substrate diffusion is limiting, K_Γ becomes directly proportional to a true diffusion coefficient; whereas at the other extreme when the catalytic step is limiting, K_Γ is proportional to k_{cat} (Keleti and Welch, 1984). For the latter case, we note that recent protein-dynamical models of enzyme action regard the catalytic process (i.e., k_{cat}) as a diffusion phenomenon in the internal conformational-coordinate space of the protein macromolecule (Kamp, Welch, and Westerhoff, 1987; Welch, 1986; Welch, Somogyi and Damjanovich, 1982). Hence, we focus specifically on diffusion as the dynamical process in our "metabolic field."

Next, we must ask, does the diffusion phenomenon yield a Lagrangian of the required form for Equations (6.24) through (6.28). We are immediately confronted with a problem; the time symmetry of Equation (6.25) (and the Lagrangian) is broken, due to the circumstance that diffusion is an irreversible, dissipative phenomenon! Morse and Feshback (1953) have addressed the general problem of the Lagrangian-variational formalism for dissipative processes. They showed that it is possible to obtain a Lagrangian which does yield the diffusion equation (i.e., the form in Equation 25, but with only the first derivative in time). The dodge is to consider the actual physical system coupled to an imaginary "mirror-image" system with "negative diffusion—"which gains as much energy as the first loses, thus conserving the total energy and giving an invariant Lagrangian function. The actual Lagrangian proposed by Morse and Feshback (1953) is as follows:

$$\mathcal{L} = -\frac{1}{2}D\nabla\phi\cdot\nabla\phi^* - \left(\phi\,\frac{\partial\phi}{\partial t} - \phi\frac{\partial\phi^*}{\partial t}\right) \qquad (6.37)$$

Where * refers to the "mirror-image" system. The justifiability of such a diffusion "pseudo-Lagrangian" must ultimately come from the results of its application. Thus far, it has been used satisfactorily to determine rate constants for certain kinds of chemical processes (see Rice, 1985). Of importance is the circumstance that, for steady-state conditions (with $\phi = \phi^*$), *the diffusion Lagrangian is precisely the dissipation function* in Equation (6.33) above (Morse and Feshback, 1953).

Buoyed by our foregoing analogical reasoning, we can now say something about the nature of the metric tensor in our "metabolic field theory." From Equations (6.26), (6.27), (6.28), (6.31) and (6.33) one is led to relate $g_{\mu\nu}$ *to a diffusion coefficient matrix*, $D_{\mu\nu}$. The fact that the diffusion matrix transforms as a tensor of rank = 2 (see DeGroot and Mazur, 1969; Risken, 1984) lends further credence to this notion.

Let us suppose that the coordinate manifold involves the spatial position of metabolite molecules. The basic diffusion equation, then, has the form:

$$\frac{\partial\mathcal{L}}{\partial t} = D\nabla^2\mathcal{L} - F_e \qquad (6.38)$$

where $\mathcal{L}$ (x, y, z, t) is the number density of the diffusion species, D the diffusion coefficient and F_e an external force. This is the well-known Smoluchowski equation. It is a special case of the more general *Fokker-Planck equation*, governing the probability density function for particle position and velocity (Risken, 1984). The diffusion matrix was shown by Graham (1977a,b) to be the necessary metric to convert the Fokker Planck equation into a manifestly covariant form on a curved Riemannian phase space. And, $D_{\mu\nu}$ can be used to generate Christoffel symbols of the form in Equation (6.18) (see Graham, 1978; Risken, 1984), allowing the construction of well-defined geodesics (see Equations (6.16) and (6.17), tensor derivatives (see Equation 6.19), etc. [*Note*: The metric tensor is actually the matrix inverse of $(D_{\mu\nu})$ (see Graham, 1977; Risken, 1984)]. Thus, a "metabolic Lagrangian" akin to Equation (6.18) (but first-order in time) seem apropos.

Graham (1977a,b) has pursued at length the "path-integral" (Lagrangian) formulation of general diffusion processes. The approach is based on an older method of Onsager, whereby the probability density W for path $x(\tau)$ in phase space is written as:

$$W[x(\tau)] \sim exp[-\int_{-8}^{+8} d\tau L(\dot{x}(\tau), x(\tau))] \qquad (6.39)$$

where, again, L is a Lagrangian-type functional and $\dot{x}$ is the time derivative of x. As expected, the condition for the most probable path is given by:

$$d\int_{t_1}^{t_2} L(\dot{x}(\tau), x(\tau))d\tau \quad 0 \qquad (6.40)$$

Equations (6.39) and (6.40) provide a formulation of macroscopic stochastic dynamics which gives the status of a *thermodynamic potential* for certain kinds of nonequilibrium regimes (Graham, 1978). Near equilibrium, W relates directly to the entropy production, with Equation 6.40 yielding the well-known "principle of minimum entropy production" (Prigogine, 1967). Graham (1977a,b) gives here the status of a "free energy," in the sense of familiar statistical thermodynamics. He has elaborated a path-integral solution for the "relativistic" (covariant) Fokker-Planck equation, whose Lagrangian is of the general form in Equation (6.26), along with a "drift" term and a Riemannian "curvature" factor (which is determined from the metric). We note that the "diffusion" (Fokker-Planck) approach has been applied to the *protein dynamics* of enzyme action by a number of workers (Gavish, 1978; Kamp, Welch and Westerhoff, 1987; Shaitan and Rubin, 1982; Welch, 1986; see also Welch, 1984). Interestingly, Berkowitz et al. (1983) have applied an extremal (Lagrangian) formalism specifically to the modelling of trajectories in reaction processes involving diffusion along a macromolecular conformational coordinate. Minimization of "friction" was considered by those workers as a key trajectory determinant for protein-based reactions (see also Somogyi, Welch and Damjanovich, 1982). A similar notion has been extended to the supermolecular level of reaction-diffusion processes in coupled multienzyme systems in organized states (Welch and Keleti, 1981).

Summarizing, we would suggest that the spatiotemporal organization of cell metabolism lends itself to a "field structure." By analogy to classical fields in physics, it would seem that the "metabolic field" should be characterized in a general tensorial form involving a tensor of rank $= 2$. The diffusion-coefficient matrix is the most logical choice here, as one may describe material transport *both* through space *and* across an enzyme-based chemical-reaction coordinate as diffusive motion. Then, it is the organizational heterogeneity of the reaction-diffusion system *in vivo* which accounts for the "curvature" of metabolic space-time. The homogeneous bulk-phase regime (which we customarily use for *in vitro* analysis) represents the "flat" (Euclidean) limit.

In the foregoing discourse we have vacillated, in a seemingly cavalier manner, between material *density* and *chemical potential* as the field quantity. (The two are, of course, related.) This reflects the rather qualitative nature of our presentation and our uncertainty at this time as to the true mathematical form of the "metabolic field." Though, biophysical intuition suggests that chemical potential is, indeed, the more appropriate entity (with the understanding that we can use this term only in a "local" sense). For, as intimated above, in the dynamics of the living state we are not concerned simply with the mere *presence* of matter, rather with its *self-activity*. And, by its very physical nature, the chemical potential of a substance is always defined *relative to* the energetic surroundings thereof.

Another reason for using the chemical potential as the field quantity lies in its relationship (viz., its gradient) to free-energy dissipation. This association immediately suggests a role of dissipation (or entropy production) as a motif in the evolutionary design of "geodesic" (nay, "biodesic") paths in metabolic space-time. Minimization of free-energy dissipation automatically lends itself to an "extremal principle" thereof. As enunciated so eloquently by Lumry (Lumry and Biltonen, 1969), the evolutionary battle in cell metabolism is ultimately fought on the reaction coordinate of combined protein-chemical subsystems. It is tempting to speculate that the "curvature" of metabolic space-time (as reflected in the organizational character of enzymatic processes) is Nature's global strategy in attempting to win this battle. Such an evolutionary trend of minimization of free-energy dissipation in the operation of cell metabolism should be contrasted to the oft-discussed role of *increasing dissipation* (intensively), as an "arrow" of biological evolution and complexity on the global scale (Nicolis and Prigogine, 1977). The two principles are seen to be complementary, but applicable at different operative levels in the living system (Welch, 1984; Welch and Kell, 1986).

As a final note to our discussion of the geometrization of metabolic space, we mention briefly the possibility of applying one of the even newer formalisms of modern physical theory to metabolizing systems—that of string theory. We shall sketch only the qualitative features. Just as a point particle is described as following a minimized geodesic curve in space-time, here we view a "string" as sweeping out a surface. The dynamics of the string's motion are then found by applying a variational principle on this surface. The string is parametrized by σ (the position along the string), and τ (the proper time). This suggests an immediate analogy to the metabolizing system. If we identify σ with the internal coordinate position along the metabolic pathway, and τ with time, then we can view a string moving through our metabolic space-time as the natural evolution of a metabolizing system subject to the constraints of cellular activity.

It is interesting to note that in order for string theory to satisfy the criteria of special relativity, the number of dimensions in which the string can exist are severely limited (26 or 10). One might speculate whether such a limitation might not also apply to organized metabolizing systems where the number of internal coordinates that can be "linked together" are similarly constrained. Certainly, the number of such coordinates does not seem to be unlimited based on the fact that microdomains do exist in the cell for the precise purpose of separating pathways and preventing close couplings. (For a review of string theory see Green, Schwartz and Witten, 1987).

6.8. THE METABOLIC MICRODOMAIN: A GAUGE FIELD

A physical system containing many degrees of freedom, such as a population of reacting-diffusing particles *or* a protein macromolecule, is never in isolation; it is always in interaction with a background field which "causes" the system to undergo a diffusion process dictated by a diffusion tensor $D_{\mu\nu}$, satisfying a variational principle $\delta \mathcal{L} = 0$. It would appear that the "background field" is just a thermal bath (reservoir). In the foregoing scheme we have seen that $D_{\mu\nu}$ serves as a Riemannian metric. Not only is it intrinsically associated with flow, but, by virtue of the *fluctuation-dissipation theorem*, it scales the local fluctuations.

The organizational character of metabolic space-time, as reflected in the affine connection, must be supplemented (or even supplanted) with something else in local metabolic domains. Here, we are aided and abetted by yet another analogy from contemporary physics—that of the *gauge field*. In the previous section, we saw a bit of the usefulness of the Lagrangian field theory, in generating "relativistically" covariant field relations. The Lagrangian formalism is also extremely valuable in discussions of invariance principles and related conservation laws. Invariance principles and their associated symmetry transformations in Nature are of many types. When the equations of motion for physical fields are derived from a variational principle, a general procedure is in place for establishing conservation conditions as a consequence of invariance properties. Accordingly, conservation laws found in Nature may be imposed as symmetries of the Lagrangian, thus defining the mathematical form of the latter. The general paradigm is provided by the famous *Noether's theorem*—which correlates a conservation law with each continuous symmetry transformation under which the Lagrangian is invariant (Agarwal, 1977; Ramond, 1981; Ryder, 1985). In particular, for gauge theories, field interactions arise as the very consequence of "local gauge symmetries." The idea of gauge invariance as a dynamical principle arose early in this century, due to Hermann Weyl's (unsuccessful) attempt to unify both gravitation and electromagnetism through the requirement of invariance under a space-time dependent change of measuring scale (or "gauge").

Gauge invariance can be applied at two levels: *global* and *local*. Let us consider first the global condition, which is applied uniformly at all points in space-time. We return to Equation (6.24) and ask what does global gauge invariance imply. Following the mathematical form of Equation (6.37), we suppose that the scalar field has two components, ϕ and ϕ^*, giving

$$\mathcal{L} = \frac{1}{2}(\nabla\phi\cdot\nabla\phi^*) - V(\phi,\phi^*) \tag{6.41}$$

(In principle, the field may have any number of components.) For familiar physical fields, conservation laws for the total energy, momentum, etc., follow from *space-time* symmetries via Noether's theorem. Considerations from the previous section suggest that we might concern ourselves with *conservation of free energy*. If there is an appropriate symmetry for this entity, it must involve internal coordinates of the two-component field variable (ϕ, ϕ^*) itself rather than space-time *per se*. Let us impose a global scale transformation on ϕ (and ϕ^*) which can be written in the general form e^α, where α is a constant. Then, we have the transformation scheme:

$$\phi \to e^\alpha\phi; \ \phi^* \to e^{-\alpha}\phi^*$$

Cytosociology: A Field-Theoretic View of Cell Metabolism

Requiring that this global transformation leave the Lagrangian unchanged leads (via Noether's theorem) to the following conserved "current:"

$$J_\mu = (\phi^* \nabla_\mu \phi - \phi \nabla_\mu \phi^*) \tag{6.42}$$

such that the following quantity, P is a constant of the motion (i.e., dP/dt = 0):

$$P = K \cdot \int \left(\phi^* \frac{2\phi}{2t} - \phi \frac{2\phi^*}{2t} \right) dV \tag{6.43}$$

where K is a dimensionality factor, and integration is performed over the spatial volume of the system (Ramond, 1981; Ryder, 1985). If our field variable is particle density, then the relation in Equation (6.43) reflects particle (e.g., mass) conservation as a "give-and-take" between the field components (probability densities) ϕ and ϕ^* (in a closed system). If the field variable denotes chemical potential, then Equation (6.43) signifies conservation of free energy. In fact, this is the spirit of Equation (6.37) wherein (as elaborated by Morse and Feshback, 1953) the "reverse-diffusion" field ϕ^* was introduced historically as a mere mathematical contrivance to insure that the "pseudo-Lagrangian" for diffusion is invariant under transformations involving energy conservation.

Of course, postulation of global free-energy conservation immediately runs afoul of the Second Law of Thermodynamics, since the reaction-diffusion processes of a metabolizing cell (when measured "globally") are dissipative! This suggests that a "metabolic gauge field" (if such exists!) must be sought *at the local level*. That is, we must abandon the stipulation that the transformation parameter, α, is a constant over all space-time. Compare this situation to that in the usual physical field theory, where limitations demanded by *special relativity* [e.g., constraint due to the speed of light] lead to the transition from a global to a local formulation (Ryder, 1985).

Thus, we write α as $\alpha(x^\mu)$ and ask what are the consequences of imposing invariance (i.e., $\delta\mathcal{L} = 0$) under a *local* transformation of the form.

$$\phi(x^\mu) \to e^{\alpha(x^\mu)} \phi(x^\mu); \quad \phi^*(x^\mu) \to e^{\alpha(x^\mu)} \phi^*(x^\mu)$$

Mathematical terms in the Lagrangian that depend only on the field variable ϕ (or ϕ^*), itself are left invariant. But such is not the case for gradient terms, which transform as:

$$\frac{\partial \phi}{\partial x^\mu} \rightarrow e^{\alpha(x^\mu)} \left[\frac{\partial \phi}{\partial x^\mu} + \phi \frac{\partial \alpha}{\partial x^\mu} \right]$$

This necessitates the introduction of a mathematical "device" to get rid of the unwanted ∇_μ term. Taking the physics to heart, one introduces a new (vector) field A_μ which couples directly to the matter flow (Equation 6.42), thereby making the Lagrangian invariant. This A_μ is a so-called *gauge field*. Accordingly, one generates a "gauge-covariant derivative"

$$D_\mu - \nabla_\mu + A_\mu(x^\mu) \tag{6.44}$$

such that transforms desirable as:

$$D_\mu \phi \rightarrow e^{\alpha(x^\mu)} D_\mu \phi$$

This will, indeed, be the situation, provided that the vector (gauge) field transforms as:

$$A_\mu(x_\mu) \rightarrow A_\mu(x^\mu) - \nabla_\mu \alpha(x^\mu)$$

The resulting locally *gauge-covariant form* of Equation (6.41) is

$$\begin{aligned} \mathcal{L} &- \nabla_\mu \phi \cdot \nabla^\mu \phi^* - J^\mu A_\mu + A^2 \phi \phi^* - V(\phi,\phi^*) \\ &- (\nabla_\mu \phi + A_\mu \phi)(\nabla^\mu \phi^* - A^\mu \phi^*) - V(\phi,\phi^*) \\ &- (D_\mu \phi)(D^\mu \phi^*) - V(\phi,\phi^*) \end{aligned} \tag{6.45}$$

where J^μ is the "Noether current" from Equation (6.42).

The nature of Equation (6.44) is quite analogous to that in Equation (6.19). (In fact, the two types of covariant derivative were defined along similar lines). In relativity theory, the additive term of the covariant derivative (see Equation 6.19) defines the local change of measuring scale as a geometric construction. Whereas, in gauge theory, invariance principles necessitate the introduction of a vector (gauge) field as the compensatory devise for local scale changes. Moreover, in quantum field theory this replacement prescribes the very form of the interaction between the gauge field and matter (e.g., the $J_\mu A^\mu$ term in Equation 6.45). All of the basic forces of nature (with the possible exception of gravitation) have been described as gauge fields, wherein, the *gauge field* (A_μ) *is mediated by mass-less bosons* (so-called "gauge bosons") of spin = 1 (except for gravitation with spin = 2).

Is there any realistic analogy between the foregoing physicomathematical interlude and putative "metabolic field?" What sort of local "balancing act" might occur, between the metabolic flow of matter and the background, to guarantee invariance of the Lagrangian function? First, let us look at the level of the individual enzyme. There are indications that the state of "perfection" of enzyme action (toward which enzymes seem to be evolving) corresponds to an *equi-energy profile of all enzyme-bound species* on the reaction path (Keleti and Welch, 1984). The microenvironment within the enzyme molecule creates this condition, interacting with the bound chemical subsystem. Moreover, the message from Section 6.3 is that the give-and-take of free energy between the protein and the chemical system is a very *dynamic process*, throughout the course of catalysis. Accordingly, most of the free-energy change arises from the entry of substrate and exit of product. During the flow of the catalytic process, the free energy of the protein matrix actually meshes with that of the bound chemical substance. This is exemplary of Lumry's "free-energy complementarity" principle in protein function (Lumry and Biltonen, 1969; Lumry and Gregory, 1986). Figure 6.3 illustrates the gauging effect of the protein matrix in smoothing out the free energy profiles.

Now, by analogy with gauge field theory, we must look for a mass-less "gauge boson" for this local field. Since much of the fluctuating character of the protein molecule (especially that relevant to enzyme catalysis—see Section 6.3, 6.4) involves hydrogen bond dynamics, we are led to suggest a role of *hydrogen-bond phonons*. The theoretical significance thereof has been emphasized by a number of workers (Fröhlich, 1980, 1986; Green and Ji, 1972; Ji, 1974, 1985, see Section 1.5, Chapter 1; Mishra, 1979; Welch and Berry, 1985). Phonons are quantized lattice vibrations (Jensen, 1964) having appropriate properties (e.g., spin = 1) for our case. Thus, one might picture enzyme catalysis as a coupling of the localized chemical subsystem to the phonon field of the protein matrix. (For a good heuristic discussion of this kind of coupling, see Bak, 1963). The dynamic hydrogen-bond phonon field in protein molecules might manifest itself in a variety of possible conformations ("conformons") (Green and Ji, 1972; Ji, 1974, 1979, 1985) (see Section 1.5, Chapter 1). For enzymes operating in bulk solution (at thermal equilibrium with a heat bath), such a dynamic "field interaction" in the enzyme-substrate complex occurs at (or near) thermodynamic equilibrium. Macroscopically, the only measurable free-energy

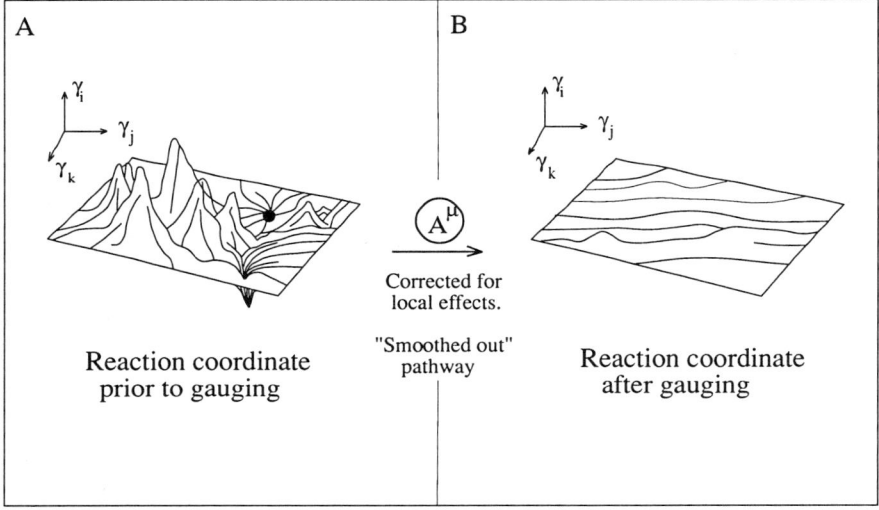

Figure 6.3. The Effect of the Metabolic Gauge Field.

change is that due to the concentration (or chemical potential) difference between the initial substrate and the final product.

Thus, locally, it seems that the Lagrangian gauge-field approach to metabolic infrastructure has some basis in physical reality. To be fully consistent with the gauge field approach one would have to take the appropriate Lagrangian and "quantize" the field (using for example, the Feynman path-integral method) (Itzykson and Zuber, 1981; Ryder, 1985)) according to the phonon modality.

The localized chemical events, occurring during the catalytic process in the enzyme active center, are vectorial in nature. For a homogeneous solution of unorganized enzyme molecules, though, there is no preferred *global* direction for material flow. The enzyme population is subject to rotational invariance. When an enzyme system is spatially organized (as those *in vivo*), the symmetry of the reaction-diffusion flow is broken. There is an interesting analogy with the behavior of the *Heisenberg ferromagnet*, a crystalline array of atomic magnetic dipoles (Quigg,

1983). Above the so-called Curie temperature, the global system manifests rotational invariance (i.e., the individual dipoles are randomly oriented). Below this critical temperature, all the dipoles line-up in parallel, resulting in global magnetization. Moreover, the individual dipoles in the ferromagnet can be aligned by application of *an external field*, breaking the symmetry in a preferred direction. A similar phenomenon may be realizable in cell metabolism, when enzymes are juxtaposed to local energy sources (e.g., electric fields, high-energy protons) at the surface of intracellular particulates (Fröhlich, 1980, 1983; Kell and Westerhoff, 1985; Mishra, Bhaumik, Mathur, and Mitra, 1979; Welch and Berry, 1983, 1985; Westerhoff, 1987). Most importantly, when an enzyme couples to these fields, the free-energy change of the chemical reaction may be "gauged" locally (Berry, 1981; Welch and Kell, 1986). One effect of such interaction is, again, to help balance the up-and-down free-energy swings in the enzyme-catalyzed reactions. Protein-protein interactions within multienzyme aggregates may participate in such free-energy conservation (Lumry and Biltonen, 1969; Welch and Kell, 1986).

In conclusion, we reiterate, that within the microenvironmental confines of organized multienzyme systems *in vivo*, there is the distinct possibility that biochemical processes are executed in a highly conservative manner—in some cases perhaps approaching thermodynamic equilibrium. Yet globally, the system is dissipative—e.g., by virtue of time-dependent boundary conditions thereon. This sort of hierarchial dichotomy in symmetry relations is often encountered in gauge fields describing the physical world. The foregoing theoretical paradigm can still be used in such situations, by splitting the Lagrangian into a symmetric part and a perturbative "symmetry-breaking" part (Quigg, 1983; Ryder, 1985). This allows us to retain at least some semblance of gauge field approach to our "metabolic field" (and its associated boson requirements), by applying *locally* the invariance principles which validate the Lagrangian formalism. As noted in Section 6.5 above, the ultimate validity of the familiar "local equilibrium assumption" in nonequilibrium thermodynamics may also be at issue here (Welch, 1985).

Additionally, we should mention that the general validity of any kind of "mechanical" approach to dissipative systems relates to basic questions of *irreversibility* in the physical world. This topic is being explored in depth at the Brussels school of thermodynamics (Prigogine, 1980)—where, interestingly, one finds the approach is to split the Liouville operator into symmetric and "symmetry-breaking" parts.

If the *local* process is actually conservative, then we can truly use a Lagrangian such as that in Equation (6.24) and the respective field equation such as Equation (6.25)—which is *second-order in time*. This is recognizable in classical physics as the basic equation for *wave motion* (under some potential). In one spatial dimension, it describes wave motion in a flexible string (subject to a deformation of the medium by the potential-energy term). By its nature, such motion can be regarded as an "equilibrium" situation corresponding to a superposition of waves moving *in both directions* (Morse and Feshback, 1953). (As above, one might introduce dissipation into this system by some condition on the boundaries.) Such an idealized ("resonance-like") state of free-energy transduction was proposed by McClare (1974), as the *modus operandi* of "molecular machines" in bioenergetics (Welch and Kell, 1986). This state of "perfection" can never be attained to completion. That is, organization of an enzyme system in conjunction with some local nonequilibrium energy source (e.g., electric field, proton-motive force) cannot "gauge" the protein-based "diffusion" process down to a single, non-dissipative mechanical degree of freedom. There is just too much degeneracy in protein-conformational states—what Cooper (1984) calls the "uncertainty principle for proteins." Nevertheless, it is tempting to speculate that this "McClare limit" represents the directional bound on enzyme evolution *in vivo*.

6.9. CONCLUDING REMARKS

In an era when many physicists (Boslough, 1985; Crease and Mann, 1986) are suggesting that the "end of physics" is in sight (what with the success of relativistic quantum field theory, gauge fields, superstrings, etc., in unifying the physical world), the time may be ripe to begin a formal embrace with biology. The rather autonomous science of biology has long resisted the kind of strict "lawfulness" so prevalent in the physicochemical sciences. The stage seems set—from a number of perspectives—for a Kuhnian paradigm transition (Kuhn, 1970). In his powerful and insightful book Mayr (1982) closes with the prophetic statement:

> When it comes to developing a truly comprehensive science of science, it can be done only by comparing the generalizations derived from the physical sciences with those of the biological and social sciences, and by attempting to integrate all three branches. I rather suspect that the raw material for such comparisons and for an integration is already available and that it is only necessary that someone adopts this as the objective of his research.

Over the years a number of eminent biophysical thinkers (e.g., A. Lotka, N. Rashevsky) have made bold attempts to bring biology under the umbrellas of defined physical lawfulness. Indeed the renowned English physiologist William Bayliss long sought to apply basic concepts of physics and chemistry to the living state, in an effort to elucidate the principles of "general physiology" by which all living things exist and owe their being. Such efforts have not always met with the approval of mainstream biologists, as evidenced by Bayliss' remarks that some found "...the introduction of mathematics into biological question mischievous" (Bayliss, 1920, page 37).

On a more positive note, in recent times the Brussels school of thermodynamics (Nicolis and Prigogine, 1977; Prigogine, 1980) has made great strides in convincing us that spatiotemporal order in the biological world is rational, and that biology has something to offer in physics (in contradistinction to the opposite directional influence so prevalent in physics, with its arrogance toward biology). With the emergence of the "anthropic principle" in cosmology (Barrow and Tipler, 1986), we find an increasing realization that there is something distinctly "biological" (or, better, "biophysical") about our universe as a whole. Thus, the heretofore naturalistic and highly empirically based science of biology may be ready for a theoretical "field structure"—above and beyond (and perhaps inclusive of) the so-called "morphogenetic fields" invoked qualitatively in developmental biology (Haraway, 1976; Sheldrake, 1981). Such theoretical constructs as relativity theory (and differential geometry), gauge fields, and symmetry-breaking dissipative structures, are too powerful and broadly-applicable in the physical world to be ignored by biology.

Perhaps a fitting place to begin a consideration thereof is at the subcellular (or metabolic) level, wherein we can more readily grasp the physicochemical principles at work (Smith et al., 1990). There are many indications that the same mathematical relations hold at various levels of biological complexity (Nicolis and Prigogine, 1977; Prigogine, 1980; Welch, 1977). [For a discussion of the application of the Fokker-Planck formalism to "diffusion" phenomena in *ecology*, for example, see Okubo (1980)]. Indeed, there may be an intrinsic symmetry at play, in this hierarchical extrapolation in biology—the "principle of relational invariance" enunciated by Rashevsky (1973) (see also Welch, 1987). As biology and physics grow ever closer to a unification, we come to a realization of the underlying significance of Bohm's "implicate order" (Bohm, 1980), whereby matter in biological systems exists in a state which is uniquely "enfolded" into the complete physical fabric of the universe. By rights, we should designate that as a distinct *state of matter*. Let us call it, aptly and simply, *bios*. Then, following Bohm, we might say that matter is "biotic" (or "alive") when it exists in a self-propagating, internally-negentropic region of space-time "curved" by global (boundary) free-energy dissipation. Mischievous, indeed....

ACKNOWLEDGMENTS

It is with much gratitude that one of us (G.R.W.) acknowledges fruitful discussions with Drs. M. N. Berry, S. Ji, T. Keleti, D. B. Kell, and H. V. Westerhoff. H. A. S. would also like to thank Dr. J. Johnson, Dr. E. Walton, Ms. C. Zimmerman, and Mr. D. Pilch for critical comments on the ideas presented herein. Our thanks also go to Ms. A. Whitley for her careful typing and editing of the manuscript.

REFERENCES

Agarwal, B. K. (1977). Quantum Mechanics and Field Theory. Asia Publications. New York.
Anderson, N. (1976). Interactive Macromolecular Sites II. Role in Prebiotic Macromolecular Selection and Early Cellular Evolution. J. theor. Biol. 60: 413-419.
Bak, T. A. (1963). Contributions to the Theory of Chemical Kinetics. Benjamin, New York.
Barrow, J. D. and Tipler, F. J. (1986). The Anthropic Principle. Oxford University Press, Oxford, New York.
Bayliss, W. M. (1920). Principles of General Physiology. 3rd ed. Longmans, Green, and Co. London.
Berkowitz, M., Morgan, J. D., McCammon, J. A. and Northrup, S. H. (1983). Diffusion Controlled Reactions: A Variational Formula for the Optimum Reaction Coordinate. J. Chem. Phys. 79: 5563-5565.
Berry, M. N. (1981). An Electrochemical Interpretation of Metabolism. FEBS Lett. 134: 133-138.
Bjorken, J. D. and Drell, S. P. (1965). Relativistic Quantum Fields. McGraw-Hill, Inc., New York.
Berry, M. N. and Welch, G. R. (1987). Unpublished Observations.
Bohm, D. (1980). Wholeness and the Implicate Order. Routledge and Gegan Paul, London.
Boslough, J. (1985). Stephen Hawking's Universe. Morrow, New York.
Clegg, J. S. (1984). Properties and Metabolism of the Aqueous Cytoplasm and Its Boundaries. Amer. J. Physiol. 246: R133.
Conrad, M. (1979). Unstable Electron Pairing and the Energy Loan Model of Enzymic Catalysis. J. theor. Biol. 79: 137-156.
Cooper, A. (1984). Thermodynamic Uncertainty Principle. Prog. Biophys. Mol. Biol. 44: 181-214.
Crease, R. P. and Mann, C. C. (1986). The Second Creation: Makers of the Revolution in Twentieth-Century Physics. Macmillan. New York.
DeGroot, S. R. and Mazur, P. (1969). Nonequilibrium Thermodynamics. Elsevier/North-Holland, Amsterdam.
Fersht, A. (1984). Enzyme Structure and Mechanism. 2nd ed. W. H. Freeman, San Francisco.
Friedrich, P. (1984). Supramolecular Enzyme Organization. Pergamon Press, New York.
Fröhlich, H. (1980). The Biological Effects of Microwaves and Related Questions. Electron Phys. 53: 85-152.
Fröhlich, H. (1986). The Fluctuating Enzyme (G. R. Welch, ed.). Wiley, New York, p. 421.
Fröhlich, H. and Kremer, F. (Eds.) (1983). Coherent Excitations in Biological Systems. Springer-Verlag, New York/Heidelberg.
Gavish, B. (1978). The Role of Geometry and Elastic Strains in Dynamic States of Proteins. Biophys. Struct. Mech. 4: 37-52.
Graham, R. (1977a). Path Integral Formulation of General Diffusion Processes. Z. Physik B 26: 281-290.

Graham, R. (1977b). Covariant Formulation of Non-Equilibrium Statistical Thermodynamics. Z. Physik B 26: 397-405.

Graham, R. (1978). In Stochastic Processes in Nonequilibrium Systems (L. Garrido, P. Seglar, and P. J. Shepherd, eds.), Springer, Verlag, New York/Heidelberg, p. 82.

Graham, R. (1981). In: Stochastic Nonlinear Systems (L. Arnold and R. Lefever, eds.). Springer-Verlag, New York/Heidelberg.

Green, D. E. and Ji, S. (1972). The Electromechanochemical Model of Mitochondrial Structure and Function. In: The Molecular Basics of Electron Transport (J. Schultz and B. F. Cameron, eds.). Academic Press, New York, pp. 1-44.

Green, M. B., Schwartz, J. H. and Witten, E. (1987). Superstring Theory Vol I, II. Cambridge University Press, Cambridge.

Haraway, D. J. (1976). Crystals, Fabrics, and Fields: Metaphors of Organicism in Twentieth-Century Developmental Biology. Yale University Press, New Haven, Connecticut.

Itzykson, C. and Zuber, J. (1980). Quantum Field Theory. McGraw-Hill, Inc. New York.

Jensen, H. H. (1964). In Phonons and Phonon Interactions (T. A. Bak, ed.). Benjamin, New York/Amsterdam, p. 1.

Ji, S. (1974). Energy and Netentropy in Enzymic Catalysis. Ann. N.Y. Acad. Sci. 227: 419-437.

Ji, S. (1979). The Principles of Ligand-Protein Interactions and Their Application to the Mechanism of Oxidative Phosphorylation. In: Structure and Function of Biomembranes (K. Yagi, ed.) Japan Scientific Societies Press, Tokyo.

Ji, S. (1985). The Bhopalator: A Molecular Model of the Living Cell Based on the Concepts of Conformons and Dissipative Structures. J. theor. Biol. 116: 399-426.

Kamp, F., Welch, G. R. and Westerhoff, H. V. (1987). "Energy Coupling and Hill Cycles in Enzymatic Processes," Cellular Biophysics, in press.

Keleti, T. (1975). In Proc. Ninth. Febs Meeting (Symposium on Mechanism of action and regulation of enzymes). (T. Keleti, ed.) 32: 3. North-Holland, Amsterdam.

Keleti, T. (1978). In: New Trends in the Description of the General Mechanism and Regulation of Enzymes (S. Damjanovich, P. Elödi and B. Somogyi, eds.) (Symp. Biol. Hung) 21: 107. Akedmian 'Kiado', Budapest.

Keleti, T. and Welch G. R. (1984). Biochem. J. 223: 299-303.

Kell, D. B. and Westerhoff, H. V. (1985). In: Organized Multienzyme Systems: Catalytic Properties (G. R. Welch, ed.) Academic Press, New York, p. 63.

Koshland, D. E. (1976). Role of Flexibility in the Specificity, Control and Evolution of Enzymes. FEBS Lett. 62(suppl) E47-E52.

Kuhn, T. S. (1970). The Structure of Scientific Revolutions (2nd Ed.). University of Chicago Press, Chicago.

Lewis, T. J. (1979). The Mechanisms of Conduction in Proteins. Ciba Found Symp. 67: 65-78.

Lumry, R. and Biltonen, R. (1969). In Structure and Stability of Biological Macromolecules (S. N. Timasheff and G. D. Fasman, eds.). Dekker, New York, p. 65.

Lumry, R. and Gregory, R. B. (1986). Free-Energy Management in Protein Reactions: Concepts, Complications, and Compensation. In: The Fluctuating Enzyme (G. R. Welch, ed.). Wiley, New York, pp. 1-190.

Mandl, F. and Shaw, G. (1984). Quantum Field Theory. John Wiley and Sons, New York.

Mayr, E. (1982). The Growth of Biological Thought. Harvard University Press, Cambridge, Massachusetts.

McClare, C. W. F. (1974). Resonance in Bioenergetics. Ann. N.Y. Acad. Sci. 227: 74-97.

McVittie, G. C. (1949). Cosmological Theory. Methuen, London.

Mishra, R. K., Bhaumik, K., Mathur, S. C. and Mitra, S. (1979). Excitons and Bose-Einstein Condensation

in Living Systems. Int. J. Quant. Chem. 169: 691-706.
Misner, C. W., Thorne, D. S. and Wheeler, J. A. (1973). Gravitation. Freeman, San Francisco.
Mitchell, P. (1979). Compartmentation and Communication in Living Systems. Eur. J. Biochem. 95: 1.
Morse, P. M. and Feshback, H. (1953). Methods of Theoretical Physics. McGraw-Hill, New York.
Nagle, J. F. and Tristram-Nagle, S. (1983). Hydrogen Bonded Chain Mechanisms for Proton Conduction and Proton Pumping. J. Membr. Biol. 71: 1-14.
Nelson, E. (1985). Quantum Fluctuations. Princeton University Press, Princeton.
Nicolis, G. and Prigogine, I. (1977). Self-Organization in Nonequilibrium Systems. Wiley, New York.
Okubo, A. (1980). Diffusion and Ecological Problems: Mathematical Models. Springer-Verlag, New York/Heidelberg.
Peters, R. A. (1930). Surface Structure in the Integration of Cell Activity. Trans. Faraday Soc. 26: 797-822.
Peusner, L. (1982). Global Reaction-Diffusion Coupling and Reciprocity in Linear Asymmetric Kinetic Networks. J. Phys. Chem. 77: 5500-5507.
Peusner, L. (1986). Studies in Network Thermodynamics. Elsevier, Amsterdam.
Porter, K. R. (1984). The Cytomatrix: A Short History of Its Study. J. Cell Biol. 99: 3s-12s.
Porter, K. R. and Tucker, J. B. (1981). Ground Substance of the Living Cell. Sci. Amer. 244: 56-67.
Prigogine, I. (1967). Introduction to Thermodynamics of Irreversible Processes. 3rd. ed. Wiley, New York.
Prigogine, I., Nicolis, G. and Babloyantz, A. (1972). Thermodynamics of Evolution. Phys. Today 25(11): 23-28.
Prigogine, I. (1980). From Being to Becoming: Time and Complexity in the Physical Sciences. Freeman, San Francisco.
Quigg, C. (1983). Gauge Theories of the Strong, Weak, and Electromagnetic Interactions. Benjamin, Reading, Massachusetts.
Ramond, P. (1981). Field Theory: A Modern Primer. Benjamin, Reading, Massachusetts.
Rashevsky, N. (1973). In: Foundations of Mathematical Biology (R. Rosen, ed). 3: 177. Academic Press, New York.
Rice, S. A. (1985). Diffusion-Limited Reactions. Elsevier, Amsterdam.
Richardson, I. W. (1980). The Metrical Structure of Aging (Dissipative) Systems. J. theor. Biol. 85: 745-756.
Richardson, I. W. and Rosen, R. (1979). Aging and the Metrics of Time. J theor. Biol. 79: 425-432.
Risken, H. (1984). The Fokker-Planck Equation. Springer-Verlag, New York/Heidelberg.
Ryder, L. H. (1985). Quantum Field Theory. Cambridge University Press, Cambridge.
Schliwa, M., Van Blerkom, J. and Porter, K. W. (1981). Stabilization of the Cytoplasmic Ground Substance in Detergent-Opened Cells and a Structural and Biochemical Analysis of its Composition. Proc. Nat. Acad. Sci. USA 78: 4329-4333.
Shaitan, K. V. and Rubin, A. B. (1982). Equation of Motion of Conformon and Primitive Molecular Machines for Electron (Ion) Transport in Biological Systems. Mol. Biol. (USSR) 16: 1004-1018.
Sheldrake, R. (1981). A New Science of Life. Tarcher, Los Angeles.
Sitte, P. (1980). In: Cell Compartmentation and Metabolic Channelling (L. Nover, F. Lynen, and K. Mothes, eds.). Elsevier/North Holland, New York, p.17.
Smith, H. A., Pilch, D. R. and Welch, G. R. (1990). Holism vs. Reductionism: A Model for Introductory Philosophy into the Biology Curriculum. In: More History and Philosophy of Science in Science Teaching (D. E. Herget, ed.). (In press).
Somogyi, B., Welch, G. R. and Damjanovich, S. (1984). The Dynamic Basis of Energy Transduction in Enzymes. Biochim Biophys. Acta. 768: 81-112.
Srere, P. A. (1981). Protein Crystals as a Model for Mitochondrial Matrix Proteins. Trends Biochem. Sci.

<u>6</u>: 4-7.
Srere, P. A. (1987). Complexes of Sequential Metabolic Enzymes. <u>Ann. Rev. Biochem.</u> <u>56</u>: 89-124.
Szent-Györgvi, A. (1941). A study of Energy-Levels in Biochemistry. <u>Nature</u> <u>148</u>: 157-159.
Thom, R. (1975). <u>Structural Stability and Morphogenesis</u>. Benjamin, Reading Massachusetts.
Van Rysselberghe, P. (1963). <u>Thermodynamics of Irreversible Processes</u>. Hermann, Paris.
Volkenstein, M. (1981). Simple Physical Presentation of Enzymic Catalysis. <u>J. theor. Biol.</u> <u>89</u>: 45-51.
Welch, G. R. (1977a). On the Role of Organized Multienzyme Systems in Cellular Metabolism: A General Synthesis. <u>Prog. Biophys. Mol. Biol.</u> <u>32</u>: 103-191.
Welch, G. R. (1977b). On the Free Energy "Cost of Transition" in Intermediary Metabolic Processes and the Evolution of Cellular Infrastructure. <u>J. theor. Biol.</u> <u>68</u>: 267-291.
Welch, G. R. (1984). In: <u>Dynamics of Biochemical Systems</u> (J. Ricard and A. Cornish-Bowden, eds). Plenum Press, New York, p. 85.
Welch, G. R. (1985). Some Problems in the Use of Gibbs Free Energy. <u>J. theor. Biol.</u> <u>114</u>: 433-446.
Welch, G. R. (ed) (1986). <u>The Fluctuating Enzyme</u>. Wiley, New York.
Welch, G. R. (1987). <u>Trends Ecol. Evol.</u> in press.
Welch, G. R. and Berry, M. N. (1983). Long-range Energy Continua in the Living Cell: Protochemical Considerations. In: <u>Coherent Excitations in Biological Systems</u> (H. Fröhlich and F. Kremer, eds.). Springer-Verlag, New York/Heidelberg, pp. 95-116.
Welch, G. R. and Berry, M. N. (1985). In: <u>Organized Multienzyme Systems: Catalytic Properties</u> (G. R. Welch, ed.). Academic Press, New York, p. 419.
Welch, G. R. and Clegg, J. S. (eds) (1987). <u>Organization of Cell Metabolism</u>. Plenum Press, New York.
Welch G. R., and Gaertner, F. H. (1980). Enzyme Organization in the Polyaromatic-Biosynthetic Pathway. <u>Curr. Top. Cell. Regul.</u> <u>16</u>: 113-162.
Welch, G. R. and Keleti, T. (1981). On the "Cytosociology" of Enzyme <u>in vivo</u>: A Novel Thermodynamic Correlate of Biological Evolution. <u>J. theor. Biol.</u> <u>93</u>: 701-735.
Welch, G. R. and Kell, D. B. (1986). Not Just Catalysts—Molecular Machines in Bioenergetics. In: <u>The Fluctuating Enzyme</u> (G. R. Welch, ed.). Wiley, New York, p. 451.
Welch, G. R., Somogyi, B. and Damjanovich, S. (1982). The Role of Protein Fluctuations in Enzyme Action: A Review. <u>Prog. Biophys. Mol. Biol.</u> <u>39</u>: 109-146.
Westerhoff, H. V., Tsong, T. Y., Chock, P. B., Chen, Y. D. and Astumian, R. D. (1986). How Enzymes can Capture and Transmit Free Energy from an Oscillating Electric Field. <u>Proc. Nat. Acad. Sci. USA</u> <u>83</u>: 4734-4738.
Westerhoff, H. V. and Kamp, F. (1987). In: <u>Organization of Cell Metabolism</u> (G. R. Welch and J. S. Clegg, eds.). Plenum Press, New York, p. 339.

Chapter 7

NON-DICHOTOMOUS, RELATIVE AND HIERARCHICAL ASPECTS OF LIFE AND DEATH

Jerome Rothstein

ABSTRACT

A fundamental physical analysis is given of how "live" and "dead" states can be characterized. Living systems are non-equilibrium and so complex in structure that the ordinary macroscopic thermodynamic state concept is totally inadequate. The microscopic quantum state is also inadequate for a variety of reasons, some of which were anticipated in part by Bohr, Elsasser and Wigner. Complexity of living matter and vital behavior force a breaking down of the total system in which the latter occurs into subsystems even to define the operationally based language needed. The organism-environment distinction is the most obvious. It is paradigmatic for subsystem definitions at all levels of description down to the quantum-molecular. It has a kind of arbitrariness resembling that of the distinction between system of interest and means for observing it long familiar in quantum mechanics. Just as physical properties of systems at the quantum level must be analyzed in the total context of system plus means of observation, so must vital properties be considered relative to an environment. The state concept needed is closer to that familiar in automata theory and computer science than to traditional state concepts. In physical embodiments this does justice to the environment as source of inputs and destination of outputs, and can treat them symmetrically (interchange interpretations of input and output on passing from system to environment). Life is then relative to an

Non-Dichotomous, Relative and Hierarchical Aspects 325

environment, and the same system can be considered as simultaneously alive with respect to one environment and dead with respect to another. The living system is a "black box" which can be further analyzed into a collection of interacting black boxes (subsystems). This reductionist strategy can be followed down to the level of microscopic quantum-molecular events and interactions. Life, though definable at almost any level, is differently characterized operationally at each level (hierarchical nature). Life at each level "emerges," by methods which generalize those used in statistical mechanics, from lower levels. A generalized entropy concept, in which information, organization, and computation are included in a consistent way, is frequently invoked.

7.1. INTRODUCTION AND SUMMARY

Anyone seriously interested in constructing a molecular theory of cell life and death can not long evade the problem of defining what is to be meant by the terms life and death. There are also problems about what molecular and theory mean in this context. However, what chemists and physicists ordinarily mean by those terms is at least a good first approximation to what is needed here, permitting us to put that problem aside temporarily. But only on the basis of clear operational concepts of life and death can the possibility (or impossibility) or building a molecular theory of them be discussed adequately.

The problem is ancient and fraught with overtones of fear, superstition, religion, philosophy, and law, as well as science. Our aim is to use logical, semantic, operational, theoretical and empirical analysis in an attempt to build conceptions of life and death robust enough to support future theory.

Are dead and alive mutually exclusive and exhaustive logical alternatives? If so there must be operationally defined methods to classify *any* physical or chemical system unambiguously under exactly one of those headings. Denying exhaustiveness admits the possibility of systems which are neither dead nor alive, while relaxing mutual exclusivity allows for systems which are both dead and alive. For scientific purposes it is thus necessary to scrutinize the operations and measurements of which such classification decisions are based.

Can dead or alive be characterized as states of a system? In what sense is the word state to be used? In thermodynamics, state is an equilibrium macroscopic concept. Living matter is not at equilibrium. In both statistical mechanics and quantum mechanics the notion of a microscopic state is central. It is not clear that operational means to prepare a microscopic "living state" can even be defined in principle. We can try to define life behaviorally, or by attributes, but then we must choose the defining attributes and specify how they are to be measured. This is no different, in essence, from the problem of state definition. In this paper we need not distinguish between the problem of defining life and that of defining "living states." These "states" are not only nonequilibrium, but are needed to characterize thermodynamically open systems of enormous complexity and heterogeneity, whose very boundaries are frequently vague, perhaps even non-existent.

Much of the later discussion is a critical examination of what is needed to replace current notions of state. The result is closer to the notion of state used by engineers and computer scientists, in many important ways, than it is to the traditional scientific ones. Reference to a context, or environment, where the boundary between it and the "living system" can be rather arbitrary, seems inevitable in general. The related concept of life (or viability) *relative to an environment* then admits a non-dichotomous life-death concept in a simple way, and has other advantages as well. The interplay between microscopic (molecular, quantum-mechanical) and macroscopic (thermodynamics with informationally-generalized entropy, plus analysis of complex systems in the spirit of engineering and computer science) plays a central role. New content is given thereby to the old insights that the mechanisms on which life depends are molecular, but life is nevertheless metamolecular in the sense of emerging as a phenomenon at a higher level of organizational complexity.

Perhaps the simplest way to introduce the idea of life as metamolecular or emergent is by means of a story about Newton's great contemporary, Leibniz. He was once walking with a friend, discussing mechanical automata, as popular then as electronic ones are now, and their application to modeling living systems. Nerves were then thought to control muscles hydraulically, which may seem amusing now, but remember that hydraulic computers can be built. We have only substituted an "electric fluid" for a mechanic one. Leibniz said (approximately) suppose these theories were really true, and we were magically shrunk and put into someone's brain while he was thinking. We would see all the pumps, pistons, gears and levers working away, and we would be able to describe their workings completely, in mechanical terms, thereby completely describing the thought processes of the brain. But that description would nowhere contain any mention of thought. It would contain nothing but descriptions of pumps, pistons, levers, and so on. Thought is "meta-automatonistic!"

Leibniz's observation applies essentially unchanged to electronic computers, where mechanical components and forces are replaced by electrical ones, or by any physico-chemical constituents whose interactions we may discover to constitute the physiological basis of thinking. Thought is thus metamolecular or metaquantal, and

Non-Dichotomous, Relative and Hierarchical Aspects   327

we submit that life is metamolecular or metaquantal in essentially the same sense.

By using the term metamolecular we do not wish to imply the necessity of believing in an unbridgeable mind-brain dualism, in a Bergsonian *e'lan vital* or in other varieties of vitalism. Rather do we use meta as in metamathematics, which discusses not what mathematics actually deals with (say numbers, geometries, etc.) but talks about the structure of mathematics itself, what its abilities, potentialities, and impossibilities might be, and so on. Similarly, to avoid things like self-referential fallacies, logicians use metalanguages to discuss particular formal languages. This carries over into theoretical computer science, which frequently deals with formal languages. We believe that the knotty problems in computer science connected with pattern, meaning (semantics), and artificial intelligence (AI) arise, at least in part, because they are "meta" to the bit strings into which computer data and commands are encoded, and even to the syntax of programming languages.

Metalanguages can say things about languages they treat which can not be expressed within those languages. We feel it likely that life and thought (also organism, species, evolution and much else) are similarly unexpressible in molecular language (or quantum mechanics), though they make use of molecular mechanisms the way an "intelligent" program uses the computer's electronics. The belief is strengthened by the fact that a real individually identifiable macroscopic physical object has no proper description in quantum mechanics. Everyday objects or events may be "meta" to quantum mechanics in the same way that thought is meta to the pumps and pistons of Leibnitz's Newtonian mechanical brain.

Though seldom expressed as in the previous paragraphs, statistical mechanics (or statistical thermodynamics), which is concerned precisely with the problem of explaining the observed properties of macroscopic objects and phenomena in terms of dynamical laws obeyed by primitive microscopic constituents, implicitly encountered such difficulties long ago. As it develops new techniques to deal with increasingly complex systems, like supercomputers and living matter, say, it will have to deal with such problems more explicitly and with increasing frequency. Much of this paper can thus be read as an exploratory attempt in this direction.

7.2. ON WHETHER LIFE AND DEATH ARE MUTUALLY EXCLUSIVE AND EXHAUSTIVE LOGICAL ALTERNATIVES

In ordinary parlance dead and alive are logical opposites. There are living things and dead (i.e., non-living) things, and the idea of something which is both or neither is likely to be dismissed as absurd. Those who have been exposed to three-valued logic (admitting the three truth-values true, false, and indeterminate) may allow for an indeterminate case, not to be identified with any of the previous four cases (living, dead, both, or neither). However, multi-valued logics have never caught on in science. Two-valued logic and probability theory have proved to be powerful enough to do what scientists need, while formalisms based on a three-valued (or n-valued) logic useful in applications have not yet materialized (despite valiant attempts). In quantum mechanics a two-valued logic for wave-functions (vectors in Hilbert space), together with suitable statistical postulates connecting them with results of measuring processes, permit construction of a powerful, elegant, and experimentally verified theory. More general logics simply never provided more to the physicist than quantum mechanics already gave him. Even in studies of the foundations of mathematics multiple-valued logics are not among the more vigorously growing fields. They shed no new light on the problems of consistency and completeness of formal systems which originally helped spur their development. Also, anything that can be said in a formal language based on n-valued logic can be said in one based on two-valued logic. The former can thus be demoted to a variant notation which might be convenient on occasion. We thus feel no need, at present, to discuss life and death in other than a two-valued logical framework.

Is ordinary parlance good enough for the purpose at hand? In every-day situations taking dead and alive as mutually exclusive and exhaustive descriptions does not lead to trouble. But Bohr taught us during the formative period of quantum mechanics, as did Einstein in his discussion of the relativity of simultaneity, that we can not regard physical alternatives as given independently of the apparatus used and the operations performed to distinguish between them. If we treat physical alternatives as clean-cut logical alternatives when the operational situation makes only a fuzzy distinction between them, then we may well encounter contradictions in the very foundations of our theory.

Long ago Einstein (1921) pointed out that any geometry was "true" or "certain" only insofar as it did not refer to reality. In referring to reality it became a physical theory, whose truth had to be checked by experiment, and thus could be falsified. This did, in fact, occur with traditional Euclidean geometry. It is now generally accepted that Riemannian geometry affords a more accurate model than the Euclidean for the structure of space-time experienced by the physicist.

From an operational point of view (OPV), whatever formalisms science uses, be they formal languages or other abstract symbolic systems, geometries, various calculi, ordinary language, or even logic itself, the same situation must occur as happens above with geometry (Rothstein, 1956 (logic), 1951, 1962 (information, measurement and quantum mechanics), 1962 (space-time)). But logic has a special place as part of the language of the operational viewpoint itself. We can not do

Non-Dichotomous, Relative and Hierarchical Aspects 329

something, e.g., prepare a system, and not do it at the same time. The concept of measurement becomes vacuous if the alternatives provided by a measuring scheme can not be distinguished. This remains the case even in quantum mechanics, which predicts probabilities of obtaining the various distinguishable results of a measurement, but retains them as logically sharp alternatives. Quantum mechanics (QM) changed the concept of state, however, in such a way that the formerly mutually exclusive and exhaustive classical cases (simultaneously sharp positions and momenta) ceased to be either, being replaced by wave functions. These are expressed in terms of a set of mutually exclusive (orthogonal) and exhaustive (complete) eigenstates or eigenfunctions, analogous to unit vectors, the states or wave functions being superpositions (more generally statistical mixtures) of states or eigenfunctions. For a closed system the set of eigenfunctions is complete in the sense that any state function describing the system can be expanded in terms of it (like a generalized Fourier expansion, where the Fourier components are the eigenfunctions). Orthogonality of subspaces (a subspace is the totality of state "vectors," i.e. possible superpositions, of some subset of the complete set of eigenfunctions) corresponds to mutual exclusivity, and a set of subspaces is exhaustive if their combination is complete. Perhaps the most celebrated example of loss of mutual exclusivity occurs with particle position. The classical state of a particle is specified by giving its position and momentum, but in QM determination of its momentum forbids saying it has a definite position. One can only calculate the probabilities that it is "here" or "there" leading to the idea that it is both "here" and "there" in the sense that neither can be precluded (let "here" and "there" be large regions, say the earth and the rest of the universe so that one might be tempted to say the particle must "be" either "here" or "there"). In QM, saying the particle is "here" means preparing it to be in a "here" state, and such a state can not be prepared by measuring momentum. States in QM are operationally meaningful, but *linguistically* distinguishable situations need not be *operationally* distinguishable.

The famous paradox of Gibbs in thermodynamics and symmetry numbers in QM also underline the importance of avoiding any ascription of physical meaning to distinctions with no operational basis. The first considers two non-reacting perfect gases, A and B, each at equilibrium in a volume V at pressure P and temperature T, and separated by a removable partition. The partition is removed, enabling both A and B to diffuse throughout volume 2V. The partial pressure of each is then P/2, the total pressure P, and the temperature still T, but the entropy has increased by $\Delta S = Nk \ln 2$, where $N = N_A + N_B$ (N_A is the number of A molecules, N_B the number of B molecules), and k is Boltzmann's constant. Now suppose A and B are chosen to be the same gas. The ΔS is zero, even though the diffusion picture might naively be thought to still imply an entropy increase. If there is any way to distinguish between A and B molecules, which can be thought of as coming from the left and right sides of the partition, say, the entropy increase is preserved. If the A and B molecules are of the same kind, but can somehow carry information about their past location accessible by measurement, the entropy increase would still occur. But to say they are of the same kind, in the context of the thought experiment, means that a left-originating molecule and a right-originating molecule are precisely as distinguishable or indistinguishable as two left ones or two right ones. If they are distinguishable, they are "individuals," the paradox remains and thermodynamics is contradicted. If they "remember history," they become individuals, again contradicting thermodynamics, by the following argument. If a molecule remembers whether it was right (say 1) or left (say 0) then we can mentally bisect the half volume (before the partition was removed) and remember which quarter it was in, i.e., it remembers a second 0 or 1. The process can obviously be continued to as many "binary decimal places" as necessary to associate a unique digit string to each molecule, which we can call its *individual* name. Note that there is no trouble with QM in sufficiently localizing the molecules, for we can take V very large to begin with. It can now be argued *thermodynamically* that no operations can exist by which individuality-determining parameters of similar molecules can be measured. Historically the identification of states differing only in linguistic, i.e., non-operational characterizations, such as ascribing individuality to particles, became prominent with the advent of QM. It is thus thought by many to have its origin in QM, but we believe, by elaboration of the foregoing, that its essential basis exists already in thermodynamics. Symmetry numbers are one usual mathematical means to eliminate spurious linguistic information from physical equations. In counting distinct states (needed in statistical mechanical calculations of thermodynamic quantities) all the spuriously distinct states are counted as one. For N indistinguishable particles in a system this reduces the count by a factor N!, and configurations with geometric symmetry can have the count reduced by an appropriate symmetry number.

Living things display individuality all the way down to a (macro) molecular level, and their behaviors reflect their history, individual and evolutionary. The foregoing considerations are therefore relevant to molecular aspects of life and death and to whether molecular theories of them are possible. See Rothstein (1990) for further discussion of individuality. It is tempting to dodge quantum difficulties by saying living things are macroscopic, but too many essential biological mechanisms have their being at what must currently be viewed as a molecular or quantum level to accept this without further discussion. The hereditary machinery (DNA or RNA in cell nuclei, organelles, or virus), immune mechanisms (protein recognition), sense

Non-Dichotomous, Relative and Hierarchical Aspects 331

organs (visual purple, and other "transducer" protein molecules) and nervous systems generally provide some of the more spectacular cases. But this seems to come out of probing *any* aspect of life deeply enough. Quantum biology and molecular biology have thus understandably become common terms, used without qualms by almost everybody.

In a companion paper to this one (Rothstein, 1990), the reservations of Bohr (1958), Elsasser (1961), and Wigner (1961) about the ability of QM to give an adequate fundamental foundation for biology are briefly summarized. Though the majority of scientists seem to reject (or be uneasy with) the conclusions to which they were led, we find their arguments ultimately compel acceptance of the view that "dead" and "alive" can not be states in the usual sense of quantum mechanics or molecular theory. In Section 7.3 below we show this by direct examination of the input step of vision. The result is to eliminate the operational basis for a simple live-dead dichotomy free of all ambiguity. This is spelled out in the next section, which also shows the operational inadequacy of conventional thermodynamic state concepts. Resolution of the conundrum (that neither microscopic (quantum) nor macroscopic (thermodynamic) states are operationally adequate) is the subject of Section 7.4.

Answers to the question in the heading of this section now clearly hinge on the operational basis for defining life and death. Linguistic, and even logical distinctions are beside the point in its absence. When a universally applicable procedure has been found which always gives an unambiguous decision between living and non-living, answers to it and many related questions will follow. Since any system is the conceptual carrier of a set of states, where the system can be recognized, in the sense of being distinguished from other systems, independent of which of those states "it is in," and since the states are defined in terms of a collection of operationally defined attributes or measurable parameters (taken as synonymous in this contest), we must consider how "living state" can be defined.

7.3. WHY "DEAD" AND "ALIVE" ARE NEITHER MOLECULAR, QUANTUM MECHANICAL, NOR THERMODYNAMIC STATES

The physical concept of state is used in two major senses. One is operational and macroscopic, the other conceptual, microscopic, essentially non-operational, and an extrapolation (with modification) of macroscopic concepts to the microscopic level. The history of both statistical and quantum mechanics chronicles the mismatch between macro concepts and micro situations and the brilliant methods used to overcome it. The states are further classifiable as equilibrium or non-equilibrium for the macroscopic case, and time-independent or time-dependent in the microscopic case (macroscopic situations can also be classified this way, as in mechanics, but in thermodynamics this is essentially the macroscopic classification already given). We now consider some possible ways to approach the problem of defining "living state."

Since the advent of quantum mechanics, molecular state in biology most often means the ground state, i.e., the lowest energy eigenstate of the time-independent Schroedinger equation. Less often, the higher energy, i.e., excited states, are involved. This can be in their own right, as in some discussions of photosynthesis, but mostly they arise from eigenfunction expansions, as in time-dependent or time-independent perturbation theory. These depend on the completeness property of the set of time-independent eigenfunctions. The following discussion is in many texts (e.g., Tolman, 1938). Write the time-independent Schroedinger equation (SE) as

$$H\Psi = E\Psi \tag{7.1}$$

where H is the Hamiltonian operator for the system of interest, Ψ is the wave function, E the energy, and q will later designate the set of dynamical variables on which H depends (here the coordinates q, the corresponding momenta p are replaced by the differential operators $(h/2\pi i)\partial/\partial q$). Then

$$\Psi(q) = \Sigma c_k u_k(q) \tag{7.2}$$

where the c_k are normalized expansion coefficients, and the u_k are the eigenfunctions of H. They satisfy

$$H u_k = E_k u_k \tag{7.3}$$

where the E_k are the corresponding energy eigenvalues (the values for which time-independent solutions of (7.1) exist).

Now the time-dependent SE, for an isolated system with Hamiltonian H, is

$$H\Psi(q, t) = -(h/2\pi i)\, \partial\Psi(q, t)/\partial t \tag{7.4}$$

Its solution can be written as

$$H\Psi(q, t) = \Sigma c_k(t) u_k(q) \tag{7.5}$$

and substituted in (7.4), which then can be integrated (from 0 to t) to find the $c_k(t)$. They turn out to be

Non-Dichotomous, Relative and Hierarchical Aspects

$$c_k(t) = c_k(0) \exp((-2\pi i/h) E_k t) \tag{7.6}$$

The probability that a measurement at time t will find the system to have energy E_k is then $|c_k(t)|^2$.

A similar procedure can be followed if the system is perturbed. Suppose, for example, that we have solved (7.1) for a system with Hamiltonian H^0, and that now it is coupled to some cause of perturbing it, which we describe by an interaction energy V, "switched on" at t = 0

$$H = H^0 + V \tag{7.7}$$

and $\quad H^0 u_k(q) = E_k^0 u_k(q) \tag{7.8}$

We can write, with a new set of $c_k(t)$'s,

$$\Psi(q,t) = \Sigma c_k(t) u_k(q) \exp(-2\pi i/h) E_k^0 t) \tag{7.9}$$

Here the $c_k(t)$ would have been constants $c_k(0)$ as in (7.5) and (7.6) if we had taken the eigenfunctions and eigenvalues of H as the basis of the expansion instead of those of H^0. But the latter are a complete set, so the expansion coefficients can be written, for any time t, as $c_k(t)$. They can be found from a set of equations derived from (7.4), subject to (7.7), (7.8) and (7.9).

Using the abbreviation

$$V_{nk} = \int u_n^*(q) V u_k(q) dq \tag{7.10}$$

the equation satisfied by c (t) can be shown to be

$$\frac{dc_n(t)}{dt} = \frac{-2\pi i}{h} \Sigma_k V_{nk} [\exp(\frac{2\pi i}{h} (E_n^0 - E_n^k)t)] c_k(t) \tag{7.11}$$

In the neighborhood of t = 0 we can write down the simple result

$$c_n(t) = c_n(0) + \Sigma_k V_{nk} \frac{\exp(\frac{2\pi i}{h}(E_n^0 - E_k^0)t) - 1}{E_k^0 - E_n^0} c_k(0) \tag{7.12}$$

We have chosen the above pedestrian and almost antiquated treatment in preference to later more elegant and powerful notations. The compact formalisms of the latter hide messy details one would have to go through in order to do more than merely pretend to discuss the wave-functions of complex systems. Mixed states, the statistical matrices connected with them, and scattering matrices will have to be discussed later. What we show now is that no operational basis can be given for constructing a Hamiltonian for a living system, either *ab initio* or by applying perturbation theory to a system built from simple interacting systems whose individual Hamiltonians are constructible. We do this by considering the transition from a simple solvable system to complex intractable ones.

The simplest system of chemical interest, the hydrogen atom, has a Hamiltonian $H = T + V$, where T is the total kinetic energy of proton and electron and V the potential energy of their interaction. In rectangular coordinates x_1, y_1, z_1 for the electron (mass m_0) and x_2, y_2, z_2 for the proton (mass M), we have, with $P_{x_1} - m_0 (dx_1/dt)$ and similarly for other momentum components

$$H - \frac{1}{2}m_0(P_{x_1}^2 + P_{y_1}^2 + P_{z_1}^2) + \frac{1}{2}M(P_{x_2}^2 + P_{y_2}^2 + P_{z_2}^2)$$
$$-e^2 / [(x_1 - x_2)^2 + (y_1 - y_2)^2 + (z_1 - z_2)^2]^{\frac{1}{2}}$$
(7.13)

The time-independent SE is a partial differential equation with six independent variables. For N protons and N electrons we would have 6N independent variables. The single Coulomb interaction term of (7.13) becomes $6N(6N - 1)/2$ such terms, corresponding to all particle pairs, and taken negative as above for opposite charges, positive for charges of the same sign. Solving (7.13) exactly is possible under certain conditions. For $N=2$ considerable progress, based on (7.1) through (7.13), and other simplifying conditions, can be made, but as N increases essentially all hope of obtaining exact solutions vanishes. Even the $N=1$ case for boundary conditions, drastically simplified compared to those relevant to biology, is intractable.

The solution of (7.13) in closed form results from the following fortunate circumstances. First one can make a coordinate transformation where the center of mass is at the origin. This replaces M by $(M + m_0)$ and m_0 by the reduced mass of the electron μ,

$$\mu = Mm_0/(M+m_0) \tag{7.14}$$

Non-Dichotomous, Relative and Hierarchical Aspects 335

Second, the chosen boundary conditions are that the system as a whole is moving freely, i.e., the only contribution to V comes from the Coulomb interaction. The motion of the center of gravity is then uniform and need not be considered further. If this were not so the potential energy would depend on the coordinates of the proton as well as those of the electron and the problem would remain six dimensional. In the three dimensional center of mass system, going to spherical polar coordinates permits solving (7.13) by the method of separation of variables. Such tricks are possible in very few coordinate systems. The hydrogen atom confined to a finite region of space allows none of these simplifications. Even if such a problem could be solved for a box with reflecting walls, say, the solution would be at best the starting point for an approximate method, for real boxes must have structure on an atomic level.

The problem for two hydrogen atoms, i.e., two protons and two electrons, can only be solved by a succession of approximations. The low energy states are those of the hydrogen molecule. The pioneering Heitler-London calculations (1927) were extended by James and Coolidge (1933) to give satisfactory agreement with experiment (Richardson, 1935; Beutler, 1936) for dissociation energy, equilibrium internuclear distance, vibrational frequency, and even anharmonicity (they also used the molecular orbital approximation to some extent). This classic work, however, plus decades of succeeding studies by scores of other brilliant investigators, still does not add up to the complete kind of understanding of the two-proton two-electron system with simple boundary conditions that has been achieved in the one-proton one-electron system (with simple boundary conditions). With increasing complexity of the system of interest, the need to make simplifying assumptions and approximations turns actual wave-functions into unattainable idealizations. One puts together a mix of empirical, theoretical, and approximational considerations, in what is hoped to be a convergent process, in order to explain empirical data (beyond a small amount which might have to be assumed anyhow).

This approach has been a great success also in quantum theory of solids, where observed crystal periodicity is assumed to carry over to the solutions of the Schroedinger equation. All but "outer shell" electrons are taken as localized in hard cores, containing nuclei and electrons tightly bound in "orbitals" in the cores. The outer electrons are then considered to move in a periodically modified "box" potential. Very often "periodic boundary conditions" are used to eliminate the box walls (it boggles the mind to interpret this physically in three dimensions; for two dimensions one has a torus, for one dimension a ring). The resulting wave-functions give rise to the well-known band theory of solids (see Wald (1968) for an introduction, including a survey of quantum background and relations to other approximations). It explained, for the first time, why solids were metals, semiconductors, or insulators and has undergone phenomenal and continuing growth. It is equivalent in the limit of large molecules to the molecular orbital (M.O.) approach, which has had great success in molecular spectroscopy.

These and other quantum methods have had, and no doubt will continue to have, a central role in clarifying structures and mechanisms in molecular biology. They have and will, of necessity, continue to have decisive roles in explaining each elementary step of complicated event sequences. Examples include photosynthesis, respiration, digestion, protein synthesis, active transport in renal or neural function, in the mechanochemistry of muscular response, in hormonal action, in the acquisition, processing and storage of information at the levels of the whole organism, of specialized subsystems like neural or immune systems, at the level of regulatory, reactive, or reproductive activity of cells, and at the DNA/RNA level of hereditary, adaptive and evolutionary mechanisms. The reader can probably write a comparable list of examples which is longer and doesn't duplicate the one just given, or can take any one of the items of either list and expand it into a similar list, perhaps longer yet.

In order to bring quantum mechanics to bear one must apply reductionist strategy on a heroic scale. As the interaction of electromagnetic radiation with matter was decisive in the discovery and development of quantum mechanics, let us consider quantum explanations of vertebrate vision, e.g., as exemplified in the fly-catching behavior of a frog or baseball outfielder. Writing an outfielder or frog wave-function is too big a problem, ditto for the eye-brain system, for the retina, or even for a retinal neuron. But we can at least begin to approach the problem of how the outfielder sees by investigating the eye's response to light. Selig Hecht and his successors, beginning more than half a century ago, made psychophysical measurements showing that human visual sensitivity extended almost to the level of individual light quanta. Rods and cones are the retinal cells responding directly to light. The other form complex neuronal circuits which process the outputs of the rods and cones in complex ways, and interact with the brain in even more complex ways. We neglect all that, and most of the complex structure of the individual rod or cone cells as well. Photon detection must occur within some macromolecular configuration. The state change produced by photon absorption somehow triggers an amplifying sequence of events culminating in visual sensation. Wald's Nobel lecture (1968) is interesting for the history and sociology of science related to the molecular basis of visual excitation as well as for its scientific material. A severely condensed version of part of his work must suffice here, due to space limitations. Before turning to it we note that a vital visual organ must be used repeatedly. The excited configuration must not only have its state change able to initiate the visual sequence of events but must also be restored to the original excitable state to detect subsequent quanta.

The approximate sequence in the baseball starts with the sensitive visual pigment rhodopsin, which consists of retinal (formerly retinene), bound as chromophore to the protein opsin. Retinal is vitamin A aldehyde. It is bound in Schiff base linkage to opsin (the aldehyde group of retinal is condensed with an amino group of opsin), to form rhodopsin, but photoresponsiveness requires the retinal to have the right shape, namely it must be 11-cis. Absorption of light makes many complications occur. These include passage through a sequence of intermediate states and electronic reconfigurations, genesis of ocular EMF's and hydrolysis of the Schiff base linkage. This last splits the rhodopsin into opsin plus reduced all trans retinal. The latter is vitamin A (now often called retinol) which must be oxidized back to retinal as part of the restoration phase of the cycle. Similarly, the rest of the system must also be "reset," and the cycles in which it is involved may have complexity comparable to or much greater than that of the rhodopsin cycle, depending on how the "rest of the system" is defined (more on this later).

This ability to recycle is essential, for an eye capable of being used only once would not be very useful. Further discussion of cycles is given below which continues the developments of Rothstein (1979a,b, 1982, 1985, 1988 and 1990). Note that the cycles are dissipative, which is important in the papers just cited and in the work of Eigen (1971, 1981), and Prigogine et al. (1972) and their schools. An empirical indication of the importance of dissipation, even in early steps in vision, is given by measurements of heat generated in bull frog retina in response to light pulses (Tasaki and Nakaye (1986)). They found an amplification of three orders of magnitude of the thermal response compared to the total light energy absorbed. Its latency was approximately 10 ms or less and localized in the photoreceptor cell layer, while the electrophysiological response occurred later (30-50 ms latency for rod polarization). It is reasonable to view this as the thermodynamic cost of detecting light, of amplifying an effect it produces and perhaps some of the cost of converting it into an electric or electrochemical signal. This last is further processed by other subsystems of the neural network. Other interpretations (which are essentially equivalent) view it as a cost of making a measurement (here determining the presence or absence of a light source), or communicating information (light channel with eye as receiver), or of preparing the input "tape" of a neural computer or data processor (preparation can sometimes be taken to include erasure of results of previous use). An analysis of dissipation in the computing process by Bennett (1982) concluded that the only source of irreversibility which could not be idealized away via conventions generally used in thought experiments was tape erasure. The reset part of the rhodopsin cycle clearly includes such an action. Further conclusions on the feasibility of reversible computation in general (Bennett, 1982) are questioned in Rothstein, (1988) because of the continuing need for subsystem measurement and preparation by other subsystems and the impossibility (arising from the required capability to perform arbitrary well-defined procedures) of setting up a system in which they are avoided. In living systems there are myriad subsystems whose cycles are coupled, the outputs of one serving as the input of another, just as there are in computers.

The reset part of the rhodopsin cycle must include isomerization back to the all-trans configuration. The retinol (vitamin A, all trans) to which the retinene was reduced must also be oxidized back to retinal. The first uses an isomerase, the second an alcohol dehydrogenase. Esterifying enzymes then condense the aldehyde group of retinal with an amino group (lysine) of the opsin, reconstituting the rhodopsin. The overall process takes energy, some of which is used to "wind up the spring" comprised by the all-trans configuration placed in the environment where it is sensitive (the combination is the spring).

Similar cycles occur with adrenergic receptors, which include the visual opsins, many neurotransmitter receptors, and receptors for peptide hormones. The alpha helices of the "G proteins" apparently form "pockets" which bind ligands like the chromophore 11-cis-retinal, and include both agonists and antagonists (the latter suppress responses). The "outputs" such systems produce can be ion flows, neurotransmitters or hormones. See Kobilka et al. (1988) for some of the structural complexities and interrelationships and how they can be elucidated using genetic techniques. The general picture of complexes of interacting subsystems, where the subsystems are ionic, molecular or macromolecular, and where the complexity of organization reminds one of the transistors on a computer chip, is ubiquitous. Can quantum mechanics deal with such computer-like cases in a meaningful way?

An early attempt to do so in a heuristic, intuitive fashion is given in Rothstein (1959). Assume now that organic evolution had a predecessor prebiotic chemical evolution. Compare DNA and its forerunners to programs stored in a computer memory and ask whether quantum mechanics admits molecules capable of performing the other functions carried out by other computer subsystems. More along these lines is found in Rothstein (1979a,b, 1982a, 1985, 1988, 1990). As logic circuits and memory suffice to build general purpose computers, it remained to be shown how to realize logic elements (switches or transistors) and their interconnections (hook-up wire). In Rothstein (1959) it was argued that solid state concepts like band theory, delocalized electron orbitals etc., applied to molecules as well, and that many molecular configurations showing chemical resonance could serve as "hook-up wire," transmitting electrical effects over macromolecular distances. The "input" and

Non-Dichotomous, Relative and Hierarchical Aspects 339

"output" sites can be utilized, for example, if an ion at one of the former could cause bond rearrangements permitting the feeling of another ion at the latter. The interactions between resonant and tautomeric forms, triad prototropy, H-bonding and ring-chain tautomerism are a few more of the rich store of chemical devices available to permit the "computation" of new configurations from initial ones. Long computations became possible, with storage of intermediate results achieved in a succession of intermediate configurations.

Recent developments in conducting polymers have spawned a veritable flood of publications. For example, a recent seventy page review by Heeger et al (1988) with almost 500 references is concentrated on the restricted field of solitons in polyacetylene and polythiophene. For surveys of many additional topics see Fumi and Tosi (1985) and Delhaes and Drillon, (1987). For technology-oriented treatments see Carter (1982, 1987). The need for new materials and devices, particularly to meet the insatiable demands for more and cheaper computational power and to convert solar energy more cheaply and efficiently, provides strong impetus to such research and development. Quantum mechanics dominates almost all theoretical discussions, suggests new experiments and explains many of the results obtained in a generally satisfactory fashion. The dominant and, in our opinion, totally justified view, in cases where theory and experiment jibe poorly, is that the fault lies not in quantum mechanics but in using an over-simplified model of the case at hand. So why doubt the ability of quantum mechanics to distinguish between dead and live states in principles?

Our discussion is essentially a continuation and generalization of Rothstein (1990) which took as one point of departure the considerations of Bohr (1958), Elsasser (1961) and Wigner (1961) (which have already been discussed briefly). The first two of these are based on generalized complementarity, namely that the deep probing necessary to distinguish between and dead and live states in a quantum mechanical sense would be incompatible with continuation of any live state. Bohr's basic idea was that the uncertainty principle, applied to determining the atomic configurations involved in living matter (for predictive purposes) to the required physical accuracies, imparts uncontrolled momenta to their constituent particles sufficient to explode those configurations. This argument seems not to have impressed biologists, or even most physicists, very much. I have determined, on a daily basis for almost half a century, that my wife was alive, and she hasn't exploded yet. But determining whether a dormant spore, or even better a virus particle, is viable could conceivably be impeded by something akin to generalized complementarity. The attitude of most scientists seems to be that there is no need to worry about this possibility until it is shown to be a real barrier to deeper understanding. After all, if one tries to locate an electron within a hydrogen atom, the probing will ionize (i.e., explode) the atom, but nobody worries about that since barriers to deeper understanding lie elsewhere. These drastic procedures are generally not needed, and the more complex the system is, the milder the procedures must be in order to be tolerated without destroying the system. Classical macroscopic measurements commute, and the perturbations they produce can generally be neglected. Drastic interventions are viewed as acts of preparing systems (e.g., preparing a metal from its ore), and acts of preparation generally do not commute. At a quantum level, of course, measurement and preparation are equivalent concepts. It is intriguing to view the "breaking" of quantum-measurement-preparation acts into meaningfully distinguishable measurement acts and preparation acts as a key element of future biophysical theory. If this is valid and if it admits a description in terms of symmetry breaking, then the evolution of life would become an aspect of cosmological evolution in an esthetic and profound way.

Throughout most of his scientific life Bohr laid great stress on the peculiar feature of "wholeness" of quantum systems, even asserting that one could not talk sensibly of physical concepts or attributes divorced from the experimental conditions under which they are measured. Under the stimulus of the Einstein-Podolsky-Rosen Paradox (hereafter called EPR) (Einstein et al.; 1935), Bohr (1935) came to view even "physical reality" as dependent on the means used to observe it. Einstein considered this view too solipsistic, and their debate never ended (see Schilpp (1957) for fascinating first-person accounts). We have no space to discuss the enormous EPR literature (see de Baere (1986) for a recent survey), but will take EPR as paradigmatic of a class of physical situations relevant to the present biophysical discussion. In the original EPR example two initially independent systems A and B are brought into interaction for a short time and then separated, becoming independent (though correlated via their interaction) once more. By measuring the state of A we can determine the state of B, and EPR asserts that one can choose any measurement at all on A "without affecting" B. In particular one can choose either of a pair of incompatible observables (like position and momentum) thereby getting either of two descriptions of the state of B whose simultaneous validity is prohibited by quantum mechanics. As this choice "can not affect" the state of distant B, according to EPR, both results are equally "real" even though we can measure only one. Making the choice "reduces the wave-packet" in quantum jargon, can be taken as describing an act of measurement (A is the instrument, B the system of interest), and introduces a discontinuity in the time dependence of the wave-function not described by Schroedinger's equation. In von Neumann's treatment of quantum measurement (Wheeler and Zurek, 1983) wave-packet reduction is irreversible (quantum mechanics, in contrast, is reversible) and accompanied by entropy increase. The same would of course be true of preparation acts, so it is not surprising that Wigner (1961) finds quantum mechanics does not account for reproduction. As emphasized in Rothstein (1990), reproduction can be viewed as a sequence of preparatory acts in which the parent assembles (or helps to assemble) the offspring. Normal metabolism can also be broken down into long chains of measurement-preparation and similar interactive acts. Our baseball player can be said to measure the radiation reflected into his eye by the baseball, and the myriad photochemical and electrochemical acts the photons initiate are complex preparations. There is literally no end to the measurement, communication, preparation and computation acts into which his overall time response can be analyzed.

Long before Wigner's paper appeared, Elsasser (1937) examined how statistical mechanics might relate to generalized complementarity. He realized that what we have called mild procedures above needed extended times for their implementation (to avoid the high-energy blasting of delicate configurations required by the uncertainty principle if the information is demanded in a short time). To determine the wave-function of an organism would take geological time, during which interactions with the environment (other than the means of observation) would have to be forbidden. This is not only impractical but generally lethal (most organisms soon die if decoupled from the environment needed to maintain them). Recourse must thus be had to quantum statistical mechanics. The sharpest state one can hope to specify is "mixed," namely a probability distribution over a large number of quantum states. We believe this is correct, but useful primarily in systems which are far less complex than biological ones. Elsasser (1961) tried to apply statistical mechanics to the foundations of biology, but concluded that special "biotonic laws," independent of quantum mechanics but consistent with it, had to be posited. Rothstein felt this to be premature surrender (1976a,b, 1985, 1988) and obtained the required supplementation of quantum mechanics by generalizing entropy. The resulting generalized entropy permits discussion of measurement, information and organization, in a (generalized) thermodynamic context, even leading to a thermodynamic rationale for the evolution of complexity and individuality in sufficiently complex non-equilibrium systems. There is even an argument that the heat engines which evolve could become not only "well-informed" (by measuring) but capable of computing and thus evolving something like intelligence.

We conclude this section with arguments that neither quantum mechanics in general (and therefore fundamental molecular approaches as well) nor current thermodynamics (including its actively growing extensions to non-equilibrium or even to non-linear thermodynamics and kinetics) suffices to characterize the difference between "live" and "dead" states. Thermodynamics fails because the complexity and heterogeneity of biological systems, as well as their non-equilibrium nature, make it impossible, in most of the crucial cases, even to define the relevant state variables needed for thermodynamic discussions. Also the number of variables (e.g., the number of macromolecular species) is enormous. Conventional thermodynamics is simply neither sufficiently detailed nor "microscopic" enough to deal with biological complexity at a sub-cellular level. This may even be true at an intercellular level. Quantum mechanics fails both because of complexity and because the only general technique we have in it for dealing with complexity, namely breaking the complex system into simpler subsystems, falls afoul of the "wholeness" property of quantum systems. As we are forced to analyze biological systems into subsystems in order to make any progress in theory at all, we are thereby forced into approximations which preclude talking as if we can define the quantum state of the whole system, or even to distinguish between large classes of states like dead and alive. Such considerations clearly are indebted to the pioneering considerations of Bohr, but they owe just as great a debt to Einstein and EPR. This is amusing, for though they were the great antagonists on the philosophy of quantum mechanics, their views are two legs of the tripod on which our argument rests. The third, generalized entropy, is the foundation of our approach to fundamental biological theory. It is made essentially inevitable by the other two.

Our earlier discussion of quantum mechanics noted how rare the solvable quantum cases are, i.e., how soon increasing complexity destroys solvability, and how solutions in time-varying cases can be integrated from assumed initial expansions explicitly only for short times. It is probably no exaggeration to say that the whole hydrogen atom problem, including all excited states, can be solved exactly for only the isolated atom. Even if one could solve the problem of the hydrogen atom in a box with perfectly reflecting walls it would correspond to solving no real problem, for the atomistic structure of any real box would have been completely idealized away. Mean field approximations, say for electron and ions in solution are suspect in principle, though possibly convergent for computing ground state energies or the like. These and other techniques, applied to subtle properties like live-dead distinctions, do not seem to be able to avoid taking averages over states which can be mutually inaccessible during meaningful averaging times. This seems tantamount to averaging the behavior of live and dead individuals and expecting the results to be informative about the properties of one or both, but they say essentially nothing about either. What is the point of averaging over a population of zebras in a game sanctuary, say, where half are contentedly grazing and half have been dead for a week? Some attributes (mass, hide color, number of limbs, number of vertebrae, etc.) are fairly insensitive to whether the animal is dead or alive (at least for times less than the time to rot), but those are not of interest here. The average heart rate, temperature, oxygen consumption, etc. of a live animal and a corpse make as much sense as averaging the parameters of the animal with those of a beer can. It is not surprising, therefore, that Elsasser (1961) tried to avoid such averaging, in effect. He allowed the representative ensemble to have few members. In Rothstein (1979a,b, 1982, 1985, 1990) this is viewed as virtually discarding ensembles altogether and an alternative view is adopted. It is to consider each individual system as characterized by internal constraints which restrict the ergodic wandering of the representative point in phase space. Averages are to be computed for each individual taking those constraints into account. These can differ enormously from unconstrained averages and between individuals. The averaging of live and dead states is avoided in principle.

Despite the problems of complexity just enumerated, the picture QM provides is nonetheless the indispensable foundation for any fundamental study of living, or other complex systems. The unavoidable and frequently the very effective strategy is to split the large system into subsystems, giving one subsystem detailed consideration while treating the remainder statistically. This is the basis of many self-consistent field approximations, for example (Hartee-Fock, Thomas-Fermi-Dirac methods) where symmetrization is imposed later to counteract special treatment of the subsystem. Alternatively one can "freeze" all but a selected subsystem (thereby constraining the overwhelming majority of variable system parameters to constant values), solve the resulting simplified problem (generally with approximations), and average and/or symmetrize over frozen configurations if necessary (Franck-Condon principle, Gurney-Condon-Gamow theory of alpha particle emission, solid state theory, etc.). At a more macroscopic level splitting into subsystems is fundamental to general systems analysis, the engineering of large systems, and compartmental analysis in physiology or biochemistry.

There is a good reason for the success of these "divide and conquer" methods, shared by both quantum and macroscopic cases: the existence of a natural conceptual cleavage between the selected subsystem and its surround. In atomic theory, for example, one often can consider the nucleus and tightly bound inner electrons as providing an environment in which the valence electrons move. The spectroscopy of alkali metal atoms can then be discussed in a manner much resembling that used for the hydrogen atom. Similarly input-output or stimulus-response treatments of networks of "black boxes" are highly successful on a macroscopic level. The condition on the black boxes is that their boundaries are stable enough to identify them as units participating in such input-output relations.

Thermodynamics can handle such systems, for example heat engines, where the input is heat from a hot source, the outputs being work and heat (rejected to the sink).

Returning to the problem of defining the living state we find the following problems. From the quantum side, complexity makes the task of specifying a "living wave-function" for the whole system hopeless, but the strategy of the previous paragraph is limited in its applicability to processes which are elementary, rather than "wholistic." For example, few would doubt the basic correctness, as applied to our ball-player's eye, of the quantum description of light absorption by matter. Similarly, one can feel reasonably confident about each step of the ensuing chain, where photon absorption excites the electron distribution, the excitation energy then migrating (using concepts like phonons, polarons, solitons, etc. to describe it) to where configurational changes are produced, chemical reactions allowed to proceed etc. While most steps are still not clear, few would doubt that their clarification will come. But this very description demands "reduction of the wave-packet" at each such elementary step, as in EPR and quantum measurement. This is not described by Schroedinger's equation, and gives information about *manufactured subsystems* carved out of the system. The "wholeness" of quantum systems does not allow us even to say that they "exist" in the original system in any consistent quantum mechanical sense. But the activities of living systems *qua* living systems consists of chains of such events eluding quantum description. The quantum evolution of the system between two such events is *not* the center of interest. Even if good quantum descriptions of the black boxes could be given, we have the problem that their boundaries and existences would be ephemeral. The descriptions might be valid between events, but events would obliterate and create boxes, probably in vast numbers and very rapidly. A completely consistent quantum treatment has a "many worlds" nature useless for discussion of a real individual system.

From the thermodynamic side we have a conceptually simpler complexity limitation, which just as effectively bars us from regarding dead and alive as thermodynamic states. Thermodynamics deals with simple systems whose states are specified by small numbers of parameters (pressure, temperature, mole fractions of various chemical species, etc.) But the complexity and heterogeneity of living matter are so enormous that an astronomical number of parameters is needed, as noted earlier. But there is also an astronomical number of subsystems to be considered which varies enormously from one discussion to the next. The subsystems appropriate at one time rapidly become inappropriate and must be replaced by others. We can not even be sure we even know a majority of the types of subsystems there are, even at a cellular level. The subsystems can go down to molecular size, and biopolymer molecules can display individuality (Rothstein, 1990).

Non-Dichotomous, Relative and Hierarchical Aspects 345

If neither quantum nor thermodynamic characterizations of dead and alive are adequate, is there another kind that might be? We turn to this problem next.

7.4. TOWARD A DEFINITION OF LIVING STATE

How can the living state be defined? If by attributes, which ones should be chosen? For what classes of system is it meaningful to ask whether it is dead, alive, both, neither? At what levels of organization is it appropriate to introduce notions of dead and alive? Probably most would find it somewhere along a sequence like

monomer - polymer - DNA, RNA, protein - virus - organelle - cell.

To what levels of organization do notions of dead and alive remain relevant, and when do they become metaphor or mysticism? Do "dead" and "alive" undergo essential changes in meaning on passage from one level to the next? The following two sequences, each starting with cell, are typical supports for questions of this type. The sequence

cell - tissue - organ - organ system - organism

strongly suggests a multi-level or hierarchical nature for life and death in a purely biological context. The sequence

cell - colony - metazoan - obligatory symbionts - society - ecology - biosphere - living earth - Spinoza's God

gets beyond science, but it is hard to say where useful generalization ends (if at all).

Are linear schemata, like the foregoing, incomplete or misleading in the sense that a partial, rather than a total ordering of terms is needed? If so, the linear graph representation would be replaced by a Hasse diagram (or a more complicated graph), whose points are terms and whose arcs correspond to the dashes used above. The term "dead" (or perhaps "unambiguously dead and surely not alive") might be taken as a "universal lower bound." There might be no universal upper bound, except as a convention, in schemata like the last one given, but there could be meaningful ones if generalization is kept to terms with well-defined scientific meanings. The boundary between dead and alive would no longer be point-like (like a point on one of the dashes, say) but more analogous to a separation hypersurface between two regions of a hyperspace whose points are terms. It is tempting to speculate that evolution of life might be viewed as a diffusion or "flow" along "paths" characterized by increasing organization as a parameter. Life would then be said to begin when some separating hypersurface is crossed. A similar description of death is crossing of the hypersurface in the opposite direction (the space here is individual rather than the global one needed for evolution). As Nature is unafraid of difficult mathematics, there is no a priori reason to reject this kind of picture in favor of linear ones other than convenience (i.e., our limitations). Lastly, it is clear that both the linear and complex schemata use gross terms capable of being analyzed into sets of more sharply defined ones. Each term thus potentially splits into a subschema, each dash or arc into a "cable," and along any arcs new terms might be found.

This last point is borne out by history, as it describes both the search, at any time, for "more fundamental" structures and mechanisms, and divide-and-conquer strategy in action. QM is the ultimate physical level for biological structure and mechanism in the sense that going further surely abandons biology for physics or metaphysics, but even going that far "overshoots," as we have seen. So let us back off to the thermodynamic-macroscopic level (which is too coarse and "undershoots") and see if we can refine it. Is the "living state" well defined macroscopically? We must be careful, as refinement proceeds, not to overlook possible changes in what we mean by dead and alive. A living organism can incorporate non-living structures (hair, teeth, water in the bladder, fat deposits, etc.) whose excision is clearly non-lethal. Our baseball player could lose eyes or limbs in an accident and still live as a human being (though he might be "dead" as a baseball player). This suggests asking whether, if a living organism is divided into parts, how far can the process continue and the parts still be called alive? If a part can regenerate the organism it is clearly alive, but what if it can't? This question is clearly not separable from the effect of level of description on the meaning of dead and alive, so trying to answer it must be postponed.

A system is, almost by definition, "in a state," but as we have seen living systems are not in equilibrium so the conventional concept of equilibrium thermodynamical state is not applicable. It is hard to get agreement on just how to generalize that concept. Living systems are open to matter and energy fluxes, and are non-uniform in structure, composition, temperature, and a host of other parameters. Their "states" are continually changing. Furthermore, the boundary between a living system and its environment is fuzzy. When do ingested food, water or oxygen become part of an organism, and, when do waste products cease to be part of it? In the sequence

Non-Dichotomous, Relative and Hierarchical Aspects 347

Vertebrate bone - cuttlefish bone - snailshell - crab carapace - hermit crab's snailshell or clothing,

where does the named entity cease being part of the organism? What of organ transplants and prostheses, and if the first but not the second be part of an organism, what about a transplant which is functioning now but will be rejected next week? If teeth be part of an organism, what about a bird's gizzard stones, without which it would starve to death? There are many similar sequences, some of which have cellular analogs. Paramecia, though single cells, have organs of locomotion, digestion, assimilation and excretion, and the questions raised at the beginning of this paragraph apply to it. Its "organs" are macromolecular.

What kind of "living state" concept is robust enough to apply to a whole organism, to its subsystems, to a single cell, to organelle systems and structures within a cell or to a virus particle, and is also compatible with both the thermodynamics above and quantum mechanics below? We submit that the abstract concept of state familiar in automata theory, computer science and system theory meets these requirements (and others), and suspect that it is unique in the sense that only something logically equivalent to it can do so. Only an abstract concept admits manifold interpretations, all of which can be discussed by means of the same formalism. If the abstraction axiomatizes the essential relations common to all cases, it does not matter too much whether the physical state it represents is a pure quantum state or mixed, stationary or time-varying, macroscopic or microscopic, equilibrium or non-equilibrium. If this seems impossible remember the infinite class of real systems to which the concept of number can be meaningfully applied. For our exposition the finite state automation (FSA) generally suffices, but mathematical generalizations to an infinite number of discrete states, or even a continuum of states, inputs, and outputs are available in the literature. We will use them on occasion, particularly the Turing machine (TM), which has a countable infinity of states. Their theory is treated in Hopcroft and Ullman (1979), whose notation we follow, and other texts.

An FSA, here called A, is a mathematical system characterized by three sets and two relations. The latter are frequently specialized further to mappings (functions). Deterministic FSA's or TM's correspond to this case, nondeterministic ones to relations. The three sets are the set of states

$$Q = \{q_0, q_1, \ldots q_m\} \tag{7.15}$$

the set of inputs

$$\Sigma = \{a_0, a_1, \ldots a_n\} \tag{7.16}$$

and the set of outputs

$$\Delta = \{b_0, b_1, \ldots b_p\} \tag{7.17}$$

The relations (or mapping) are the next state relation

$$\delta = Q \times \Sigma \to Q \tag{7.18}$$

and one of the two output relations

$$\lambda = Q \times \Sigma \to \Delta \quad \text{or} \tag{7.19}$$

$$\lambda = \Sigma \to \Delta \tag{7.20}$$

It is an important theorem that deterministic and nondeterministic FSA's (and TM's) are equivalent in the sense that given an arbitrary one of either kind, one can construct one of the other kind with the same input-output behavior. The two classes have exactly the same descriptive or modeling power; either provides adequate "black box" characterizations. It is also a theorem that choice of (7.19) or (7.20) is similarly immaterial. The first is called a *Mealy machine*, where an output is generated on receipt of an input, depending both on what input is received and what the state is when the input is received. The second is called a *Moore machine*, and the output depends only on the state. The Mealy machine directly models the behavior of transducers, the Moore machine that of machines with memory, like shift registers.

FSA's are useful for modeling input-output behavior. Their physical embodiments presuppose an environment in which they act, for inputs come from that environment and outputs go to that environment. The situation must also be non-equilibrium, for at equilibrium no input signals would come in nor would output signals go out. The inputs can be read from a tape or be received as signals (electrical, acoustic or other), with similar diversity possible for the outputs. An FSA model can be used to represent measurement (or communication) systems, for the inputs bear information about their source, state changes record that information, in effect, and the outputs embody readout of the state-stored information. Similarly state preparation (or manufacture) is modeled by the state change, the input representing an external agent engendering that change.

Irreversibility arises naturally in the abstract FSA for the following reason.

Non-Dichotomous, Relative and Hierarchical Aspects 349

In dynamics, both classical and quantum, the changes in state of a system can be described as a group of transformations T(t), T(0) being the identity transformation, composition of transformations expressed by

$$T(t_1) \, T(t_2) = T(t_1 + t_2) \tag{7.21}$$

while reversibility is expressed by the existence of inverse transformations T^{-1} satisfying

$$T^{-1}(t) = T(-t) \tag{7.22}$$

Irreversibility, which comes from thermodynamics, means that the time history of a state is expressed by time-displacement operators with no inverses. The transformation group is replaced by a semigroup (or more precisely, a monoid, which is a semigroup with identity). In FSA's state histories consist of successions of states entered into as a result of the string of successive inputs, i.e., transformations or operators on the states forming a monoid, generally with no inverses and generally not commutative either. This last is noteworthy, for a characteristic feature of QM, not shared with classical physics, is that operators representing observables do not generally commute. The compatibility of FSA models with reversible mechanics, irreversible thermodynamics and QM is striking. Receipt (detection, absorption) of an input or generation (emission, creation) of an output is an irreversible act in physical automata. It is accompanied by entropy increase in general. At a quantum level wave-packet reduction of the compound source-destination system is involved.

Two additional characteristics of FSA models of special attractiveness for theoretical biology are their upward and downward extensibility. By the first we mean the fact that an FSA results from combining any finite number of FSA's into an interacting network, in which selected outputs can be inputs of other FSA's (inputs are allowed to be "vectors" whose components are outputs of other FSA's or external (vector) inputs, similar generality in form is permitted for the outputs). The theory is preserved under "complexification." If Nature were to evolve FSA's, then the same tendencies might be expected to evolve them further into coalitions of FSA's. This is discussed further in Rothstein (1979a,b, 1985, 1990). The second characteristic is a kind of converse of the first. An FSA can be "refined" by splitting its states into a multiplicity of states, or more generally, any state can be replaced by an FSA (with additional "microscopic" inputs and outputs, if desired, and both internal transitions and couplings to external states). The result is still an FSA. An initial FSA model can be made more detailed (in the reductionist tradition) by invoking "more fundamental" mechanisms, without abandoning the class of FSA models. Hierarchical structures can thus often be conveniently analyzed or designed using FSA techniques.

The class of computations or data processings performable by FSA's is limited. It is now the consensus that the most general class possible is that which Turing machines can do, for despite valiant attempts, nobody has ever exhibited a well-defined procedure, algorithm, program, deductive scheme, or finitely axiomatizable formal system that can not be implemented on a TM. If TM is substituted for FSA in the above, all statement remain valid. The TM is an abstract model of a general purpose computer, or rather the universal TM (UTM) is. Turing showed that a UTM could be constructed, which, given an appropriately encoded description of an arbitrary TM and the data it is to process, would simulate that TM, i.e., produce an encoded version of what that TM would have produced. The reason the FSA is limited in the class of computations it can perform is that its memory is bounded. The standard TM construction is simply an FSA with an infinite tape (unbounded memory), divided into squares which are blank or contain a symbol. The FSA also called a finite control or read/write head reads a symbol (input) from a square (its current address), prints a symbol on that square (output), and, in a new state, moves left or right to an adjacent square. In a finite time the FSA can access only a finite amount of tape, but that amount is unbounded, given an arbitrarily long access time. If the tape is finite or the FSA is constrained to move in one direction only, the TM degenerates to an FSA. Multiple tapes, multidimensional tapes, complicated repertoires of moves, multiple heads and other increases in resources do not increase the generality of what the machine can do, though they may make it faster. The state of the FSA is only part of the configuration needed to determine the next configuration. A configuration is a triple (q, α, n) where q is the state of the FSA, α the string of symbols on the tape, and n is the address (an integer) of the FSA on the tape. In physical embodiments, reading or writing symbols clearly constitute measurement or preparation acts. They are thus irreversible, and at the quantum level correspond to wave-packet reductions and which are not described by Schroedinger's equation. Every such event corresponds to a new initial condition, after which the time-dependent Schroedinger equation must be integrated again. Wave-functions are computed after boundary or initial conditions are chosen. The irreducibility of life to QM can thus be expressed, in a ubiquitous class of cases, by saying that QM doesn't choose what problem is to be solved. The choice is "meta" to QM. The QM machinery takes over after that choice is made.

The TM overcomes FSA limitations by indefinitely extending the latter's state set, in effect, via passive memory. The same thing can be done by using an

Non-Dichotomous, Relative and Hierarchical Aspects 351

unbounded network of automata, which is done in cellular automata (CA's) (Wolfram, 1986), neural nets (NN's) (Denker, 1986; Computer, 1988), and bus automata (BA's) (Rothstein, 1970-1988). Extensive bibliographies on CA's and NN's are given in Wolfram (1986), Denker (1986) and Computer (1988), but as BA's have a relatively small and unknown literature, the original papers are cited here. Developments in CA's and NN's were originally inspired by biology, but both have recently become of computational interest where a high degree of parallelism is needed. The origin of BA's was simultaneously from biology and computer science. The specific stimulus from the latter was a desire to understand the limits of parallelism and distributed processing, for they seemed to provide the only visible means to increase computational speed in large problems whose ultimate limitations on that increase were unclear in principle. Such large problems abound in fields like pattern recognition, artificial intelligence, modeling of biological, economic and other complex systems, and in many scientific or engineering problems.

In CA's each cell has a set of neighbors (which includes itself). Its input is a "vector" whose components are the states of the whole set of neighbors, and its output is its state. Information transfer is thus by "bucket brigade" from neighbor to neighbor which prevents full utilization of the potential parallelism of the whole array. The set of neighbors is fixed (not extensible), so that the networking of cells can not adapt to varying needs arising in the course of a computation. Neural nets were historically prior both to CA's and modern computation and many concepts of automata and logic circuits were first encountered there. Until recently their language was thus almost totally alien to that of computer science, in which theory of automata and logic circuits had in the meantime become highly developed. The NN's have been enjoying a revival, but their language is still struggling to evolve to where they can properly assimilate and exploit the intervening computer science advances. An important advantage of NN's over CA's is their greater network plasticity, but their disadvantage is a language in which the general theory is hard to develop.

Bus automata achieve up to the full potential parallelism of CA's, the plasticity of NN's, do so by extending the concepts of computer science, and appear to have a bright future in large problems (modeling complex systems, etc.). The CA cells are generalized to include both a resident local FSA and a module of a global interconnection network controlled by the automaton. The input and output links between neighbors can serve to carry inputs and outputs as in CA's, or input links can be switched directly to output links (both can be done in parallel). With appropriate switching an arbitrary global communication network between FSA's can be realized in a BA. Indeed, the entire architecture of the BA can be changed on each clock pulse. The state of a cell is a couple (Q,B) where Q is the state of the

FSA and B is the state of the switching module. In the most general binary signal case these are the Boolean (logic) functions from the set of link inputs to the set of link outputs. The resulting class of machines is so powerful that we have been unable to conceive of any way to increase that power beyond what it already has in principle. In this sense we suspect it may be the TM of parallel/distributed computation. We have achieved the ultimate speed-ups possible in principle for many important cases of pattern recognition, computation, data processing, formal language processing etc., and the end is not visible. It has been suggested for neural, visual, and brain modeling, and for use in robots (Rothstein, 1979, 1982, 1986, 1988).

The compatibility of BA's with computer and information science both accelerated BA research and opened new areas in computer science. As the "brain" of a "well-informed heat engine" (WHE) it contributes both to biological modeling (Rothstein, 1979a,b, 1982b, 1985, 1988b,c, 1990) and to a program of attempting to reformulate philosophical problems as scientific ones for a world of WHE's (Rothstein, 1964).

A WHE is a physical system with measurement/communication means to acquire information about its environment and about itself, with effector apparatus to perform operations on its environment (or part of itself), a computer, memory, programs, etc., whereby it can compute the "proper" operations to perform in response to the information acquired, and reservoirs of energy and matter to which it has access in order to fuel its activities which include construction, repair, or even reproduction (anything programmable). I believe the WHE is as far as physical science can go, in principle, in providing a model of a living system. If it becomes possible to model intelligence adequately on a computer, then I believe a WHE using a BA with "intelligent" programs as its computer would be as far as physical science can go in providing a model of intelligent living systems. Networks of WHE's would model populations, ecologies, and so on. Advances in biochemistry, heredity etc. would induce continuing modifications of the current best WHE model, but those modifications, I believe, would always be WHE models themselves (as suspected by Leibniz).

Let us abbreviate a WHE model of a living system as WHEM, and suppose that the presence of some operationally defined set of attributes and behaviors is taken as the definition of "life-like." The mechanist will identify life processes with the life-like processes of the "ultimate" WHEM. But FSA, WHE, or WHEM behaviors and attributes can not even be defined without reference to an environment. No external stimuli, energy, or matter come in except from it, and no responses, waste products or waste heat can leave except by being exported to the environment. Even in thermodynamics a non-equilibrium system with continued, as opposed to transient,

Non-Dichotomous, Relative and Hierarchical Aspects 353

non-equilibrium behavior can not be specified without specification of the environmental constraints maintaining the non-equilibrium behavior. We maintain that the definition of life-like behavior can be displayed. In an unsuitable environment that behavior will not occur. I believe this forces the conclusion that life is a relative phenomenon. One should talk of life relative to an environment, and there is then no necessary contraction in calling similar systems, in the same state, alive in one environment and dead in another.

An example may be helpful. Regular stacking of virus particles into "crystals" is well known. In all environments in which such crystals are stable they are as dead as salt crystals in all environments in which salt crystals are stable. Now dissolve them in water, and add cells which the virus particles can attack. They will be alive relative to that environment, as shown by viral attack on those cells and the production (or reproduction) of new virus particles. Dissolved salt would be inert. It is not alive relative to the environment in which the virus is alive.

A somewhat paradoxical situation now arises, for given any system whatsoever, it seems possible to construct a complicated environment relative to which that system "comes alive," like virus, and "reproduces itself." An automatic factory, which given a widget as template will copy it and make a bunch of widgets, can be called an environment relative to which widgets live in the same sense virus does! Though startling at first, there is no logical problem here, and I think it must be accepted if physical models of life are. The same is true *a fortiori* for physical models of the origin of life. The "fitness of the environment" is tautologously present as part of any attempt at a physical definition of life.

As WHEM can scale up or down, there is no objection in principle to defining different kinds of life (different criteria relative to different possible kinds of behavior in different environments) at different levels. Life at "lower levels" is necessary but not sufficient for life at "higher levels." A legally dead (brain-dead) individual can have transplantable organ systems which thus can be regarded as still alive at the organ system level, dead organs can still have some live tissues, and the tissues some live cells. Can a dead cell contain live DNA? Presumably yes in the sense that the DNA can be removed from the cell (thereby killing the cell), inserted in a different freshly enucleated, and thus "dead," similar cell, and the new cell will be viable. Such transplantations have been done successfully between zygotes, and the situation is basically similar to the earlier virus discussion. It is even closer to virus RNA or DNA vis-a-vis its protein coat. Neither molecule is "molecularly alive" without the other, and the combination is not "reproductively alive" without the host cell, as previously noted. Molecular is here potential reproductive aliveness (a predicted behavior) with respect to the host cell environment. Aliveness thus can not be considered an absolute property of a molecular assemblage defined without reference to an environment. We therefore suggest that "absolute life" can not be defined as a property of any closed system at the level of that closed system.

Nothing is lost operationally, compared to conventional usage. Physicians or biologists generally use behavioral tests for the presence of life, e.g., respiration, metabolism, irritability, emission of signals like EEG's, and so on. These are environmental interactions. The analogy of the view proposed to Bohr's insistence on the indivisibility of quantum phenomena is so striking that we propose our view as an appropriate generalization of Bohr's. It is also responsive, at least in part, to the intent of Wigner's call for a generalization of quantum mechanics sufficient in principle for biology. Relativity of life is a term evoking Einstein, whose presence also looms via EPR and wave-packet reduction. It is thus aesthetically satisfying that a relativistic argument (in the sense of the special theory of relativity) has been given in which irreversibility shows up as a necessary condition for the consistency of QM in a generalized EPR-like situation (Rothstein, 1982).

The physics of selective systems (SS's) which includes all physical systems for which FSA or computer models are appropriate, has been studied by Rothstein (1982a), as has the state concept espoused here. The SS is endowed with a generally large set of discrete macroscopic states, called B-states, which are *behaviorally* distinguishable from one another. By this we mean that for any two such states there exist well-defined sequences of inputs for which different (operationally distinguishable) output behaviors result. Many different microscopic states can be assigned, in theory, to a particular B-state, as with macroscopic states in statistical mechanics, but the B-states are not the usual gross states considered in that discipline. In thermodynamic systems, observables like temperature, pressure, and amounts of different constituents dominate. Here they are often irrelevant. For example, the temperature of a computer and ambient pressure can vary over wide tolerable ranges with essentially no effect on behavior., Equilibrium itself can be irrelevant. Similarly my weight and relative composition (due to changes in dietary or water intake, obesity, etc.) can vary with little effect on some behavior of interest. Higher level behavioral alternatives, though macroscopically and operationally defined, need correspond to no conventional quantum observables or even to intermediate level observables. It is not always clear that there is any relation at all between concepts at different levels. Averaging processes in statistical mechanics bridge the conceptual gap between molecular impacts and mechanical pressure, and similarly with other thermodynamic observables, but such connections are traceable in detail only in simple systems.

Music is more than acoustics, architecture cannot be reduced to the quantum

Non-Dichotomous, Relative and Hierarchical Aspects 355

bonding in cement or bricks, and no studies of transistor physics or circuitry will determine what programs to run or in what computer language they should be written. The B-state idea captures the essence of the black box at any level. Analysis into subsystems permits reductionist refinements in principle, all the way down to the level of fundamental quantum mechanisms. There can be many levels, with each reductionist transition to a lower level affecting the semantics (interpretation) of inputs and outputs. But the input-output class of model is preserved under such transitions. The complexity of living systems forces quantum level descriptions to surrender attempts to describe the whole system. Splitting the system into simple interacting subsystems seems to be the only way to make progress at the quantum level. It is illegitimate in the sense that it replaces the real quantum system by one frequently inconsistent with it in principle. It has the advantage of giving a good picture nevertheless, perhaps in loose analogy to the way mathematicians use divergent series (asymptotic expansions) to compute quantities of interest more accurately than they can be computed by convergent series (for comparable computing time).

7.5. CONCLUDING DISCUSSION

To sum up, we propose that the problem of defining the living state for a system is that of defining a set of suitable B-state orbits relative to a specified environment. Such states are behavioral, often at the level of an organism, and can, at that level, correspond to conventional metabolic criteria. With each refinement in the level of description the meanings of the inputs and outputs involved can change. The B-states are usually non-equilibrium, generally at macroscopic operational levels of description, and are conveniently discussed with the aid of computer science and statistical thermodynamics-like concepts. Attempts to reduce the discussion to the quantum-molecular level must stop short of a complete QM description. Life is metamolecular and metaquantal, but not metaphysical. The relative nature proposed for life makes the live-dead distinction non-dichotomous in the sense that a system can both be alive with respect to one environment and dead with respect to another without contradiction. It can be multiply hierarchical, with life definable at many levels. Death at any level ensues when the B-states appropriate to that level become inaccessible. Life could be restored if they are made accessible again.

The introduction summarizes additional views and conclusions relevant here. It is suggested that the reader review it before continuing. Its last paragraph will now be amplified.

Statistical mechanics (SM) can be viewed as the physical methodology for dealing with "emergent properties," to borrow a biological term. These are characteristics of an individual complex system which may be difficult, perhaps impossible, to describe satisfactorily in terms completely definable in the language of theory relevant at a level of "more fundamental" constituents of that complex system. In the Boltzmann-Gibbs SM, this means augmentation of mechanics with statistical averaging, representative ensembles, measure theory, ergodic theory etc. These are elements of a mathematical theory whose axioms are completely independent of those of a formal theory of the underlying mechanics. The SM representations of thermodynamic variables like pressure, temperature, entropy etc. "emerge" only within a formalism broad enough to include both of those theories. It is thus a "metamechanics," and this situation is preserved when QM is adopted as the fundamental lower level theory.

Generalizing SM to bridge the gaps, over many levels, from QM to biology poses new problems. These can be met and include clarification of the relation between information, entropy, measurement and QM (Rothstein, 1951), and showing both that the relation between the first three survives the transition in level from microscopic QM to macroscopic thermodynamics (Rothstein, 1952a). It gives an adequate viewpoint for handling organization at thermodynamic (Rothstein, 1952b) and operational levels (Rothstein, 1962b). The generalized entropy concept to which these concepts led clarified the paradoxes of SM (irreversibility, paradoxes of Loschmidt and Zermelo) (Rothstein, 1957a, 1974) provided an entropy measure for the amount of information or organization provided by theory, and thus established on a unified basis many criteria for preferring one theory to another. The correspondingly generalized second law of thermodynamics admitted these novel formulations as the operational undecidability of some physical questions (Rothstein, 1964).

All formal theories, complicated enough to include arithmetic, and which claim operational verifiability must be incomplete in Goedel's sense (Goedel, 1931; Hopcroft and Ullman 1979) lest they be logically inconsistent. This means that statements exist which are meaningful within the system, and which can even be true in the system, but which are impossible to prove in the system. If such a proof existed the system would be inconsistent. One such statement is that the system is consistent. Such statements might be proved in a metasystem strong enough to include the original system, but the metasystem would then have undecidable (unprovable, noncomputable) questions of its own (Goedel, 1931; Hopcroft and Ullman, 1979; Arbib, 1987). The famous Einstein-Bohr debates on the nature of QM (Einstein et al., 1935; Bohr, 1935; Schilpp, 1957) were declared in a tie (Rothstein, 1982a), because EPR said QM was incomplete, while Bohr defended QM as being

Non-Dichotomous, Relative and Hierarchical Aspects 357

logically consistent; Goedel and generalized entropy allow both views to be correct without contradiction. The generalized EPR (called EPR^2) there discussed shows that QM becomes inconsistent unless a question closely relating to wave packet reduction is undecidable. It becomes so if there is irreversibility in measurement. This supported the idea that the second law might somehow be connected with Goedel's theorem, which almost suggested itself in Rothstein (1964), and was explicitly opined in footnote reference 10 of Rothstein (1967). Later work on WHE's and WHEM's strengthened this view and exhibited additional noncomputable (undecidable) behaviors (Rothstein, 1979a, 1982a). The classical diagonalization argument proving the non-denumerability of the continuum is transformed to show both how WHE's can refute all finitely axiomatizable theories of their behavior, and that evolution is non-predictable (Rothstein, 1979a). It may well be that almost all behavior of real physical systems is undecidable, for the class of all possible behaviors over all time has the cardinality of the continuum, while the subclass of computable behaviors is countable. Computable is here taken to mean computable within the logical system of quantum mechanics, while behavior must be describable in operational terms. The rules of correspondence between the microscopic language of QM and the macroscopic language of behavior are not logically part of the first, so we doubt that a valid proof of the second law can ever be given by quantum mechanics.

We can view the intrusion of FSA, or more generally CIS (computer and information science) concepts in another way. Splitting a quantum system into subsystems, though a transgression against fundamental quantum principles, is forced on us by complexity. In atonement we try to put the pieces back together by having them interact. The interaction can be characterized as a set of input-output relations between subsystems. This makes the introduction of FSA or CIS concepts seem more natural, and perhaps even inevitable. A corollary may be physical irreversibility in situations where the tradition of theoretical physics has been to view a purely quantum dynamical treatment as the only proper one. Scattering matrix techniques, creation and annihilation processes and phase change (creation and destruction of embryos in evaporation, condensation, melting or crystallization) might be worth examining from this point of view.

In conclusion we note that though we had to bring in a wide range of disciplines and subject matter into this discussion, and though the complexities seemed sufficiently forbidding to deter all but the foolhardy, the final picture seems coherent and reasonable. Like divergency in quantum electrodynamics, much of the complexity can be put in boxes, so to speak, and a meaningful discussion given which doesn't need to open those boxes very often. When it does happen the discussion continues in terms of the interactions of smaller boxes within a box. Life as a non-equilibrium, non-dichotomous, hierarchical and relative concept will, we hope, contribute to biophysical systems research both by organizing known data, suggesting synthesizing viewpoints and suggesting new experiments.

REFERENCES

Arbib, M. A. (1987). Brains, Machines and Mathematics. Second edition. Springer-Verlag, New York.
Bennett, C. H. (1982). The Thermodynamics of Computation—A Review, Internat. Jour. Theor. Physics 21: 905-940.
Beutler, H. (1936). Z. Physik 101:304.
Bohr, N. (1935). Can Quantum-Mechanical Description of Physical Reality be Considered Complete? Physical Review, Ser. 2 48: 696-702.
Bohr, N. (1958). Atomic Physics and Human Knowledge. Wiley, New York.
Carter, F. L., ed. (1982, 1987). Molecular Electronic Devices Volumes I, II. Marcel Dekker, New York.
Champion, D. and Rothstein, J. (1987). Immediate Parallel Solution of the Longest Common Subsequence Problem. Proc. 1987 Int. Conf. on Par. Proc. Penn State Univ. Press (University Park, PA 16802), pp. 70-77.
Computer 21 (1988). Special issue devoted to artificial neural systems.
de Baere, W. (1986). Einstein-Podolsky-Rosen Paradox and Bell's Inequalities Advances in Electronics and Electron Physics 68: 245-336.
Delhaes, P. and Drillon, M., eds. (1987). Organic and Inorganic Low-Dimensional Crystalline Materials, NATO ASI Series B, Vol. 168, Plenum Press New York, London.
Denker, J. S., ed. (1986). Neural Networks for Computing, AIP Conference Proceedings Volume 151, New York.
Eigen, M. (1971). Self-Organization of Matter and the Evolution of Biological Macromolecules. Naturwiss. 58, 465-522.
Eigen, M. and Winkler, R. (1981). Laws of the Game. A. A. Knopf, New York.
Einstein, A. (1921). Geometrie und Verfahrung. Sitzungsberichte, Preussische Akademie d. Wiss. pt. I, pp. 123-30. English translation in Sidelights on Relativity (London, Methuen, 1922; Dutton, New York 1923). Reprinted in Methods of the Sciences, 2nd ed., Univ. of Chicago, Chicago 1947.
Einstein, A., Podolsky, B. and Rosen, N. (1935). Can Quantum-Mechanical Description of Physical Reality be Considered Complete? Phys. Rev. Series 2 47: 777-780.
Elsasser, W. M. (1937). On Quantum Measurements and the Role of the Uncertainty Relations in Statistical Mechanics. Phys. Rev. 56:987.
Elsasser, W. M. (1961). Quanta and the Concept of Organismic Law. J. theor. Biol., 1: 27-58.
Fumi, F. and Tosi, M. P., editors (1985). Proceedings of the International School of Physics, "Enrico Fermi," Course LXXXIX, Highlights of Condensed-Matter Theory (Bassani, F., ed) North-Holland, Amsterdam, Oxford, New York, Tokyo.
Goedel, K. (1931). Ueber formal unentscheidbar Saetze der Principia Mathematica und verwandter Systeme I. Monatschefte f. Mathematik u. Physik 38, 173-198.
Heeger, A. J., Kivelson, S., Schrieffer, J. R. and Su, W. P. (1988). Solitons in Conducting Polymers Rev. Mod. Physics 60: 781-850.
Heitler, W. and London, F. (1927). Z. Physik 44: 455.
Hopcroft, J. E. and Ullman J. D. (1979). Introduction to Automata Theory, Languages and Computation. Addison-Wesley, Reading, Mass.
James, H. M. and Coolidge, A. S. (1933) J. Chem. Phys. 1: 825, 3 129.

James, H. M. and Coolidge, A. S. (1935). ibid. 3: 129.
James, H. M. and Coolidge, A. S. (1938). ibid. 6: 730.
Kobilka, B. K., Kobilka, T. S., Daniel, K., Regan, J. W., Caron, M. G. and Lefkowitz, R. J. (1988). Chimeric α_2-, β_2-Adrenergic Receptors: Delineation of Domain Involved in Effector Coupling and Ligand Binding Specificity. Science 240: 1310-1316.
Moshell, J. M. and Rothstein, J. (1979). Bust Automata and Immediate Languages. Information and Control 40: 88-121.
Prigogine, G., Nicolis, G. and Babloyantz, A. (1972). Thermodynamics of Evolution. Physics Today 25(11): 23-28; ibid. 25(12): 38-44.
Richardson, W. (1935). On the Ground State of (H_2), the Molecular Ion (H_2^+) and Wave Mechanics. Proc Roy Soc Lond 152: 503-514.
Rothstein, J. (1951). Information, Measurement and Quantum Mechanics. Science 114: 171-175.
Rothstein, J. (1952a). Information and Thermodynamics. Phys. Rev. 85: 135.
Rothstein, J. (1952b). Organization and Entropy. Jour. Applied Phys. 23: 1281-1282.
Rothstein, J. (1956). Information, Logic and Physics. Philosophy of Science 23: 31-35.
Rothstein, J. (1957a). Nuclear Spin Echo Experiments and the Foundations of Statistical Mechanics. Jour. Applied Physics 25: 510-518.
Rothstein, J. (1957b). A Physicist's Thoughts on the Formal Structure and Psychological Motivation of Theory and Experiment. Revue International de Philosophie 211-226.
Rothstein, J. (1959). Heuristic Application of Solid State Concepts to Molecular Phenomena of Possible Biological Interest. In: Proceedings of the First National Biophysics Conference, (H. Quastler and H. J. Morowitz, eds.) Yale University Press, New Haven, pp. 77-85.
Rothstein, J. (1962a). Wiggleworm Physics. Physics Today, pp. 28-32, 34, 36, 38.
Rothstein, J. (1962b). Information and Organization as the Language of the Operational Viewpoint. Philosophy of Science 29: 406-411.
Rothstein, J. (1964). Thermodynamics and Some Undecidable Physical Questions. Philosophy of Science 31: 40-48.
Rothstein, J. (1967). Chapter XIX in Communication: Concepts and Perspectives (L. Thayer, ed.). Spartan Books, pp. 397-423.
Rothstein, J. (1970). Patterns and Algorithms. Proc. 1970 IEEE Symp. on Adaptive Processes (9th) Pub. No. C58-AC, paper II-4.
Rothstein, J. (1974). Loschmidt's and Zermelo's Paradoxes Do Not Exist. Foundations of Physics 4: 83-89.
Rothstein, J. (1976). On the Ultimate Limitations of Parallel Processing. Proc. 1976 Int. Conf. on Parallel Processing, IEEE. Cat. No. 76CH1127-OC, 206-212.
Rothstein, J. (1977a). Toward an Arithmetic for Parallel Processing. Proc. 1977 Int. Conf. on Par. Proc., IEEE. Cat. No. 77CH 1253-4C, 224-233.
Rothstein, J. (1977b). Transitive Closure, Parallelism, and the Modeling of Skill Acquisition. Proc. Int. Conf. on Cybernetics and Society IEEE. Cat. No. 77CH 1259-1AMC, 232-236.
Rothstein, J. (1978). Topological Pattern Recognition in Parallel and Neural Models on Bus Automata. Proc. 1978 Int. Conf. on Par. Proc., IEEE. Cat. No. 78CH 1321-9C, 92-107.
Rothstein, J. (1979a). Generalized Entropy, Boundary Conditions and Biology. In: The Maximum Entropy Formalism (R.D. Levine and M. Tribus, eds.). The MIT Press Cambridge, Mass pp. 423-468.
Rothstein, J. (1979b). Generalized Life. Cosmic Search 1: 35-8, 44-6.
Rothstein, J. (1982a). Physics of Selective Systems: Computation and Biology. Internat. Jour. of Theoretical Physics 21: 327-350.
Rothstein, J. (1982b). Toward Pattern-Recognizing Visual Prostheses. In: Proc. IFAC Symp. "Control Aspects of Prosthetic and Orthotics". Pergamon Press, New York, pp. 87-95.

Rothstein, J. (1985). On the Scientific Validity and Utility of the Living Earth Concept and its Further Generalization. In: Proceedings of the Symposium "Is the Earth a Living Organism?" National Audubon Society Expedition Instutitue, Northeast Audubon Center, Sharon, Connecticut 06069. Also available as OSU technical report OSU-CISRC-TR-86-4.

Rothstein J. (1986). Bus Automata, Brains and Mental Models. IEEE Transactions on Systems, Man and Cybernetics 18(4): 522-531.

Rothstein, J. (1988a). On Ultimate Thermodynamic Limitations in Communication and Computation. In: Performance Limits in Communication Theory and Practice. NATO ASI Series E, Vol. 142 (J.K. Skwirzynski, ed.), Kluwer Academic Publishers, Dordrecht, Boston, London, pp. 43-58.

Rothstein, J. (1988b). Bus Versus Cellular Automata and Ultimate Limitations of Parallel Processing. Mathl Comput. Modeling, 11, 357-362.

Rothstein, J. (1988c). Bus Automata for Intelligent Robots and Computer Vision, In: Intelligent Robots and Computer Vision: Sixth in a Series, (David P. Casasent, Ernest L. Hall, eds.) Proc. SPIE 848, pp. 299-309.

Rothstein, J. (1990). Entropy and the Evolution of Complexity and Individuality. In: Prebiological Self-Organization of Matter, (C. Ponnamperuma and F.R. Eirich, eds), A. Deepak Publishing, Hampton, VA, pp. 51-98..

Rothstein, J. and Davis, A. (1979). Parallel Recognition of Parabolic and Conic Patterns by Bus Automata. In: Proc. 1979 Int. Conf. on Par. Proc., IEEE Cat. No. 79CH 1433-2C, 288-297.

Schilpp, P. A., ed. (1957). A. Einstein: Philosopher-Scientist. Dover Publications, New York.

Takasi, I. and Nakaye, T. (1986). Heat Produced by the Dark-Adapted Bullfrog Retina in Response to Light Pulses. Biophys. J. 50: 285-293.

Tolman, R. C. (1938). The Principles of Statistical Mechanics. Oxford University Press, Oxford.

Wald, G. (1968). Molecular Basis of Visual Excitation. Science 162: 230-239. For a recent survey of retina science, see J.E. Dowling, The Retina, Harvard University Press, Cambridge, Mass., 1987.

Wheeler, J. A. and Zurek, W. H., eds. (1983). Quantum Theory and Measurement. Princeton University Press. Rather than referring to the original publication we cite an anthology which includes it and much more of interest here.

Wigner, E. P. (1961). The Probability of Existence of a Self-Reproducing Unit. In: The Logic of Personal Knowledge (Essays presented to M. Polyanyi). Routledge and Kegan Paul Ltd., London.

Wolfram, S. (1986). Theory and Applications of Cellular Automata. World Scientific, Singapore.

Chapter 8

EMERGENT PROPERTIES IN NEURAL NETWORKS

Steven Finette

ABSTRACT

Living systems possess a high degree of spatial and temporal order among their constituent parts. In addition, they have the ability to reliably code, decode, store and transfer information among different hierarchial structures. Since nonlinear interactions among the components of a biological system are quite common, it is natural to ask whether cooperative effects can play a role in processing information and achieving order. We consider here a brief discussion of collective phenomena at the cellular level, using as an example the interactions among neurons.

8.1. INTRODUCTION

While both inanimate and living matter are composed of the same fundamental components, atoms, and obey the laws of quantum mechanics, they differ dramatically in the collective behavior exhibited at the macroscopic level. Two properties commonly associated with living systems and dependent on the coordinated motions of many components are (1) a high degree of spatial and temporal order and (2) the ability to reliably code, decode, store, and transfer information. These two properties are not independent and are also shared by diverse inanimate systems such as nonlinear chemical reactions and digital computers, respectively. Living systems, however, differ from their inanimate counterparts in the degree of complexity and organization associated with these properties. When both properties are absent, it would be reasonable to conclude that the system was not living, or at least lacked important attributes normally associated with biological systems.

Property (1) can be understood, independent of the underlying mechanisms, by recognizing that any living organism is an open thermodynamic system maintained far from equilibrium by an external flow of energy. While the theory of dissipative structures (Nicolis and Prigogine, 1977) forms a very plausible framework for describing the emergence of spatial and temporal coherence in living systems, complementary microscopic models are necessary in order to understand the molecular mechanisms responsible for the (macroscopic) low entropy structures so prevalent in biological systems. The "information" referred to in the second property may be obtained internally through DNA transcription and externally through sensory stimuli. The interpretation and decoding of external stimuli ultimately depends on the genetic code which, in turn, prescribes the transduction mechanisms, neural anatomy, and physiology used by the organism to process externally driven stimuli. DNA also provides the code for determining of which dissipative structures will be formed (Ji, 1985).

Both spatial-temporal coherence and information processing appear to be organized in hierarchial structures within living systems. We tend to associate these structures with levels of complexity: subcellular, cellular, multicellular and so on up to the highest level, that of an ecological system. The term complexity is used rather loosely here to refer to some measure of both connectivity and coupling among components at a given level of organization and among hierarchial structures. The coupling may be viewed as an input/output relation for any pair of components that share information through chemical or electrical signalling. The connectivity or topology at one system level may be static or dynamic and the coupling strength may be linear, non-linear and/or time dependent.

Traditionally, biologists have successfully analyzed living systems within a reductionist framework. It is known, however, that a multicomponent system may possess qualities that are absent or meaningless at the component level. These emergent (collective or holistic) properties often appear as a natural consequence of complexity. With respect to biological systems, there are a number of structures or tasks for which such phenomena can be of use. From an engineering standpoint these tasks include hardware design (exquisite cell and organ structures), software design (efficient and reliable compression and coding of information) and communication links (neural and chemical networks) for transferring information through noisy channels. If we make the reasonable working hypothesis that nature is an excellent engineer, then it is quite plausible for nature to use collective phenomena as fundamental processing tools. One must be aware, however, that because evolution is a dynamic process, the biological structures that one studies at the present time are, in some sense, representative of state of the art design and are not necessarily the finished product. Therefore, while the above hypothesis is quite reasonable, using it to ask questions about *optimal* biological systems design must be approached with caution.

A recent report (Science, 1985) by the National Academy of Sciences on models for biomedical research states "...in every hierarchial level from molecules to ecosystems, common hardware, common programs and common strategies are used to achieve diverse ends." Since interactions among biological system components are universal, we might conclude from this statement and the above comments that a study of collective phenomena at one level could yield insights into the dynamics of other levels of structure and that common themes may emerge. With this motivation, we will consider here a brief discussion of collective phenomena at the cellular level in the context of a specific example, that of interaction among groups of neurons.

8.2. THE NETLET AS A PARADIGM FOR NEURAL SELF-ORGANIZATION

A set of interconnected neurons is defined to be a neural network. The fundamental component of the network, the neuron, has been the subject of extensive anatomical and physiological research, and its dynamics is fairly well understood. The electrophysiological response of single neurons can be isolated and measured, as can the bulk electrical response of the brain using the electroencephalogram. The latter, however, is very difficult to interpret in terms of underlying neural responses because it is an indirect measure of cell activity. Therefore, most of our understanding of basic brain function is due to the interpretation of single cell responses. Practical considerations pertaining to electrophysiological recording techniques have forced biologists to formulate models of the nervous system based on single cell physiology, involving a limited knowledge of cell group interactions.

Unfortunately, neural responses can only be partially interpreted in this manner—a cortical neuron may synapse with 10^2-10^4 other neurons, forming a densely connected structure where interaction among components is the rule rather than the exception. While the dynamical response of the network will be a function of the responses of its individual components, emergent phenomena may exist which may be best studied at the network level, rather than the component level. Several quantitative approaches have been proposed to study the dynamics of neural networks (for example, see Harth, et al., 1970; Amari, 1971; Wilson and Cowen, 1972; Grossberg, 1980; Little, 1974; Hopfield, 1982). A rather common assumption is that the macroscopic variable of interest is related to the number of action potentials generated by the network per unit time. This assumption may be criticized as overly simplistic, yet our ignorance concerning events, such as local membrane changes throughout the dendritic tree of each neuron, makes this assumption a reasonable first approximation. The approach discussed here describes network response similar to those obtained using other methods and these results are representative of some of the types of non-linear dynamics that can occur.

We consider a neural network as being composed of one or more *netlets*—discrete populations of neurons in which the formation of a synapse between any two cells is equally probable (Harth et al., 1970). Netlets are assumed to form functional units, where the fractional number of neurons firing at a given time is called the activity a. The activity is assumed to be the macroscopic correlate of sensory and motor events. Because of the large number of microscopic parameters characterizing the structure and function of a single neuron and our lack of knowledge concerning the distributions of these parameters, most researchers choose lumped parameter models of single neurons when describing the network dynamics. In this approach, the netlet microstates are of secondary importance. Netlets are coupled in such a way as to specify randomness-in-the-small, structure-in-the-large where each netlet is characterized by its own set of statistical parameters (Harth et al., 1970). To some extent, this design requirement is due to our ignorance concerning the microscopic connectivity of a neural network. While there is no doubt that the nervous system has a great deal of *global* specificity, our knowledge of *local* specificity is less well understood. This lack of knowledge is incorporated in the random character of the connectivity, which is characterized by probability functions within each netlet. The assumption is not as restrictive as one might think. It can be relaxed to include distance-dependent, anisotropic connectivity distributions (Finette, 1978, 1979).

We can introduce two sets of parameters describing the structural and functional characteristics of a netlet. The internal structural parameters include the total number of neurons N, the fraction of inhibitory neurons h, and the average number of neurons μ^+ and μ^- with excitatory (+) or inhibitory (-) connections from a given excitatory or inhibitory neuron in the same netlet. The coupling coefficients w_{ij} are lumped parameters which describe the strength of the synaptic connections from cell i to cell j. They are implicit functions of the cable properties of the dendrites and the distance of the synapse from the axon hillock of cell j. Coupling coefficients describe the weighted potential variations at the axon hillock due to one or more post-synaptic inputs distributed over the dendritic tree. The afferent input

Emergent Properties In Neural Networks

is an external structural parameter and is interpreted as axonal input from another netlet. It is partially characterized by ξ, the number of afferent fibers entering the netlet. These excitatory or inhibitory afferent axons are assumed to uniformly distribute their synapses with neurons in the netlet (Anninos, et al., 1970).

The netlet is considered to be a discrete dynamical system where the smallest unit of time is τ. Since the activity at time $n\tau$ is a function of the activity in previous time increments, netlets are related to stochastic (Vichniac, 1986). Functional parameters include the synaptic delay d, absolute and relative refractory periods, t_{abs} and t_{rel}, respectively, and the temporal summation time λ. Each of these quantities is a multiple of the time quantization parameter τ. The neuronal threshold voltage, θ, controls the transfer of information among neurons by regulating the generation of action potentials. The fractions of active afferent fibers in a given iteration, σ^+ or σ^-, are external functional parameters which, as we will see below, may be interpreted as setting the operating point of the network. Additional functional properties that can be taken into account are cumulative hyperpolarization and post-inhibitory rebound (Harth et al., 1975; Finette, 1979). The former is closely related to neural accommodation; physiologically, it causes the threshold θ to increase in direct proportion to the neuronal firing rate. Post-inhibitory rebound is a lowering of the neuronal threshold after a period of hyperpolarization.

The time quantization is necessary for digital simulations, but can introduce a rather artificial synchronization of neural activity. Extensive simulations show that the network is effectively desynchronized by allowing τ to be smaller than the synaptic delay. Under this condition, the discrete time system approximates the continuous time system and the network properties appear robust (Wong and Harth, 1973). The similarities between netlet dynamics and the dynamics obtained using continuous-time approaches also supports this conclusion (Wilson and Cowen, 1973; Amari, 1977).

We can define a microstate vector $X = (x_1, x_2...x_N)$ where x_i denotes the firing state of a neuron: $x_i = 1$ if the neuron has fired an action potential in the previous time iteration, and $x_i = 0$ if it has been quiescent. The condition for the occurrence of an action potential in the ith neuron at the time $n\tau$ is given by $\Sigma_{j=1}^{N} w_{ij} x_j > \Theta_i(n\tau)$ with n an integer. Note that the threshold is a function of time due to the absolute and relative refractory periods. The activity of the net can be determined by probabilistic arguments to be of the general form (Wong and Harth, 1973)

$$<\alpha_{n+1}> = (1-\alpha_n')P(\alpha_n', \sigma) \qquad (8.1)$$

where $<\alpha_{n+1}>$ is the expectation value for the fraction of neurons firing at

$(n+1)\tau$. The quantity $(1-\alpha_n')$ is the fraction of neurons available for firing at $(n+1)\tau$ where α_n' depends on the activity in the previous k time intervals: $\alpha_n' = \Sigma_{i=1}^{k} \alpha_{n+1-i}$. At $(n+1)\tau$, the probability that a neuron receives input equal to or exceeding θ is given by P, a non-linear function of α_n'. The parameter σ is specified explicitly as is now considered to be a variable which changes slowly with time. The responses of the network for variations of this parameter are discussed below. Equation (8.1) describes a non-Markovian difference equation which is non-linear in the macroscopic variable of interest. Since computer simulations indicate that the dynamical responses are robust with respect to the order of the difference Equation (8.1), it is convenient to discuss the netlet dynamics using a first order difference equation $\alpha_{n+1} = f(\alpha_n, \sigma)$ where $f(\alpha_n, \sigma)$ corresponds to the right hand side of Equation (8.1) for the special case $\alpha_n' = \alpha_n$ (Anninos et al., 1970). This implies $\lambda < \tau$, $\tau_{rel} = 0$, and that τ_{abs} satisfies the inequality $t < \tau_{abs} < 2\tau$.

One can show that several solutions exist for the non-linear eigenvalue problem $\alpha_{ss} = f(\alpha_{ss})$ describing steady-state activity in the netlet. For $\sigma = 0$, stable stationary states are determined by the condition $|(\frac{\partial f}{\partial \alpha})|_{\alpha=\alpha_{ss}} < 1$. There is a trivial stationary point at $\alpha_n = 0$ common to all netlets. The existence of fixed points other than $\alpha_n = 0$ indicates that steady-state activity patterns can occur. Note that the function $f(\alpha_n)$ depends implicitly on all the parameters of the netlet. If one or more of those parameters are allowed to vary, the fixed points will change and this can introduce bifurcation points where the steady-state activity jumps from one dynamical mode to another. The steady state response patterns, as well as the change in response due to the emergence of new bifurcation points, are examples of functional or structural self-organization, depending on which set of parameters is changing. In the general case both sets could vary simultaneously leading to very complicated behavior. The self-organization is due to cooperative interaction among many neurons. We will consider here only the effect of functional self-organization, in which the afferent input σ is given the status of an independent variable (Equation (8.1)). An example of structural self-organization might involve learning through neuronal plasticity or synaptic facilitation, where the coupling constants are functions of α. While structural self-organization is of considerable interest (Denker, 1986), we wish to mention here that interesting properties of the network can emerge without invoking any structural learning schemes.

The fixed points α_{ss} can be determined by finding solutions of $\alpha_{ss} = f(\alpha_{ss})$. Alternatively, a neural network can be simulated with the statistical properties described by this equation to obtain an estimate of the fixed points. The agreement between the theory and Monte Carlo simulations is quite good (Anninos, 1970). One obtains the results sketched in Figure (8.1) if we plot the activity trajectories α_{n+1} vs.

Emergent Properties In Neural Networks

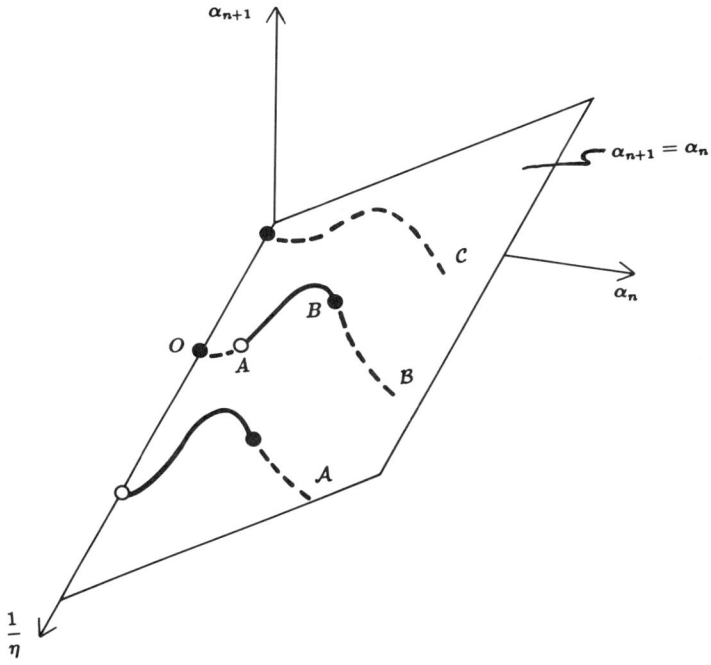

Figure 8.1. Schematic illustration of activity curves for three classes of netlet. Steady-state activity occurs when an activity curve intersects the plan $\alpha_{n+1} = \alpha_n$.

α_n for netlets parametrized by η, where η is a measure of the average number of excitatory post-synaptic potentials that must occur in a neuron in a given iteration to cause an action potential. In other words, it is a measure of netlet excitability. In this figure, three activity trajectories are drawn to illustrate the three classes of netlet behavior. Fixed points are denoted by ● (stable) and ○ (unstable). They correspond to the intersections of the activity curves with the plane $\alpha_{n+1} = \alpha_n$. A class A netlet has an unstable fixed point at $\alpha_n = 0$ since the activity will converge to a sustained level for any non-zero value of α. On the other hand, a class B netlet has two attractors at points O and B, respectively, separated by an unstable fixed point at A. In this case a threshold activity exists above which the netlet will approach a sustained activity level. Below the threshold point A, the netlet activity will be damped out. For class C netlets, the origin is the only fixed point, and for any non-zero value of α_n, the activity will damp out. The above results are valid provided that the netlet consists of only two groups of neurons, excitatory and inhibitory. If, for example, the excitatory group consists of several subpopulations characterized by different mean threshold values, then other models predict that additional attractors and repellers can exist (Wilson and Cowen, 1972). If a slowly varying external input σ is now provided to the netlet, the fixed points are modified in such a way that one class of netlet can be transformed into another. In other words, new fixed points can be created and old ones shifted or destroyed due to functional self-organization. Figure (8.2) shows the stationary behavior as a function of σ^+, σ^-. The figure shows three phase curves parametrized by σ^- and indicates that both phase transitions and hysteresis can occur as the level of afferent input to the netlet is changed. Both effects emerge spontaneously from interactions of groups of neurons and it is found through simulation that such behavior can occur for groups as small as thirty neurons (Harth et al., 1975; Shaw et al., 1982). Below this number of cells, coherence effects are not as well defined because of large fluctuations about steady-state activity values.

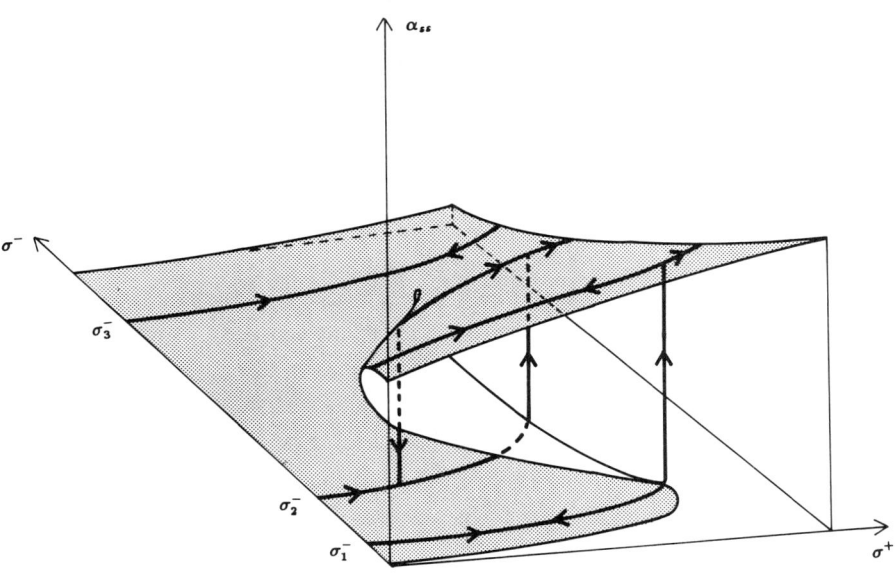

Figure 8.2. Illustration of three qualitatively different steady state trajectories, parametrized by three different values of external inhibitory input. The figure illustrates both phase transitions and hysteresis. The surface corresponds to steady state activity as a function of afferent input. Reproduced by permission from Finette et al., 1978.

Other cooperative effects occur when additional functional network properties are included. For example, the introduction of cumulative hyperpolarization can have an interesting effect on the network behavior. Consider the netlet response under excitatory input σ^+ chosen so that the netlet exhibits steady-state activity α_1. Cumulative hyperpolarization causes a down-shift in the steady-state activity to other phase curves corresponding to larger values of η until a bifurcation point is reached. At this point, a phase transition to lower activity α_2 will occur. Since cumulative hyperpolarization is proportional to the activity of the netlet, η will now decrease, moving the activity level back toward the original stationary point corresponding to α_1. The time-dependent behavior thus exhibited is an example of non-linear oscillation induced by phase transitions between two stable stationary states. Coupling netlets together and incorporating cumulative hyperpolarization allows one to control the oscillation on/off times, amplitude and frequency by choosing appropriate values for the functional parameters. These emergent properties were used to provide a neural mechanism for the swimming escape sequence of the mollusk Tritonia. An important point brought out by this application is the following. By exploring some aspects of emergent behavior, one can accurately simulate a complex stereotyped response pattern using both a minimum of knowledge concerning the precise neural circuitry and some basic principles of single-cell physiology. The information about the escape sequence is stored globally among groups of neurons, rather than locally in any given cell.

The introduction of a distance metric allows the netlet concept to be generalized to include spatially inhomogeneous and anisotropic connectivity distributions among neurons. When distance dependent coupling is introduced, the netlet is found to exhibit the property of local addressability (Finette, 1978). For example, if the external excitatory input to an initially quiescent netlet is spatially confined to a particular domain R, the resulting netlet activity pattern remains within or near R. The dynamical responses of this region show the same qualitative behavior as the class A, B and C netlets described previously. Emergent properties such as spatially localized phase transitions and hysteresis have been demonstrated by computer simulation and shown to be sensitive to the particular spatial distribution of σ^+ for a specified netlet connectivity distribution. These spatially localized activity regions can be interpreted as pattern recognition units, selectively tuned for specific input patterns. In this sense, the netlet serves as a template in a matched filtering system where the recognition of a particular pattern is indicated by a spatially localized phase transition. The concept of local addressability, in conjunction with anisotropic connectivity distributions, has been used to provide an explanation for orientation sensitivity in the visual cortex (Finette, 1978).

8.3. RELIABILITY AND COOPERATIVE COMPUTATION

The dynamics we have briefly outlined above are inherently cooperative. The fact that these properties depend upon interaction among neurons is of importance for the design of a reliable system capable of neural computation and information processing. The significance of cooperativity in this context is related to the known variability of single neuron responses to identical physical stimuli. Neurons are noisy devices—most neurons maintain some level of activity even in the absence of external stimuli. The purpose of this endogenous firing rate is unknown, but any neural signal must have characteristics which allow it to be recognized by the network in the presence of noise. Provided that a certain critical netlet size is reached, neuronal cooperativity implies a "built-in" reliability factor because the netlet activity is not significantly affected by the presence of noise and/or failure of a small fraction of the neurons comprising the system. Simulation has determined that statistical fluctuations in netlet activity around a stable steady-state have standard deviations $\approx 1 / \sqrt{N_a}$, where N_a is the number of active neurons. For small N_a, these fluctuations could drive a class B netlet to an unstable steady-state causing a phase transition. However, for $N \geq 50$ the reliability of the system, defined here as a measure of the probability of maintaining the sustained steady-state level for a specified length of time, was approximately 95% (Shaw et al., 1982). While this result may vary depending on the specific task to be performed, it is clear that cooperative interactions can provide a plausible mechanism for enhancing the reliability of neural processing relative to that of single neuron responses.

The speed at which individual neurons transmit information through action potentials is several orders of magnitude slower than a typical electronic circuit which simulates neural behavior. Physiological constraints place an upper bound of approximately 1 Khz. on the frequency of action potentials in a single axon. Yet, for many complex applications such as visual pattern and speech recognition, the computational efficacy of neural circuits associated with these important biological functions far exceeds that of sophisticated algorithms run on modern supercomputers. Anatomical evidence implies a considerable opportunity for parallel processing of sensory information in mammalian brain, and the recent trend toward concurrent computer architectures for high speed computation has led to considerable interest in the computational properties of neural networks (Shastri, 1987). In this respect, emergent phenomena can play an important role (Hopfield, 1982; Hogg and Huberman, 1984).

An example is provided by the design of a content addressable (associative) memory based on a simplified analog view of neural network dynamics. The memory can be implemented in hardware using operational amplifiers, resistive and capacitive elements (Hopfield, 1982). Such a circuit is analog rather than digital, and deterministic rather than probabilistic in its connectivity. A content addressable memory (CAM) differs from a typical computer memory in that a memory stored in the latter is addressed by location rather than content. A CAM does not locate a memory by knowing an address, but rather from partial (incomplete) information about the memory. Such a system is capable of error-correcting, since only partial information concerning the meaning or content of the memory is necessary to retrieve the full memory. The basis for the design of a CAM utilizing neural network behavior is related to the inclusion of phase space attractors which code the memory or memories of interest (Hopfield, 1982). In this view, the memory is mapped onto a fixed point of the system. This fixed point occupies a known position in phase space. A state point "near" such a fixed point will contain partial information about the memory and in time will converge to that memory if it is in the basin of that attractor. Convergence can be guaranteed by specifying a Liapunov function for the system. An important attribute of such a system design is its fault-tolerant behavior.

8.4. CONCLUSIONS

We have briefly considered some cooperative properties of neural networks, associated with non-linear interactions among groups of neurons. Some emergent properties were illustrated in the context of the netlet model. While a number of different quantitative approaches have been used to study neural networks, the qualitative behavior exhibited by these models is quite similar when one considers functional self-organization. This tends to support the hypothesis that emergent behavior is a natural consequence of both neural connectivity and non-linear interactions, and that it might play a role in neural information processing. The introduction of structural self-organization, via synaptic plasticity, will increase the types of neuronal group responses beyond those that are considered here. The enhanced reliability of groups of interacting neurons, relative to single cell reliability, has been demonstrated by simulation for some simple cases and deserves further investigation.

Emergent properties such as phase transitions, hysteresis and limit cycles can be viewed as forms of spatial/temporal order which have the ability to store and process information. In such a view, information can be stored globally among cells rather than locally in a given cell. It is not appropriate to perform a simple extrapolation of emergent phenomena associated with neural systems to cells without excitable membranes. However, it is plausible that some form(s) of cooperativity exist in other cell populations which are tightly coupled either through physical contact, chemical networks or possibly long-range electric forces (Fröhlich, 1968).

REFERENCES

Amari, S. (1971). Characteristics of Randomly Connected Threshold Element Networks and Network Systems. Proc. IEEE. 59: 35-47.

Amari, S. (1977). Competition and Cooperation in Neural Nets. In: Systems Neuroscience (J. Metzler, ed.), Academic Press, New York.

Anninos, A., Beek, B., Csermely, T. J., Harth, E. M. and Pertile, G. (1970). Dynamics of Neural Structures. J. theor. Biol. 26: 121-148.

Denker, J. S., ed. (1986). Neural Networks for Computing. American Institute of Physics: New York.

Finette, S. (1979). Anisotropic Connectivity and Cooperative Phenomena as a Basis for Orientation Sensitivity in the Visual Cortex. Ph.D. dissertation, Syracuse University, Syracuse, N.Y..

Finette, S., Harth, E. and Csermely, T. J. (1978). Anisotropic Connectivity and Cooperative Phenomena as a Basis for Orientation Sensitivity in the Visual Cortex. Biol. Cybernetics 30: 231-140.

Fröhlich, H. (1968). Long Range Coherence and Energy Storage in Biological Systems. Int. J. Quantum Chem. 2: 641-649.

Grossberg, S. (1980). How Does the Brain Build a Cognitive Code? Psych. Rev. 87: 1-51.

Harth, E. M., Csermely, T. J., Beek, B. and Lindsay, R. D. (1970). Brain Functions and Neural Dynamics. J. theor. Biol. 26: 93-120.

Harth, E., Lewis, N. S. and Csermely, T. J. (1975). The Escape of Tritonia: Dynamics of a Neuro-Muscular Control Mechanism. J. theor. Biol. 55: 201-228.

Hogg, T. and Huberman, B. A. (1984). Understanding Biological Computation: Reliable Learning and Recognition. Proc. Natl. Acad. Sci. USA 81: 6871-6875.

Hopfield, J. J. (1982). Neural Networks and Physical Systems with Emergent Collective Computational Abilities. Proc. Natl. Acad. Sci. USA 79: 2554-2558.

Ji, S. (1985). The Bhopalator: A Molecular Model of the Living Cell based on the Concepts of Conformons and Dissipative Structures. J. theor. Biol. 116: 399-426.

Little, W. A., (1974). Existence of Persistent States in the Brain. Math. Biosci. 19: 101-120.

Nicolis, G. and Prigogine, I. (1977). Self-organization in Non-equilibrium systems. Wiley, New York.

Science (1985). An Omnifarious Data Bank for Biology? 228: 1412-1413.

Shastri, L. (1987). Massive Parallelism in Artificial Intelligence. Appl. Opt. 26: 1829-1844.

Shaw, G. L., Harth, E. and Scheibel, A. B. (1982). Cooperativity in Brain Function: Assemblies of Approximately 30 Neurons. Exp. Neurol. 77: 324-358.

Vichniac, G. Y. (1986). Cellular Automata Models of Disorder and Organization. In: Disordered Systems and Biological Organization (E. Bienenstock, ed.), Springer-Verlag, Heidelberg, pp. 3-20.

Wilson, H. R. and Cowen, J. S. (1972). Excitatory and Inhibitory Interactions in Localized Populations of Model Neurons. Biophys. J. 12: 1-24.

Wilson, H. R. and Cowen, J. S. (1973). A Mathematical Theory of the Functional Dynamics of Cortical and Thalamic Nervous Tissue. Kybernetik 13: 55-80.

Wong, R. and Harth, E. (1973). Stationary States and Transients in Neural Populations. J. theor. Biol. 40: 77-106.

II. EXPERIMENTS

Chapter 9

MOLECULAR MECHANISMS OF CELL DEATH

Sten Orrenius

ABSTRACT

Molecular mechanisms involved in the development of lethal cell injury have attracted increased interest in recent years. In particular, the possible existence of a final common pathway in toxic cell killing has been addressed in a number of studies. This presentation summarizes recent work from our laboratory, suggesting that a disruption of intracellular calcium ion homeostasis may represent a common step in the development of toxic damage to hepatocytes, and that activation of Ca^{2+}-dependent degradative enzymes caused by a sustained increase in cytosolic free Ca^{2+} concentration may be responsible for cell killing by agents causing intracellular Ca^{2+} accumulation.

9.1. INTRODUCTION

For many years now our laboratory has been actively engaged in studies of mechanisms of toxic cell injury. Our interest in this area originated from early work on drug biotransformation in isolated, intact cells and on the role of reduced glutathione (GSH) in the cellular defense against electrophilic drug metabolites. In our studies we have used several experimental model systems, including perfused organs, isolated cells and subcellular organelle fractions, and purified enzymes as well as a variety of potentially toxic agents, such as bromobenzene, carbon tetrachloride, t-butyl hydroperoxide, and the redox active compounds, diquat (1,1'-ethylene-2,2'-bipyridilium dibromide) and menadione (2-methyl-1,4-naphthoquinone). Our findings have emphasized the role of glutathione as the predominant intracellular defense mechanism against the toxicity of these agents and suggested that a disruption of intracellular Ca^{2+} homeostasis may trigger their cytotoxicity, once intracellular GSH has been depleted.

The aim of this presentation is to briefly summarize available evidence for a role of Ca^{2+} in toxic cell injury and to discuss possible mechanisms of Ca^{2+}-mediated cytotoxicity.

9.2. CHARACTERISTICS OF TOXIC INJURY TO HEPATOCYTES

It is now generally recognized that most chemicals require metabolic activation to produce toxicity. As a result of their biotransformation reactive intermediates are formed which, normally, are readily inactivated by intracellular defense systems. Whereas the cytochrome P-450-linked monooxygenase system plays a predominant role in the formation of such reactive intermediates, their interaction with intracellular GSH represents the most important defense mechanism. If the balance between metabolic activation and inactivation is perturbed, however, the reactive intermediates can accumulate intracellularly and cause toxicity. Depending on the nature of the reactive intermediates, depletion of intracellular thiols, covalent binding to cellular macromolecules, and lipid peroxidation may result from their intracellular accumulation. A subsequent disruption of intracellular Ca^{2+} homeostasis seems to play an important role in the further development of irreversible cell damage.

9.3. MECHANISMS OF GLUTATHIONE DEPLETION

Glutathione plays a unique role in the cellular defense against toxic chemicals. It is present at high concentrations in most mammalian cells and almost entirely in its reduced form, with glutathione disulfide (GSSG), mixed disulfides, and thioethers constituting minor fractions of the total glutathione pool (Larsson et al., 1983). Depletion of intracellular GSH is known to be one of the most detrimental effects of toxic injury, since loss of glutathione protection against reactive intermediates rapidly leads to the failure of vital cell functions and to cell death.

Conjugation with GSH represents the most important cellular defense mechanism against the toxicity of a wide variety of reactive electrophiles (Figure

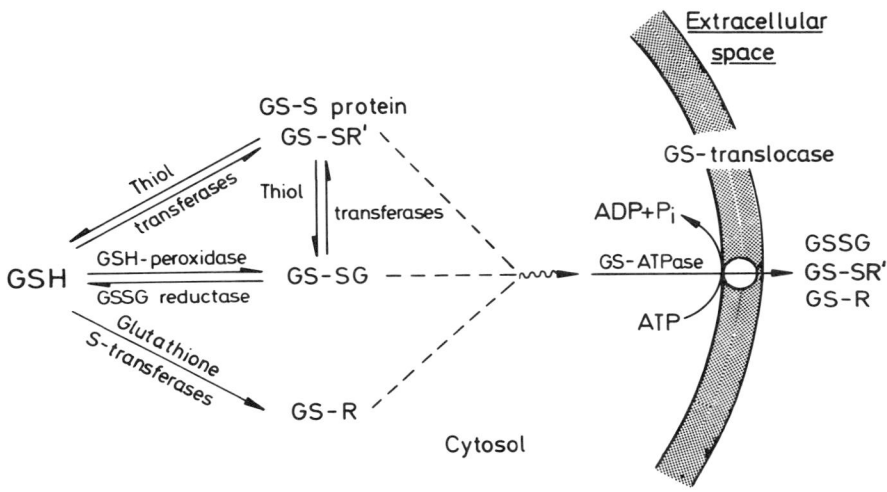

Figure 9.1. Schematic illustration of intracellular reactions leading to GSH depletion. GS-R, glutathione conjugate; GS-SR', soluble mixed disulfides containing glutathione.

9.1). Although GSH can react non-enzymatically with many electrophilic compounds, this process is normally catalyzed by the glutathione transferases (Jakoby, 1980; Larsson et al., 1983). In the liver there are multiple cytosolic glutathione transferases which are all inducible and display overlapping substrate specificities.

The selenoprotein glutathione peroxidase provides protection against H_2O_2 and a variety of organic hydroperoxides (Jakoby, 1980). GSH serves as the electron donor, and the GSSG formed in the reaction is subsequently reduced back to GSH by glutathione reductase at the expense of NADPH. Under conditions of oxidative stress, when the cell must cope with large amounts of H_2O_2 and/or organic hydroperoxides, the glutathione reductase is unable to keep up with the rate of glutathione oxidation, and GSSG accumulates. In an apparent effort to avoid the detrimental effects of increased intracellular levels of GSSG, the cell activity excretes the disulfide which can lead to a depletion of the intracellular glutathione pool (Figure 9.1). The increase in intracellular GSSG concentration, which seems to be a prerequisite for the efflux to occur, is most likely due to insufficient regeneration of NADPH (Eklöw et at., 1984).

9.4. ROLE OF PROTEIN THIOL MODIFICATION

Although the presence of free sulfhydryl groups in proteins has been recognized since the beginning of this century, interest in the role they may play as highly reactive functional groups in biological systems arose only after the discovery of glutathione. A large number of enzymes, catalyzing a wide variety of reactions, are now known to depend on free sulfhydryl groups for their activity, and it is therefore not surprising that modification of protein thiol groups can result in severe functional damage in biological systems.

Thiol groups are highly reactive and participate in several different reactions, such as alkylation, arylation, oxidation, thiol-disulfide exchange, etc. All of these reactions may be involved in the modification of protein thiols resulting from the interaction with reactive intermediates formed during the metabolism of toxic chemicals. Thus, although the toxicological implications of "covalent binding" (e.g., alkylation, arylation) of reactive intermediates to various proteins have often been emphasized, it is clear that thiol oxidation and mixed disulfide formation may also interfere with normal protein function. Such modifications of protein thiols may be particularly important during oxidative stress (Di Monte et al., 1984).

9.5. APPEARANCE OF SURFACE BLEBS IN HEPATOCYTES EXPOSED TO TOXIC AGENTS

Incubation of hepatocytes with a variety of toxic agents results in alterations in surface morphology characterized by a loss of microvilli and appearance of multiple blebs on the surface of the hepatocytes (Jewell et al., 1982). These changes, which are illustrated in Figure 9.2, usually appear before any signs of increased plasma membrane permeability are observed and seem to be reversible initially, *i.e.* they disappear when the toxic agent is removed from the incubation. The formation of surface blebs does not seem to be related to cell swelling caused by increased plasma membrane permeability, since it in not affected by changes is the osmolarity of the incubation medium (Smith et al., 1984).

Cell surface morphology is thought to be determined by the organization of cortical microfilaments associated with the plasma membrane (Cheung, 1980 and references therein). This assumption is supported by the finding that two classes of compounds, the cytochalasins and phalloidins, which disrupt cortical microfilament

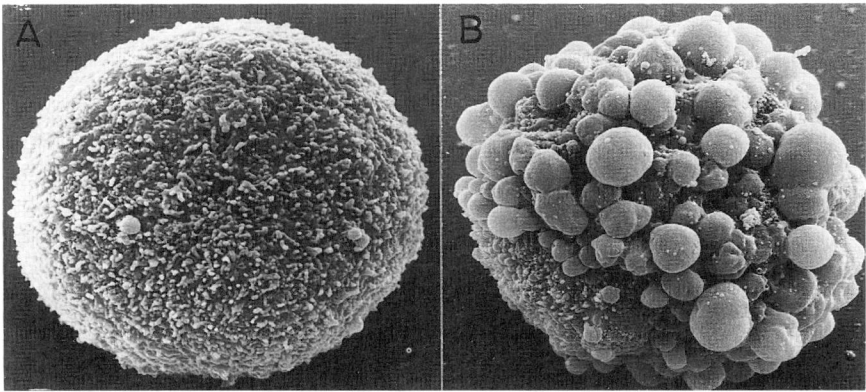

Figure 9.2. Scanning electron micrographs of hepatocytes incubated in absence (A) or presence (B) of menadione. Magnification X 3000. The photomicrograph illustrate the formation of numerous surface blebs as result of menadione toxicity. See Jewell et al., 1982 for experimental details. Copyright 1982 by the AAAS.

structure, cause bleb formation on the surface of hepatocytes similar to that observed with other toxic agents. Since there is no evidence for direct interaction of the toxic metabolites with cytoskeletal components, it appears likely that the observed abnormalities are produced indirectly by alterations in levels of regulatory cofactors or ions.

The polymerization of monomeric to filamentous form of actin is dependent upon ATP; one molecule of bound ATP is converted to ADP for every monomeric actin subunit polymerized. Thus, one would expect that, in the hepatocytes, ATP depletion would result in actin depolymerization, breakdown of the actomyosin network, and plasma membrane blebbing. Inhibition of ATP synthesis by treatment of isolated hepatocytes with antimycin A is associated with extensive plasma membrane blebbing which precedes loss of cell viability. However, these alterations in surface morphology occur before ATP depletion does and are better correlated with alterations in intracellular Ca^{2+} distribution following antimycin A-induced release of mitochondrial Ca^{2+} (Smith et al., 1984). A change in intracellular Ca^{2+} distribution can alter hepatocyte cytoskeletal structure because Ca^{2+} and its associated binding proteins play a pivotal role in regulating cytoskeletal structure (Cheung, 1980). Furthermore, the observation that the calcium ionophore A23187 produces plasma membrane blebbing in isolated hepatocytes supports the assumption that the alterations in surface morphology observed during oxidative stress are related to perturbation of intracellular Ca^{2+} homeostasis (Jewell et al., 1982).

9.6. DISRUPTION OF INTRACELLULAR Ca^{2+} HOMEOSTASIS BY TOXIC AGENTS

The low resting concentration of Ca^{2+} in the cytosol of hepatocytes is maintained by active compartmentation processes and by calcium binding to specific proteins, including calmodulin (Figure 9.3). Mitochondrial Ca^{2+} homeostasis is regulated by a cyclic mechanism, involving Ca^{2+} uptake by an energy-dependent pathway and Ca^{2+} release which is probably mediated by a Ca^{2+}/H^+ antiporter. The latter appears to be regulated by the redox level of intramitochondrial pyridine nucleotides (Lehninger et al., 1978), although membrane-bound protein thiols may also be important in modulating mitochondrial Ca^{2+} fluxes. The active transport of calcium ions into the endoplasmic reticulum and through the plasma membrane is mediated by Ca^{2+}-stimulated, Mg^{2+}-dependent ATPases which appear to depend on free sulfhydryl groups for activity (Moore et al., 1975; Bellomo et al., 1983).

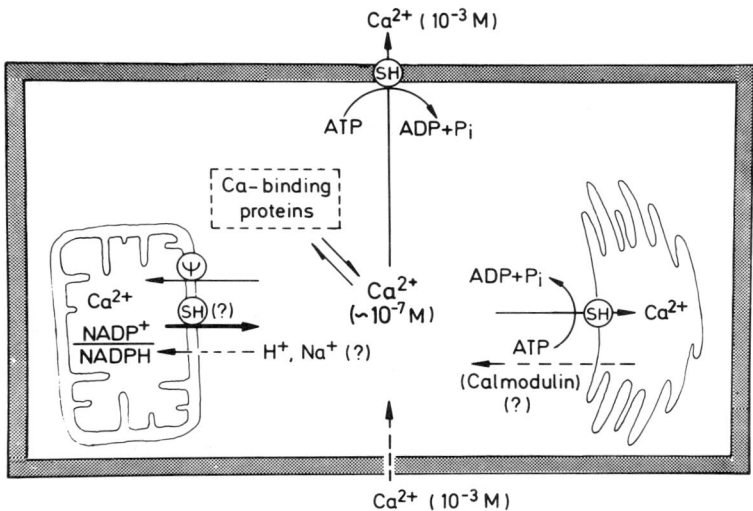

Figure 9.3. Schematic representation of the regulation of Ca^{2+} compartmentation in hepatocytes.

Molecular Mechanisms of Cell Death

The availability of non-invasive techniques to measure Ca^{2+} content in intracellular compartments has made it possible to monitor alterations in Ca^{2+} compartmentation during the development of toxicity in hepatocytes (Bellomo et al., 1982). Under normal conditions freshly isolated rat hepatocytes contain approximately 3 nmol/10^6 cells of exchangeable Ca^{2+}; ~60 % of this Ca^{2+} is sequestered in the mitochondria and ~40 % in the endoplasmic reticulum. The concentration of cytosolic free Ca^{2+} is normally very low (~150 nM).

As shown in Table 9.1, incubation of hepatocytes with various toxic agents is associated with alterations in the intracellular Ca^{2+} pools. Common to all agents listed in the table is their ability to produce a sustained increase in cytosolic free Ca^{2+} concentration. This effect precedes the appearance of surface blebs and the loss of cell viability and seems to be a critical event in the development of toxicity.

As briefly mentioned above, the hepatic plasma membrane Ca^{2+}-ATPase contains functional sulfhydryl group(s), the modification of which is associated with inhibition of both Ca^{2+} ATPase activity and Ca^{2+} extrusion from hepatocytes (Nicotera et al., 1985). In fact, it has been speculated that the impairment of Ca^{2+} efflux resulting from the interaction between electrophilic intermediates and a critical pool of plasma membrane protein thiols may represent a critical biochemical lesion in the development of hepatotoxicity (Nicotera et al., 1986a).

9.7. MECHANISMS OF Ca^{2+}-MEDIATED TOXICITY

To further investigate the relationship between perturbation of intracellular Ca^{2+} homeostasis and cytotoxicity, we have recently exposed isolated hepatocytes to agents that selectively interfere with the normal Ca^{2+} influx-efflux balance, thereby causing a sustained increase in cytosolic Ca^{2+} level (Nicotera et al., 1986b). As illustrated schematically in Figure 9.4, incubation of hepatocytes with either extracellular ATP or cystamine is associated with an inhibition of Ca^{2+} efflux, whereas treatment of the cells with the calcium ionophore A23187 results in enhanced influx of extracellular Ca^{2+}. All treatments result in a sustained increase in cytosolic Ca^{2+} concentration which is followed by the appearance of surface blebs and, subsequently, the loss of cell viability (Figure 9.5). A previous study has shown that the effects of cystamine are caused by the selective inhibition of Ca^{2+} efflux associated with the formation of mixed disulfides between cystamine and a small pool of superficially located plasma membrane protein thiols (Nicotera et al., 1986a). The mechanism by which extracellular ATP inhibits Ca^{2+} extrusion from hepatocytes is not yet well characterized (Bellomo et al., 1984).

Table 9.1. Effects of Toxic Agents on Ca^{2+} Homeostasis in Hepatocytes[*]

Agents	Mitochondrial Pool	Endoplasmic recticular pool	Cytosolic Pool
Menadione	Decrease	Decrease	Increase
t-Butyl hydroperoxide	Decrease	Decrease	Increase
N-Acetyl-p-benzoquinone imine	Decrease	Decrease	Increase
Carbon tetrachloride	-	Decrease	Increase
Bromobenzene	-	Decrease	Increase
Extracellular ATP	Increase	Increase	Increase
Cystamine	Increase	Increase	Increase

[*] See Bellomo et al., 1982 and 1984; Jewell et al., 1982; Nicotera et al., 1986a and b for experimental details.

The finding of a relationship between a sustained increase in cytosolic Ca^{2+} concentration and toxicity of several agents in hepatocytes, has led to a search for mechanisms by which the increased cytosolic Ca^{2+} could trigger cytotoxicity. In the case of extracellular ATP, cystamine, and ionophore A23187, this appears to occur by the activation of a group of Ca^{2+}-dependent non-lysosomal proteases, since inhibitors of these proteases abolish the stimulation of proteolytic activity and prevent both bleb formation and loss of cell viability in hepatocytes exposed to either of the three agents (Nicotera et al., 1986a,b).

Thus, it appears that the toxicity of certain agents in hepatocytes is related to a disruption of intracellular Ca^{2+} homeostasis and the subsequent activation of Ca^{2+}-dependent degradative processes (Figure 9.6). The applicability of this model to

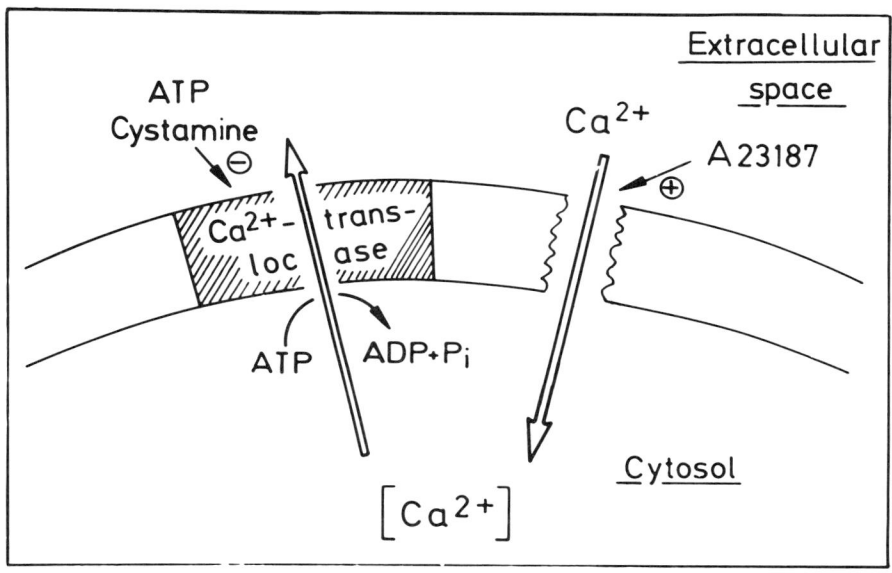

Figure 9.4. Schematic illustration of the modulation of cytosolic Ca^{2+} concentration in hepatocytes by various agents. The figure illustrates the inhibitory effects of extracellular ATP and cystamine on active Ca^{2+} extrusion and the permeabilization of the plasma membrane to extracellular Ca^{2+} by ionophore A23187.

other toxins and tissues is presently being investigated in our laboratory.

9.8. FUTURE PERSPECTIVES

In this overview I have briefly discussed possible mechanisms of chemical toxicity derived from studies with isolated hepatocytes and various model toxins. The importance of a perturbation of intracellular thiol and calcium ion homeostasis for the development of toxicity in this experimental system is obvious. As mentioned above, future investigations will focus on the general applicability and *in vivo* significance of our findings. It will be equally important to establish whether a perturbation of thiol and calcium ion homeostasis is also important in the development of subchronic and chronic toxicity. Should it be possible to demonstrate that a similar sequence of events is responsible for the development of toxicity *in vivo*, this will increase our understanding of the nature of toxic tissue injury and make new therapeutic approaches possible.

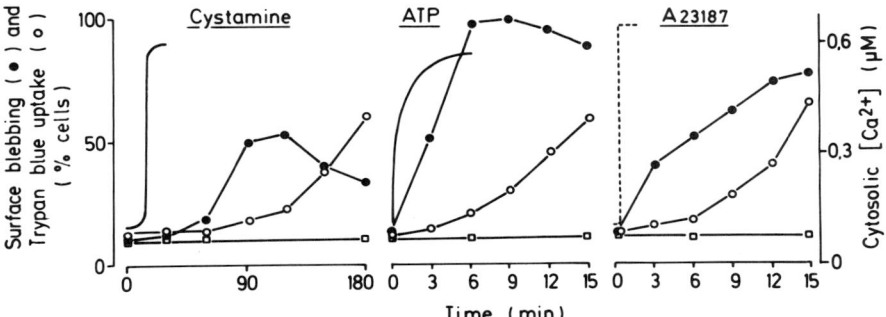

Figure 9.5. Relationship between increase in cytosolic Ca^{2+}, surface blebbing, and cell death in hepatocytes exposed to cystamine, ATP, and ionophore A23187. See Nicotera et al., 1986a and b for experimental details.

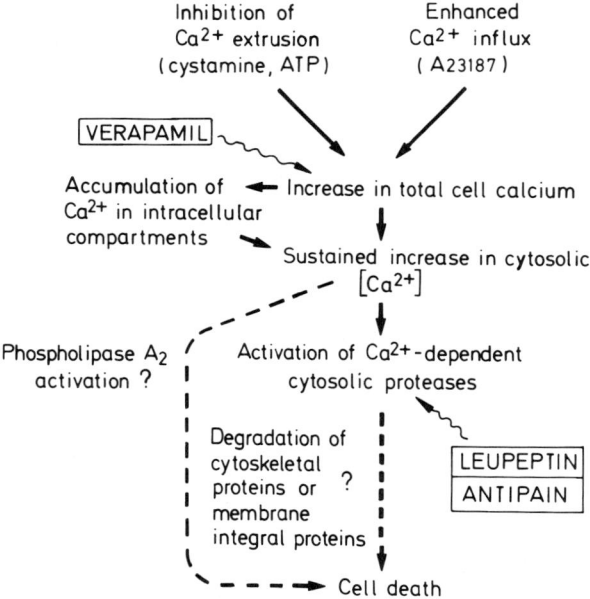

Figure 9.6. Schematic illustration of proposed mechanisms of Ca^{2+}-mediated cytotoxicity in isolated hepatocytes. Probable sites of action of protective agents are also shown. DTT, dithiothreitol.

REFERENCES

Bellomo, G., Jewell, S. A., Thor, H. and Orrenius, S. (1982). Regulation of Intracellular Calcium Compartmentation; Studies with Isolated Hepatocytes and t-butyl Hydroperoxide. Proc. Natl. Acad. Sci. USA 79: 6842-6846.

Bellomo, G., Mirabelli, F., Richelmi, P. and Orrenius, S. (1983). Critical Role of Sulfhydryl Group(s) in ATP-dependent Ca^{2+} Sequestration by the Plasma Membrane Fraction from Rat Liver. FEBS Letters 163: 136-139.

Bellomo, G., Nicotera, P. and Orrenius, S. (1984). Alterations in Intracellular Ca^{2+} Compartmentation Following Inhibition of Ca^{2+} Efflux from Isolated Hepatocytes. Eur. J. Biochem. 144: 19-23.

Cheung, W. Y. (1980). Calmodulin Plays a Pivotal Role in Cellular Regulation. Science 207: 19-27.

Di Monte, D., Bellomo, G., Thor, H., Nicotera, P. and Orrenius, S. (1984). Menadione-induced Cytotoxicity is Associated with Protein Thiol Oxidation and Alteration in Intracellular Ca^{2+} Homeostasis. Arch. Biochem. Biophys. 235: 343-350.

Eklöw, L., Moldéus, P. and Orrenius, S. (1984). Oxidation of Glutathione During Hydroperoxide Metabolism. A Study Using Isolated Hepatocytes and the Glutathione Reductase Inhibitor 1,3-bis(2-chloroethyl)-1-nitrosourea. Eur. J. Biochem. 138: 459-463.

Jakoby, W. B. (1980). Enzymatic Basis of Detoxication. Academic Press, New York, Vol. I & II.

Jewell, S. A., Bellomo, G., Thor, H., Orrenius, S. and Smith, M. T. (1982). Bleb Formation in Hepatocytes During Drug Metabolism is Caused by Disturbances in Thiol and Calcium Ion Homeostasis. Science 217: 1257-1259.

Larsson, A., Orrenius, S., Holmgren, A. and Mannervik, B. (1983). Functions of Glutathione: Biochemical, Physiological, Toxicological and Clinical Aspects. Raven Press, New York.

Lehninger, A. L., Vercesi, A. and Bababunmi, E. (1978). Regulation of Ca^{2+} Release From Mitochondria by the Oxidation-reduction State of Pyridine Nucleotides. Proc. Natl. Acad. Sci. USA 75: 1690-1694.

Moore, L., Chen, T., Knapp, H. R. and Landon, E. (1975). Energy-dependent Calcium Sequestration Activity in Rat Liver Microsomes. J. Biol. Chem. 250: 4562-4568.

Nicotera, P., Moore, M., Mirabelli, F., Bellomo, G. and Orrenius, S. (1985). Inhibition of Hepatocyte Plasma Membrane Ca^{2+}-ATPase Activity by Menadione Metabolism and its Restoration by Thiols. FEBS Letters 181: 149-153.

Nicotera, P., Hartzell, P., Baldi, C., Svensson, S.-Å., Bellomo, G. and Orrenius, S. (1986a). Cystamine Induces Toxicity in Hepatocytes Through the Elevation of Cytosolic Ca^{2+} and the Stimulation of a Non-lysosomal Proteolytic System. J. Biol. Chem. 261: 14628-14636.

Nicotera, P., Hartzell, P., Davies, G. and Orrenius, S. (1986b). The Formation of Plasma Membrane Blebs in Hepatocytes Exposed to Agents that Increase Cytosolic Ca^{2+} is Mediated by the Activation of a Non-lysosomal Proteolytic System. FEBS Letters 209: 139-145.

Smith, M. T., Thor, H., Jewell, S. A., Bellomo, G., Sandy, M. S. and Orrenius, S. (1984). In: Free Radicals in Molecular Biology, Aging and Disease. Raven Press, New York, pp. 103-118.

Chapter 10

DAMAGE TO THE NUCLEUS AND ACUTE CELL DEATH PRODUCED BY ALKYLATING HEPATOTOXINS

Sidhartha D. Ray and George B. Corcoran

ABSTRACT

Hepatocellular necrosis arises under a relatively wide range of pathological conditions. In most cases, acute cell death occurs indirectly over a finite span of time, delayed from the point of initial cell injury and amidst a backdrop of diverse and complex cellular responses. Present strategies to discover the critical steps in cell death stem from the belief that lethal damage results from the loss of one or more compartmentalized cellular function. Much recent progress in understanding hepatocellular necrosis and acute lethal damage has come from studies that examine the plasma membrane, the endoplasmic reticulum, the cytoplasm, and the nucleus. This chapter emphasizes investigations that concentrate on these compartments and particularly those reports that consider the nucleus for changes that may be critical to acute lethal injury produced by alkylating hepatotoxins.

10.1. INTRODUCTION

A greater understanding of toxic reactions that injure the cell and of the succeeding responses mounted by the cell in reaction to injury will ultimately give us the means to grasp fundamental limitations imposed upon all living systems, including those of disease and senescence. Cells of broadly different types and organization

Damage to the Nucleus and Acute Cell Death

show underlying similarities in how they respond to injurious stimuli. This includes cells ranging from plant to animal in origin, and from unicellular organisms to highly differentiated cells dedicated to specialized functions. As a result of this general uniformity in cellular response to injury, leading theories of cell death are reductionist in nature. Theories emphasize conceptual simplification and seek unifying mechanisms that can extend to most living cell types.

Cell death arises under a wide variety of physical and biological conditions. From an operational standpoint, the killing of some cells can be viewed as occurring via direct means whereas the killing of others takes place via indirect means. Instances of cell death are known in which the initial injury kills cells directly, such as when physical or mechanical forces disrupt the cell. However, most acute cell death occurs indirectly, over a finite span of time. Death is usually delayed from the point of initial injury and occurs later amidst a backdrop of cellular responses that are as diverse as they are complex. After damage by a toxin which will ultimately result in cell death, changes that are responsible for cell death are joined and effectively concealed by a host of powerful but futile stimuli from a network of control systems that attempt to return the cell to homeostasis. While the control network strives to restore basal condition, systems that are devoted to cellular repair and regeneration also become engaged through mechanisms that remain largely undefined. Thus, it is apparent why most cell responses during lethal injury are believed to be unrelated to the ultimate cause of cell death. Only a select response or subset of responses is considered to be capable of producing permanent irreversible damage and the ultimate loss of cell viability.

There has been a great deal of progress in characterizing specific *secondary* responses that could be integral to indirect cell death. Advances have occurred despite the sizable number of unrelated biochemical and morphological changes that are provoked by lethal insult. Theoretical and mechanistic experiments to identify critical secondary responses and to demonstrate cause-and-effect relationships have grown in number and substance over the past several decades. This is particularly true for the period of the 1980s, when *in vitro* models of acute lethal injury have matured in quality and opened new avenues of sophistication in hypothesis testing. These *in vitro* systems remain at the forefront of research into cell death and allow for a highly systematic evaluation of specific molecular sites during cell killing. The driving force behind the need to understand cell death comes from related but far-reaching medical disciplines. The benefits of a comprehensive knowledge of cell death are not easily predicted but clearly stand to be immense.

Research into lethal cell injury has offered many new insights as well as raised new questions about this fundamental process. Recent studies that are casting new light on the mechanism(s) of cell death have attempted to relate damage of a putative critical function to the loss of cell viability, most often *in vitro*. In light of the breadth of this topic and the large amount of data addressing concepts in this field, the remainder of this discussion will consider primarily the liver, because this organ is a principal target of injury, and consequently of research in this area. The focus will be on toxins that kill hepatic parenchymal cells presumably due to their ability to produce alkylation damage. For the purpose of this discussion, the term alkylation will be used to indicate covalent modification of tissue through the addition of alkyl or aryl groups.

10.2. COMPARTMENTATION OF MOLECULAR DAMAGE LEADING TO ACUTE CELL DEATH IN THE LIVER

A large number of agents with sparingly few similarities damage the liver acutely. The effects of these agents have been studied carefully, both *in vivo* and *in vitro*. Common to both past and present research is the careful consideration given to compartmentation of cell function. Many research strategies have been founded upon the notion that lethal damage results from the loss of a specific compartmentalized function. Given that compartmentation is vital to cell organization and the hierarchical nature of cell function, researchers have attempted to implicate events from particular organelles in the search for answers to the vexing question of acute cell death. Table 10.1 summarizes information on the subcellular compartmentation of damage by selected drugs and chemicals that alkylate liver cells and produce acute hepatocellular necrosis. Due to increasing efforts to relate acute lethal damage to events in the plasma membrane, the endoplasmic reticulum, the cytoplasm, and the nucleus, the discussion that follows will highlight developments that concentrate upon these subcellular compartments. The reader is referred to the large number of summaries of past work addressing mechanisms and cellular sites not developed here (Keppler, 1975; Popper and Schaffner, 1960; Slater, 1978; Zimmerman, 1978; Farber and Fisher, 1979; Zakim and Boyer, 1982; Arias et al., 1988).

10.2.1. Plasma Membrane

Changes in the plasma membrane are among the most striking to be documented during acute cell injury leading to cell death *in vitro* and *in vivo*. The plasma membrane delimits the contents of the mammalian cell from a markedly different extracellular fluid that bears a close resemblance to plasma. The selective

Table 10.1. Subcellular compartmentation of damage by selected drugs and chemicals that alkylate liver cells and produce hepatocellular necrosis

Cell compartment	Hepatotoxin	References
Plasma membrane	Acetaminophen	Moore et al., 1985; Corcoran et al.,1987a
	CCl_4	Berger et al., 1986
	Ethanol	Lieber & Rubin, 1968; Gonzalez-Calvin, 1985
	Menadione	Nicotera et al., 1985
Endoplasmic reticulum and cytoplasm	Acetaminophen	Corcoran et al., 1987b, 1988; Moore et al., 1985; Farber et al., 1988; Mitchell et al., 1981
	CCl_4	Recknagel, 1967; Moore & Ray, 1983
	Dimethylnitrosamine	Magee & Swann, 1969; Hong & Yang, 1985; Reitman et al., 1988
	Ethanol & C_1-C_{20} Alcohols	Lieber, 1988; Ray & Mehendale, 1989
Mitochondrion	Acetaminophen	Moore et al., 1985; Meyers et al., 1988; Esterline et al., 1989; Burcham & Harman, 1990
	CCl_4	de Castro et al., 1984
	1,1-Dichloroethylene	Reynolds et al., 1975

(Table 10.1 continued)

	Dimethylnitrosamine	Reitman et al., 1988
	Ethanol	Ray & Ji, 1986
Lysosome	Acetaminophen	Moore et al., 1985; Orrenius et al.,1989
	CCl_4	Slater & Greenbaum, 1966
	Cystamine	Nicotera et al., 1986
Nucleus	Acetaminophen	Corcoran et al., 1990; Holme et al., 1988; Harris et al., 1989
	Dimethylnitrosamine	Naslund & Decken, 1981
	Menadione	McConkey et al., 1989
	Trichloroethylene	Vamvakas et al., 1989

permeability and specialized transport functions of the plasma membrane combine to create the unique intracellular environment required by living cells. Because of its physical location and its role in creating conditions relied upon by control systems that preserve homeostasis, the plasma membrane is thought by some to be the weak link in cell viability. Understandably, the status of the plasma membrane has moved to become a leading consideration for research into acute lethal injury.

Experiments examining menadione toxicity in cultured hepatocytes offer evidence that plasma membrane blebbing precedes the death of liver cells in this *in vitro* model (Jewell et al., 1982). Although the mechanism(s) of deteriorating plasma membrane structure is not clear, blebbing could be related to rising Ca^{2+} concentrations and attack of Ca^{2+}-dependent proteases upon cytoskeletal anchor proteins (see Chapter 9) (Nicotera et al., 1985). Hepatotoxins typified by acetaminophen and menadione deplete the cytosol of soluble thiols, disrupt Ca^{2+} homeostasis, and increase protein S-thiolation (Moore et al., 1985; Nicotera et al., 1985). Faltering plasma membrane Ca^{2+}-ATPase activity following exposure to these and other toxins may stem from membrane protein alkylation and/or thiol oxidation, particularly if this damage occurs to the ATP-dependent Ca^{2+} translocase enzyme itself (Bellomo et al., 1985; Tsokos-Kuhn et al., 1988). Disruption of plasma membrane function by alkylating toxins is not limited to Ca^{2+} transport activity. For example, acetaminophen inhibits the Na^+/K^+-ATPase ion pump very early in the course of hepatic necrosis in the rat (Corcoran et al., 1987a). Transporters of ions other than Ca^{2+} may be important in acute lethal injury to hepatocytes by alkylating drugs.

Not all agree that uncontrolled and therefore rising cytosolic Ca^{2+} is required for initial alkylation or peroxidation damage to be translated into cell surface dysfunction. In studying hypoxic injury to the perfused liver, Lemasters and colleagues find plasma membrane changes and membrane blebbing are early consequences of injury, much like after alkylation (Anundi et al., 1987). Cultured hepatocytes subjected to cyanide and iodoacetate, a chemical model of hypoxia, also show membrane blebbing prior to rupture, but without large changes in cytosolic Ca^{2+} (Lemasters et al., 1987). It is also possible for membrane blebbing to result from accelerated membrane lipolysis. Observations of Farber and colleagues show that the loss of membrane lipids precedes cell death in cultured hepatocytes *in vitro*. Lipid loss is paralleled by activation of lipases such as phospholipase A_2, and is primarily dependent on Ca^{2+} and calmodulin. However, additional data from this group would now suggest that lipolysis and cell death can occur largely independent of changes in cellular Ca^{2+}. Findings from these *in vitro* models of acute lethal injury not involving alkylation are provocative and may have substantial implications for how cell death occurring under other conditions is viewed. The significance of individual *in vitro* observations to the overall process of cell death remains unclear, due in part to the difficult challenge of extrapolating such findings to the *in vivo* setting.

The chemicals listed in Table 10.1 impair functioning of the plasma membrane. It is clear that the mammalian cell membrane is not an independent entity but is engaged in intimate interactions with macromolecular components of the cell cytosol and endoplasmic reticulum. Severe impairment of the plasma membrane should also drastically perturb important intra-organ functions including intercellular communication and conductance via gap junctions. Toxin-induced damage to the plasma membrane remains a leading candidate to explain irreversible changes culminating in acute hepatocellular necrosis.

10.2.2. Endoplasmic Reticulum and Cytoplasm

The comparative ease of isolating the endoplasmic reticulum and cytoplasm has resulted in extensive study of these compartments and their successful exploitation to explain fundamental relationships governing many cell processes. These two subcellular components carry out important functions ranging from basic intermediary metabolism and protein synthesis to the elimination of xenobiotics and detoxification of toxic byproducts from normal cell reactions. Of equal or perhaps greater importance, these compartments preserve various cell conditions, such as a favorable redox potential, which range from permissive to obligatory for the cellular functions that are performed within these locations.

Many changes take place within the endoplasmic reticulum and cytoplasm during acute cellular injury leading to hepatic necrosis. These organelles sustain a high frequency of damage, in part because they contain the enzymes responsible for bioactivation of hepatotoxins to alkylating agents. Of agents the listed in Table 10.1, all are metabolized and detoxified by these organelles. For example, carbon tetrachloride and dichloroethylene damage the endoplasmic reticulum and, notably, the Ca^{2+} pump early en route to producing hepatic necrosis *in vivo* and isolated hepatocyte killing *in vitro* (Moore et al., 1976; Long and Moore, 1988). This may represent an important event in faltering Ca^{2+} regulation and, in conjunction with changes in other organelles, a potentially critical action of this hepatotoxin leading to lethal Ca^{2+} overload. A range of other functions and structures are attacked or altered during lethal injury of parenchymal cells by alkylating hepatotoxins. Among the targets of carbon tetrachloride, acetaminophen, menadione, ethanol, and other agents are basal processes such as protein synthesis and glycogen regulation (Smuckler et al., 1962; Kasbekar et al., 1959; Hinson et al., 1983), and integral cell structures such as the endoplasmic reticulum membrane (Recknagel, 1967) and the cytoskeleton (Baraona and Lieber, 1982; Thor et al., 1988).

A final selected example of how alkylating hepatotoxins may produce potentially lethal damage within the endoplasmic reticulum and cytoplasm involves acetaminophen. Cohen, Khairallah and colleagues have recently prepared an antibody that detects acetaminophen-protein adducts. They have used the antibody to show how protein alkylation by acetaminophen is highly selective (Bartolone et al., 1987). A limited number of hepatic proteins in these fractions can account for a high percentage of covalent binding. Of particular note is a 58 kD cytosolic protein that represents one of the major, preferentially alkylated targets. This protein is alkylated to a much greater extent in older mice which are susceptible to acetaminophen toxicity than in younger mice which are resistant (Beierschmitt et al., 1989). The function of the 58 kD protein is not as yet known. However, it is interesting to speculate that the protein may be involved in one or another critical cell function, or may physically resemble proteins that are critical to cell viability. Should either of

Damage to the Nucleus and Acute Cell Death

these speculations prove to be accurate, the 58 kD protein would be extremely valuable in understanding and predicting lethal cellular injury.

It is conceivable that inhibition of protein synthesis with detachment of ribosomes from the endoplasmic reticulum, or the prolonged stimulation of glycogenolysis plays a role in cellular injury. However, it is currently considered more likely that such changes are unrelated secondary responses rather than critical events in acute cell injury leading to necrosis. Although the same could also be said of damage to the endoplasmic reticulum Ca^{2+} pump, cytosolic proteins, or cytoskeletal macromolecules, there continues to be intense interest in these and various other sites within these compartments as possible molecular keys to unlocking the mystery of acute cellular necrosis.

10.3. CELL DEATH AND THE HEPATOCYTE: THEORIES OF ACUTE CELLULAR NECROSIS

Despite the number of compartmentalized functions within the cell and the diversity in mechanisms that regulate these functions, current efforts to understand acute cell death are converging upon a single, common conceptual framework. This body of thinking stresses the global role that ion regulation plays in continued cell vitality and emphasizes, in particular, the large number of regulatory systems that rely upon the actions of calcium ion. Judah and Rees (1959) and Reynolds and co-workers (Thiers et al., 1960) pioneered the early investigation of acute liver injury. They discovered that Ca^{2+} accumulated during toxic insult by the alkylating hepatotoxin, carbon tetrachloride, and drew attention to a possible link between losses in Ca^{2+} regulation and in cell viability. Current research into the specific vital targets of initiation damage again focuses on ion regulation, particularly on intracellular Ca^{2+}. Trump and colleagues have mapped the stereotypical changes that occur during lethal insults such as ischemia, and conclude that shifts in both Na^+ and Ca^{2+} may be required to bring cells beyond the "point of no return" to acute cell death (Trump et al., 1980, 1981, 1984, 1989).

Thus, thinking about acute cell death over the past 15 years has centered increasingly upon the **"Calcium Hypothesis"** of acute lethal injury. Intracellular Ca^{2+} is strictly regulated within $100\text{-}200 \times 10^{-9}$ M and controls a diverse number of processes. Lost Ca^{2+} regulation would be expected to result in cell death if one or more of these processes is required for cell viability. The calcium hypothesis of cell death is based on a series of complementary observations. Ca^{2+} overload accompanies alkylation-induced liver necrosis *in vivo* (Judah and Rees, 1959; Thiers et al., 1960) and both overload and necrosis are prevented by the calcium-calmodulin antagonists promethazine and imipramine (Rees and Spector, 1961; McLean et al., 1965; de Ferreyra et al., 1985). *In vitro*, removal of extracellular Ca^{2+} prevents the killing of cultured hepatocytes by diverse toxins and points to calcium influx as being required for injury (Schanne et al., 1979). And finally, Ca^{2+} and calmodulin blockers including nicardipine, nifedipine, verapamil, and chlorpromazine prevent experimental liver injury *in vivo* (Garay et al., 1984; Landon et al., 1986).

Reports that different Ca^{2+} control sites are damaged early during liver injury have produced several variations on the Ca^{2+} hypothesis. All emphasize the cytotoxicity of unregulated and therefore elevated Ca^{2+} concentrations. Theories differ on whether rising Ca^{2+} originates from extracellular or intracellular pools. *In vivo* data show damage to the microsomal Ca^{2+} pump after carbon tetrachloride administration and suggest the source may be mainly intracellular (Moore et al., 1976; Lowrey et al., 1981). Orrenius et al. identify the mitochondria as an important source for oxidizing hepatotoxins (Thor et al., 1984). However, substantial evidence implicates extracellular Ca^{2+} as the cytotoxic pool, and Farber and others have proposed that loss of Ca^{2+} extrusion at the plasma membrane is a common final pathway of cell death (Schanne et al., 1979; Farber, 1979, 1981; Schiessel et al., 1984; Tsokos-Kuhn et al., 1988). Subsequent work suggests that microsomal, mitochondrial and extracellular pools of Ca^{2+} may each contribute to cytotoxicity *in vitro* (Bellomo and Orrenius, 1985; Moore et al., 1985).

The breadth of intellectual appeal and the ongoing development of calcium hypotheses are evidenced in numerous recent reviews (Thomas and Reed, 1989; Orrenius et al., 1989; Boobis et al., 1989; Trump et al., 1989). Nonetheless, many remaining gaps need to be closed before this theory can be considered conceptually or functionally complete. Most of the challenges and further development of the hypothesis relate not to the premise that Ca^{2+} changes are important, but to details of lost Ca^{2+} regulation and subsequent effects. The identity and the location of failed Ca^{2+} regulators that are critical to lethal cell damage remain to be established, and the ultimate molecular target of the pathogenic effect of excessive Ca^{2+} activity has not been explained.

10.4. THE NUCLEUS AS A CRITICAL SITE OF LETHAL CELL INJURY

Advances in our quest to understand acute lethal injury at the molecular level have built on discrete, novel observations. Each successive observation has stimulated a flurry of activity related to that biochemical or regulatory change, and has led to heightened interest in the implicated subcellular organelle. Over time, attention has fallen on mitochondria, lysosomes, and most other subcellular

Damage to the Nucleus and Acute Cell Death

organelles. Notably absent from this list has been the nucleus. Although it is clear that this organelle is critical to such important events as mutations and to the development of cancers, the nucleus has not received serious consideration in the past as a site of critical injury leading to necrosis. This situation has changed within just the past several years. During this time, several discoveries have placed the nucleus and genomic DNA at the center of activity and speculation. The study of acute cell death in model systems that are not necessarily not related to alkylation insult, such as the regression of immature lymphoid tissue and the oxidative damage of vascular endothelial cells, has offered several new insights that have been instrumental in the turnabout. Cell killing in these *in vitro* systems appears to rely upon the early and extensive fragmentation of genomic DNA or upon the activation of nuclear enzymes that modify proteins involved in repairing DNA or maintaining nucleosomal structure. In addition, it has been known for some time that hepatotoxins have the capacity to alkylate DNA directly and, more recently, that extensive changes in the mobility including the possible loss of large genomic DNA precede liver necrosis *in vivo*.

In considering the events described above, an most important stimulus for increased interest in the role of the nucleus in acute lethal injury has come from the *in vitro* models of thymocyte and endothelial cell death. Wyllie (1980) and Cohen and Duke (1984) show rather convincingly that chromatin undergoes cleavage quite early during the killing of immature thymocytes by glucocorticoids. Activation of a Ca^{2+}-sensitive endonuclease appears to be an obligatory step in the appearance of DNA cleavage. Investigators highlight the role that Ca^{2+} plays in the process and propose that faltering Ca^{2+} regulation at other sites within the cell is needed for glucocorticoids to produce cell killing in this model. In looking at killing of thymocytes by the immunotoxin 2,3,7,8-tetrachlorodibenzo-*p*-dioxin (TCDD), Orrenius and co-workers conclude that this environmental chemical also promotes endonuclease-mediated DNA damage in a Ca^{2+}-dependent manner (McConkey et al., 1988a, 1989). These discoveries raise a number of important questions. How do the mechanisms of acute cell killing gleaned from these *in vitro* models relate to cell death 1) from other forms of cell insult such as alkylation, 2) in other cell types such as the hepatocyte, and 3) under more realistic conditions such as in the living animal *in vivo* ?

Recent experiments with the hepatotoxin acetaminophen offer some insights into these important but unresolved questions. Dybing and colleagues show that acetaminophen directly alkylates genomic DNA both *in vivo* and *in vitro*, and that the analgesic produces genotoxic effects in a variety of test systems (Dybing et al. 1984). These findings have now been extended to show inhibition of DNA synthesis, and increases in single strand breaks and sister chromatid exchanges following exposure of V79 chinese hamster cells to acetaminophen (Hongslo et al., 1988). Although some of these events appear relatively late and may relate more to the carcinogenic rather than the hepatotoxic potential of acetaminophen, they are important in that they demonstrate the ability of acetaminophen to damage DNA directly or to inhibit systems that maintain DNA integrity.

In recent years, our laboratory has been examining whether the highly integrated process of acetaminophen-induced cell necrosis *in vivo* resembles aspects of cell death *in vitro*, particularly as described for the lymphoid model using immature thymocytes. A 600 mg/kg dose of acetaminophen in mice produces > 100 fold increases in alanine aminotransferase activity in plasma from 4 hours onward, significant accumulation of bound plus free Ca^{2+} in the nucleus at 2-3 hours, and a sustained increase in the fragmentation of nuclear DNA from 2 hours onward (Ray et al., 1990). Ca^{2+} accumulation, DNA fragmentation, and decreased nuclear DNA recovery all appear prior to extensive enzyme leakage into plasma and cellular necrosis. The timing of these events suggests that faltering Ca^{2+} regulation throughout the cell, but particularly in the nucleus, leads to activation of an endonuclease very early during necrosis, as has been described for thymocyte killing *in vitro*. Regression analysis to examine the relatedness of these events shows strong linear correlations between nuclear Ca^{2+} accumulation and alanine aminotransferase activity ($r = 0.886$, $p < 0.05$), between DNA fragmentation and alanine aminotransferase activity ($r = 0.871$, $p < 0.05$), and an association between DNA fragmentation and nuclear Ca^{2+} accumulation ($r = 0.624$, $p < 0.05$). Results of DNA separation by electrophoresis show changes that are qualitatively similar to but quantitatively different from sedimentation analysis of DNA. By 2 hours, slowly migrating large DNA is virtually absent from isolated nuclei and whole liver homogenates while accumulating DNA fragments often appear in a ladder-like pattern (Corcoran et al., 1990). These findings suggest that during necrosis *in vivo*, there is activation of a constitutive endonuclease in mouse liver nuclei by supraphysiologic Ca^{2+} concentrations prior to attack upon intranucleosomal regions of genomic DNA. McConkey and colleagues describe a similar stimulation of endogenous endonuclease activity in isolated hepatocytes killed by oxidative stress (McConkey et al., 1988b).

In our studies, specific knowledge of nuclear Ca^{2+} activity would be more informative than our measurement of total nuclear Ca^{2+} concentration during acetaminophen injury. Nonetheless, reproducible early increases in total nuclear Ca^{2+} coincide with evidence of endonuclease activation reflected in DNA fragmentation. Ca^{2+} homeostasis seems to be under as careful regulation in the nucleus as elsewhere throughout the cell. The Ca^{2+} concentration gradient present between nucleus and cytoplasm (Williams et al., 1988) implies the existence of

Damage to the Nucleus and Acute Cell Death

structures that transport Ca^{2+} across the nuclear membrane. Skeletal muscle nuclei demonstrate Ca^{2+}-stimulated ATPase activity (Kulikova et al., 1982). Nuclei isolated from rat liver accumulate Ca^{2+} *in vitro* via an ATP and calmodulin dependent process, and contain a constitutive endonuclease that responds to pathophysiological Ca^{2+} concentrations in the submicromolar range by cleaving DNA into periodic intranucleosomal fragments (Jones et al., 1989). In order for significant accumulation of Ca^{2+} and fragmentation of DNA to occur in the nucleus, as observed during our studies, it may be necessary for hepatotoxins such as acetaminophen to impair both plasma membrane and nuclear pumps that provide for Ca^{2+} extrusion.

We do not find overall DNA fragmentation to be extensive at 2-4 hours after acetaminophen overdose. This probably reflects several factors: 1) the number of target cells undergoing DNA fragmentation is small relative to uninjured nontarget cells, 2) the declining yield of nuclei from 2 hours onward produces a downward bias in estimates of fragmentation due to loss of the most severely damaged nuclei, and 3) the relatively insensitive DNA sedimentation assay fails to disclose the full nature of changes in large genomic DNA. The latter limitations are overcome in large part by isolation of DNA from whole liver homogenates and evaluation of this fraction by agarose gel electrophoresis.

Before concluding a discussion of nuclear events that may contribute to acute lethal injury, it is important not to overlook a nuclear enzyme system that appears to participate in cell killing following oxidative insult. The enzyme poly(ADP-ribose) polymerase [EC 2.4.2.30] is thought to participate in a range of functions in the nucleus, including modulating changes in chromatin architecture and in nuclear enzyme activities, and producing lethal depletion of cellular NAD when DNA damage is so great that cell replication might be deleterious to the organism (Gaal et al., 1987). This enzyme system has been implicated in macrophage killing by hydrogen peroxide (Schraufstatter et al., 1986) and in isolated hepatocyte killing by selenium (Garberg et al., 1988). It appears that this enzyme may also play a role in damage of vascular endothelial cells by other oxidants (Autor, 1990). However, this enzyme would appear to make a negligible contribution to thymocyte cell killing *in vitro* by 2,3,7,8-tetrachlorodibenzo-*p*-dioxin (McConkey et al., 1989a). What role this enzyme may play in acute cellular necrosis produced *in vivo* by alkylating hepatotoxins remains to be established.

10.5. OUTLOOK

During acute chemical injury, sustained increases in cytosolic Ca^{2+} activity can inflict a variety of potentially lethal cellular lesions. These include cytoskeletal alterations leading to plasma membrane blebbing (Jewell et al., 1982), activation of phospholipases with resultant membrane damage (Chien et al., 1979), stimulation of Ca^{2+}-dependent neutral proteases (Nicotera et al., 1986), and activation of a constitutive endonuclease in the nucleus (Hewish and Burgoyne, 1973; Chien et al., 1979; Vanderbilt et al., 1982). Which of these lesions or ones yet undiscovered is critical to lethal cellular damage *in vivo* remains unclear. Orrenius and colleagues believe that DNA fragmentation is the lesion responsible for thymocyte killing *in vitro* by glucocorticoids and 2,3,7,8-tetrachlorodibenzo-*p*-dioxin (McConkey et al., 1988, 1989). Interestingly, Long et al. (1989) find no early breakdown of hepatic DNA *in vivo* or *in vitro* during carbon tetrachloride and dichloroethylene toxicity, and Coleman et al. (1989) show that they can dissociate single strand breaks in DNA from cell killing by oxidative stress *in vitro*. This has prompted the former authors to propose that Ca^{2+} concentration probably increases in the cytosol of liver cells but does not rise sufficiently in the nucleus to activate endonucleases. Our results with acetaminophen in mice differ from those observed with halocarbons in rats. It is possible that DNA fragmentation represents little more than a bystander event during lethal cell injury *in vivo*. However, these contrasting data may be an indication that certain hepatotoxins such as acetaminophen, while producing cellular necrosis *in vivo*, activate mechanisms common to those that occur during cell death *in vitro*. Although it is generally believed that sharp biochemical and morphologic distinctions exist between the processes of necrosis and programmed cell death, it may well be that the actual lines of demarcation are less well defined during lethal injury in the whole animal setting.

ACKNOWLEDGMENTS

For work originating from our own laboratory, we wish to express our appreciation for the valuable collaborative input of Dr. Judy L. Raucy and for the excellent technical contributions of Mr. Christopher L. Sorge and Mr. Asadollah Tavacoli. This work was supported in part by National Institute of General Medical Sciences Grant # GM 41564. The authors appreciate suggestions made by Drs. Scott W. Burchiel and William M. Hadley to improve the manuscript.

REFERENCES

Anundi, I., King J., Owen, D.A., Schneider, H. and Lemasters, J.J. (1987). Fructose Prevents Hypoxic Cell Death in Liver. Amer. J. Physiol. 253: G390-G396.

Arias, I.M., Popper, H., Jakoby, W.B., Schachter, D. and D.A. Shafritz, eds. (1988). Liver: Biology and Pathobiology, 2nd Ed., Raven Press, New York.

Autor, A. P. (1990). Intracellular responses in oxidant-exposed vascular endothelial cells. In: Biological

Reactive Intermediates. Molecular and Cellular Effects and Their Impact on Human Health, (Jollow, D. J., Snyder, R. and Sipes, I. G. eds.) Plenum Press, New York (In Press).

Baraona, E. and Lieber, C. S. (1982). Effects of Alcohol on Hepatic Transport of Proteins. Ann. Rev. Med. 33: 281-292.

Bartolone, J. B., Sparks, K., Cohen, S. D. and Khairallah, E. A. (1987). Immunochemical Detection of Acetaminophen-bound Liver Proteins. Biochem. Pharmacol. 36: 1193-1196.

Beierschmitt, W. P., Brady, J. T., Bartolone, J. B., Wyand, D. S., Khairallah, E. A. and Cohen, S. D. (1989). Selective Protein Arylation and the Age Dependency of Acetaminophen Hepatotoxicity in Mice. Toxicol. Appl. Pharmacol. 98: 517-529.

Bellomo, G. and Orrenius, S. (1985). Altered Thiol and Calcium Homeostasis in Oxidative Hepatocellular Injury. Hepatology 5: 876-882.

Berger, M. L., Bhatt, H., Burton, C. and Estabrook, R. W. (1986). CCl_4-Induced Toxicity in Isolated Hepatocytes: The Importance of Direct Solvent Injury. Hepatology 6: 36-45.

Boobis, A. R., Fawthrop, D. J. and Davies, D. S. (1989). Mechanisms of Cell Death. TIPS 10: 275-280.

Burcham, P. C. and Harman, A. W. (1990). Mitochondrial Dysfunction in Paracetamol Hepatotoxicity: In Vitro Studies in Isolated Mouse Hepatocytes. Toxicol. Lett. 50: 37-48.

Chien K. R., Pfau, R. G. and Farber, J. L. (1979). Ischemic Myocardial Cell Injury: Prevention by Chlorpromazine of an Accelerated Phospholipid Degradation and Associated Membrane Dysfunction. Amer. J. Pathol. 97: 505-530.

Cohen, J. J. and Duke, R. C. (1984). Glucocorticoid Activation of a Calcium-dependent Endonuclease Leads to Cell Death. J. Immunol. 132: 38-42.

Corcoran, G. B., Chung, S. -J. and Salazar, D. E. (1987a). Early Inhibition of the Na^+/K^+-ATPase Ion Pump During Acetaminophen-induced Hepatotoxicity in Rat. Biochem. Biophys. Res. Comm. 149: 203-207.

Corcoran, G. B., Wong, B. K. and Neese, B. L. (1987b). Early Sustained Rise in Total Liver Calcium During Acetaminophen Hepatotoxicity in Mice. Res. Comm. Chem. Pathol. Pharmacol. 58: 291-305.

Corcoran, G. B., Bauer, J. A. and Lau, D. T. -W. (1988). Immediate Rise in Intracellular Calcium and Glycogen Phosphorylase a Activities Upon Acetaminophen Covalent Binding Leading to Hepatotoxicity in Mice. Toxicology 50: 157-167.

Corcoran, G. B., Ray, S. D., Sorge, C. L., Braun, E. L., Tavacoli, A. and Raucy, J. L. (1990). Nuclear Ca^{2+} Accumulation and DNA Fragmentation In Vivo During Acetaminophen-induced Liver Injury in Mice. Toxicologist 10: 294.

de Castro, C. R., Bernacchi, A. S., Villarruel, M. C., Fernandez, G. and Castro, J. A. (1984). Carbon Tetrachloride Activation by Highly Purified Liver Mitochondrial Preparations. Agents and Actions 15: 664-667.

de Ferreyra, E. C., de Fenos, E. C. and Castro, J. A. (1985). Late Protective Effects of Several Anti-calmodulin Drugs on Galactosamine-induced Liver Necrosis. Res. Comm. Chem. Pathol. Pharmacol. 47: 289-292.

Dybing, E., Hølme, J. A., Gordon, W. P., Søderlund, E. J., Dahlin, D. and Nelson, S. D. (1984). Genotoxicity Studies with Paracetamol. Mutat. Res. 138: 21-32.

Esterline, R. L., Ray, S. D. and Ji, S. (1989). Reversible and Irreversible Inhibition of Hepatic Mitochondrial Respiration by Acetaminophen and its Toxic Metabolite, N-acetyl-p-benzoquinoneimine (NAPQI). Biochem. Pharmacol. 38: 2387-2390.

Farber, E., and Fisher, M. M. eds. (1979). Toxic Injury of the Liver, Parts A and B, Marcel Dekker, Inc., New York.

Farber, J. (1979). Reactions of Liver to Injury: Necrosis. In: Toxic Injury of the Liver, Part A, (E. Farber and M. M. Fisher, eds.) Marcel Dekker, New York, pp. 215-241.

Farber, J. (1981). The Role of Calcium in Cell Death. Life Sci. 29: 1289-1295.

Farber, J. L., Leonard, T. B., Kyle, M. E., Nakae, D., Serroni, A. and Rogers, S. (1988). Peroxidation-dependent and Peroxidation-independent Mechanisms by Which Acetaminophen Kills Cultured Rat Hepatocytes. Arch. Biochem. Biophys. 267: 640-650.

Gaal, J. C., Smith, K. R. and Pearson, C. K. (1987). Cellular Euthanasia Mediated by a Nuclear Enzyme: A Central Role for Nuclear ADP-ribosylation in Cellular Metabolism. TIPS April: 129-130.

Garay, G. L., Annesley, P. and Burnette, M. (1984). Prevention of Experimental Liver Injury in Rats by Nicardipine, a Calcium Entry Blocker. Gastroenterology 86: 1319.

Garberg, P., Ståhl, A., Warholm, M. and Högberg, J. (1988). Studies on the Role of DNA Fragmentation

in Selenium Toxicity. Biochem. Pharmacol. 18: 3401-3406.
Gonzalez-Calvin, G. L., Saunders, J. B., Crossley, I. R., Dickenson, C. J., Smith, H. M., Tredger, M. and Williams, R. (1985). Effects of Ethanol Administration on Rat Liver Plasma Membrane-bound Enzymes. Biochem. Pharmacol. 34: 2685-2689.
Harris, C., Stark, K. L., Luchtel, D. L. and Juchau, M. R. (1989). Abnormal Neurulation Induced by 7-hydroxy-2-acetylaminofluorene and Acetaminophen: Evidence for Catechol Metabolites as Proximate Dysmorphogens. Toxicol. Appl. Pharmacol. 101: 432-446.
Hewish, D. R. and Burgoyne, L. A. (1973). The Calcium Dependent Endonuclease Activity of Isolated Nuclear Preparations. Relationships Between its Occurrence and the Occurrence of Other Classes of Enzymes Found in Nuclear Preparations. Biochem. Biophys. Res. Comm. 52: 475-481.
Hinson, J. A., Mays, J. B. and Cameron, A. M. (1983). Acetaminophen-induced Hepatic Glycogen Depletion and Hyperglycemia in Mice. Biochem. Pharmacol. 32: 1979-1988.
Holme, J. A., Hongslo, J. K., Bjornstad, C., Harvison, P. J. and Nelson, S. D. (1988). Toxic Effects of Paracetamol and Related Structures in V79 Chinese Hamster Cells. Mutagenesis 3: 51-56.
Hong, J. and Yang, C. S. (1985). The Nature of Microsomal N-nitrosodimethylamine Demethylase and its Role in Carcinogen Activation. Carcinogenesis 6: 1805-1809.
Hongslo, J. K., Christensen, T., Brunborg, G., Bjørnstad, C. and Hølme, J. A. (1988). Genotoxic Effects of Paracetamol in V79 Chinese Hamster Cells. Mutat. Res. 204: 333-341.
Jewell S. A., Bellomo, G., Thor, H., Orrenius, S. and Smith, M. T. (1982). Bleb Formation in Hepatocytes During Drug Metabolism is Caused by Disturbances in Thiol and Calcium Homeostasis. Science 217: 1257-1259.
Jones D. P., McConkey, D. J., Nicotera, P. and Orrenius, S. (1989). Calcium-activated DNA fragmentation in Rat Liver Nuclei. J. Biol. Chem. 264: 6398-6403.
Judah, J. D. and Rees, K. R. (1959). Mechanism of Action of Carbon Tetrachloride. Fed. Proc. 18: 1013-1020.
Kasbekar, D. K., Labate, W. V., Rege, D. V. and Sreenivasan, A. (1959). A Study of Vit-B_{12} Protection in Experimental Liver Injury to the Rat by Carbon Tetrachloride. Biochem. J. 72: 384-389.
Keppler, D., ed. (1975). Pathogenesis and Mechanisms of Liver Cell Necrosis University Park Press, Baltimore, MD.
Kulikova O. G., Savostianov, G. A., Beliavsteva, L. M. and Razumovskaia, N. I. (1982). ATPase Activity and ATP-dependent Accumulation of Ca^{2+} in Skeletal Muscle Nuclei. Effects of Denervation and Electrical Stimulation. Biokhimiia 47: 1216-1221.
Landon, E. J., Naukam, R. J. and Rama Sastry, B. V. (1986). Effects of Calcium Channel Blocking Agents on Calcium and Centrilobular Necrosis in the Liver of Rats Treated with Hepatotoxic Agents. Biochem. Pharmacol. 35: 697-705.
Lemasters J. J., DiGuiseppi, J., Nieminen, A. L. and Herman, B. (1987). Blebbing, Free Ca^{2+} and Mitochondrial Membrane Potential Preceding Cell Death in Hepatocytes. Nature 325: 78-81.
Lieber, C. S. and Rubin, E. (1968). Alcoholic Fatty Liver in Man on a High Protein and Low Fat Diet. Amer. J. Med. 44: 200-206.
Lieber, C. S. (1988). Biochemical and Molecular Basis of Alcohol-induced Injury to Liver and Other Tissues. N. Engl. J. Med. 319: 1639-1650.
Long, R. and Moore, L. (1988). Biochemical Evaluation of Rat Hepatocyte Primary Cultures as a Model for Carbon Tetrachloride Hepatotoxicity: Comparative Studies *In Vivo* and *In Vitro*. Toxicol. Appl. Pharmacol. 92: 295-306.
Long, R. M., Moore, L. and Schoenberg, D. R. (1989). Halocarbon Hepatotoxicity is not Initiated by Ca^{2+}-stimulated Endonuclease Activation. Toxicol. Appl. Pharmacol. 97: 350-359.
Lowrey, K., Glende, E. A. and Recknagel, R. O. (1981). Destruction of Liver Microsomal Calcium Pump Activity by Carbon Tetrachloride and Bromotrichloromethane. Biochem. Pharmacol. 30: 135-140.
Magee, P. N. and Swann, P. F. (1969). Nitroso Compounds. Brit. Med. Bull. 25: 240-244.
McConkey, D. J., Hartzell, P., Duddy, S. K., Hakasson, H. and Orrenius, S. (1988). 2,3,7,8-Tetrachlordibenzo-p-dioxin Kills Immature Thymocytes by Ca^{2+}-mediated Endonuclease Activation. Science 243: 256-258.
McConkey, D. J., Hartzell, P., Nicotera, P., Wyllie, A. H. and Orrenius, S. (1988). Stimulation of Endogenous Endonuclease Activity in Hepatocytes Exposed to Oxidative Stress. Toxicol. Lett. 42: 123-130.
McConkey D. J., Hartzell, P., Nicotera, P. and Orrenius, S. (1989). Calcium-activated DNA Fragmentation Kills Immature Thymocytes. FASEB J. 3: 1843-1849.

McLean, A. E. M., McLean, E. and Judah, J. D. (1965). Cellular Necrosis in the Liver Induced and Modified by Drugs. Internat. Rev. Exp. Pathol. 4: 127-157.

Meyers, L. L., Beierschmitt, W. P., Khairallah, E. A. and Cohen, S. D. (1988). Acetaminophen-induced Inhibition of Hepatic Mitochondrial Respiration in Mice. Toxicol. Appl. Pharmacol. 93: 378-387.

Mitchell, J. R., Corcoran, G. B., Smith, C. V., Hughes, H. and Lauterburg, B. H. (1981). Alkylation and Peroxidation Injury From Chemically Reactive Metabolites. In: Biological Reactive Intermediates - II, Snyder, R., Parke, D. V., Kocsis, J. J., Jollow, D. J., Gibson, G. G. and Witmer, C. M. eds., Plenum Press, New York, pp. 199-223.

Moore, L., Davenport, G. and Landon, E. G. (1976). Calcium Uptake of a Rat Liver Microsomal Subcellular Fraction in Response to an In Vivo Administration of Carbon Tetrachloride. J. Biol. Chem. 251: 1197-1201.

Moore, L. and Ray, P. (1983). Enhanced Inhibition of Hepatic Microsomal Calcium Pump Activity by CCl_4 Treatment of Isopropanol-pretreated Rats. Toxicol. Appl. Pharmacol. 71: 54-58.

Moore, M., Thor, H., Moore, G., Nelson, S., Moldeus, P. and Orrenius, S. (1985). The Toxicity of Acetaminophen and N-acetyl-p-benzoquinone Imine in Isolated Hepatocytes is Associated with Thiol Depletion and Increased Cytosolic Ca^{2+}. J. Biol. Chem. 260: 13035-13040.

Naslund, B., and von der Decken, A. (1981). Chain-length Heterogeneity of Nucleosomal DNA in Mouse Liver After Dimethylnitrosamine Administration. Toxicology 47: 169-177.

Nicotera, P., Moore, M., Mirabelli, F., Bellomo, G. and Orrenius, S. (1985). Inhibition of Hepatocyte Plasma Membrane Ca^{2+}-ATPase Activity by Menadione Metabolism and its Restoration by Thiols. FEBS Lett. 181: 149-153.

Nicotera P., Hartzell, P., Baldi, C., Svenson, S. -A., Bellomo, G. and Orrenius, S. (1986). Cystamine Induces Toxicity in Hepatocytes Through the Elevation of Cytosolic Ca^{2+} and the Stimulation of a Non-lysosomal Proteolytic System. J. Biol. Chem. 261: 14628-14635.

Orrenius, S., McConkey, D. J., Bellomo, G. and Nicotera, P. (1989). Role of Ca^{2+} in Toxic Cell Killing. TIPS 10: 281-285.

Popper, H. and Schaffner, F. eds. (1960). Progress in Liver Disease, Volumes 1-9 Grune and Stratton, Inc., New York.

Ray, S. D. and Ji, S. (1986). Extracellular Calcium-dependent and -independent Mechanisms of Acetaminophen Hepatotoxicity. Toxicologist 6: 187.

Ray, S. D. and Mehendale, H. M. (1989). Potentiation of Carbon Tetrachloride Hepatotoxicity and Lethality by Various Alcohols. Toxicologist 9: 59.

Ray, S. D., Sorge, C. L., Tavacoli, A., Raucy, J. L. and Corcoran, G. B. (1990). Extensive Alteration of Genomic DNA and Rise in Nuclear Ca^{2+} In Vivo Early After Hepatotoxic Acetaminophen Overdose in Mice. In: Biological Reactive Intermediates. Molecular and Cellular Effects and Their Impact on Human Health (Jollow, D. J., Snyder, R. and Sipes, I. G. eds.) Plenum Press, New York (In Press).

Recknagel, R. O. (1967). Carbon Tetrachloride Hepatotoxicity. Pharmacol. Rev. 19: 145-208.

Rees, K. R. and Spector, W. G. (1961). Reversible Nature of Liver Cell Damage Due to Carbon Tetrachloride as Demonstrated by the Use of Phenergan. Nature 190: 821-822.

Reitman, F. A., Berger, M. L., Minnema, D. J. and Shertzer, H. G. (1988). Calcium Transport, Thiol Status, and Hepatotoxicity Following N-Nitrosodimethylamine Exposure in Mice. J. Toxicol. Environ. Health 23: 321-331.

Reynolds, E. S., Moslen, M. T., Szabo, S., Jaeger, R. J. and Murphy, S. D. (1975). Hepatotoxicity of Vinyl Chloride and 1,1-dichloroethylene. Amer. J. Pathol. 81: 219-232.

Schanne, F. A. X., Kane, A. B., Young, E. E. and Farber, J. L. (1979). Calcium Dependence of Toxic Cell Death: A Final Common Pathway. Science 206: 700-702.

Schiessel, C., Forsthove, C. and Keppler, D. (1984). ^{45}Calcium Uptake During Transition from Reversible to Irreversible Liver Injury Induced by D-galactosamine In Vivo. Hepatology 4: 855-861.

Schraufstatter, I. U., Hyslop, P. A., Hinshaw, D. B., Spragg, R. G., Sklar, L. A. and Cochrane, C. G. (1986). Hydrogen Peroxide-induced Injury of Cells and its Prevention by Inhibitors of Poly(ADP-ribose) Polymerase. Proc. Nat. Acad. Sci. USA 83: 4908-4912.

Slater, T. F., ed. (1978). Biochemical Mechanisms of Liver Injury. Academic Press, New York.

Slater, T. F. and Greenbaum, A. L. (1966). Changes in Lysosomal Enzymes in Acute Experimental Liver Injury. Biochem. J. 96: 484-491.

Smuckler, E. A., Iseri, O. A. and Benditt, E. P. (1962). An Intracellular Defect in Protein Synthesis Induced by Carbon Tetrachloride. J. Exp. Med. 116: 55-71.

Thiers, R. E., Reynolds, E. S. and Vallee, B. L. (1960). The Effect of Carbon Tetrachloride Poisoning on Subcellular Metal Distribution in Rat Liver. J. Biol. Chem. 235: 2130-2133.

Thomas, C. E. and Reed, D. J. (1989). Current Status of Calcium in Hepatocellular Injury. Hepatology 10: 375-384.

Thor, H., Hartzell, P. and Orrenius, S. (1984). Potentiation of Oxidative Cell Injury in Hepatocytes Which Have Accumulated Calcium. J. Biol. Chem. 259: 6612-6615.

Thor, H., Mirabelli, F., Salis, A., Cohen, G. M., Bellomo, G. and Orrenius, S. (1988). Alterations in Hepatocyte Cytoskeleton Caused by Redox Cycling and Alkylating Quinones. Arch. Biochem. Biophys. 266: 397-407.

Trump, B. F., Berezesky, I. K., Laiho, K. U., Osornio, A. R., Mergner, W. J. and Smith, M. W. (1980). The Role of Calcium in Cell Injury. A Review. In: Scanning Electron Microscopy (Becker, R.P. and Johari, O. eds.) Scanning Electron Microscopy, Inc., Chicago, pp. 437-462.

Trump, B. F., Berezesky, I. K. and Phelps, P. C. (1981). Sodium and Calcium Regulation and the Role of the Cytoskeleton in the Pathogenesis of Disease. A Review and Hypothesis. Scan. Electron Microsc. 2: 435-454.

Trump, B. F., and Berezesky, I. K. (1984). Role of Sodium and Calcium Regulation in Toxic Cell Injury. In: Drug Metabolism and Drug Toxicity (Mitchell, J. R. and Horning, M. G. eds.) Raven Press, NY, pp. 261-300.

Trump, B. F., Berezesky, I. K., Smith, M. W., Phelps, P. C. and Elliget, K. A. (1989). The Relationship Between Cellular Ion Deregulation and Acute and Chronic Toxicity. Toxicol. Appl. Pharmacol. 97: 6-22.

Tsokos-Kuhn, J. O., Hughes, H., Smith, C. V. and Mitchell, J. R. (1988). Alkylation of the Liver Plasma Membrane and Inhibition of the Ca^{2+} ATPase by Acetaminophen. Biochem. Pharmacol. 37: 2125-2131.

Vamvakas, S., Dekant, W. and Henschler, D. (1989). Genotoxicity of Haloalkane and Haloalkane Glutathione S-conjugates in Porcine Kidney Cells. Toxicol. In Vitro 3: 151-156.

Vanderbilt, J. N., Bloom, K. S. and Anderson, J. N. (1982). Endogenous Nuclease. Properties and Effects on Transcribed Genes in Chromatin. J. Biol. Chem. 257: 13009-13017.

Williams D. A., Becker, P. L. and Fay, F. S. (1988). Regional Changes in Calcium Underlying Contraction of Single Smooth Muscle Cells. Science 235: 1644-1648.

Wyllie, A. H. (1980). Glucocorticoid-induced Thymocyte Apoptosis is Associated with Endogenous Endonuclease Activation. Nature 284: 555-556.

Zakim, D., and Boyer, T. D. eds. (1982). Hepatology. A Textbook of Liver Disease. W. B. Saunders Co., Philadelphia.

Zimmerman, H. L. (1978). Hepatotoxicity. The Adverse Effects of Drugs and Other Chemicals on the Liver. Appleton-Century-Crofts, Inc., New York.

Chapter 11

ELECTRONICALLY EXCITED STATES IN CELLS AND ORGANS: RELATION TO PROOXIDANT/ANTIOXIDANT BALANCE

Helmut Sies

ABSTRACT

Electronically excited states in biological molecules have recently been found to occur. Several sources of excited state conditions have been characterized, including membrane-localized processes such as lipid peroxidation. Since membrane integrity is one of the prerequisites of intactness of cells, the process of lipid peroxidation may become crucial. Electronically excited states include singlet molecular oxygen ($^1O_2$) or excited carbonyls. Methods for detecting these compound have been recently employed with intact cells. It is concluded that, in biology, electronically excited states are of interest, both in physiological and pathophysiological states.

11.1. INTRODUCTION

Reactive oxygen species are formed in normal metabolism and can be considered a normal attribute of aerobic life (Chance et al., 1979; Sies, 1986). The exposure of cells to oxidative conditions of a diverse nature can be accompanied by an elevated production of free radicals. This, in turn, is expressed as an enhanced generation of electronically excited states. Electronically excited states, therefore, can be formed not only photochemically using appropriate photosensitizers, but also in the absence of excitation by light, and this has been called "photochemistry in the dark" (Cilento, 1982). Electronically excited states of biological interest include those of molecular oxygen and carbonyl functions. In terms of the electronic structure, this would be singlet molecular oxygen ($^1\Delta_g O_2$) and triplet carbonyls (RO*). The energy above ground state is sufficient for initiation and maintenance of chemical reactions. Once excited, the compound undergoes what is called photochemical reactions, regardless of whether the excited state was initially formed by light excitation or by dark chemical processes.

Research into the excited states in biological samples like cells and tissues has been greatly advanced by the development of new equipment, in particular photon-counting techniques. The return to the respective ground states leads to the emission of low-level chemiluminescence, with the wavelength at 634 nm and 703 nm for $^1O_2$ and in the visible range (340-460 nm) for excited carbonyls (Figure 11.1). At the present state of knowledge, it appears that the monitoring of photoemission from such electronically excited states can provide useful information during oxidative challenge.

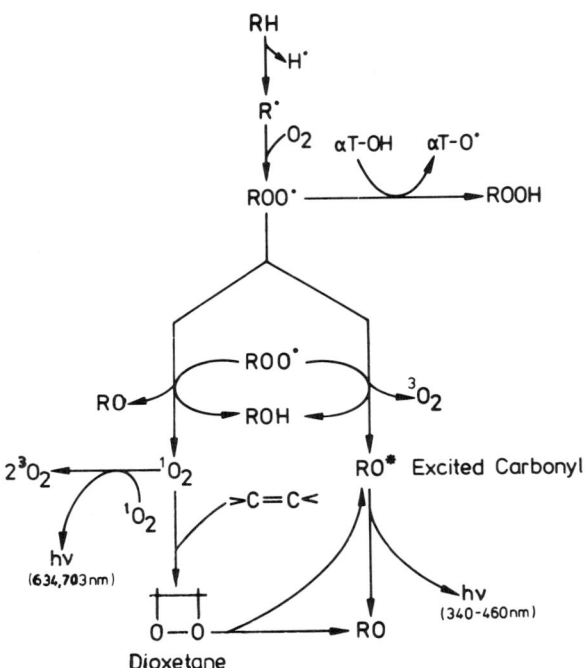

Figure 11.1. Pathway of generation of excited oxygen (singlet oxygen; left-hand branch) or excited carbonyls (triplet carbonyls; right hand branch) from lipid peroxy radicals according to Russell's mechanism. αT-OH signifies α tocopherol.

This means that cell damage is associated with the generation of electronically excited states under aerobic conditions. Whether such excited states can also be a cause of cell damage will be open for further study. In this article, some recent work from our group will be presented. It is clear to the author that the potential role of electronically excited states in cell life is an area of research still in its early stage of development (Slawinska and Slawinski, 1983; Cadenas, 1984). Therefore, the emphasis will be restricted to cell damage and associated processes. Many other aspects, e.g., the excited states of nucleic acids (Daniels, 1983; Piette et al., 1986) or the potential role in mutagenesis (Sargentini and Smith, 1985) will not be treated here. Recently, the current state of knowledge on the biochemistry of oxidative stress has been surveyed (Sies, 1986). Oxidative damage extends to almost all major classes of compounds, DNA, protein, lipids and carbohydrates.

11.2. LOW-LEVEL CHEMILUMINESCENCE

The detection of light emission from biological samples is a useful method for studying oxidative reactions in intact systems (Boveris et al., 1979; Cadenas and Sies, 1984, 1985a, 1985b; Sies and Cadenas, 1985; Wefers, 1987). As indicated above, the generation of electronically excited states during intracellular oxidative conditions can result from free-radical interactions that may or may not be associated with the peroxidation of membrane fatty acids (lipid peroxidation). Alternatively, the direct generation of excited states can occur in enzyme-catalyzed reactions (Cilento, 1982) as demonstrated in model systems using peroxidases. The excited states discussed are singlet molecular oxygen as measured by photoemission during its decay to the triplet state in the monomol (Reaction 11.1) or dimol (Reaction 11.2) reactions:

$$^1O_2 \rightarrow {^3O_2} + h\nu(1270nm), \tag{11.1}$$

$$2\ {^1O_2} \rightarrow 2\ {^3O_2} + h\nu(634\ and\ 703nm) \tag{11.2}$$

and (b) excited triplet carbonyls (RO*) exhibiting a weak emission in the blue-green region of the spectrum (Reaction 11.3) or an indirect emission after energy transfer to a suitable acceptor A (Reaction (11.4a)), thus eliciting sensitized emission (Reaction (11.4b)):

$$RO^* \rightarrow RO + h\nu \tag{11.3}$$

$$RO^* + A \rightarrow RO + A^* \tag{11.4a}$$

$$A^* \rightarrow A + h\nu \tag{11.4b}$$

Interactions of lipid peroxy radicals can be a source of excited state(s) as in Reactions (11.5a,b) (Russell 1957; Howard and Ingold 1968). Because peroxy radicals are produced at the final stages of lipid peroxidation (Figure 11.1), they might be considered as the common mechanisms for chemiluminescence (Reactions 11.5a,b) shared by different oxidative conditions which promote lipid peroxidation: CCl_4 poisoning, iron overload, oxidative breakdown of hydroperoxides, hyperoxia, etc.

$$ROO\cdot + ROO\cdot \rightarrow ROH + RO + {}^1O_2 \tag{11.5a}$$

$$ROO\cdot + ROO\cdot \rightarrow ROH + RO^* + O_2 \tag{11.5b}$$

Free-radical interactions supporting redox cycling, however, elicit chemiluminescence that is not associated with lipid peroxidation. Although $O_2^{\doteq}$ and HO· are generally produced during the activation of different xenobiotics by redox cycling, a direct link between them and the electronically excited state(s) generated remains to be determined. The fact that redox cycling-supported photoemission can be inhibited by superoxide dismutase indicates that the $O_2^{\doteq}$ generated is required at some stage for the generation of photoemission (Wefers and Sies, 1986).

The molecular mechanisms for the production of singlet oxygen during several peroxidase- or peroxidase-like reactions seems to follow the dismutation of hydroperoxides without involving free radical intermediates. The production of triplet excited carbonyl compounds during the peroxidase-catalyzed oxidation of aliphatic aldehydes to their lower analogs (Cilento, 1982) is thought to proceed via the formation of a dioextane intermediate.

11.3. REDOX CYCLING

The molecular mechanism for activation of certain xenobiotics (often of quinone structure) involves the univalent reduction of the compound via formation of the superoxide anion radical and, subsequently, other oxygen radical-derived species (Kappus & Sies, 1981; Doroshow & Hochstein 1982; Wefers & Sies, 1986). This reduction is accomplished by different enzymatic activities present in the endoplasmic reticulum, mitochondria and nuclei of different tissues. O_2 is required for cytotoxicity; $O_2^{\doteq}$, H_2O_2, HO· and ${}^1O_2$ were thought to be the species responsible for cell damage, whereas the radical form of the xenobiotic appeared less likely to be responsible for cytotoxicity. The relative contribution by the quinone itself remains to be evaluated. The one-electron reduction of the redox cyclers that leads to $O_2^{\cdot-}$ and subsequent hydrogen peroxide generation provides substrate for GSH peroxidase. Hence, there is an increase in cellular GSSG and an increase in mixed disulfides. These observations suggest that substantial losses of GSH are caused by intracellular redox cycling. Thus, the cellular glutathione system is related to the oxidative challenge occurring during redox cycling. Some special aspects of GSH involvement have recently been studied and will not be discussed here (Medeiros et al., 1987; Napetschnig and Sies, 1987; Wefers et al., 1985).

As one example of cell damage through redox cycling, we have studied menadione and the generation of photoemissive species. When menadione was infused into the isolated perfused rat liver via the portal vein, there was an increase in red photoemission that was attributed to singlet oxygen (Wefers and Sies, 1983). The intensity of photoemission was increased when the two-electron reduction capacity that is exhibited by NADPH: quinone oxidoreductase (DT diaphorase) was restricted. This can be effected by the addition of an inhibitor, dicoumarol. As shown in similar experiments in isolated cell fractions (S 9), the effect of menadione is shown to be increased when dicoumarol is present (Figure 11.2).

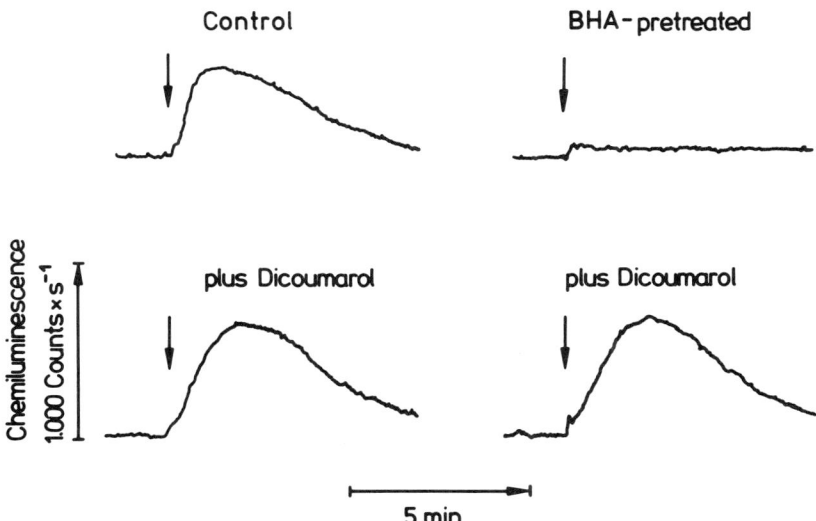

Figure 11.2. Menadione-induced low-level chemiluminescence (beyond 610 nm) of postmitochondrial supernatant fractions of mouse liver. Controls (left-hand side) and animals treated with butylated hydroxy anisole (right-hand side). Menadione (100 μM) was added at the time indicated by the arrows. NADPH (0.4 mM) has been added 1 min prior to menadione. Protein content: 1.5 mg/ml; dicoumarol 30 μM. From Wefers et al. (1984).

The protective effect of the enzyme is substantiated in the right-hand side of Figure 11.2 where the level of NADPH: quinone oxidoreductase was increased more than 10-fold (Wefers et al., 1984). This induction of the level of the protective enzyme was obtained by feeding an antioxoidant, BHA, in the diet. More recently, the addition of pure enzyme to the measuring cuvette provided final proof. The full protection was obtained by the readdition of enzyme as predicted from the difference between control animals and BHA-infused animals (Prochaska et al., 1987). Further, the protection by the enzyme was abolished upon the addition of antibody against the enzyme (Prochaska et al., 1987). These experiments show that quinone toxicity and protection against it can be monitored by photoemissive species. However, it is not clear whether the photoemissive species themselves are damaging agents or whether they are accompanying the more aggressive species which cause the cell damage.

It should be noted that other parameters of cell damage include those of enzyme leakage and parameters for lipid peroxidation. The latter include the evolution of ethane and pentane as well as malondialdehyde (TBA-reactive material). In several instances, there was a parallel rise in the parameters of lipid peroxidation just mentioned and the low-level chemiluminescence. As shown in Figure 11.3, the generation of excited states in lipid peroxidation can parallel other parameters. However, in the temporal relationships between these different parameters, there are distinct differences. We suggested in a detailed study of comparison of time courses for different parameters of lipid peroxidation that sub-sets of polyunsaturated fatty acids may be recruited sequentially during the process of lipid peroxidation (Noll et al., 1987). In states of very low oxidative challenge, therefore, there may be selective depletions of specific fatty acids within specialized parts of the biological membrane.

11.4. ANTIOXIDANT ACTIVITIES

The protection against generation of excited species is not only exerted by NADPH: quinone oxidoreductase as mentioned above, but also by a large variety of other antioxidants. These have been recently discussed in survey articles (Ishikawa et al., 1986; Sies, 1986). We have been interested in employing different antioxidants in a model system consisting of isolated rat liver microsomes. The lag-time before the onset of lipid peroxidation is a useful measure for antioxidant activity (Bartoli et al., 1983). For example, the evaluation of alpha-tocopherol as an antioxidant in microsomal lipid peroxidation was performed by detecting low-level chemiluminescence (Cadenas et al., 1984a). It was found that the vitamin E added to microsomes was much less effective as antioxidant as compared to a similar

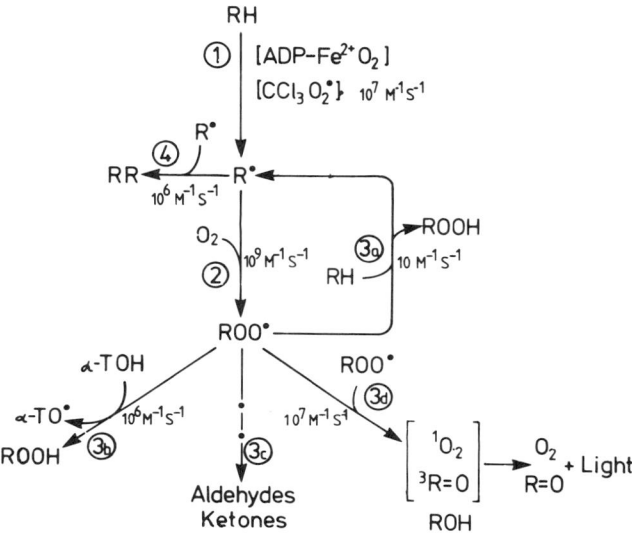

Figure 11.3. Relationship between the process of lipid peroxidation and the generation of some products, including photoemissive species. Rate constants are given as approximate numbers collected from the literature. From Noll et al. (1987).

amount incorporated into vitamin E-deficient animals by pretreatment of the intact animal one day before the experiment. In other words, the orientation of vitamin E in the membrane appears crucial in its effectiveness. The difference between *in vitro* addition and *in vivo* incorporation was about 50-fold in terms of increasing the lag-time before the onset of the rise in chemiluminescence.

Such studies were also extended to intact cells. Using an organic selenocompound capable of carrying out the glutathione peroxidase reaction (PZ 51; ebselen), it was found in intact cells that the oxidative challenge exerted by the addition of ADP/iron could be counteracted (Figure 11.4). As seen in Figure 11.4A, there is a decrease in chemiluminescence as well as an increase in the lag-time upon the addition of ebselen (PZ 51), being half-maximal at about 8 μmol/l (Müller et al., 1985). As the compound ebselen is capable of catalyzing the glutathione peroxidase reaction (Müller et al., 1984; Wendel et al., 1984), the depletion of glutathione in the cells was also studied. As shown in Figure 11.4B, in cells pretreated with phorone, thus having been essentially depleted of glutathione, ebselen has only negligible protective effect. In Figure 11.4C, the sulphur analog, PZ25, has been studied under conditions similar to Figure 11.4A; this compound exerts much less protective effect than the selenoorganic compound.

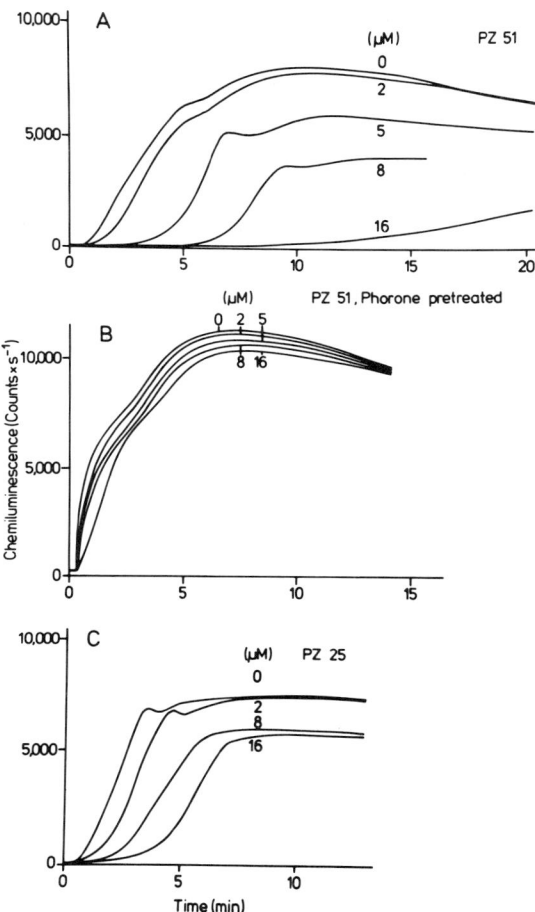

Figure 11.4. Time course of ATP-Fe-induced generation of low-level chemiluminescence, and the effect of different concentrations of ebselen (PZ 51) and PZ 25 in isolated hepatocytes. (a) and (c), hepatocytes from untreated rats; (b) hepatocytes from phorone-pretreated rats. From Müller et al. (1984).

11.5. DNA-DAMAGE BY SINGLET OXYGEN

Radiation-induced and much of the chemically induced DNA-damage is attributed to free radical reactions involving notably the hydroxyl radical (see Simic et al., 1986; Schulte-Frohlinde and von Sonntag, 1985). However, non-radical reactions of electronically excited species are also important, e.g. in DNA-damage by photoxidation, and can explain the effects of photosensitizers which generate singlet molecular oxygen. $^1O_2$ was demonstrated recently by its monomol emission in tetracycline photosensitization (Hasan and Khan, 1986) and enzymatically with the lactoperoxidase reaction (Kanofsky, 1984), and by its dimol emission in enzyme reactions such as cycloxygenase (Cadenas et al., 1983). Although $^1O_2$ has long been known to react with constituents of nucleic acids, there has been uncertainty as to its importance in eliciting DNA-damage. Using a physical source of $^1O_2$ by employing a microwave discharge system, we recently examined the transforming activity of the plasmid pBR 322 in E.coli. Taking care to exclude O atoms and O_3, it is shown that $^1O_2$ leads to a loss of biologically active DNA (Wefers et al., 1987). As shown in Figure 11.5, there is an increase in the loss of transforming activity in deuterium oxide as compared to water. In agreement with such results, Lafleur et al. (1987) using $^1O_2$ generated chemically from the thermodissociable endoperoxide of 3,3'-(1,4-napthy-lidene)dipropionate, observed a loss of plaque-forming capacity of ØX174DNA. Likewise, preliminary work on the generation of $^1O_2$ by irradiation of rose bengal absorbed on silicagel beads adhered to a glass slide and measuring transforming activity of pBR 322 (Hildebrand et al., 1986) is in support of the capability of $^1O_2$ as a non-radical excited species to damage biologically active DNA.

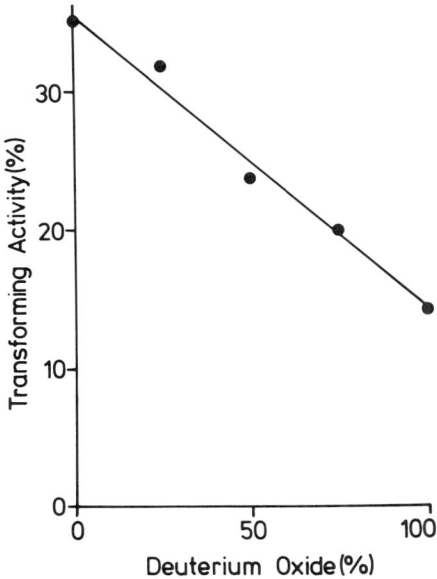

Figure 11.5. Dependence of the transforming activity of plasmid DNA (pBR322) in *E. coli* caused by singlet molecular oxygen as a function on D_2O concentration. The inhibition was obtained by 20 min exposure to singlet molecular oxygen generated in the microwave discharge-system. 100% in H_2O corresponds to 4.9 x 10^{-3} transformants x(ug DNA)$^{-1}$ x (total number of cells)$^{-1}$. From Wefers et al. (1987)

In the literature, there has been much further work on the reactions of electronically excited states with DNA, but this field is beyond the scope of the present article.

11.6. EXCITED CARBONYLS

Recent work, notably from the group of Cilento, has established the potential biological importance of excited carbonyls. This was achieved mainly by generating triplet acetone-related compounds. For example, there has been energy transfer from enzyme-generated triplet carbonyls to the thylakoid membrane fractions enriched in photosystems I and II (Nassi and Cilento, 1985). Intracellularly, polymorphonuclear leukocytes were shown to generate excited carbonyls in myeloperoxidase-catalyzed reaction. This was found to be the cause of considerable cell damage (Nascimento et. al., 1986). Likewise, the generation of excited carbonyls in liver microsomal membranes was detected. For this, the sensitized photoemission of chlorophyll as an indicator of triplet carbonyls was found most useful (Cadenas et. al., 1984b).

11.7. CONCLUDING REMARKS

The detection of electronically excited states in biological material so far mainly rests on the photoemission that occurs upon the return to the ground state. The association with cell life and death probably is greater than is currently known. In our own work, the photoemission during the process of cell damage and lipid peroxidation was the major focus. The so-called mitogenic radiation, initially described by Gurvich (1926), is a phenomenon associated with fundamental processes in cell life.

ACKNOWLEDGMENTS

Studies carried out in our laboratory were supported by Deutsche Forschungsgemeinschaft, Bonn, and by the National Foundation for Cancer Research, Washington.

REFERENCES

Bartoli, G. M., Müller, A., Cadenas, E. and Sies, H. (1983). Antioxidant Effect of Diethyldithiocarbamate on Microsomal Lipid Peroxidation Assessed by Low-Level Chemiluminescence and Alkane Production. FEBS Lett. 164: 371-374.

Boveris, A., Cadenas, E., Reiter, R., Filipkowski, M., Nakase, Y. and Chance, B. (1979). Organ

Chemiluminescence: Noninvasive Assay for Oxidative Radical Reactions. Proc. Natn. Acad. Sci. U.S.A. 77: 347-351.
Cadenas, E. (1984). Biological Chemiluminescence. Photochem. Photobiol. 40: 823-830.
Cadenas, E. and Sies, H. (1984). Low-Level Chemiluminescence as an Indicator of Singlet Molecular Oxygen in Biological Systems. In: Oxygen Radicals in Biological Systems, Methods in Enzymology 105: 221-231 (L. Packer, ed.) Academic Press, New York.
Cadenas, E. and Sies, H. (1985a). Oxidative Stress: Excited Oxygen Species and Enzyme Activity. Advances in Enzyme Regulation 23: 217-237.
Cadenas, E. and Sies, H. (1985b). Detecting Singlet Oxygen by Low-Level Chemiluminescence in: Handbook of Methods for Oxygen Radical Research (R. A. Greenwald, ed.) pp. 191-195. CRC Press, Inc., Boca Raton, Florida.
Cadenas, E., Ginsberg, M., Rabe, U. and Sies, H. (1984a). Evaluation of Alpha-Tocopherol Antioxidant Activity in Microsomal Lipid Peroxidation as detected by Low-Level Chemiluminescence. Biochem. J. 223: 755-759.
Cadenas, E., Sies, H., Campa, A. and Cilento, G. (1984b). Electronically Excited States in Microsomal Membranes: Use of Chlorophyll-$\underline{a}$ as an Indicator of Triplet Carbonyls. Photochem. Photobiol. 40: 661-667.
Cadenas, E., Sies, H., Nastainczyk, W. and Ullrich, V. (1983). Singlet Oxygen Formation Detected by Low-Level Chemiluminescence During Enzymatic Reduction of Prostaglandin G_2 to H_2. Hoppe-Seyler s Z. Physiol. Chem. 364: 519-528.
Chance, B., Sies, H. and Boveris, A. (1979). Hydroperoxide Metabolism in Mammalian Organs. Physiol. Rev. 59: 527-605.
Cilento, G. (1982). Electronic Excitation in Dark Biological Processes. In: Chemical and Biochemical Generation of Excited States (W. Adam & G. Cilento, eds.) Academic Press, New York, pp. 277-307.
Daniels, M. (1983). Recent Developments in the Excited States of Nucleic Acids. Photochem. Photobiol. 37: 691-693.
Doroshow, J. and Hochstein, P. (1982). Redox Cycling and the Mechanism of Action of Antibiotics in Neoplastic Diseases. In: Pathology of Oxygen (A.P. Autor, ed.) Academic Press, New York, pp. 245-260.
Gurvich, A. G. (1926). Das Problem der Zellteilung. J. Springer, Berlin.
Hasan, T. and Khan, A. U. (1986). Phototoxicity of the Tetracyclines: Photosensitized Emission of Singlet Delta Dioxygen. Proc. Natl. Acad. Sci. USA 83: 4604-4606.
Hildebrand, E. L., Midden, W. R. and Murr, B. L. (1986). Singlet Oxygen Mutagenesis of Plasmid DNA. Photochem. Photobiol. 43: 14S, Abstr. MAM-D10.
Howard, J. A. and Ingold, K. U. (1968). Rate Constants for the Self-reactions of n- and t-butyl Peroxy Radicals and Cyclohexylperoxy Radicals. The Deuterium Isotope Effects in the Termination of Secondary Peroxy Radicals. J. Am. Chem. Soc. 90: 1058-1059.
Ishikawa, T., Akerboom, T. P. M. and Sies H. (1986). Role of Key Defense Systems in Target Organ Toxicity. In: Target Organ Toxicity, Vol. I (G.M. Cohen, ed.) CRC Press pp. 129-143.
Kanofsky, J. R. (1984). Singlet Oxygen Production by Lactoperoxidase: Halide Dependence and Quantification of Yield. J. Photochem. 25: 105-113.
Kappus, H. and Sies, H. (1981). Toxic Drug Effects Associated with Oxygen Metabolism: Redox Cycling and Lipid Peroxidation. Experientia 37: 1233-1241.
Lafleur, M. V. M., Nieuwint, A. W. M., Aubry, J.M., Kortbeek, H., Arwert, F. and Joenje, H. (1987). DNA Damage by Chemically Generated Singlet Oxygen. Free Rad. Res. Comm. 2: 343-350.
Medeiros, M. H. G., Wefers, H. and Sies, H. (1987). Generation of Excited Species Catalyzed by

Horseradish Peroxidase or Hemin in the Presence of Reduced Glutathione and H_2O_2. J. Free Rad. Biol. Med. 3: 107-110.

Müller, A., Cadenas, E., Graf, P. and Sies, H. (1984). A Novel Biologically Active Selenoorganic Compound I. Glutathione Peroxidase-like Activity In Vitro and Anti-oxidant Capacity of PZ 51. Biochem. Pharmacol. 33: 3235-3239.

Müller, A., Gabriel, H. and Sies, H. (1985). A Novel Biologically Active Selenoorganic Compound IV. Protective Glutathione-Dependent Effect of PZ 51 (Ebselen) Against ADP-Fe Induced Lipid Peroxidation in Isolated Hepatocytes. Biochem. Pharmacol. 34: 1185-1189.

Napetschnig, S. and Sies, H. (1987). Generation of Photoemissive Species by Mitomycin C Redox Cycling in Rat Liver Microsomes. Biochem. Pharmacol. 36(10): 1617-1621.

Nascimento, A. L. T. O., da Fonseca, L. M., Brunetti, I. L. and Cilento, G. (1986). Intracellular Generation of Electronically Excited States. Polymorphonuclear leukocytes Challenged with a Precursor of Triplet Acetone. Biochim. Biophys. Acta 881: 337-342.

Nassi, L. and Cilento, G. (1985). Energy Transfer from Enzyme-generated Triplet Carbonyls to Thylakoid Membrane Fractions Enriched in Photosystem I and II. Photochem. Photobiol. 41: 195-201.

Noll, T., de Groot, H. and Sies, H. (1987). Distinct Temporal Relation Among Oxygen Uptake, Malondialdehyde Formation, and Low-Level Chemiluminescence During Microsomal Lipid Peroxidation. Arch. Biochem. Biophys. 252(1): 284-291.

Piette, J., Merville-Louis, M. P. and Decuyper, J. (1986). Damages Induced in Nucleic Acids by Photosensitization. Photochem. Photobiol. 44: 793-802.

Prochaska, H. J., Talalay, P. and Sies, H. (1987). Direct Protective Effect of NAD(P)H:Quinone Reductase against Menadione-induced Chemiluminescence of Post-mitochondrial Fractions of Mouse Liver. J. Biol. Chem. 262(5): 1931-1934.

Russell, G. A. (1957). Deuterium-isotope Effects in the Autoxidation of Aralkyl Hydrocarbons. Mechanism of Interaction of Peroxy Radicals. J. Am. Chem. Soc. 79: 3871-3877.

Sargentini, N. J. and Smith, K. C. (1985). Spontaneous Mutagenesis. Mutat. Res. 154: 1-27.

Schulte-Frohlinde, D. and Von Sonntag, C. (1985). Radiolysis of DNA and Model Systems in the Presence of Oxygen. In: Oxidative Stress (H. Sies, ed.) Academic Press, London, pp. 11-40.

Sies, H. (1986). Biochemistry of Oxidative Stress. Angew. Chem. Int. Ed. 25: 1058-1071.

Sies, H. and Cadenas, E. (1985). Oxidative Stress: Damage to Intact Cells and Organs. Phil. Trans. R. Soc. Lond. B. 311: 617-631.

Simic, M. G., Grossmann, L. and Upton, A. C. (eds.) (1986). Mechanisms of DNA Damage and Repair. Plenum, New York.

Slawinska, D. and Slawinski, J. (1983). Biological Chemiluminescence. Photochem. Photobiol. 37: 709-715.

Wefers, H. (1987). Singlet Oxygen in Biological Systems. Bioelectrochem. Bioenergetics 18: 91-104.

Wefers, H. and Sies, H. (1983). Hepatic Low-Level Chemiluminescence During Redox Cycling of Menadione and the Menadione-Glutathione Conjugate: Relation to Glutathione and NAD(P)H:Quinone Reductase (DT-Diaphorase) Activity. Arch. Biochem. Biophys. 224: 568-578.

Wefers, H., Komai, T., Talalay, P. and Sies, H. (1984). Protection Against Reactive Oxygen Species by NAD(P)H:Quinone Reductase Induced by the Dietary Antioxidant Butylated Hydroxyanisole (BHA). Decreased Hepatic Low-level Chemiluminescence During Quinone Redox Cycling. FEBS Lett. 169: 63-66.

Wefers, H., Riechmann, E. and Sies, H. (1985). Excited Species Generation in Horseradish Peroxidase-Mediated Oxidation of Glutathione. J. Free Radicals Biol. Med. 1: 311-318.

Wefers, H. and Sies, H. (1986). Generation of Photoemissive Species During Quinone Redox Cycling. Biochem. Pharmacol. 35: 22-24.

Wefers, H., Schulte-Frohlinde, D. and Sies, H. (1987). Loss of Transforming Activity of Plasmid DNA (pBR 322) in E. coli (CMK) Caused by Singlet Molecular Oxygen. FEBS Lett. 211: 49-52.

Wendel, A., Fausel, M., Safayi, H. and Otter, R. (1984). A Novel Biologically Active Selenoorganic Compound. II. Activity of PZ-51 (Ebselen) in Relation to Glutathione Peroxidase. Biochem. Pharmacol. 33: 3241-3245.

Chapter 12

HYPOXIA AND MOLECULAR MECHANISMS OF CELL DEATH

Herbert de Groot and Thomas Noll

ABSTRACT

Hypoxia, i.e., O_2 deficiency, is a frequent cause of cell death, implicating clinical medicine, e.g., stroke or myocardial infarction. Cell death can occur by impairment of the energy (ATP)-producing capacity of the cell due to O_2 limitation of mitochondrial cytochrome oxidase. Under certain conditions, however, cell death may also result from pathological cell functions which occur only under hypoxia. The PO_2 distribution in tissues and the O_2 dependence of cell functions are depicted. Subsequently, the current hypotheses on the molecular mechanism of hypoxic cell death are presented. The underlying mechanisms of cell death caused by misonidazole, a nitroimidazole used in tumor therapy, and by CCl_4, a well known hepatotoxin, are described as examples of those pathological functions which are aggravated under hypoxia. Experimental examples given mainly refer to liver parenchymal cells.

12.1. INTRODUCTION

In mammals, certain tissues like erythrocytes, white skeletal muscle fibers, and the cornea and lens of the eye can maintain their functions in the absence of O_2. Most mammalian cells, however, depend for their functions on the presence of O_2. Examples include liver parenchymal cells, heart muscle cells, and neurons. Hypoxia

Hypoxia and Molecular Mechanisms of Cell Death

is present when there are deviations in the functions of these cells from their normal values owing to a subnormal oxygen partial pressure (PO_2) (Figure 12.1). This condition comes about when cellular O_2 consumption surpasses O_2 supply; for example, due to an increased work load on the cell or a deterioration in perfusion. Usually decreases in cell functions are observed under hypoxia. Among these decreases the diminution of mitochondrial cytochrome oxidase activity plays a central role. There are, however, some cell functions, such as glycolysis, which actually increase under hypoxia.

The present article is concerned with hypoxic cell death and its relation to the decreased mitochondrial energy (ATP) producing capacity due to O_2 limitation of cytochrome oxidase. Further, recently identified pathological cell functions are introduced which occur only under hypoxia. These particular cell functions can lead under certain circumstances to cell death. They are independent of the cytochrome oxidase activity and may occur at PO_2 where the latter function is not yet impaired. Before illustrating both of these, however, a short overview on the PO_2 distribution in mammalian tissues will be given and the O_2 dependence of mammalian cell functions will be described in more detail.

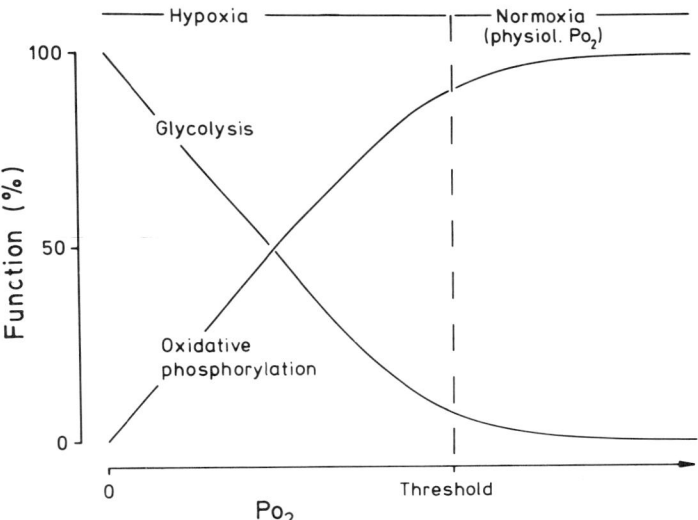

Figure 12.1. Schematic representation of the O_2 dependence of mammalian cell functions. The diagram stresses the fact that under hypoxia decreases (such as oxidative phosphorylation) as well as increases (such as glycolysis) in cell functions occur. It is a simplification in the sense that a certain cell function must not necessarily reach maximal values under normoxia or hypoxia but may also show a continuous O_2 dependence over the whole PO_2 range.

12.2. OXYGEN PARTIAL PRESSURES IN MAMMALIAN TISSUES

In situ measurements of PO_2 in mammalian tissues by multiwire surface electrodes and PO_2 needle electrodes reveal bell-shaped distributions of PO_2 (Kessler et al., 1985). In the various tissues two characteristic types of PO_2 distribution curve exist (Figure 12.2). The first type is found in liver, brain, pancreas, and skeletal muscle. Its characteristics are mean PO_2 between 20-40 mm Hg and lowest PO_2 values within the range of 1-5 mm Hg. The second type of PO_2 histogram which can be recorded from the lung, the heart and the outer cortex of the kidney is characterized by mean PO_2 values of significantly above 40 mm Hg and lowest PO_2 values higher than 20 mm Hg.

In solid mammalian tumors relatively large areas of very low PO_2 typically exist (Vaupel, 1977) (Figure 12.2). These areas may represent up to 50% of the solid tumor mass. They are due to insufficient blood supply of the rapidly growing tumor tissue. Tumor cells in those hypoxic areas are an obstacle to a successful tumor therapy since they are several times more resistant to radio therapy and chemotherapy than tumor cells of higher PO_2.

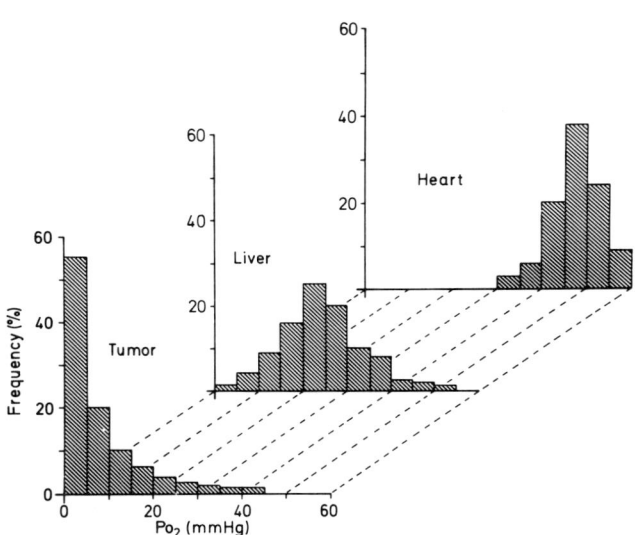

Figure 12.2. PO_2 frequency distribution in rat liver, rat heart and a solid tumor (mouse mammary carcinoma). The in situ measurements were performed with PO_2 multiwire surface and PO_2 needle electrodes (Kessler et al., 1985; Vaupel, 1977).

Due to O_2 consumption intercellular and intracellular O_2 gradients are formed within tissues. For instance in liver there are steep intercellular O_2 gradients along the sinusoids (liver capillaries) from the portal to the central regions of the liver lobules (Sies, 1977; Ji et al., 1982). As indicated by the PO_2 distribution curves (Figure 12.2) these intercellular O_2 gradients may be as high as 50 mm Hg. Intracellular O_2 gradients mainly occur towards the mitochondrial compartment and, again in liver, they have been estimated to be in the order of 2-4 mm Hg (Jones and Mason, 1978; Jones, 1984; de Groot and Noll, 1987). The O_2 gradients occur predominantly immediately at the surface of the mitochondria and there are only very flat gradients to other parts and hence to other O_2-consuming sites in the cell such as the endoplasmic reticulum and the peroxisomes.

12.3. OXYGEN DEPENDENCE OF CELL FUNCTIONS

Oxidases and oxygenases are those cellular enzymes that require O_2 as a substrate. Among these enzymes a unique role is played by cytochrome oxidase of the mitochondrial respiratory chain. The energy and the oxidation-reduction status of the O_2-dependent cells directly depend on the proper function of this enzyme. Limitation of its function by O_2 availability leads to a decrease in the ATP/ADP concentration ratio and increases in the reduction state of the electron transferring components of the respiratory chain and in the $NADH/NAD^+$ concentration ratio (Chance, 1952, 1965; Starlinger and Lübbers, 1973; Sugano et al., 1974; Jones and Mason, 1978; Wilson et al., 1979).

In isolated liver cells incubated in a special incubation apparatus, the oxystat system (Noll et al., 1986), the PO_2 where the cytochrome oxidase activity was limited by O_2 appeared to be about 2 mm Hg (Figure 12.3). At this critical PO_2, the O_2 uptake rate, which is more than 90% due to cytochrome oxidase activity (de Groot and Noll, 1985) and therefore directly reflects the energy flux through the cell, became dependent upon O_2 and rapidly decreased, being half-maximal at a PO_2 of about 0.7 mm Hg (P_{50} value). Likewise, the ATP/ADP concentration ratio, which only slightly decreased from 10 at a PO_2 of 100 mm Hg to 8 at the PO_2 of 2 mm Hg, rapidly declined, reaching a value of about 1 under anaerobic conditions. In line with the decreased energy flux through the liver cell and presumably mediated by the decline of the ATP/ADP concentration ratio, gluconeogenesis from lactate started to decrease at the critical PO_2 as well.

Values of similar magnitude for the critical PO_2 of the cytochrome oxidase function in the liver have been reported by other groups (Longmuir, 1957; Jones and Mason, 1978). As indicated by the cellular O_2 uptake, and the oxidation-reduction state of the pyridine nucleotides and the mitochondrial respiratory chain, they appear

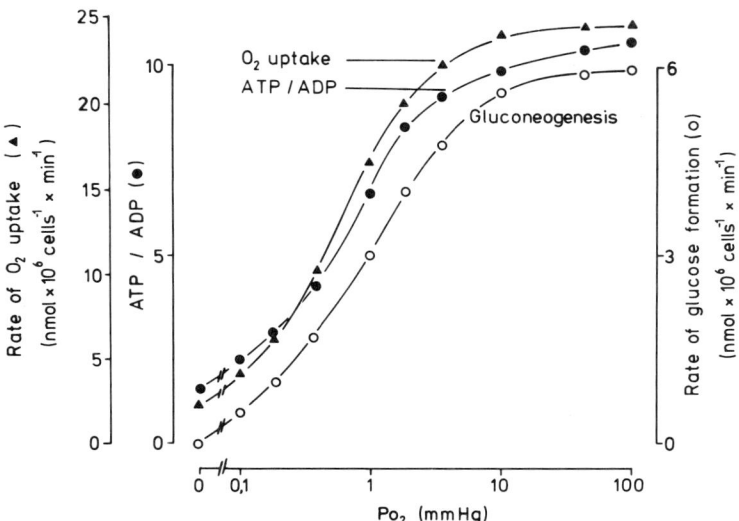

Figure 12.3. O_2 dependence of hepatocellular O_2 uptake, gluconeogenesis and the ATP/ADP concentration ratio (de Groot, H., Noll, T., Hummerich, H., Romero, F., and Soboll, S., unpublished results). The incubations were performed with isolated hepatocytes in the oxystat system (Noll et al., 1986). Glucose formed from lactate, ATP and ADP were determined with enzymatic methods (Bergmeyer et al., 1974a, b; Gruber et al., 1974).

to apply also to other O_2-dependent mammalian cells such as the heart muscle cell (Froese, 1962; Chance, 1965; Sugano et al., 1974; Jones and Kennedy, 1982).

Oxidases and oxygenases other than cytochrome oxidase usually exhibit significantly lower affinities for O_2. Examples include liver microsomal cytochrome P-450 with P_{50} values of up to 150 mm Hg, depending on the isoenzyme of cytochrome P-450 and the substrate of the monooxygenase reaction, mitochondrial amine oxidase with P_{50} values of 160 mm Hg and above and placental diamine oxidase with a P_{50} value of about 200 mm Hg (see Jones, 1981, for recent review).

12.4. HYPOXIC CELL DEATH

In liver parenchymal cells, at PO_2 below the critical value of 2 mm Hg, there is obviously an imbalance between energy (ATP) demand and the energy (ATP) producing capacity of the cell. It is the depression of the ATP synthesis rate due to O_2 limitation of the cytochrome oxidase activity which is considered to be the decisive functional lesion responsible for hypoxic cell death not only in the liver parenchymal cell but also in other hypoxia-sensitive cells such as heart muscle cells

Hypoxia and Molecular Mechanisms of Cell Death

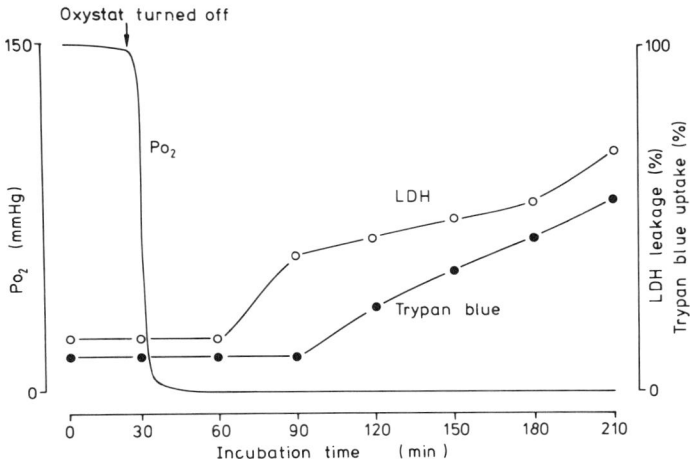

Figure 12.4. Lactate dehydrogenase leakage and trypan blue uptake during a typical incubation of isolated hepatocytes under anaerobic conditions. As indicated by the PO_2 trace, anaerobic conditions were reached after about 10 min following turning off the oxystat system due to O_2 consumption by the hepatocytes. Lactate dehydrogenases (LDH) activity was measured according to Bergmeyer and Bernt (1974a, b). LDH leakage is the percentage of extracellular to total (extra plus intracellular) LDH activity. Trypan blue uptake was determined by incubating an aliquot of cells with 0.25 % trypan blue for 3 min and subsequent microscopic counting of the percentage of stained cells.

and neurons (see Jennings, 1978; Venkatachalam, 1983; Hochachka, 1986, for recent reviews).

In Figure 12.4 a typical incubation of isolated rat liver cells is shown where the PO_2 is not maintained constant by special procedures. Starting from a PO_2 of 150 mm Hg, the PO_2 rapidly dropped due to O_2 uptake by the hepatocytes reaching anaerobic conditions after about 10 min. The decrease of PO_2 was paralleled by a decrease in the ATP/ADP concentration ratio (compare Figure 12.3). While the ATP/ADP concentration ratio responded within seconds to the altered PO_2, the leakage of lactate dehydrogenase and the uptake of trypan blue, both indicators of severe cell injury, only started to increase after 30 and 60 min, respectively. At present those events connecting the early impairment of the energy status of the cell with the loss of cell viability are the subject of intensive research.

Among those cell functions which depend on ATP, the disturbance of the ion homeostasis is usually supposed to be the critical further step towards cell death (Figure 12.5) (see also Chapter 9). The intracellular concentrations of K^+, Na^+ and Ca^{2+} tend to equilibrate with the extracellular concentrations of these ions (Farber et al., 1981; Fleckenstein, 1983; Nakaya et al., 1985; Hochachka, 1986). The

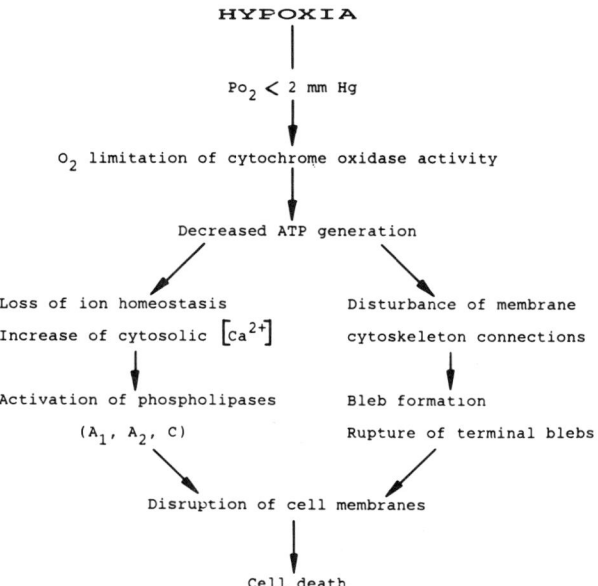

Figure 12.5. Possible mechanisms of hypoxic liver cell death.

increase in the cytosolic Ca^{2+} concentration is regarded as especially critical. Depending on cell type it may be mediated by several factors: a) ATP-dependent Ca^{2+} transporters in the plasma membrane and the endoplasmic (sarcoplasmic) reticulum are not efficient due to lack of ATP, b) mitochondria do not accumulate Ca^{2+}, a process which directly depends on the proton electrochemical potential across the inner mitochondrial membrane, c) in excitable cells voltage-dependent Ca^{2+} channels open because of a change of the membrane potential resulting from alterations in the intracellular concentrations of K^+ and Na^+, and d) increases in inositol triphosphate and diacylglycerol, due to continuous activity of phospholipases but decreased renewal of phosphoinositols from diacylglycerol, may cause an increased Ca^{2+} efflux from the endoplasmic (sarcoplasmic) reticulum and a decreased Ca^{2+}-Na^+ exchange across the plasma membrane.

There are several cell functions which may be disturbed by an increased cytosolic Ca^{2+} concentration. Once again with regard to cell death, the activation of phospholipases of the A class is hypothesized to represent the decisive subsequent step (Farber et al., 1981) (Figure 12.5). In line with this assumption there is a decrease in the cellular phospholipids in liver and heart muscle cells during hypoxia (Chien et al., 1978) while in kidney cells increasing concentrations of lysophosphatidylcholine, diglycerides, and free fatty acids have been detected (Matthys et al., 1984). Without control the continuous hydrolysis of membrane phospholipids may lead to disruption of cellular membranes and hence to cell death.

It should be mentioned, however, that the actual mechanism of hypoxic cell death may be more complex than outlined above. For instance, the degradation of phospholipids during hypoxia proceeds to a significant degree within the mitochondrial fraction (Chien et al., 1978). However, even though several mitochondrial lesions are known to occur during hypoxia, following reoxygenation these lesions are reversed and the mitochondria resume their functions (Farber et al., 1981; Fleckenstein et al., 1983). Among other things they start to accumulate Ca^{2+} ions. This accumulation of Ca^{2+} occurs to an excessive extent. Actually, since it leads to mitochondrial inactivation, it is one of the major problems which arise following reoxygenation of hypoxic tissues (Nayler et al., 1979; Nakanishi et al., 1982; Nayler 1983; Fleckenstein et al., 1983). In liver cells the reason for this excessive accumulation of Ca^{2+} is probably a shift in the kinetics of the mitochondrial Ca^{2+} uniporter from sigmoidal to near hyperbolic kinetics triggered by the increased cytosolic Ca^{2+} concentration (Kröner, 1986, 1987). This alteration results in a marked increase in the activity of the Ca^{2+} uniporter and an increased mitochondrial Ca^{2+} uptake.

In addition, even the assumption of an increase in the cytosolic Ca^{2+} may be invalid. In a very recent paper Lemasters et al. (1987), using the Ca^{2+} indicator fura-2 and a digitized low-light video microscope, found no indication for an increase in the cytosolic Ca^{2+} concentration in hepatocytes where cell death was induced by inhibition of oxidative phosphorylation and glycolysis by cyanide and iodoacetate, respectively. The authors proposed an alternative pathway to irreversible cell injury where the formation of cell surface blebs due to decreased ATP formation plays a decisive role (Figure 12.5).

12.5. PATHOLOGICAL CELL FUNCTIONS UNDER HYPOXIA

There are drugs which exert selective toxicity against hypoxic cells. Examples include nitroimidazoles such as misonidazole and haloalkanes such as carbon tetrachloride and the anesthetic halothane. Several lines of evidence suggest that their selective toxicity against hypoxic cells is closely related to their metabolism to reactive intermediates, a process which proceeds with greater likelihood under hypoxia.

Nitroimidazoles, such as misonidazole, are used as so-called hypoxic cell sensitizers in the chemo- and radiotherapy of tumors. A prerequisite for their cytotoxicity is the reductive metabolism to reactive intermediates, such as the nitro radical anion, the nitroso and the respective hydroxylamine derivative (Adams and Stratford, 1986; Biaglow et al., 1986; de Groot and Noll 1987) (Figure 12.6). This activation is catalyzed by flavoenzymes such as cytochrome P-450 reductase and xanthine oxidase. It only takes place at decreased PO_2. It is assumed that the first one-electron reduced product, the nitro radical anion, reacts with O_2 to yield $O_2^{-\bullet}$ and that the reductive activation of the parent compound is inhibited by O_2 leading to an inhibition of the cytotoxicity of these drugs (Mason and Holtzman, 1975). Under hypoxia, reactions of intermediates of the reductive metabolism of misonidazole with thiol compounds occurs (Varnes and Biaglow, 1982; Biaglow, 1986). In addition, DNA single strand breaks and the formation of DNA adducts following administration of misonidazole have been described (Palcic and Skarsgard, 1978; Chapman et al., 1983). It is the reaction with particular cellular constituents which is believed to be responsible for cell injury caused by these nitro compounds (Figure 12.6).

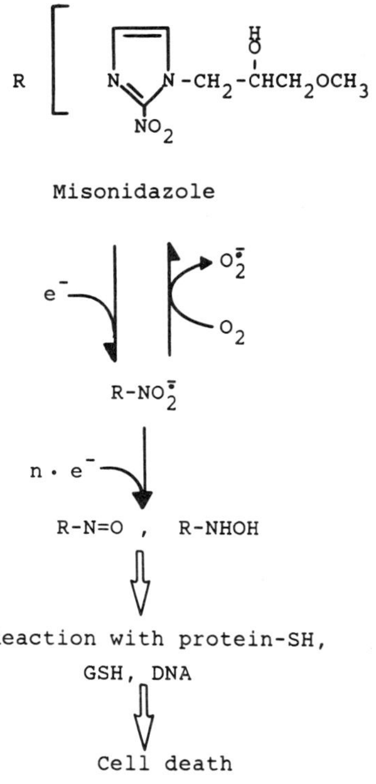

$$R \left[\underset{NO_2}{\underset{|}{N\!\!\diagup\!\!\diagdown\!\!N}} - CH_2 - \overset{OH}{\underset{|}{C}}HCH_2OCH_3 \right]$$

Misonidazole

$$e^- \searrow \nearrow O_2^{-\bullet}$$
$$ \nwarrow O_2$$

$$R-NO_2^{-\bullet}$$

$$n \cdot e^- \searrow$$

$$R-N=O \; , \; R-NHOH$$

$$\Downarrow$$

Reaction with protein-SH, GSH, DNA

$$\Downarrow$$

Cell death

Figure 12.6. Possible mechanisms of misonidazole cytotoxicity. The reaction of the reduced forms of misonidazole primarily the nitro radical anion, with O_2 to yield $O_2^{-\bullet}$ and the respective parent compound, are considered responsible for the inhibitory effects of O_2 on the reductive metabolism of misonidazole and hence on its cytotoxicity.

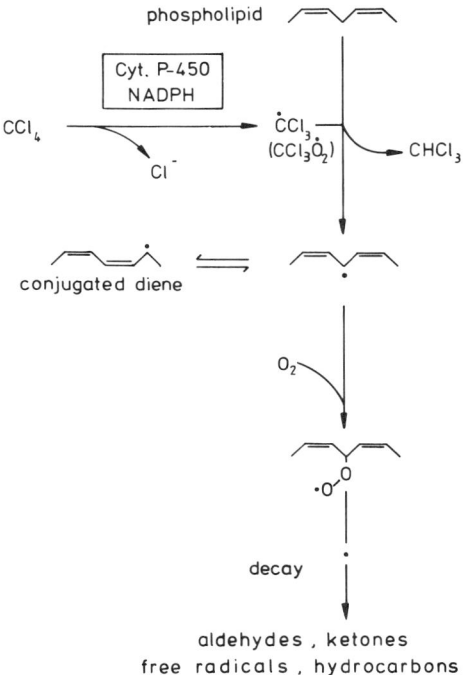

Figure 12.7. Mechanism of CCl_4-induced lipid peroxidation. Note that O_2 plays a dual role. It inhibits the formation of CCl_3 radicals but it is necessary for the formation of lipid hydroperoxides.

Halogenated alkanes such as CCl_4 are well known hepatotoxins. Their hepatotoxicity depends on their reductive activation to free radicals by particular isoenzymes of cytochrome P-450, the terminal oxidase of the microsomal monooxygenase system (Slater 1984; Brattin et al., 1986; de Groot and Noll, 1986) (Figure 12.7). As concluded from the difference absorption spectra of enzyme-substrate complexes and from synthetic heme ligand complexes the activation takes place at the heme moiety of the cytochrome (Ullrich et al., 1979). Since it is at that locus that O_2 becomes activated normally during the monooxygenase cycle, there is a competition between O_2 and the haloalkane for reducing electrons and hence the activation proceeds at maximal rate under anoxia. The reductive activation of haloalkanes is accompanied by covalent binding of reactive metabolites, such as the CCl_3 radical, to cellular macromolecules (Reynolds, 1967; Gordis, 1969; Uehleke et al., 1973). However, this apparently plays only a minor role for cell injury. Under anaerobic conditions where the covalent binding is maximal no significant effect of CCl_4 on the viability of hepatocytes was detectable (de Groot et al., unpublished results). The only cellular damage which was found to occur under anaerobic conditions was an inactivation of cytochrome P-450 (de Groot and Haas, 1980, 1981). Since this enzyme is the site where the reactive metabolites are formed, the loss of its enzymatic function may even be considered beneficial for the cell. However, cell death occurred, as indicated by lactate dehydrogenase leakage and trypan blue uptake, when small amounts of O_2 were present. In fact it was found that cell death was maximal when the experiments were performed at PO_2 between 1 and 60 mm Hg. At PO_2 of 70 mm Hg and above the haloalkane again was almost without effect on cell viability (Noll et al., 1987). This complex O_2 dependence of cell injury caused by CCl_4 and related compounds (de Groot and Noll, 1985, 1986) can be explained when it is assumed that cell injury is mediated by lipid peroxidation. As we could demonstrate in experiments with NADPH-reduced rat liver microsomes and isolated hepatocytes (Noll and de Groot, 1984; de Groot and Noll, 1986; Noll et al., 1987) this damaging process, which is induced by haloalkane free radicals, exhibits a similar O_2 dependence as already described for the haloalkane-mediated cell injury (Figure 12.8). The complex O_2 dependence of the haloalkane free radical-induced lipid peroxidation is due to the fact that O_2 plays a dual role. It inhibits the formation of free radical metabolites at the active site of cytochrome P-450 but it is necessary for the propagation steps of lipid peroxidation (de Groot and Noll, 1983, 1986). As indicated in Figure 12.7, fatty acid radicals are formed following interaction of the free radicals with unsaturated fatty acids of membrane phospholipids. They readily add O_2 to yield the respective peroxy radical. Following rearrangements and decomposition a great variety of reactive products are formed. These products, such as 4-hydroxynonenal (Esterbauer, 1985) but also the marked disturbance of the structure of the membrane are considered to mediate the further steps of cell injury leading ultimately to cell death. The earliest signs of damage to the liver cell are found at the endoplasmic reticulum. They include inactivation of glucose-6-phosphatase (de Groot et al., 1985, 1986) and the endoplasmic reticulum Ca^{2+} pump (Lowrey et al., 1981a, 1981b; Long and Moore, 1986). It is the inactivation of this Ca^{2+} pump which is held responsible for the marked loss of the endoplasmic reticulum-associated Ca^{2+} following CCl_4 intoxication (Reynolds and Moslen, 1980). As indicated by an increase in phosphorylase a activity (Long and Moore, 1986), the loss of the endoplasmic reticulum Ca^{2+} is paralleled by an increase in the Ca^{2+} concentration in the cytosol. This increase in the cytosolic Ca^{2+} concentration, however, is only intermediate presumably due to the fact that it triggers the change of the mitochondrial Ca^{2+} uniporter from sigmoidal to hyperbolic kinetics (Kröner, 1986, 1987) as already outlined for the situation of the reoxygenation of the

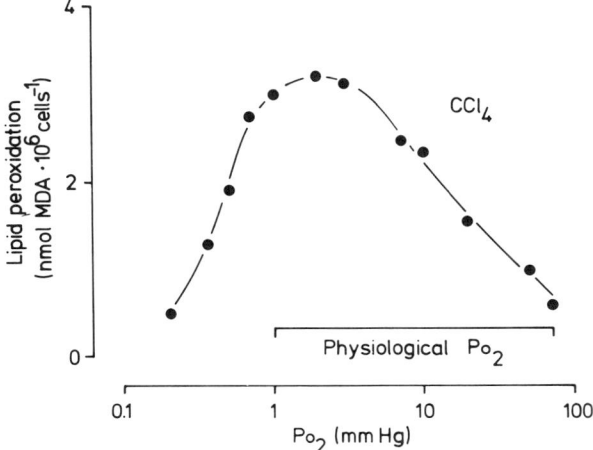

Figure 12.8. Oxygen dependence of CCl_4-induced lipid peroxidation in isolated hepatocytes (de Groot and Noll, 1986).

previously hypoxic liver cell. Nevertheless it is assumed that activation of phospholipases may also play a decisive role in this kind of cell damage so that the steps subsequent to lipid peroxidation may be similar to those described for the hypoxic cell death (Figure 12.5).

12.6. CONCLUSIONS

In mammalian cells hypoxia can lead to cell death due to O_2 limitation of the energy (ATP) producing capacity of the cell. In addition, hypoxia can impair the viability of mammalian cells by increasing certain pathological cell functions. These possible effects of hypoxia on cell viability differ in their critical PO_2.

In liver cells the critical PO_2 of the medium surrounding the cells for the impairment of their energy producing capacity appears to be 2 mm Hg. It is notable that this PO_2 is found just at the borderline of the physiological PO_2 in liver (Figure 12.2). Already a small shift of the PO_2 distribution curve to the left, as has been observed in experiments with rats following acute and chronic ethanol consumption (Ji et al., 1982), would expose significant centrolobularly located portions of the liver to hypoxic PO_2, leading to cell injury in these areas (Lemasters et al., 1981). A critical PO_2 of similar magnitude for the impairment of the energy producing capacity appears to exist also in other O_2-dependent mammalian cells.

The critical PO_2 for the pathological cell functions is not only different from the critical PO_2 for the cytochrome oxidase activity but also varies with the respective pathological function. For example, the reductive activation of misonidazole proceeds at half-maximal rate at a PO_2 of about 2 mm Hg (Koch et al., 1984) while CCl_4-mediated lipid peroxidation is already maximal at that PO_2 and proceeds at a significant rate even at a PO_2 of 60 mm Hg (Figure 12.8). In the former case severe hypoxia would be necessary before misonidazole exerts its cytotoxicity. In the latter case already under normal conditions large areas of the liver lobule, especially those around the central vein, possess the optimum PO_2 for the activation of CCl_4 and the subsequent decisive step of lipid peroxidation. Hypoxia would enlarge that area and hence increase the extent of tissue damage.

ABBREVIATIONS

PO_2, oxygen partial pressure (a PO_2 of 7 mm Hg is equivalent to about 1% O_2 in air, 10 μM O_2 in water, and about 60-100 μM in biological membranes at 37°C); P_{50}, oxygen partial pressure of half-maximal alteration of a function.

ACKNOWLEDGMENTS

The authors would like to thank Professor Dr. H. Sies for his critical discussion of the manuscript. Work at the authors' laboratory was generously supported by the Deutsche Forschungsgemeinschaft, Schwerpunktprogramm "Mechanismen toxischer Wirkungen von Fremdstoffen" and Ministerium für Wissenschaft und Forschung, Nordrhein-Westfalen.

REFERENCES

Adams, G. E. and Stratford, I. J. (1986). Hypoxia-Mediated Nitro-Heterocyclic Drugs in the Radio- and Chemotherapy of Cancer. Biochem. Pharmacol. 35: 71-76.

Bergmeyer, H. U. and Bernt, E. (1974a). Lactat-Dehydrogenase. UV-Test mit Pyruvat und NADH: In Methoden der enzymatischen Analyse (Bergmeyer, H. U., ed.). Verlag Chemie, Weinheim, Vol. 1, pp. 607-612.

Bergmeyer, H. U., Bernt, E., Schmidt, F. and Stork, H. (1974b). D-Glucose Bestimmung mit Hexokinase und Glucose-6-phosphate-Dehydrogenases, In: Methoden der enzymatischen Analyse (Bergmeyer, H. U., ed.). Verlag Chemie, Weinheim, Vol. 2, pp. 1241-1250.

Biaglow, J. E., Varnes, M. E., Roizen-Towle, L., Clark, E. P., Epp, E. R., Astor, M. B. and Hall, E. J. (1986). Biochemistry of Reduction of Nitro Heterocycles. Biochem. Pharmacol. 35: 77-90.

Brattin, W. J., Glende, Jr., E. A. and Recknagel, R. O. (1985). Pathological Mechanisms in Carbon Tetrachloride Hepatotoxicity. J. Free Rad. Biol. Med. 1: 27-38.

Chance, B. (1952). Spectra and Reaction Kinetic of Respiratory Pigments of Homogenized and Intact Cells. Nature 169: 215-221.

Chance, B. (1965). Reaction of Oxygen with the Respiratory Chain in Cells and Tissue. J. Gen. Physiol. 49: 163-188.

Chapman, J. D., Baer, K. and Lee, J. (1983). Characteristics of the Metabolism-Induced Binding of Misonidazole to Hypoxic Mammalian Cells. Cancer Res. 43: 1523-1528.

Chien, K. R., Abrams, J., Serroni, A., Martin, J. T. and Farber, J. L. (1978). Accelerated Phospholipid Degradation and Associated Membrane Dysfunction in Irreversible, Ischemic Liver Cell Injury. J. Biol. Chem. 253: 4809-4817.

de Groot, H. and Haas, W. (1980). Oxygen-Independent Damage of Cytochrome P450 by Carbon Tetrachloride-Metabolites in Hepatic Microsomes. FEBS Lett. 115: 253-256.

de Groot, H. and Haas, W. (1981). Self-Catalyzed, Oxygen-Independent Inactivation of NADPH- or Dithionite-Reduced Microsomal Cytochrome P-450 by Carbon Tetrachloride. Biochem. Pharmacol. 30: 2343-2347.

de Groot, H. and Noll, T. (1983). Halothane Hepatotoxicity Relation Between Metabolic Activation, Hypoxia, Covalent Binding, Lipid Peroxidation and Liver Cell damage. Hepatology 3: 601-606.

de Groot, H. and Noll, T. (1984). The Crucial Role of Hypoxia in Halothane-Induced Lipid Peroxidation. Biochem. Biophys. Res. Commun. 119: 139-143.

de Groot, H. and Noll, T. (1985). Haloalkane Free Radicals and Lipid Peroxidation under Low Steady-State Oxygen Partial Pressures. In: Free Radicals in Liver Injury (Poli, G., Cheeseman, K. H., Dianzani, M. U., and Slater, T. F., eds.). IRL Press, Oxford, pp. 185-189.

de Groot, H. and Noll, T. (1986). The Crucial Role of Low Steady State Oxygen Partial Pressures in Haloalkane Free-Radical-Mediated Lipid Peroxidation. Biochem. Pharmacol. 35: 15-19.

de Groot, H. and Noll, T. (1987). Oxygen Gradients: The Problem of Hypoxia. Biochem. Soc. Trans. (in press).

de Groot, H., Ling, L. L. and Sutherland, R. M. (1987). The Role of Glycolysis and Hexose Monophosphate Pathway in the Hypoxic Toxicity of Misonidazole. Free Rad. Res. Comms. 3: 93-98.

de Groot, H., Noll, T. and Rymsa, B. (1986). Alterations of the Microsomal Glucose-6-phosphatase System Evoked by Ferrous Iron- and Haloalkane Free-Radical-Mediated Lipid Peroxidation. Biochim. Biophy. Acta. 881: 350-355.

de Groot, H., Noll, T. and Sies, H. (1985). Oxygen Dependence and Subcellular Partitioning of Hepatic Menadione-Mediated Oxygen Uptake. Arch. Biochem. Biophys. 243: 556-562.

de Groot, H., Noll, T. and Tölle, T. (1985). Loss of Latent Activity of Liver Microsomal Membrane Enzymes Evoked by Lipid Peroxidation. Studies of Nucleoside Diphosphatase, Glucose-6-phosphatase, and UDP Glucoronyltransferase, Biochim. Biophys. Acta. 815: 91-96.

Esterbauer, H. (1985). Lipid Peroxidation Products: Formation, Chemical Properties and Biological Activities. In: Free Radicals in Liver Injury (Poli, G., Cheeseman, K. H., Dianzani, M. U., and Slater, T. F., eds.). IRL Press, Oxford, pp. 29-47.

Farber, L. J., Chien, K. R. and Mittnacht, Jr., S. (1981). The Pathogenesis of Irreversible Cell Injury in Ischemia. Am. J. Pathol. 102: 271-281.

Fleckenstein, A., Frey, M. and Fleckenstein-Grün, G. (1983). Consequences of Uncontrolled Calcium Entry and its Prevention with Calcium Antagonists. Eur. Heart J. 4: 43-50.

Froese, G. (1962). The Respiration of Ascites Tumor Cells at Low Oxygen Concentrations. Biochim. Biophys. Acta 57: 509-519.

Gordis, E. (1969). Lipid Metabolites of Carbon Tetrachloride. J. Clin. Invest. 48: 203-209.

Gruber, W., Möllering, H. and Bergmeyer, H. U. (1974). Analytische Differenzierung von Purin- und Pyrimidin-Nucleotiden. In: Methoden der enzymatischen Analyse (Bergmeyer, H. U., ed). Verlag Chemie, Weinheim Vol. 2, pp. 2128-2137.

Hochachka, P. W. (1986). Defense Strategies Against Hypoxia and Hypothermia. Science 231: 234-241.

Jennings, R. B., Hawkins, H. K., Lowe, J. E., Hill, M. L., Klotman, S. and Reimer, K. A. (1978). Relation Between High-Energy Phosphate and Lethal Injury in Myocardial Ischemia in the Dog. Am. J. Pathol. 92: 187-214.

Ji, S., Lemasters, J. J., Christenson, V. and Thurman, R. G. (1982). Periportal and Pericentral Pyridine Nucleotide Fluorescence from the Surface of the Perfused Liver: Evaluation of the Hypothesis that Chronic Treatment with Ethanol Produces Pericentral Hypoxia. Proc. Natl. Acad. Sci. USA 79: 5415-5419.

Jones, D. P. (1981). Hypoxia and Drug Metabolism. Biochem. Pharmacol. 30: 1019-1023.

Jones, D. P. (1984). Effect of Mitochondrial Clustering on Oxygen Supply in Hepatocytes. Am. J. Physiol. 247: 83-89.

Jones, D. P. and Kennedy, F. G. (1982). Intracellular Oxygen Gradients in Cardiac Myocytes. Lack of a Role for Myoglobin in Facilitation of Intracellular Oxygen Diffusion. Biochem. Biophys. Res. Commun. 105: 419-424.

Jones, D. P. and Mason, H. S. (1978). Gradients of Oxygen Concentration in Hepatocytes. J. Biol. Chem. 253: 4874-4880.

Kessler, M., Höper, J., Harrison, D. K., Skolasinska, K., Klövekorn, W. P., Sebening, F., Volkholz, H. J., Beier, I., Kernbach, C., Rettig, V. and Richter, H. (1985). Tissue Oxygen Supply under Normal and Pathological Conditions. In: Oxygen Transport to Tissue-V (Lübbers, D. W., Acker, H., Leniger-Follert, E., and Goldstick, T. K., eds.). Plenum Press, New York pp. 69-80.

Koch, C. J., Stobbe, C. C. and Baer, K. A. (1984). Metabolism-Induced Binding of 14C-Misonidazole to Hypoxic Cells: Kinetic Dependence on Oxygen Concentration and Misonidazole Concentration. Int. J. Radiat. Oncol. Biol. Phys. 10: 1327-1331.

Kröner, H. (1986). Calcium Ions, an Allosteric Activator of Calcium Uptake in Rat Liver Mitochondria. Arch. Biochem. Biophys. 251: 525-535.

Kröner, H. (1987). "Allosteric Regulation" of Calcium-Uptake in Rat Liver Mitochondria. Biol. Chem. Hoppe Seyler 367: 483-493.

Lemasters, J. J., Ji, S. and Thurman, R. G. (1981). Centrilobular Injury Following Hypoxia in Isolated, Perfused Rat Liver. Science 213: 661-663.

Lemasters, J. J., DiGuiseppi, J., Nieminen, A. -L. and Herman, B. (1987). Blebbing, Free Calcium and Mitochondrial Membrane Potential Preceding Cell Death in Hepatocytes. Nature 325: 78-81.

Long, R. M. and Moore, L. (1986). Inhibition of Liver Endoplasmic Reticulum Calcium Pump by Carbon Tetrachloride and Release of a Sequestered Calcium Pool. Biochem. Pharmacol. 35: 4131-4137.

Longmuir, I. S. (1957). Respiration Rate of Rat-Liver Cells at Low Oxygen Concentrations. Biochem. J. 65: 378-382.

Lowrey, K., Glende Jr., E. A. and Recknagel, R. O. (1981a). Destruction of Liver Microsomal Calcium Pump Activity by Carbon Tetrachloride and Bromotrichloromethane. Biochem. Pharmacol. 30: 135-140.

Lowrey, K., Glende Jr., E. A. and Recknagel, R. O. (1981b). Rapid Depression of Rat Liver Microsomal Calcium Pump Activity after Administration of Carbon Tetrachloride or Bromotrichloromethane and Lack of Effect after Ethanol. Toxicol. Appl. Pharmacol. 59: 389-394.

Mason, R. P. and Holtzman, J. L. (1975). The Mechanism of Microsomal and Mitochondrial Nitroreductase. Electron Spin Resonance Evidence for Nitroaromatic Free Radical Intermediates. Biochemistry 14: 1626-1632.

Matthys, E., Patel, Y., Kreisberg, J., Stewart, J. H. and Venkatachalam, M. (1984). Lipid Alterations Induced by Renal Ischemia: Pathogenic Factor in Membrane Damage. Kidney Int. 26: 153-161.

Nakanishi, T., Nishioka, K. and Jarmakani, J. M. (1982). Mechanism of Tissue Ca^{2+} Gain During Reoxygenation after Hypoxia in Rabbit Myocardium. Am. J. Physiol. 242: H437-H449.

Nakaya, H., Kimura, S. and Kanno, M. (1985). Intracellular K^+ and Na^+ Activities under Hypoxia, Acidosis, and no Glucose in Dog Hearts. Am. J. Physiol. 249: H1078-H1085.

Nayler, W. G. (1983). Calcium and Cell Death. Eur. Heart J. 4: 33-41.

Nayler, W. G., Poole-Wilson, P. A. and Williams, A. (1979). Hypoxia and Calcium. J. Mol. Cell. Cardiol.

11: 683-706.
Noll, T. and de Groot, H. (1984). The Critical Steady-State Hypoxic Conditions in Carbon Tetrachloride-Induced Lipid Peroxidation in Rat Liver Microsomes. Biochim. Biophys. Act 795: 356-362.
Noll, T., de Groot, H. and Wissemann, P. (1986). A Computer-Supported Oxystat System Maintaining Steady-State Oxygen Partial Pressures and Simultaneously Monitoring Oxygen Uptake in Biological Systems. Biochem. J. 236: 765-769.
Noll, T., Hugo-Wissemann, D., Littauer, A., de Sagara, R. M. and de Groot, H. (1987). The Decisive Oxygen-Levels in Haloalkane-Mediated Liver Cell Injury. Free Rad. Res. Comms. 3: 293-298.
Palcic, B. and Skarsgard, L. D. (1978). Cytotoxicity of Misonidazole and DNA Damage in Hypoxic Mammalian Cells. Br. J. Cancer 37: 54-59.
Reynolds, E. S. (1967). Liver Parenchymal Cell Injury. IV. Pattern of Incorporation of Carbon and Chlorine from Carbon Tetrachloride into Chemical Constituents of Liver in Vivo. J. Pharmacol. Exp. Ther. 155: 117-126.
Reynolds, E. S. and Moslen, M. T. (1980). Free-Radical Damage in Liver. In: Free Radicals in Biology (Pryor, W. A., ed.). Academic Press, New York, Vol. 4, pp. 49-90.
Sies, H. (1977). Oxygen Gradients During Hypoxic Steady States in Liver. Biol. Chem. Hoppe-Seyler 358: 1021-1032.
Slater, T. F. (1984). Free-Radical Mechanisms in Tissue Injury. Biochem. J. 222: 1-15.
Starlinger, H. and Lübbers, D. W. (1973). Polarographic Measurements of the Oxygen Pressure Performed Simultaneously with Optical Measurements of the Redox State of the Respiratory Chain in Suspensions of Mitochondria under Steady-State Conditions at Low Oxygen Tensions. Pflügers Arch. 341: 15-22.
Sugano, T., Oshino, N. and Chance, B. (1974). Mitochondrial Functions under Hypoxic Conditions. The Steady States of Cytochrome c Reduction and of Energy Metabolism. Biochim. Biophys. Acta 146: 340-358.
Uehleke, H., Hellmer, K. H. and Tabarelli, S. (1973). Binding of 14C-Carbon Tetrachloride to Microsomal Protein in vitro and Formation of CHC13 by Reduced Liver Microsomes. Xenobiotica 3: 1-11.
Ullrich, V. (1979). Cytochrome P450 and Biological Hydroxylation Reactions. Top. Curr. Chem. 83: 67-104.
Varnes, M. E. and Biaglow, J. E. (1982). Inhibition of Glycolysis of Mammalian Cells by Misonidazole and Other Radiosensitizing Drugs. Biochem. Pharmacol. 31: 2345-2351.
Vaupel, P. (1977). Hypoxia in Neoplastic Tissue. Microvasc. Res. 13: 399-408.
Venkatachalam, M. A., Kreisberg, J. I., Stein, J. H. and Lifschitz, M. D. (1983). Salvage of Ischemic Cells by Impermeant Solute and Adenosinetriphosphate. Lab. Invest. 49: 1-3.
Wilson, D. F., Owen, C. S. and Erecinska, M. (1979). Quantitative Dependence of Mitochondrial Oxidative Phosphorylation on Oxygen Concentration: A Mathematical Model. Arch. Biochem. Biophys. 195: 494-504.

Chapter 13

POPULATION DYNAMICS OF TUMORS ATTACKED BY IMMUNOCOMPETENT KILLER CELLS

Jacques R. Hiernaux, Patricia Meyers and René Lefever

ABSTRACT

We have developed a model describing the competition between the process of tumor progression due to tumor cell replication, and the process of tumor regression due to the host's immune responses. We show that, depending on the value of immune parameters characterizing effector-target binding and target lysis, various tumoral states may arise. We discuss the relation of this behavior with well known paradoxical and physiological situations such as tumor dormancy.

13.1 INTRODUCTION

Tumors *in vivo* usually present a complex nonmonotonic dynamical behavior. Notably: (i) the phases of progression, during which the tumor size increases, may be interrupted by regression phases during which the tumor size decreases; (ii) in some tumors, after phases of rapid growth and regression, a dormant state (Wheelock, 1981) occurs such that the number of cancer cells remains rather low and nearly constant; (iii) the establishment of a progressive tumoral state after the injection of an initial inoculum of cancer cells may involve intuitively hard to grasp threshold phenomena (Old et al., 1962; Klein, 1972; Bonmassar et al., 1974): small-sized inoculum progressively grow (the so-called "sneaking-through phenomenon"), medium-sized inoculum are rejected and large inoculum quickly proliferate; (iv) inoculum of identical size can grow differently in different animals: total regression, progressive growth or dormancy have been observed in different animals subjected to an identical treatment.

There is no doubt that at least part of these phenomena originate from the interactions between the neoplasm and the host's defense mechanisms. These interactions essentially rely on the immune system and can be decomposed into four main steps: (i) neoplastic proliferation independent of the immune system; (ii) recognition of immunogenic cancer cells by the immune system; (iii) triggering of an anti-tumor immune response; (iv) development of a regulated anti-tumor response. Once triggered, the immune response leads to the production of immune effector cells. The latter proliferate, differentiate and migrate to the tumoral site. Different types of anti-tumor effectors have been identified (Gorlzynski, 1974; Adams and Snyderman, 1979; Heberman and Ortaldo, 1981; Adams et al., 1984): cytotoxic T lymphocytes (CTL), activated macrophages and natural killer (NK) cells. They are involved in distinct killing pathways which have the same ultimate goal: the cytolysis of cancer cells. Hence, tumor growth *in vivo* results from a competition between the proliferative capacity of the neoplastic cells and the efficiency of the anti-tumor immune response.

The proliferative capacity of cancerous cells depends on growth factors activating or inhibiting mitosis. The immune system is a highly regulated network of specialized cells and molecules: any immune response is controlled by positive and negative feedbacks which either amplify or suppress the production and the activity of immunocompetent effector cells. In the case of an anti-tumor immune response, the regulatory processes thus determine the quantity of killer cells. Moreover, in the case of a specific response (i.e. mediated by CTL), it might be assumed that immunocompetent effectors are selected on the basis of their affinity for tumor-associated antigens. So, the regulation of the anti-tumor immune response could control the proliferation or regression of the tumor. The last assumption seems reasonable because tumor regression is related to the production of killer cells in numerous experimental systems. In some systems, a switch from regression to progression is related to immunosuppression (Broder et al., 1978; Greene, 1980; North, 1984). Interestingly, in one experimental system, the occurrence of immunosuppression has been correlated with the size of the tumor (Bear, 1986). Tumor escape has also been linked to antigenic modulation which induces the loss of key tumor-associated antigens involved in the stimulation of the immune system (Uyttenhove et al., 1983). Those facts illustrate the interdependence of neoplastic growth and of the immune system.

Nevertheless, paradoxical situations where tumor cells escape an active immune response have been described. Such situations challenge the idea that the goal of the anti-tumor immune response is the destruction of malignant cells. We distinguish at least three types of such paradoxical situations. (i) Immunostimulation of tumor growth (Prehn, 1977) has been observed, i.e. a situation where immunocompetent cells stimulate tumoral proliferation rather than destruction. (ii) A dormant state arises after nearly complete rejection of a progressive tumor mediated by cytotoxic T cells recognizing tumor-associated antigens still displayed on the surface of dormant cells (Wheelock, 1981; Hiernaux et al., 1986). (iii) Concomitant immunity is characterized by the rejection of a second implant of the original tumor whereas the progressive growth of the initial implant persists (Tuttle et al., 1983). At the time of rejection, the size of the second implant is smaller than the size of the initial one.

13.2. THE CYTOLYTIC REACTION AND THE SWITCHING PARAMETER β

Carcinogenesis or the interaction between a killer cell and its target can be studied at the single cell level. By contrast, the dynamics of tumor growth must be investigated at the population level because the onset of the various growth patterns appears as a multicellular property. Indeed, the sensitivity to the initial conditions reflected by the dependance of tumor progression or regression on the size of the initial inoculum as well as the multiplicity of dynamical behaviors following identical treatment typically illustrate population dynamics properties. In the case of cancer growth, the latter essentially rely on the competition between two cellular populations: the malignant cells and the cytotoxic cells able to kill them. We developed a mathematical modeling (Garay and Lefever, 1978; Lefever and Erneux, 1984) of this competition. In this modeling, we focus on the tumor dynamical properties resulting from the type of kinetic interactions which govern the cytolytic cycle by which effector cells destroy cancer cells rather than on the multiple immunoregulatory processes of the anti-tumor immune response. The purpose of this approach is to investigate the complex dynamical behaviors resulting from the effector-target kinetic competition, independently from the immunoregulation of killer cell production. The role of the latter in our point of view is to account for the switching between the various dynamical behaviors allowed by the kinetics of effector-target interactions. This switching is determined by the quantity (i.e., the actual number) and the quality (i.e. the binding affinity and the lytic capacity) of the immunocompetent effectors. In the sequel, we present the model and discuss the possible modes of tumoral growth.

Broadly speaking, the interaction between a cytolytic cell and a cancerous cell can be schematized as follows:

$$E_O + T \xrightarrow{b} E_1 \xrightarrow{\ell} E_O + D$$

A free effector E_0 binds a target T leading to the formation of a complex E_1. This binding step is characterized by a binding constant b. The bound effector lyses the target; this lytic step, characterized by a lytic constant ℓ, produces a free effector E_0 and a dead target D. According to this scheme, cytolysis is treated as an enzymatic reaction where the effector is equivalent to the enzyme. This analogy is based on the fact that a single effector can kill several targets (Berke et al., 1972; Rothstein et al., 1978). It has been discussed by various authors (Garay and Lefever, 1978; Thorn and Henney, 1976; Callewaert et al., 1978; Merrill, 1982).

Let us now consider a volume element of tissue containing E_T effector cells ($E_T = E_0 + E_1$). We suppose that this volume element is sufficiently large to contain a great number of cells ($E_T >> 1$) but on the other hand small enough for the target and effector cellular populations to be homogeneous, i.e. the scale of dimensions of interest here is much smaller than that over which the heterogeneity of tumors manifest itself. On this local scale, we consider the outcome of the competition between the replication of target cells and their cytolysis. We expect that when *locally* the balance of this competition is in favor (or disfavor) of target population, this is a necessary condition for the overall progression (or regression) of the tumor to be possible.

This local dynamics essentially involves four kinetic parameters. Namely, E_T, the binding and lytic constants b, ℓ and the replication constant λ describing the process of division and growth of the target cells. Formally, it corresponds to a step which can be schematized as:

$$T \xrightarrow{\lambda} 2T$$

The kinetics of this reaction is obviously more complicated than that of a simple autocatalytic reaction. In particular, it is essential in its modeling to account for the saturation exhibited by the growth of the T population in any finite volume element.

The value of those parameters depends on regulatory processes controlling the dynamics of the competition between the tumor and the immune system. Growth promoting or growth inhibiting substances controlling the mitosis of cancer cells modify the value of λ. Immunosuppression can be reflected in decreasing the value of E_T or in modifying b and/or ℓ. Selection processes resulting from immune regulation can alter the values of b and ℓ. Antigenic modulation of the tumor reflected by a decreasing density (or a lack) of tumor associated antigens leads to smaller values of b.

In order to grasp how such variations affect the local dynamics of tumors, it is useful to formulate this question as a stability problem and to introduce an adimensional switching parameter controlling the exchanges of stability taking place in the system between the normal and tumoral states of the tissue. More precisely, we say that the normal state of a tissue is stable if the level of surveillance which the immune system exerts is sufficient to eliminate the neoplastic cells. On the contrary, if the appearance of a single cancerous cell is the starting point of a tumor, we say that the normal tissue is unstable. This notion of stability can be quantified in term of the following switching parameter:

$$\beta = \frac{b\,E_T}{\lambda}$$

This parameter compares the efficiency of the immune response and the proliferative capacity of the malignant cells. Clearly, if the value of β is small, the immune response is weak and the tumor can proliferate: the tumorless normal state is unstable. By contrast, in presence of a strong immune response, the neoplasm is rejected: the normal state is stable. It can be shown, as depicted on Figure 13.1, that the exchange of stability occurs when the value $\beta = 1$ is crossed.

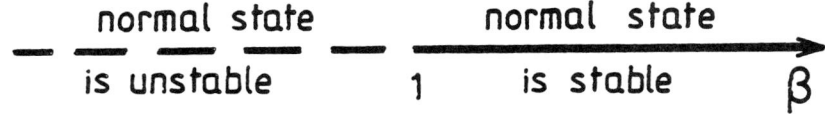

Figure 13.1. Presentation of the switching parameter β.

Population Dynamics of Tumors

Those intuitive stability considerations in term of the switching parameter β are however not sufficient to predict the final outcome of the balance between neoplastic proliferation and cytolysis, i.e., to predict if the tumor is in a dynamical state of progression or regression. Indeed, the analysis of our model (Lefever and Erneux, 1984; Lefever, et al., 1987), shows that tumor rejection is not automatically insured once the normal state is stable, i.e., β larger than 1 which implies that the stability of tissue's normal state is not automatically sufficient to insure the rejection of an established tumoral state. The latter may be stable as well. This happens when the conditions are such that the evolution equations of the system admit as solution, besides a stable normal state of the tissue, another stable solution corresponding to a cancerous state. In other words, a bistability phenomenon occurs: the tumoral and the normal state are stable simultaneously.

In fact, during the unfolding of the immune response two situations with markedly different properties may be encountered when the switching parameter β increases, e.g., as a result of the production of immunocompetent killer cells which increases the values of E_T and b. These situations are represented in Figure 13.2a and 13.2b where the density of proliferative target cells in the tissue is plotted against

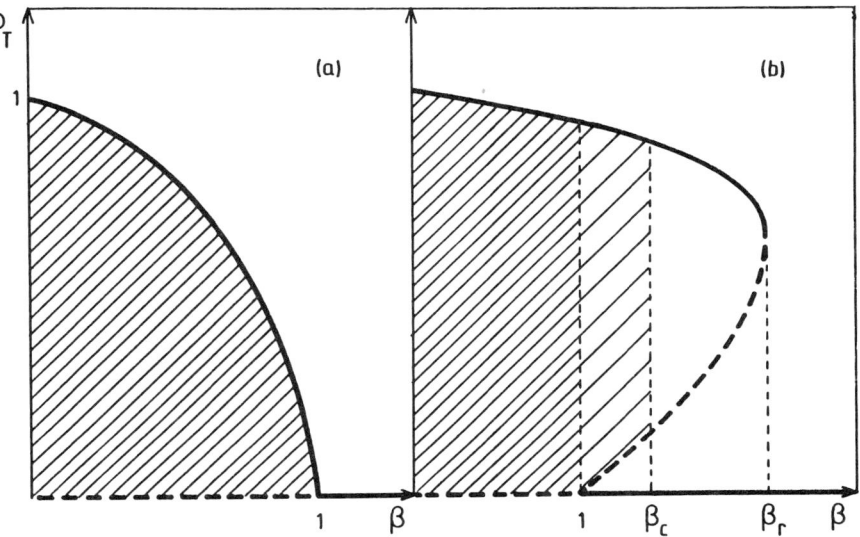

Figure 13.2. Variation of the density ρt of tumoral cells as a function of the switching parameter β. (a) For $\beta < 1$, the neoplasm proliferates, and for $\beta > 1$, it is rejected. (b) For $\beta > \beta_c$, the tumor is rejected. For $\beta < \beta_c$, tumor progression is observed.

β. Obviously, in both cases, when $\beta = 0$, i.e., in the absence of immune attack, the tumoral cells proliferate freely and hence their density is maximal (i.e., $\rho t = 1$ is some suitable normalized units). The normal state of the tissue corresponding to $\rho t = 0$ is then unstable: one cancer cell which appears somewhere in the tissue will proliferate and develop into a stable tumoral state corresponding to $\rho t = 1$. The distinction between the case 2a and 2b then concerns the mechanism of exchange of stability between the normal and the tumoral state. Indeed, in case 2a, the density ρt simply is a monotonously decreasing function of β which, as expected intuitively, crosses the axis $\rho t = 0$ (normal state) at the point $\beta = 1$. In the interval $0 < \beta < 1$, the tumor is progressive; rejection occurs as soon as $\beta > 1$. On the contrary, in the case 2b, the tumoral branch $\rho t \neq 0$ extends and remains stable for $1 < \beta < \beta_r$. Notably, the transition from progression to regression does not occur at the point $\beta = 1$ where the normal state $\rho t = 0$ becomes stable, but at some larger value β_c as indicated on the figure.

It is interesting from the point of view of the immune response to examine how the transition between these two situations takes place in term of the parameters of the cytolytic process. This is summarized in Figure 13.3. It presents the variation of the density of tumor cells as a function of the value of λ and bE_T, ℓ being kept constant. If there is no immune responses (i.e., if $bE_T = 0$), tissue stability only depends on the value of λ. If λ is smaller than 0, the tumorless state is stable. By contrast, if λ is larger than 0, the tumoral state is the only stable attractor. This means that the tumor will progressively invade the normal tissue through local growth. Now, as soon as there is an immune response (i.e., bE_T larger than 0), the transition between the normal and the tumoral state depends on the value of the switching parameter β as explained in Figure 13.2. If we progressively increase the intensity of the anti-tumor immune response (i.e., we consider larger values of bE_t), we observe a progressive transition from this simple situation to the case presenting multiple stationary states which is illustrated in Figure 13.2b.

The possible existence of bistable states is interesting. It allows the maintenance of a tumoral state in presence of a strong immune response, i.e., when the switching parameter β is larger than 1. Such a situation is compatible with the paradoxical immunostimulation of tumor growth. Indeed, once a tumoral tissue is in a state corresponding to the upper branch, no regression will be observed as long as β is smaller than β_c.

Our data also have some broader implications concerning the cell-mediated anti-tumor immune response. In order to be rejected, a tumor must stimulate the immune system. This results in increasing the value of β because the number of killer cells increases. Now, it is well established that lysis and binding are distinct steps. We can assume that immune selection will choose killer cells binding more

Population Dynamics of Tumors

strongly the target (i.e., b and β increase) and presenting comparable lytic capacity (i.e., comparable ℓ). In the course of this process, b might become larger than ℓ and this would induce a switch from the situation of Figure 13.2a to the situation of Figure 13.2b. So improving the immune response by selecting clones of higher affinity might finally favor tumor growth. In the sequel, we describe dynamical situations compatible with tumor dormancy.

It has been shown that an effector can bind several targets (Zagury et al., 1979). Let us examine the new dynamical situations observed as a result of this property. If a killer cell can bind one or two targets, the cytotoxic reaction can be schematized as follows:

$$E_O + T \xrightarrow{2b_0} E_1 \xrightarrow{b_1} E_2$$
$$\downarrow \ell_1 \qquad \downarrow 2\ell_2$$
$$E_0 + D \quad E_1 + D$$

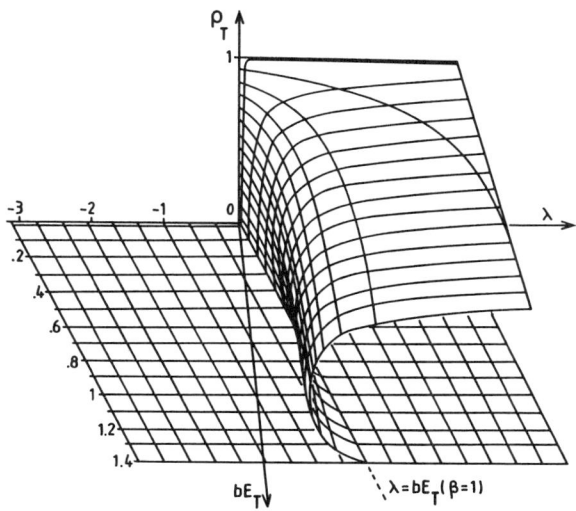

Figure 13.3. Variation of the density of tumoral cells as a function of λ and the product bE_T (ℓ being kept constant). As the value of bE_T increases, one observes a transition from the situation depicted in Figure 13.2a to the one presented in Figure 13.2b.

A free effector cell binds a target T. This step is characterized by a binding constant b_0; 2 is a combinational factor indicating that E_0 can bind two targets. The bound effector E_1 can either bind a second target T (this step is characterized by a binding constant b_1) or lyse the bound target (lysis is characterized by the lytic constant ℓ_1). When a killer cell has bound two targets (E_2), it lyses them sequentially (Zagury et al., 1979). The first step of this lytic sequence is characterized by a lytic constant ℓ_2 (the factor 2 indicates that two targets can be potentially lysed). It leads to E_1 and dead target D.

We now have six kinetic parameters: λ, b_0, ℓ_1, b_1, ℓ_2, and E_T ($E_T = E_0 + E_1 + E_2$). Four types of solutions can be obtained. The first two solutions correspond to the situations illustrated on Figures 13.2a and 13.2b. Figures 13.4 and 13.5 present new solutions which can also occur. Each situation corresponds to distinct relative values of the parameters. If the ratio ℓ_1/b_0 is smaller than 1, the solution of the kinetic equation describing the time evolution of the cellular density of tumor cells is presented on Figure 13.2b. If this ratio is greater than 1, either one of the three other situations can occur. The curves presented on Figures 13.4 and 13.5 reflect interesting dynamical properties. In Figure 13.4, for values of β smaller than 1 and fairly different from 0, the system evolves to a "microcancer" state characterized by a low density of tumoral cells. We propose to establish an analogy between this state observed for values of β corresponding to a weak immune response and the tumor dormant state. Figure 13.5 presents a more complex type of solution where multiple steady states again are observed for β bigger than β_P and smaller than β_r. Once more, the cellular densities corresponding to the intermediate branch represent unstable attractors which means that the system does not evolve to a state corresponding to those values for the cellular density. When β is smaller than β_P, the immune response is weak and the tumor progresses. When β is greater than β_r, a very strong immune response induces tumor regression. If β is smaller than 1 and larger than β_P, the tumor will proliferate either strongly (upper branch) or weakly (lower branch). The cellular densities of the lower branch are compatible with the establishment of dormancy. Now, if, in the course of the dynamics of tumor growth, a state characterized by a value of β between β_P and 1 and by a tumor cell density within the strippled area is reached, the system evolves to a "microcancer" state corresponding to the lower branch. Such a state will be maintained as long as β is smaller than 1 and larger than β_P. When β is greater than 1 and smaller than β_r, the tumor either regresses (lower branch corresponding to $\rho t = 0$) or progresses (upper branch). If the system reaches a dynamical state corresponding to the hatched area, i.e., $0 < \beta < \beta_P$, neoplastic proliferation is observed. Once more, a tumor can grow in the presence of a strong immune response. Broadly speaking, the situation presented in Figure 13.5, will be obtained when ℓ_1 is greater than b_0 and when ℓ_2

Population Dynamics of Tumors 439

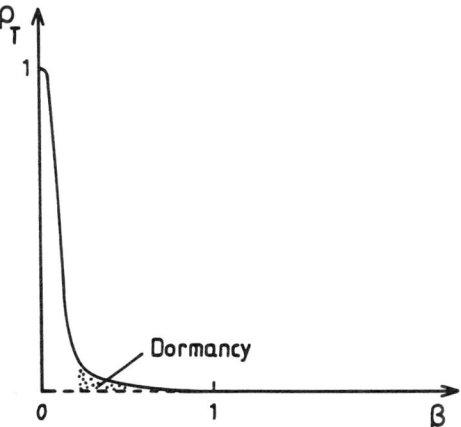

Figure 13.4. Variation of the density of tumoral cells as a function of β. The strippled area corresponds to a "micro cancer state" (i.e., to the maintenance of a small number of tumor cells, which can be assimilated to dormancy); for $\beta > 1$, the tumor regresses.

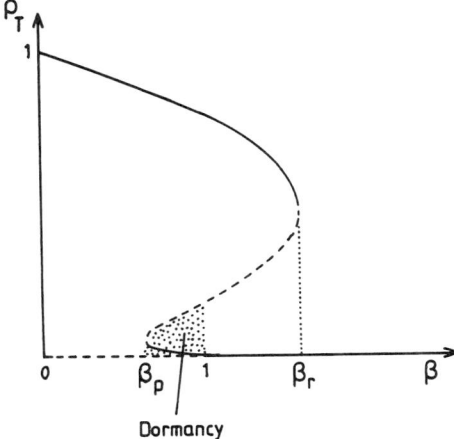

Figure 13.5. Multiple steady states are observed for $\beta_p < \beta < \beta_r$. The strippled area corresponds to the establishment of a dormant state and the hatched area corresponds to tumor progression.

is much smaller than b_1.

13.3. CONCLUSION

The summary of our theoretical analysis presented here indicates that various dynamical situations can occur if we look at tumor growth as a simple competition between proliferating malignant cells and immunocompetent killer cells. Dormancy and immunostimulation of tumor growth are compatible with our description. Our ideas could be tested experimentally by evaluating the values of the different kinetic parameters during tumor growth *in vivo*. We are currently developing experimental techniques allowing the determination of binding and lytic constants (Meyers et al., 1990).

ACKNOWLEDGMENTS

We thank Professor Prigogine for providing continuous support and interest in this research. This work is conducted pursuant to a contract with the National Foundation for Cancer Research. We thank J. Thierie for drawing Figure 13.3.

REFERENCES

Adams, D. O., Hall, T., Steplewski, Z. and Koprowski, H. (1984). Tumors Undergoing Rejection Induced by Monoclonal Antibodies of the IgG_{2a} Isotype Contain Increased Numbers of Macrophages Activated for a Distinctive Form of Antibody-Dependent Cytolysis. Proc. Natl. Acad. Sci. USA 81: 3506-3510.

Adams, D. O. and Snyderman, R. (1979). Do Macrophages Destroy Nascent Tumors? J. Natl. Cancer Inst. 62: 1341-1345.

Bear, H. D. (1986). Tumor-Specific Suppressor T-cells Which Inhibit the In Vitro Generation of Cytotoxic T-cells from Immune and Early Tumor-bearing Host Spleens. Cancer Res. 46: 1805-1812.

Berke, G., Sullivan, K. A. and Amos, D. B. (1972) Tumor Immunity In Vitro: Destruction of Mouse Ascites Tumor Through a Cycling Pathway. Science 177, 433-435.

Bonmassar, E., Menconi, E., Goldin, A. and Cudkowicz, G. (1974). Escape of Small Numbers of Allogeneic Lymphoma Cells from Immune Surveillance. J. Natl. Cancer Inst. 53: 475-479.

Broder, S., Muul, L. and Waldmann, T. A. (1978). Suppressor Cells in Neoplastic Disease. J. Natl. Cancer Inst. 61: 5-11.

Callewaert, D., Johnson, D. F. and Kearney, J. (1978). Spontaneous Cytotoxicity of Cultured Human Cell Lines Mediated by Normal Peripheral Blood Lymphocytes. J. Immunol. 121: 710-717.

Garay, R. and Lefever, R. (1978). A Kinetic Approach to the Immunology of Cancer: Stationary States Properties of Effector-Target Cell Reactions. J. theor. Biol. 73: 417-438.

Gorczynski, R. M. (1974). Evidence for In Vivo Protection Against Murine-Sarcoma Virus-Induced Tumors by T Lymphocytes from Immune Animals. J. Immunol. 112: 532-536.

Greene, M. I. (1980). The Genetic and Cellular Basis of Regulation of the Immune Response to Tumor Antigens. Contemp. Top. Immunobiol. 11: 81-116.

Heberman, R. B. and Ortaldo, J. R. (1981). Natural Killer Cells: Their Role in Defenses Against Disease. Science 214: 24-30.

Hiernaux, J. R., Lefever, R., Uyttenhove, C. and Boon, T. (1986). Tumor Dormancy as a Result of Simple Competition Between Tumor Cells and Cytolytic Effector Cells. In: Paradoxes in Immunology (Hoffmann et al., eds). CRC Press, Boca Raton, pp. 95-109.

Klein, E. (1972). Tumor Immunology: Escape Mechanisms. Ann. Inst. Pasteur (Paris) 122: 593-602.

Lefever, R. and Erneux, T. (1984). On the Growth of Cellular Tissues Under Constant and Fluctuating Environmental Conditions. In: Nonlinear Electrodynamics in Biological Systems (Adey, W.R. and Lawrence, A.F., eds) Plenum Press, New York; Pp 287-305.

Lefever, R., Hiernaux J., and Meyers, P. (1989). Evolution of Tumors Attacked by Immune Cytotoxic Cells: the Immune Response Dilemma. In: Cell to Cell Signalling: From Experimental to Theoretical Models (Goldbeter, A., ed). Academic Press, London, pp. 315-333.

Merrill, S. J. (1982). Foundations of the Use of an Enzyme-Kinetic Analogy in Cell-Mediated Cytotoxicity. Math. Biosci. 62: 219-235.

Meyers, P., Urbain, J., Hiernaux, J., and Lefever, R. (1990). Kinetics of Cytotoxic Reaction mediated by T-Lymphocyte Clones and Influence of Effector-Target Specificity on the Stabilization of Tumors. (Submitted).

North, R. J. (1984). The Murine Antitumor Immune Response and Its Therapeutic Manipulation. Adv. Immunol. 35: 89-155.

Old, L. J., Boyse, E. A., Clarke, D. A. and Carswell, F. A. (1962). Antigenic Properties of Chemically Induced Tumors. Ann. NY Acad. Sci. 101: 80-106.

Prehn, R. T. (1977). Immunostimulation of the Lymphodependent Phase of Neoplastic Growth. J. Natl. Cancer Inst. 59: 1043-1049.

Rothstein, T. L., Mage, M., Jones, G. and McHugh, L. L. (1978). Cytotoxic T Lymphocyte Sequential Killing of Immobilized Allogenic Tumor Target Cells Measured by Time-Lapse Microcinematography. J. Immunol. 121, 1652-1656.

Thorn, R. M. and Henney, C. S. (1976). Kinetic Analysis of Target Cell Destruction by Effector Cells. J. Immunol. 117: 2213-2219.

Tuttle, R. L., Knick, V. C., Stopford, C. R. and Wolberg, G. (1983). In Vivo and in Vitro Antitumor Activity Expressed by Cells of Concomitantly Immune Mice. Cancer Res. 43: 2600-2605.

Uyttenhove, C., Maryanski, J. and Boon, T. (1983). Escape of Mouse Mastocytoma P815 After Nearly Complete Rejection is Due to Antigen-Loss Variants Rather than Immunosuppression. J. Exp. Med. 157: 1040-1052.

Wheelock, E. F. (1981). The Tumor Dormant State. Adv. Cancer Res. 34: 107-140.

Zagury, D., Bernard, J., Jeannesson, P., Thiernesse, N. and Cerottini, J.C. (1979). Studies on the Mechanism of T Cell-Mediated Lysis at the Single Effector Cell Level. J. Immunol. 123: 1604-1609.

Chapter 14

ORGANIZATION OF HISTONE GENES AND REGULATION AND CONTROL OF THEIR EXPRESSION IN EARLY DEVELOPMENT

Hans V. Westerhoff, Johanna G. Koster, and Oliver H.J. Destrée

ABSTRACT

In the life strategy of the cell histones are crucial proteins. They are needed in large amounts to allow eukaryotic DNA to attain its chromatin structure. Because of their proximity to the transcription machinery, histones have also been implicated in regulation of cellular differentiation.

In many organisms the genes encoding histones are reiterated and are organized in "complete" clusters (each cluster encoding a linker histone and the four histones constructing the complete nucleosome, the basic structural unit for chromatin). The possible function and evolutionary origin of this organization are discussed.

In the early development of many organisms there is a phase of rapid DNA synthesis, with a concomitant need of excessive amounts of histones. With the aid of a recently developed quantitative model for histone gene expression in *Xenopus laevis*, we review how this and other organisms cope with the challenge of providing a sufficient amount of histones to structure the new DNA of their early embryos. Cell life here depends on storage in the oocyte, as well as well-tuned regulation of, the rate of cell division.

Later in the development of *Xenopus laevis*, unregulated histone gene expression would lead to overproduction of histones. In terms of a new and exact terminology for the control and regulation of gene expression, we discuss how a regulatory step (at the pretranslational level) some 11 hours after fertilization removes this problem.

14.1. INTRODUCTION

One of the central problems in developmental biology is how cells manage to maintain the concentrations of their proteins at the proper magnitudes throughout periods of rapid growth and differentiation. For the basic proteins that structure the DNA, this problem has two aspects. Firstly, because they determine the three dimensional structure of the DNA, the histones might well play a role in the regulation of the expression of other genes. Thus, the occurrence of histone variants might determine differentiation. We have recently (Koster, 1987; Koster et al., 1988, 1990) discussed (i) if histone gene variants are expressed differentially, (ii) how such a differential expression may be regulated, and (iii) whether the expression of histone gene variants plays a regulatory role in the expression of other genes (i.e., does it regulate genome activity by structural changes in the chromatin?).

In this chapter we shall dwell more on the second aspect of the problem: How is the expression of histone genes themselves regulated? Knowledge of how the histone-RNA and protein concentrations vary with time, is a prerequisite for understanding the second aspect of the problem. The canonical structure of chromatin requires approximately equal masses of DNA and histones (McGhee and Felsenfeld, 1980). At periods of rapid DNA replication, this poses the rather serious problem to the cell of how to ensure the availability of sufficient histones. Part of this chapter (Section 14.4) will therefore review (i) how the total amount of DNA (and therewith the amount of histone-encoding DNA) and histone mRNA vary during the early development of *Xenopus laevis* and (ii) if the observed concentrations (of histone-encoding DNA and mRNA) would be sufficient for the synthesis of the histone proteins necessary to structure the total genomic DNA. In the final part of this chapter, we will discuss at which points the early embryo of *Xenopus laevis* may control the expression of its histone genes in this time period (Section 14.5).

The histone genes of many organisms are reiterated. In Section 14.4 we shall discuss whether this serves the function of providing sufficient template for transcription during early embryogenesis. The reiterated histone genes are not randomly dispersed throughout the genome, but organized in a rather peculiar fashion. In Section 14.3 we shall discuss this organization and the possible functional importance thereof.

As a basis for Sections 14.3-14.5, we shall now begin (Section 14.2) to review roles and properties of histone proteins.

14.2. THE HISTONES AND THE ORGANIZATION OF EUKARYOTIC GENES IN CHROMATIN

Ever since the histones were first detected as part of the chromatin structure (Stedman and Stedman, 1950), experiments have been aimed at revealing the structure of chromatin. Supported by data from electron microscopy, circular dichroism and X-ray diffraction, the first order of chromatin structure is now known in considerable detail (Richmond et al., 1984) (cf. Figure 14.1). Stretches of 146 basepairs (bp) of a double helical DNA molecule are wrapped in 1.8 left-handed turns around a disc-like "nucleosome core" consisting of two each of the histones H2A, H2B, H3 and H4. The "linker" DNA in between these stretches is of variable length (20-100 bp per subunit). Together with the nucleosome core it constitutes the so-called nucleosome. Since DNA is continuous, the result is a "beads (or, rather, discus) on a string" structure. In some cases it has been possible to demonstrate that the nucleosomes bind preferentially to certain positions along the string of DNA (such that in a population all nucleosomes will tend to be in phase with a certain sequence on the DNA) (Simpson and Stafford, 1983; Thoma and Simpson, 1985; Ramsay, 1986).

A second step in the packaging of the DNA is the association of histone H1 to DNA outside the nucleosome core as the DNA enters and exists the nucleosome. The central globular domain of H1 is able to induce the further winding of DNA to 2 turns of DNA per histone octamer. The corresponding particle containing 166 bp of DNA has been termed chromatosome (Simpson, 1978). H1 also interacts with the linker DNA of 20-100 bp per nucleosome. Variation in the H1 (or "linker") histone

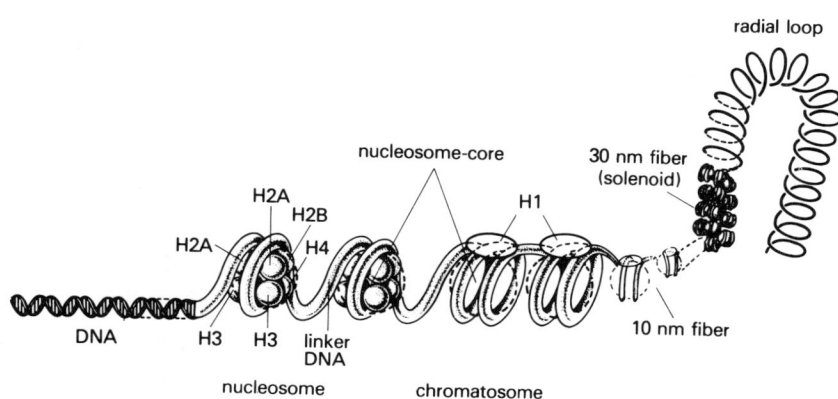

Figure 14.1. Histones and the organization of eukaryotic DNA.

Organization of Histone Genes

has been correlated to the length of the linker DNA, as well as to terminal differentiation (Weintraub, 1978). On the other hand, the absence of linker histones from lower eukaryotes (e.g., yeast) (Smith, 1984), suggests that they may not play a universal role in eukaryotic gene expression.

The next step of structural organization is (Figure 14.1) the folding of the nucleosome array, the 10 nm chromatin fiber, to form the 30 nm fiber, the precise structure of which is still under discussion (Eissenberg et al., 1985; Williams et al., 1986; Felsenfeld and McGhee, 1986). In one of the models for the higher order structure of chromatin (Thoma et al., 1979; Widom and Klug, 1985), the string with discs forms (at physiological ionic strength) a single helix. In this and in the other proposed structures, the linker histones appear to be involved in the interaction of the discs. Indeed, histone H1 has been shown to be instrumental in the condensation of the nucleosomes into the 30 nm fiber (e.g., Thoma et al., 1979; Dimitrov et al., 1986; Boulikas, 1986; Thomas et al., 1985; Jin and Cole, 1986).

The 30 nm fiber is again folded into a higher order structure. Eventually the chromatin is organized in loops that are attached to a matrix of protein (Berezney and Coffey, 1974; Cook and Brazell, 1976; Alberts et al., 1983; Eissenberg et al., 1985; Bureau et al., 1986; Gasser and Laemmli, 1986): in metaphase this structure is observable as a nuclear scaffold (Lewis et al., 1984; Mirkovitch et al., 1984). One function of the organization of eukaryotic DNA in terms of chromatin, is obvious: it allows the DNA of considerable length (e.g. 2 m for the human genome) to reside within a single nucleus of the cell (Laskey, 1986).

On the basis of their presence in chromatin the histones can be divided into two groups: the nucleosomal histones, occurring in the nucleosome core, and the linker histones, bound to the linker DNA. The group of nucleosomal histones is divided into four classes, historically named H2A, H2B, H3 and H4. The class of "linker" histones is distinct in that its member bind to linker DNA, have a significantly higher molecular mass than the core histones, as well as a rather different amino acid composition. Because of the high amount of lysine present in these linker histones they are often called "very lysine rich" histones. The characteristics of the different classes are summarized in Table 14.1.

Each class of histones can be subdivided in subclasses and or subtypes called histone variants. Variants within each class can replace each other in the chromatosome, but differ in amino acid sequence. Variants for the different classes of nucleosomal histones are best described in the sea urchin, the chicken and mammals. In the sea urchin the variants are separated as different protein bands on polyacrylamide gels. The variants appearing at consecutive times in early development are encoded by different genes located in different parts of the genome (for reviews see Kedes, 1979; Maxson et al., 1983a and c).

Table 14.1. Characteristics of the five classes of histone proteins

Class	Classification based on amino acid composition	Molecular weight in *Xenopus*	Classification based on position in chromatin	Degree of conservation
H1	Very lysine rich	21,000	Linker	Varies markedly between tissues and species
H2A	Lysine rich	13,908	Nucleosomal	Moderate variation between tissues and species
H2B	Lysine rich	13,994	Nucleosomal	Moderate variation between tissues and species
H3	Arginine rich	15,335	Nucleosomal	Highly conserved
H4	Arginine rich	11,234	Nucleosomal	Highly conserved

In the chicken, Urban and Zweidler (1983) recognized, by gel electrophoretic analysis, variants for the H2A, H2B, and H3 classes of nucleosomal histones, the expression of which appeared to be related to the rate of cell division. Detailed research of the various nucleosomal histones of mouse (Zweidler, 1984) showed that the variants could be distinguished into strictly replication-dependent and replication-independent variants. In HeLa cells, Wu and Bonner (1981) detected similar variants: "basic" variants expressed independently of the phase of the cell cycle and "replication" variants expressed predominantly during the S phase of the cell cycle.

A wide variation in expression has been observed for the class of the linker histones in different organisms, such as sea urchin (Newrock et al., 1978; Kedes, 1979), *Xenopus* (Byrd and Kasinsky, 1973; Koster et al., 1979; Risley and Eckhardt, 1981), chicken (Harbourne and Allan, 1986) and mammals (Cole, 1984; Lennox and Cohen, 1984), depending on the time in development, the differentiation status of the cell and the activity of the chromatin. Correlated with the variation in linker histone in the chromatin in various tissues, a variation in the length of the linker DNA is found (McGhee and Felsenfeld, 1980). A well studied variant in the linker histone class is the presumably erythrocyte-specific H5. There is a relatively large difference in molecular weight between H5 and most of the other linker histone variants. The composition of the amino acids shows that the H5 histones are very rich in arginine and lysine. The presence of particular linker histones has been correlated, in some instances, to the particular structure of the chromatin, e.g., H5 abounds in the very condensed erythrocyte chromatin (Neelin et al., 1964; Weintraub, 1978; Destrée et al., 1979). Testis-specific $H1^t$ appears in the pachytene spermatocytes (Seyeden and Kistler, 1980). $H1^0$ is predominant in quiescent cells (Pehrson and Cole, 1981), or particular regions of the chromatin that are not actively transcribed (Roche et al., 1985).

Histones can undergo several postsynthetic modifications: acetylation, phosphorylation, methylation, ADP-ribosylation, and, in the case of H2A and H2B, covalent linkage to the protein ubiquitin (for review see Isenberg, 1979; McGhee and Felsenfeld, 1980). Although these postsynthetic modifications provide ample opportunities for structural alterations in the chromatin, causal connections between these modifications and chromatin structure or function have not yet been elucidated (for review see Wu et al., 1986).

From the above it may be clear that the histones, especially the class of linker histones, might play a role in the regulation of gene expression. This role could consist of a direct effect of the expression of a specific histone gene variant on the chromatin structure, which in turn could affect the transcription of the gene itself or of other genes. However, up to this date, little definitive evidence exists for such a direct role of histones in the regulation of transcription.

14.3. THE ORGANIZATION OF HISTONE GENETIC INFORMATION

14.3.1. Sets of "Complete" Clusters That May Lie in Tandem

The histone genes of *Xenopus laevis* are largely organized in clusters (Destrée et al., 1984). Each of the assayed clusters seems to be "complete," i.e., each contains the information for five histones, one out of each of the five histone classes.

The eight *Xenopus laevis* histone-gene clusters cloned and analyzed by Destrée et al. (1984) differed in the order in which the histone genes were present. This is in line with the results obtained by Old et al. (1982) who concluded that there was homology between different histone gene clusters of *Xenopus laevis*. However, the clones obtained by Old et al. (1982) were mere fragments of complete clusters, so that these authors were not yet able to distinguish between different variant linker histone genes. In agreement with the results of Zernik et al., (1980) who detected H1-gene variation between clusters, we grouped the "complete" clusters into three categories, based on whether they encoded the H1A, H1B, or H1C linker histone. We noted that within each category the length of the "complete" cluster and the order of the genes might be homogeneous, at least for the few clusters (3 for H1B, 3 for H1A) that we had analyzed. Within each category the restriction maps of the clusters were similar, though not identical.

Some of our clones contained parts of what seemed to be neighboring clusters, such that it is likely that the clusters are arranged in tandem. Indeed, part of the clusters show some tandem repetition as can be concluded from the restriction enzyme mapping of genomic DNA (Van Dongen et al., 1981). The cluster encoding H1A is most likely organized in tandem repeats as judged from Southern blot analysis of genomic DNA (Perry et al., 1985). Because the H1B genes appeared to be more heterogeneous, Perry et al. (1985) doubted that the H1B containing clusters would be arranged in tandem. Comparison of the heterogeneity between three H1A clusters with that between three H1B clusters (cf. Destrée et al., 1984) does not lend much support for such a distinction between H1A and H1B clusters. The information concerning the H1C cluster type is too little to allow speculation on organization in the genome. Its repetition frequency is not known and until now only two clones of this cluster type have been isolated (Turner et al., 1983; Destrée et al., 1984; H. Schilthuis, personal communication).

It is not certain if the tandem arrangement would have the H1A, H1B, H1C clusters in random order; it remains possible that the *Xenopus laevis* histone genes are organized in three (or perhaps more) sets of tandemly repeated complete clusters, each set carrying its own linker variant. Within each set the clusters might all have the same gene order, but would exhibit differences in nucleotide sequence at least to the extent that their restriction maps would differ. These differences would be more frequent than within the homogeneous major histone gene cluster arrangement in the genome of *Xenopus borealis* (Turner and Woodland, 1983) or within the tandemly repeated clusters in sea urchin (for review see Kedes, 1979; Hentschel and Birnstiel, 1981; Maxson et al., 1983a) *Drosophila* (Lifton et al., 1978) and the newt (Stephenson et al., 1981). Indeed, the sequence data from Perry et al. (1985) of two gene clusters suggest that the variation between histone genes lying in different types of clusters (e.g., two H3 genes one lying in an H1A-type and the other lying in an H1B-type cluster) is much higher than the variation between histone genes lying in the same type of cluster (e.g., two H3 genes lying in different H1A-type clusters) (see also the restriction enzyme maps of clusters in Destrée et al., 1984).

In sea urchin the histone genes are organized mainly in three different sets expressed consecutively during the development of the sea urchin. The so-called "early" histone genes of sea urchin are organized in "complete" clusters. Typically, such "complete" clusters are repeated in tandem for up to several hundred times (Hentschel and Birnstiel, 1981; Maxson et al., 1983a). In some sea urchin species there are two such distinct tandem arrays, differing in the number of clusters they contain. There is an extensive homology between the individual clusters of "early" sea urchin genes; all five histone genes are present in the same order and orientation and the clusters have very similar restriction maps (Kedes, 1979).

A second set of histone genes are expressed at the protein level before the "early" genes, i.e., during cleavage stages. The genes encoding this set of histone proteins are still to be isolated and characterized. The third set of histone genes expressed during adult stages in sea urchin has been better characterized. These "late" histone genes have a low reiteration level (approximately 10 fold) and the sequence variation between the "late" histone gene variants is high. The organization of these "late" sets of genes shows no clustering and no head to tail arrangement on the same chromosome (Maxson et al., 1983a).

There is a functional difference between the "early" and the "late" sets of histone genes in sea urchin. The "early" genes are needed in high amounts to secure the high demand for histones during rapid cleavage stages and are not expressed during adult life. The "late" genes are primarily expressed later in development and during adult life and are therefore not needed in high amounts (Maxson et al., 1983b). The histone proteins of the "early," "cleavage-stage" and "late" genes can be distinguished by polyacrylamide gel electrophoresis (Newrock et al., 1978; Maxson et al., 1983a). The extensive differences between them are probably reflected in differences in structural details in the encoded chromatin (Simpson, 1981). The switch between the expression of the "early" and the "late" genes (Knowles and Childs, 1984; Busslinger and Barberis, 1985) suggests a functional role for the switch between the two kinds of chromatin.

The histone genes of chicken, mouse, and man also appear in multiple copies in the genome, although the repetition frequency is smaller than in *Xenopus laevis*. However, there is considerable heterogeneity among the different genes belonging to the same histone class (Maxson et al., 1983a, b, c). In view of the multitude of different genomic restriction fragments of genomic DNA hybridizing with a single histone, it is considered unlikely that there is much order in the arrangement of histone genes in these species (Maxson et al., 1983a,c). In fact, many of the isolated histone gene clones of these species, as well as yeast, do not encode all histone gene classes (cf. Maxson et al., 1983a; Hentschel and Birnstiel, 1981): in these species most histone genes are not organized in "complete" clusters, although some incomplete clustering may occur. Thus, they are more comparable to the "late" histone genes of sea urchin.

The question of whether or not the gene coding for the linker-histone H5 variant of *Xenopus laevis* is also clustered with nucleosomal genes remains unanswered. The lack of a proper DNA hybridization probe precluded the isolation of a genomic clone. However, in none of the 15 genomic clones isolated and checked for the presence of histone genes by hybridization selection procedures, an H5 gene was detected. The existence of solitary ("orphon," Childs et al., 1981) histone genes has been reported for other organisms than *Xenopus*. Of the more than 200 histone gene clones of *Xenopus laevis* that we isolated through cross hybridization with nucleosomal histone genes, more than 97% contained more than one histone gene, indicating that few, if any, nucleosomal histone genes of *Xenopus laevis* are orphons. In the chicken at least one, the H5 gene, is surrounded by 41,000 basepairs that do not encode nucleosomal histone genes (Krieg et al., 1983).

14.3.2. What Might be the Function of the Organization of Histone Genes in Clusters?

One may ask if the organization of histone genes in clusters plays a role in the regulation of histone gene expression, serves some other function, or is perhaps just an evolutionary relic. Before discussing possible functions of the histone gene organization, we recall the point made by Scherrer (1980) and others that a strategy of regulation of gene expression in which each individual gene would have its own, unique regulator, is impracticable. Hence, organisms will tend to reduce the number of messengers involved in the regulation of gene expression.

In organisms in which the rate of transcription of histone genes is a limiting factor at some point of the life cycle, the optimal situation would be that in which all histone genes would have the most active promoter that is possible. In such a situation, the rate of synthesis of the protein part of chromatosomes is limited by the copy number of the histone gene class that is the least reiterated; in as far as the histone genes of the other classes have higher copy numbers, they will be overproduced. Thus, the optimal genome would have the genes for the different histone classes present at stoichiometries corresponding to the composition of chromatin. With the exception of the linker histone class, the histone gene copy numbers in *Xenopus laevis* are in line with this scenario: they are equal (50) (Van Dongen et al., 1981; Perry et al., 1985). The total number of linker-histone genes exceeds 35 per haploid *Xenopus laevis* genome (Perry et al., 1985), whereas a number of 25 would be expected in the present scenario.

"Early" sea urchin H1, H2B and H3 mRNAs are present and synthesized in approximately a 1:2:2 ratio, i.e., the ratio at which the corresponding proteins occur in the chromatosome (Maxson et al., 1983c). In exponentially growing HeLa cells, the transcription of histone genes, as determined through Northern blot hybridization, has been shown to be stoichiometric (Baumbach et al., 1984).

Although this accounts for the fact that the nucleosomal genes are reiterated to the same extent, it does not account for their organization in clusters. However, if at other points in time, maximal rates of histone synthesis are not needed, the first possible function of the cluster organization we shall consider here would be "stoichiometric regulation," i.e., allow for a simple strategy to achieve synthesis of the histones at slower but equal rates: leaving the transcription rate of all histone genes in all the transcribed clusters maximal, the number of transcribed clusters could be reduced. The equimolarity of nucleosomal histone gene expression would remain.

This mechanism is by no means universal in the insurance of equimolarity of gene expression. In hemoglobin equal amounts of α-globin and β-globin are present, whereas the genes coding for these two different globin chains are present on different chromosomes (Deisseroth et al., 1977, 1978). In that case we do see a different type of clustering of the genes. The α-like globin genes (which are expressed alternatedly rather than stoichiometrically) are clustered and the β-like globin genes are clustered. The cluster probably evolved by gene duplication and subsequent divergence (Jeffreys, 1982).

With respect to the clustering of the histone genes in *Xenopus laevis*, a transcriptional expression study (Koster et al., 1988) showed that, independently of the type of cluster expressed and independently of the time in development, the nucleosomal histone genes are not always transcribed into equal amounts of mRNA. We conclude that the organization of histone genes in clusters does not serve the purpose of guaranteeing equimolar transcription. Presumably, some sort of translational regulation reinstates the equimolar expression of the nucleosomal histone genes that has been disrupted at the level of mRNA. Of course, this conclusion depends on the assumption that throughout early embryogenesis chromatin requires nucleosomes of the canonical composition. Though a likelihood, this assumption may not be absolutely safe, since it has been shown that nucleosome-like particles can be formed from H3 and H4 histones alone (Simon et al., 1978; Stockley and Thomas, 1979). On the other hand H2B repression in yeast eliminates cell division (Han et al., 1987).

Although the cluster organization does not seem to enforce equimolar expression of the nucleosomal histone classes, it might still serve the related purpose of keeping the transcription rates of the histone genes constant relative to one another. If the transcription of histone genes were to be reduced at a given time in development, this might be achieved by reducing the transcriptional activity of each cluster as a unit, or by completely shutting down transcription of some of the clusters. In either scenario, the expression ratios of the histone genes of the five classes would not be affected by the transcriptional regulation.

We found indications for a transition in the rate of histone gene transcription approximately 11 hours after fertilization (Koster et al., 1988). Around this transition there was little indication of a change in the transcription rates of the nucleosomal histone genes relative to one another. This would be consistent with regulation by turning off a number of clusters, or by reducing the transcriptional activity of each cluster as a unit. Of course, it cannot be excluded that all nucleosomal genes would be regulated individually to produce the same reduction in transcription rate; such a type of regulation would, however, be more costly in terms of the number of required regulatory elements.

The cluster type of organization of the histone genes could have a second function, i.e., that of variant coordination. In this scenario each cluster would contain a set of histone gene variants together coding for a specific combination of histone variants producing chromatin with special properties. By transcribing different clusters at different times in its development, an organism would be able to vary the properties of its chromatin and perhaps also the transcription pattern of the genes enclosed therein, during development. In *Xenopus laevis* it would seem possible that at some time in development, the H1A clusters would be uniquely expressed, at other times the H1B clusters, and at yet other times the H1C clusters. However (Perry et al., 1986; Koster, 1987) in *Xenopus laevis'* early development there is little evidence for unique expression of certain histone-gene variants at certain points in time.

The third possible function of the clustering of histone genes that serve high rates of histone synthesis in early development might lie in ease of regulation. The physical proximity would allow the expression of a large number of genes to be shut off by a single event that would repress a region of the genome. This however would

Organization of Histone Genes

not seem to require clustering in the scene of units of five genes, one for each histone class.

The cluster type of organization could also be a consequence of the evolutionary origin of the genes. In organisms with rapid replication cycles (high rates of DNA synthesis) like sea urchin, *Drosophila*, and *Xenopus* the need for multiple histone genes is evident (see Section 14.4). In these organisms the individual in possession of multiple histone genes would be favored during rapid growth. For organisms in which a major drive for the evolution of histone genes has been towards more rapid growth of the early embryos, it is understandable that histone genes have been amplified, starting from one or two single histone gene clusters. The existence of tandem repeats of the "early" histone genes in several species of sea urchin suggests this. The same is true for *Drosophila*, in which at least two tandem repeats of histone genes have been detected (Lifton et al., 1978). This then may explain the tandem repetition of histone gene clusters in *Xenopus* (see above).

As described by several authors (Dover, 1982; Maxson et al., 1983; Busslinger et al., 1983), several types of recombination, such as unequal cross over and gene conversion, can be involved in the maintenance of the homogeneity among repeated gene families. The extreme homogeneity in the tandem repeats of histone gene clusters in sea urchin (the early histone gene sets) and *Drosophila* could be explained in this way.

The reason that the three cluster types identified in *Xenopus laevis* exhibit larger heterogeneity between each other than between the elements of each cluster type, might be the consequence of different origins of the different cluster types at different times in the evolutionary past.

Although, repetition of histone genes is favorable for the organisms in which DNA replication is very rapid in early development, these organisms could do with fewer histone genes as soon as the cell-cycle time lengthens. Then histone gene sets located at other places in the genome (far from the tandem-repeated clusters) might be selectively expressed, causing changes in chromatin structure. In *Drosophila* no such second, non-tandem repeated histone gene set has been described. Sea urchin and *Xenopus* might fit in this scenario. In the former organism the late genes are not present in complete clusters, although semi-complete clusters (comprising H3+H4 genes) occur. Northern blot experiments reveal 4 different "late" H3 mRNAs, which may occur at somewhat different points in time (e.g., Kaumeyer and Weinberg, 1986).

In organisms that lack a rapid early development in terms of DNA synthesis, one would not expect tandemly reiterated histone genes. Indeed, human histone genes, which are reiterated some 15 fold, map to different chromosomes (Tripputi et al., 1986). Some of the histone genes do occur in (single) "complete" clusters

(Carozzi et al., 1984).

In yeast, the amount of DNA per cell is not high enough to cause significant capacity problems for the histone synthesizing apparatus (Maxson et al., 1983c). Yeast has merely two gene copies per haploid genome for each histone class (Smith, 1984). The histone genes are sometimes organized in semi-complete clusters (a cluster containing H2A and H2B and one containing H3 and H4) (see however Choe et al., 1985; Matsumoto and Yanagida, 1986).

14.4. KINETICS OF HISTONE SYNTHESIS IN *XENOPUS LAEVIS*: THE RACE AGAINST TIME

14.4.1. Exponentially Dividing Cells

In many organisms, fertilization is followed by a lag phase, in which no DNA synthesis occurs and subsequently by a number of cleavages with concomitant DNA synthesis. In invertebrates and amphibia these early cleavages are very rapid and the time in between them tends to be constant. This suggests that, after the initial lag phase, the DNA content of the embryos of these organisms increases exponentially with time. If this phase of exponential growth would continue, the amount of DNA would become extremely high at times long after fertilization, and so would the need for histones to package the DNA into chromatin. It seems questionable if the capacity of the embryos to synthesize histones supplemented by the inherited maternal store of histones would be sufficient to satisfy this need.

At any point in time, the concentration of histone protein relative to the concentration of DNA depends on a number of factors (Figure 14.2). These include the rate of transcription of histone genes, the stability of histone mRNAs, the rate of translation of histone mRNAs, the rate at which these mRNAs are translated to protein, as well as the amounts of DNA, histone mRNA and histone protein at zero time. As a consequence it is difficult to estimate in a qualitative argument whether there will be sufficient histone proteins throughout early embryogenesis. We therefore developed a quantitative analysis of histone gene expression in early development (Koster et al., 1988, Koster, 1987), essentially by integrating the relevant differential equations at the appropriate boundary conditions. A special case of this analysis was the hypothetical case in which there would be no end to the exponential growth early in development.

The first question one may ask is whether or not, assuming maximum rate constants for transcription and translation, the histone genes present in the early embryo would provide sufficient template for synthesis of histone mRNA directing synthesis of sufficient histone protein. The lower dashed line in Figure 14.3 shows the development of the calculated histone to DNA (mass/mass) ratio (for the canonical chromatin structure this ratio should be 1) (note the logarithmic scale for the mass/mass ratio). It is clear that there is a problem for early *Xenopus* embryos: even with high rate constants for transcription and translation (Koster et al., 1988) the ratio of histone protein to DNA would be far below 1.

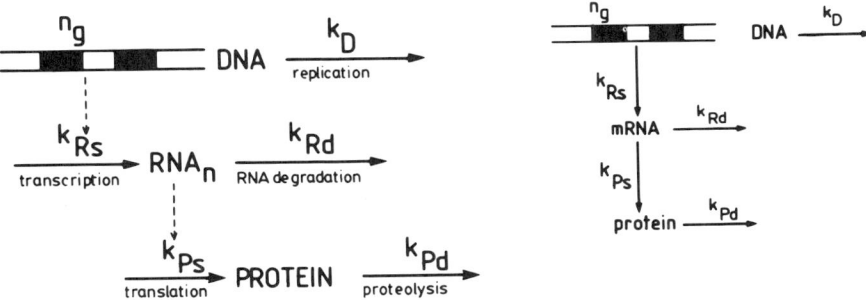

Figure 14.2. Simplified scheme of gene expression in a eukaryotic cell (For a detailed description of the parameters, see Koster et al., 1988).

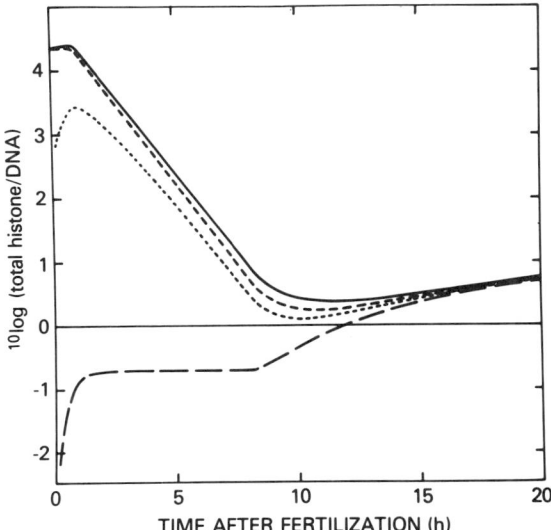

Figure 14.3. Histone to DNA (mass/mass) ratios as a function of time expected for exponentially dividing cells. Full line: with protein and RNA storage. Upper dashed line: only protein storage. Dotted line: only RNA storage. Lower dashed line: no storage (From Koster et al., 1988, Koster, 1987).

In *Xenopus laevis* embryos, there are histones synthesized on maternal mRNA, and there are residual maternal histones in addition to histones synthesized since fertilization on *de novo* synthesized histone mRNAs. Taking these two forms of storage into account, we calculated (cf. the full line in Figure 14.3) that there is a clear (approximately 22 thousand fold) excess of histones over DNA at fertilization. Because of the translation of maternal mRNA, this excess slightly increases for as long as the lag phase in DNA synthesis lasts. Then however, DNA synthesis is sufficiently rapid to cause a continuous decrease in this ratio such that it would drop below the critical ratio of 1 at about 9.9 hours after fertilization.

If maternal histone mRNA (but not maternal histone protein) would have been absent (the upper dashed line), the picture would hardly have been different. Apparently, for the first few hours, the maternal mRNA directs some histone synthesis, but this soon stops, because of the limited life time of these mRNAs (3 h, (Woodland and Wilt, 1980a,b)). In the absence of the inherited mRNAs the histone to DNA ratio would have dropped below the critical ratio of 1 at about 9.5 hours after fertilization only some 20 min earlier than in the actual case described above. In the absence of stored histone protein, but in the presence of maternal mRNAs (dotted line in Figure 14.3), the simulation predicts that the (invisible in Figure 14.3) initial shortage of histones would be rapidly overcome through translation of maternal mRNA. Also here however, the ratio would begin to decrease once DNA replication starts. The critical ratio would be reached at 9.0 h after fertilization.

Thus, for the early exponential growth phase, *de novo* synthesis of histone mRNA and protein is insufficient, but the early embryo may solve this problem by making use of histone mRNA (cf., Woodland, 1980, 1982) and/or protein stored in the oocyte. However, in the case that growth would continue to be exponential, storage of histone mRNA and protein would fail to help out around 10 hours after fertilization (cf. Figure 14.3).

Evolutionarily, a possibility to ensure a histone to DNA ratio of 1 in exponentially dividing cells, would have been to increase the number of genes coding for histone. In Figure 14.4 the histone to DNA ratio attained for exponential growth at times long after fertilization, is given as a function of the number of genes encoding histones for *Xenopus laevis* (full line) and two other species (i.e., sea urchin and *Drosophila*, represented by the dotted line and the dashed and dashed/dotted lines respectively). As was already concluded from Figure 14.3. in *Xenopus laevis* the 50 genes per haploid genome coding for each nucleosomal histone can not supply sufficient histones in order to structure the DNA in a 1:1 mass ratio with histones. In order to reach the critical value of 1, the *Xenopus laevis* embryo would need at least 268 gene copies for each histone. Due to the longer time in between two cell divisions (90 versus 35 minutes) and the lower amount of DNA per haploid genome

Organization of Histone Genes

(0.89 versus 3.15 pg) the sea urchin embryo needs only 15 copies of each histone gene to attain a histone to DNA ratio of 1 for exponentially growing embryos (assuming the same rates for transcription and translation as in the case of *Xenopus laevis*) (see also Maxson et al., 1983c). The latter number is much lower than the real reiteration of the early histone genes in the sea urchin (i.e., 400 fold (Hentschel and Birnstiel, 1981)). In *Drosophila*, 83 copies of each histone gene per haploid genome are needed when we assume that the cell cycle time is 14 minutes (the average cell cycle time early in development of this species), whereas 4 genes are sufficient assuming a cell cycle time of 75 minutes (the average nuclear doubling time during gastrulation (Anderson and Lengyel, 1984). In both cases this number is lower than the number of histone genes per haploid genome determined in *Drosophila* (approximately 100 genes for each histone) (Lifton et al., 1978). A cell-cycle time as short as 8 min (the length of the first nine cycles in *Drosophila* development (Edgar et al., 1986) would bring the need of histone genes above the 100 genes available (i.e., 254).

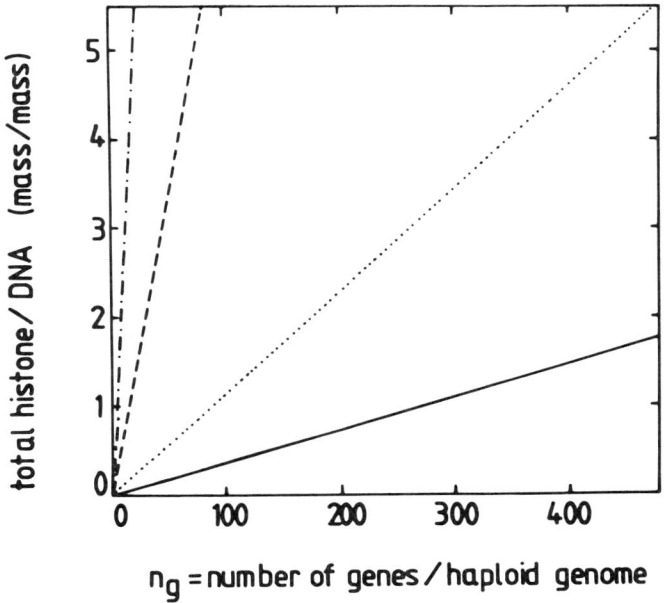

Figure 14.4. Ultimate histone to DNA (mass/mass) ratio as a function of the number of genes encoding the histones for exponentially dividing cells. The full line gives the case for *Xenopus laevis*, the dotted line for sea urchin (90 min cell cycle time), the dashed line (14 min cell cycle time) and the dashed/dotted line (75 min cell cycle time) for *D. melanogaster* (cf., Koster, 1987).

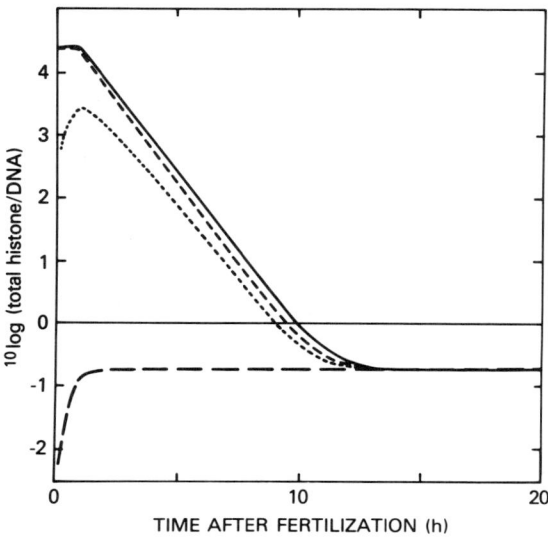

Figure 14.5. Histone to DNA (mass/mass) ratios as a function of time calculated for early development of *Xenopus laevis* as determined by use of empirical DNA contents (cf. Figure 14.1). Full line: with protein and mRNA storage. Dashed line: only protein storage. Dotted line: only mRNA storage. Full/dashed: no storage. DNA (t) was taken from Dawid (1965) and Newport and Kirschner (1982) from Koster et al. (1988).

From the above it can be concluded that with its reiteration of the histone genes, sea urchin does not require any maternal storage to provide sufficient histones to form chromatin with the newly synthesized DNA. In *Xenopus*, however, exponential growth for an infinite time would cause a problem regarding the histone to DNA ratio: its histone genes are not reiterated 268 fold. (In *Drosophila* the same problem would occur with extension of the extremely rapid growth to infinite time.)

14.4.2. The Actual Growth of *Xenopus* Embryos

In the absence of a sufficient number of histone genes to support histone synthesis during continued exponential growth, there is an alternative strategy: slow down DNA synthesis when synthesis and storage no longer keep up. Indeed, the growth of *Xenopus* embryos in terms of DNA content is not exponential for an infinite time. After the period of exponential growth (Newport and Kischner, 1982), the cell-cycle time increases (Dawid, 1965). The question is whether the slowdown of the DNA synthesis after 8.1 h is sufficient to prevent the shortage of histones that occurred in Figure 14.3. Indeed (Figure 14.5), due to the slowdown of the DNA

Organization of Histone Genes 459

synthesis the histone/DNA ratio does not cross the critical value of 1. Depending on the type of storage, histone protein, histone mRNA or both, the histone/DNA ratio reaches a minimum (of 1.6, 1.2, and 2.1 respectively) at different times (i.e., 11 h, 10 h and 11.5 h, respectively) and then increases again. In (Koster et al., 1988) we checked that these calculations are realistic by comparing the implied concentrations of histone mRNA to those measured experimentally Koster et al. (1988). Apparently, *Xenopus* solves its problem of providing sufficient histone protein for its new DNA by storing histone protein or histone mRNA in its eggs and be decelerating the rate at which its embryo divides around 8 hours after fertilization.

14.5. THE CONTROL OF HISTONE-GENE EXPRESSION IN *XENOPUS LAEVIS*

14.5.1. Principles: Control Versus Regulation

Biological chemistry has entered the arena of quantitative sciences (Van Dam, 1986; Westerhoff and Van Dam, 1987). Quantification of the extent to which gene expression is controlled or regulated, has, however, lagged behind quantifications of metabolic processes. In this section we shall attempt to reduce this lag. For discussions about control it is important to distinguish two aspects. One aspect asks the question above which steps limit (or "control" in the narrower sense of the word used in the discussions of the control of metabolism (Groen et al., 1982)) the rate at which a gene is expressed. The other aspect asks at which step the expression is actually regulated by the cell.

To illustrate the two aspects and the difference between them, we consider the scheme of gene expression of Figure 14.2. In this scheme a gene present at copy number n_g in the DNA is transcribed into RNA which is degraded at some rate. The RNA directs protein synthesis. Protein concentrations are limited by proteolysis. To simplify the present discussion (In fact too little is known about the systems we shall discuss, to warrant a more detailed model), the scheme neglects a number of additional, potentially relevant steps, such as RNA processing and transport, and polypeptide processing and transport. In the scheme the concentrations of DNA, histone mRNA and histone protein, as well as the rates of DNA replication, transcription, translation, RNA degradation and protein degradation are variables: they vary with time. At any point in time their value is determined by the magnitudes of the parameters in the system and by time itself. These parameters are invariant unless their values are changed by some act of regulation. They include the activities (characterized by their respective unimolecular rate constants) of the five processes in Figure 14.2: DNA replication (k_D), transcription (K_{Rs}), RNA degradation (k_{Rd}), translation (k_{Ps}), and proteolysis (k_{Pd}). Five additional parameters are the DNA (DNA(0)), histone mRNA(0)), and histone protein (PROT(0)) concentrations at fertilization, the number (n_g) of genes per haploid genome encoding the histone in question and the fraction of the cell cycle that is available for DNA and histone synthesis.

14.5.1.1. Control of Gene Expression

The first type of question one may ask in discussions of control is to what extent the magnitude of each of these parameters influences the concentration of a certain protein, or the magnitude of a certain parameter. To make the answer independent of the magnitude of the parameter change, the limit to (infinitesimally) small changes is considered. For the "Control Coefficient" indicating to what extent the concentration of histone H3 would be controlled by the rate constant of transcription, this definition turns out to be:

$$C_{Rs}^{[H3]} = \frac{d[H3]}{[H3]} \bigg/ \frac{dk_{Rs}}{k_{Rs}} = \frac{d(\ln[H3])}{d(\ln[k_{Rs}])}$$

In this differentiation, the magnitudes of the other parameters (but not of variables, such as the RNA concentration) should be held constant. The time at which (H3) is considered should be specified if non-steady states are considered, but may differ from the time at which k_{Rs} was changed. With this definition, this concentration control coefficient is virtually equal to the percentage increase in H3 concentration resulting from a sudden 1% increase solely in the transcription rate constant. Analogous definitions pinpoint the extent of control of the concentration of H3 by DNA replication, RNA degradation, translation and proteolysis. It may be noted that the latter control coefficient would be negative, as an increase in proteolysis would lead to a decrease in the concentration of H3.

14.5.1.2. Regulation of Gene Expression

In case $C_{Rs}^{[H3]}$ would be close to 1, the histone concentration would depend proportionately on the rate constant of transcription; one might call transcription completely limiting. It is important to note that this does not imply that the cell actually regulates the concentration of H3 by modifying the rate constant of transcription. Parameters with high control coefficients are potential points for regulation, but not more than that. Thus, the second type of control question, i.e.,

Organization of Histone Genes

at which points the organism regulates its gene expression, differs from the former (section 14.5.1.1) type of question, which asks which points have the highest control coefficients.

It is also important to define what we will mean if we speak of regulation at the translational level. It might seem that, if not due to a decrease in the rate constant for proteolysis, an increase in H3 concentration would have to be due to regulation at the translational level: the rate of translation would have to be increased. However, we shall only call this translational regulation if the increase in translation rate and consequently in H3 protein, is not due to an increase in the concentration of H3 mRNA, but due to an increase in k_{ps}. If the change in translation rate is due to a change in the concentration of mRNA and not due to a change in k_{Ps}, we shall speak of regulation at the pretranslational level, which may be at the transcriptional or at the RNA degradation level.

With respect to questions of regulation, it is important to realize that the magnitude of a concentration (or a rate) at one point in time cannot be said to be regulated. For instance, the question as to what regulates the concentration of the histone H3 6 hours after fertilization cannot be asked (the question as to what controls this concentration can be asked, see above). The question of regulation only comes in when one considers changes in concentrations or reaction rates that would not have occurred if no parameter value would have been changed. Here it should be noted that in non-stationary situations, concentrations can change in the absence of any regulation. For instance, although the histone/DNA ratio in Figure 14.3 varies strongly with time, these changes occur at constant magnitudes of all the parameters in the system and are therefore not due to regulation.

Below we shall discuss the control and the regulation of histone gene expression in the early development and erythrogenesis of *Xenopus laevis*.

14.5.2. Control and Regulation of Histone Gene Expression in the Early Development of *Xenopus Laevis*

14.5.2.1. Which Parameters have the Greatest Impact on the Histone/DNA Ratio?

With respect to the control of the concentration of the nucleosomal histones (We shall here again focus on histone H3, cf., Section 14.4.1), Figure 14.5 is illuminating. It reveals that until some 10 hours after fertilization, time is an important factor (full line in Figure 14.5). Later, the histone to DNA ratio becomes rather independent of time. In fact, when more than 10 hours have passed since fertilization, the histone to DNA ratio is virtually constant. Using the definitions given in Section 14.5.1, and the values for the kinetic constants used in our model

(Koster et al., 1988b), one can calculate the magnitudes of the different control coefficients with respect to the histone/DNA ratio late in embryogenesis. For the number of histone genes, transcription, the fraction of the cell cycle the cell spends in S phase, and translation the control coefficients are equal to 1. For RNA degradation and proteolysis the control coefficients are equal to -0.87 and -0.01, respectively, whereas DNA replication has a control coefficient of approximately -1.12. For obvious reasons, the amounts of histone mRNA and histone protein stored in the oocyte no longer determine the histone/DNA ratio late in embryogenesis; their control coefficients equal zero. In these calculations we used a cell cycle time of 20 h (this is the actual cell-cycle time 30 and more hours after fertilization (Dawid, 1965)). These results indicate that proteolysis is rather irrelevant for the histone to DNA ratio and stress the easily-overlooked importance of the rate at which the DNA replicates. Also noteworthy is the importance of the stability of the histone mRNAs.

With respect to the control of the histone to DNA ratio earlier in embryogenesis we note (cf. Figure 14.5) that especially before $t=11$ h, both the inherited histone mRNA and the inherited histone protein are significant factors in determining the histone/DNA ratio. Also for the minimum histone/DNA ratio both these parameters as well as the translation rate constant, and to a lesser degree the transcription rate constant and the number of genes are important, whereas proteolysis is irrelevant (Koster et al., 1988).

14.5.2.2. Some 12 Hours After Fertilization Regulation Occurs

Figure 14.5 shows that in the absence of any regulation other than the regulation of DNA replication the histone to DNA ratio at tailbud stage would amount to some 34 (w/w) rather than the approximate 1 (or slightly below 1) that has been measured in and after blastula stage (Destrée et al., unpublished). Consequently, some regulation must take place to prevent the increase in histone to DNA ratio predicted after 10 h post fertilization in the absence of any regulation. Although the above magnitudes of the control coefficients tell us which processes are influential with regards to the histone to DNA ratio late in embryogenesis, they do not indicate at which point the organism actually intervenes to regulate the histone/DNA ratio.

A distinction between regulation at the translation or protein degradation level and regulation at any pretranslational level is that in the former case the histone mRNA concentration would have become much higher than in the latter. The dashed line in Figure 14.6 represents the maximal amount of mRNA that can be synthesized from the available DNA plus what is left of the maternal mRNA storage at that time. For the case that the transcription rate (k_{Rs}) would be decreased 14 fold by 11.35 h, the RNA(t) (i.e., number of H3-mRNAs at any time t after fertilization) is given by

Organization of Histone Genes

the solid line. The triangles in this figure represent the measured values for the number of H3-mRNAs (Koster et al., 1988).

At times more than 12 h after fertilization, the mRNA concentrations calculated for the case in which there would be no regulation (- - - in Figure 14.6) greatly exceed the observed H3-mRNA concentrations (triangles). On the other hand, if a single regulatory event (at the level of transcription) is assumed to occur around 11 h after fertilization (the full line in Figure 14.6), the calculated mRNA concentrations fit the experimental RNA(t) well. Based on the observed data a possible gene expression scenario (though not the only one, see below) would be, that in the early *Xenopus* embryo all histone genes are transcribed until 11.35 h. After that time the embryo can manage with the transcription of 4 histone genes (of each class) assuming the same constants for transcription as before, or with transcription of the same number of genes at a 50/4 times lower rate, or with transcription of the same number of genes at the same rate, but only during 4/50th of the cell-cycle. An alternative scenario, in which at t=12.75 the stability of histone mRNA would suddenly be decreased by a factor of 16.5 (dotted line in Figure 14.6), fits the experimental data equally well as the scenario in which the transcription was suddenly decreased.

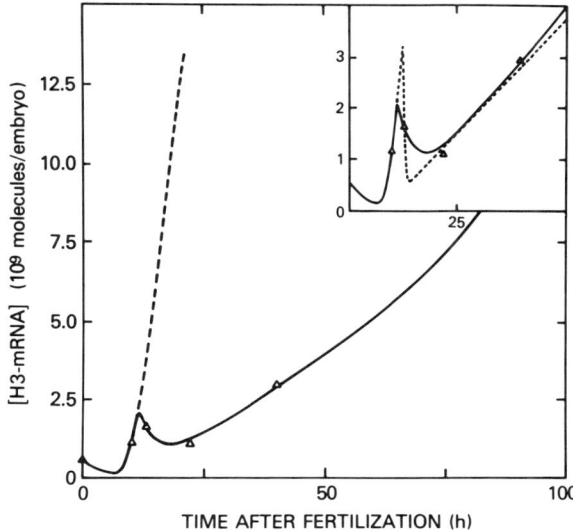

Figure 14.6. Number of H3 mRNA molecules per embryo as a function of time. ▵:Experimental results from Koster et al., (1988). Dashed line: As predicted by the standard model, with constant transcription rate per cell. Full line: As predicted by the modified model, in which the number of transcribed H3 genes (or the transcription rate constant) is suddenly reduced by a factor of 14 at t=11.35 h. Dotted line in inset: As predicted by the modified model, in which the RNA degradation constant (kRd) is suddenly increased by a factor of 16.5 at t=12.75. From Koster et al. (1988).

When we used the RNA(t) function of either scenario that fit the observed mRNA data (cf. Figure 14.6) and calculated the histone to DNA ratio, we obtained ratios close to 1 for times later in development (Koster et al., 1988). Thus, the single regulatory event proposed above accounts simultaneously for the observed RNA concentrations and deals with the problem that if unregulated, there would be a wide excess of histone protein late in development.

We conclude that approximately 12 hours after fertilization one or a combination of the following regulatory events must occur: (i) the fraction of the cell cycle in which histone gene transcription occurs is reduced by a factor of 14, (ii) the number of transcribed histone genes (n_g) is reduced 14 fold, (iii) the rate constant of histone gene transcription (k_{Rs}) is reduced by a factor of 14, (iv) the stability of the histone mRNAs is reduced 16 fold (k_{Rd} is increased 16 fold).

In cultured somatic cells, the largest part of histone gene transcription is limited to the S phase. Thus, a reduction of the fractional time the embryonic *Xenopus* cell spends in the S phase, might lead to a reduction in the average rate of transcription of histone genes. Around 12 hours after fertilization the cell-cycle time has increased from 35 minutes to approximately 3 h (Dawid, 1965). If we would assume that the S phase would have retained its duration of approximately 30 minutes, then the fraction of the cell cycle taken by the S phase would have been reduced by a factor of 6 only. Thus, by itself this falls short of the required factor of 14. Moreover, as the cell cycle time increases, also the duration of the S phase tends to lengthen (Graham and Morgan, 1966), further reducing the above factor of 6. We conclude that although mechanism (i) may occur additional regulation must be operative.

In HeLa cells fewer variant histone genes are transcribed outside the S phase than during the S phase (Wu and Bonner, 1981; Wu et al., 1984). In the early development of sea urchin the transcription of a special class of "early" histone genes dwindles just after the hatching blastula stage (Childs et al., 1979; see however, Angerer et al., 1985). These two cases correspond to mechanism (ii), mentioned above. In *Xenopus laevis* embryos we found no indications that certain histone genes are shut off specifically about 12 hours after fertilization; the time dependence of the mRNA concentrations was rather similar for the (limited) number of histone-gene variants studied. To this the only exception seemed to be that the reduction in amount of mRNA per embryo around gastrula stage, seemed to take place earlier for H1A than for H1B type mRNA (Koster et al., 1988). Of course the possibility cannot be excluded that similar fractions of the H1A and the H1B genes are inactivated at around 12 hours after fertilization.

Option (iv), though apparently wasteful would not be without parallel: it occurs at the end of the S phase in somatic tissue culture cells (Heintz et al., 1983;

Organization of Histone Genes 465

Alterman et al., 1984; Schümperli, 1986) in *Drosophila* at the syncytial blastoderm stage (Anderson and Lenglyel, 1980; 1984) and in sea urchin where the early mRNAs are rapidly degraded after the hatching blastula stage (Maxson and Wilt, 1982). As one may estimate (Koster, 1987), a combination of options (i) and (iv) in the sense that histone gene transcription be largely limited to the S-phase and histone mRNA be rapidly degraded thereafter, could constitute the regulatory event. Here it is relevant that Kimelman et al. (1987) attributed much of what happens at the midblastula transition to changes in the cell cycle.

14.5.2.3. Between Maturation of the Egg and 11 Hours After Fertilization the Only Regulation of the Histone/DNA Ratio is at the Level of the Cell-Cycle Time

For the discussion of the regulation of histone gene expression shortly after fertilization, the distinction we make between control ("limitation") and actual regulation is important: translation seems to be the process with the highest control on the minimum histone/DNA ratio attained during development. Yet, the organism does not regulate translation to save itself from a shortage of histone around ten hours after fertilization. In fact the cessation of the decrease in the histone to protein ratio around ten hours after fertilization seems to occur in the absence of any regulation of gene expression. The major regulation that does occur is the regulation of the cell-cycle time at approximately 9 hours after fertilization (Koster et al., 1988). Woodland (1980) has discussed the regulation of histone gene expression during oogenesis and subsequent embryogenesis. Adamson and Woodland (1974, 1977) had observed that, as the oocyte converts to egg, the rate of incorporation of injected radiolabelled lysine into nucleosomal histones increased. Between 5 and 8 hours after fertilization the embryo staged another increase. Because *in vitro* translation of isolated mRNA (Ruderman et al., 1979) had suggested that the amount of histone mRNA differed little between oocyte and blastula stage embryo, Woodland (1980) concluded that there is translational regulation (translational "control" in his terminology) of histone gene expression both around oocyte maturation and very early in embryogenesis.

It may be noted that in drawing this conclusion Woodland assumed that the absence of a significant difference in histone mRNA between oocyte and blastula reflected that the histone mRNA concentration remains constant in between. This assumption is however at odds with the observation of Woodland and Wilt (1980b) that (sea urchin) histone mRNAs injected into *Xenopus laevis* eggs or early embryos have a half-life of only some 3 h. Experiments with "androgenetic haploids" obtained by U.V. irradiation of the laevis nucleus of the fertilized eggs resulting from crosses of a female *Xenopus laevis* with a male *Xenopus borealis* confirmed that the histone mRNAs inherited from the oocyte exhibited this type of half-life. Since the time span and amount of DNA between fertilization and the stage-7 embryo are too small for an appreciable amount of histone mRNA to be synthesized, these results imply that after maturation the concentration of mRNA in the egg and the embryo should decrease (with a $t_{1/2}$ of some 3 h), such that its concentration 5 hours after fertilization would have decreased to 20% of that in the oocyte (40% of that in the egg). Our calculations (cf. Figure 14.6) confirmed this, but also showed that *de novo* synthesis of histone mRNA would cause a subsequent increase in histone mRNA concentration restoring it to more than 50% (on a per embryo basis) of the oocyte level by the late blastula stage. This could well explain why Ruderman et al. (1979), who took data points only for the oocyte and at the late blastula stage, observed little difference between the histone mRNA concentrations in these two stages (That they missed the predicted 50% decrease may be the result either of the inaccuracy of our calculations, or an inaccuracy in their conclusion; they based their conclusion on visual impression rather than on a quantitative analysis of the fluorographic bands). In fact, more recent measurements by the group of Woodland (Flynn and Woodland, 1980) revealed that the increase in synthetic rates of nucleosomal histones immediately upon fertilization they had suggested earlier (Adamson and Woodland, 1977; Woodland, 1980) may not occur until 5 hours after fertilization (That these experiments do not exhibit a drop in histone synthesis in going from the egg to the stage 7-stage 8 embryo may well be due to incomplete mixing of the injected radiolabelled lysine with the lysine pool of the early embryo).

Since molecular clones of histone genes have become available, it has become possible to measure the concentrations of histone mRNA more specifically and accurately. We (Koster et al., 1988) found not only that the concentration of histone mRNA dropped by a factor of 2-3 between oocyte and blastula stage embryo, but also that this drop is preceded by a more extensive drop and a subsequent rise, in line with what one would calculate (see above) from the known half-lives of the histone mRNA molecules and a reasonable transcription rate constant for histone mRNA synthesis. Consequently, we conclude that, although the increase in synthesis of nucleosomal histones (Adamson and Woodland, 1974; 1977) accompanied by a decrease in the concentration of histone mRNA diagnoses translational regulation of histone gene expression at maturation of the oocyte. There is no such indication for subsequent translational regulation early in embryogenesis. Rather, calculations on the basis of known or estimated properties of *Xenopus laevis* embryos would suggest that after maturation of the oocyte, histone gene expression may well be unregulated (except for regulation at the level of the cell-cycle time; see above) until some 12 hours after fertilization when some pretranslational regulation seems to occur.

14.6. CONCLUDING REMARKS

Histone genes in *Xenopus laevis* tend to be organized in complete clusters. The functional significance of this organization is not yet clear.

If its early exponential growth continued, *Xenopus* would not be able to sustain sufficient synthesis of histones; its newly synthesized DNA could not then be packaged into chromatin. However, the rate of DNA synthesis slows down some ten hours after fertilization. The initial period is bridged by making use of histone protein stored in the oocyte or by translating histone mRNA stored in the oocyte.

During early development the regulation of histone gene expression consists of three parts. First, there is translational regulation at maturation of the oocyte. Thereafter there is only regulation at the level of the rate of DNA replication, until at about 12 hours after fertilization there is repression of histone gene expression at some pretranslational step.

More empirical information is necessary to distinguish between the different pretranslational possibilities of regulation.

REFERENCES

Adamson, E. D. and Woodland, H. R. (1974). Histone Synthesis in Early Amphibian Development: Histone and DNA syntheses are not Coordinated. J. Mol. Biol. 88: 283-285.

Adamson, E. D. and Woodland, H. R. (1977). Changes in the Rate of Histone Synthesis During Oocyte Maturation and Very Early Development of *Xenopus Laevis*. Dev. Biol. 57: 136-149.

Alberts, B., Bray, D., Lewis, J., Raff, M., Roberts, K. and Watson, J. D. (1983). Molecular Biology of the Cell. Garland, New York.

Alterman, R. B., Ganguly, S., Schulze, D. H., Marzluff, W. F., Schildkraut, C. L. and Skoultchi, A. I. (1984). Cell Cycle Regulation of Mouse H3 Histone mRNA Metabolism. Mol. Cell. Biol. 4: 123-132.

Anderson, K. V. and Lengyel, J. A. (1980). Changing Rates of Histone mRNA Synthesis and Turn-over in *Drosophila* Embryos. Cell 21: 717-727.

Anderson, K. V. and Lengyel, J. A. (1984). Histone Gene Expression in *Drosophila* Development: Multiple Levels of Gene Regulation. In: Histone Genes: Structure, Organization and Regulation (Stein, G.S., Stein, J.L. and Marzluff, W.F., eds.). Wiley and Sons, New York, pp. 135-163.

Angerer, L., DeLeon, D., Cox, K., Maxson, R., Kedes, L., Kaumeyer, J., Weinberg, E. and Angerer, R. (1985). Simultaneous Expression of Early and Late Histone Messenger RNAs in Individual Cells During Development of the Sea Urchin Embryo. Dev. Biol. 112: 157-166.

Baumbach, L. L., Marashi, R., Plumb, M., Stein G. and Stein, J. (1984). Inhibition of DNA Replication Coordinately Reduces Cellular Levels of Core and H1 Histone mRNAs: Requirement for Protein Synthesis. Biochemistry 23: 1618-1625.

Berezney, R. and Coffey, D. (1974). Identification of a Nuclear Protein Matrix. Biochem. Biophys. Res. Commun. 60: 1410-1419.

Boulikas, T. (1986). Nucleosomes are Assembled into Discrete Size Structures by Histone H1 In Vitro. Biochem. Cell. Biol. 64: 463-473.

Bureau, J., Hubert, J. and Bouteille, M. (1986). Attachment of DNA to Nucleolar and Nuclear Skeletal

Structures as Visualized by Kleinschmidt Molecular Spreading. Biology of the Cell 56 7-16.

Busslinger, M. and Barberis, A. (1985). Synthesis of Sperm and Late Histone cDNAs of the Sea Urchin With a Primer Complementary to the Conserved 3' Terminal Palindrome: Evidence for Tissue-specific and More General Histone Gene Variants. Proc. Natl. Acad. Sci. U.S.A. 82: 5676-5680.

Busslinger, M., Rusconi, S., Rohrer, U. and Birnstiel, M. L. (1983). A Horizontal Gene Transfer Between Sea Urchin Species? In: Perspectives on Genes and the Molecular Biology of Cancer (Robertson, D.L. and Saunders G.F., eds.) Raven Press, New York, pp. 105-

Byrd, E. W. and Kasinsky, H. E. (1973). Nuclear Accumulation of Newly Synthesized Histones in Early *Xenopus* Development. Biochim. Biophys. Acta 331: 430-441.

Carozzi, N., Marashi, F., Plumb, M., Zimmerman, S., Zimmerman, A., Coles, L. S., Wells, J. R., Stein, G. and Stein, J. (1984). Clustering of Human H1 and Core Histone Genes. Science 4: 1115-1117.

Childs, G., Maxson, R., Cohn, R. H. and Kedes, L. H. (1981). Orphons: Dispersed Genetic Elements Derived From Tandem Repetitive Genes of Eukaryotes. Cell 23: 651-663.

Childs, G., Maxson, R. and Kedes, L. H. (1979). Histone Gene Expression During Sea Urchin Embryogenesis: Isolation and Characterization of Early and Late Messenger RNAs of *Strongylocentrotus Purpuratus* by Gene-Specific Hybridization and Template Activity. Dev. Biol. 73: 153-173.

Choe, J., Schuster, T. and Grunstein (1985). The Organization, Primary Structure and Evolution of Histone H2A and H2B Genes of the Fission Yeast *Schizosacharomyces Pombe*. Mol. Cell. Biol. 5: 3261-3269.

Cole, R. D. (1984). A Minireview of Microheterogeneity of H1 Histone and Its Possible Significance. Anal. Biochem. 136: 24-30.

Cook, P. and Brazell, I. (1976). Conformational Constraints in Nuclear DNA. J. Cell Sci. 22: 287-302.

Dawid, I. (1965). Deoxyribonucleic Acid in Amphibian Eggs. J. Mol. Biol. 12: 581-599.

Deisseroth, A., Nienhuis, A., Turner, P., Velez, R., Anderson, W. F., Lawrence, J., Creagan, R. and Kucherlapati, R. (1977). Localization of the Human α-Globin Structural Gene to Chromosome 16 in Somatic Cell Hybrids by Molecular Hybridization Assay. Cell 12: 205-218.

Deisseroth, A., Nienhuis, A., Lawrence, J., Giles, R., Turner, P. and Ruddle, F.H. (1978). Chromosomal Localization of Human β-globin Gene on Human Chromosome 11 in Somatic Cell Hybrids. Proc. Natl. Acad. Sci. USA 75: 1456-1460.

Destrée, O. H. J., Bendig, M. M., De Laaf, R. T. M. and Koster, J. G. (1984). Organization of *Xenopus* Histone Gene Variants Within Clusters and Their Transcriptional Expression. Biochim. Biophys. Acta 782: 132-141.

Destrée, O. H. J., Hoenders, H. J., Moorman, A. F. M. and Charles, R. (1979). Histones of *Xenopus Laevis* Erythrocytes. Purification and Characterization of the Lysine-rich Fractions. Biochim. Biophys Acta 577: 61-70.

Dimitrov, S. I., Apostolova, T. M., Makarov, V. L. and Pashev, I. G. (1986). Chromatin Superstructure. A study with an Immobilized Trypsin. FEBS Lett. 200: 322-326.

Dover, G. (1982). Molecular Drive: A Cohesive Mode of Species Evolution. Nature 299: 111-117.

Edgar, B. A., Kiehle, C. P. and Scubiger, G. (1986). Cell Cycle Control by the Nucleo-Cytoplasm Ratio in Early *Drosophila* Development. Cell 44: 365-372.

Eissenberg, J. C., Cartwright, I. L., Thomas, G. H. and Elgin, S. C. R. (1985). Selected Topics in Chromatin Structure. Annu. Rev. Genet. 19: 485-536.

Felsenfeld, G. and McGhee, J. D. (1986). Structure of the 30 nm Chromatin Fiber. Cell 44: 375-377.

Flynn, J. M. and Woodland, H. R. (1980). The Synthesis of Histone H1 During Early Amphibian Development. Dev. Biol. 75: 222-230.

Gasser, S. M. and Laemmli, U. K. (1986). Cohabitation of Scaffold Binding Regions with Upstream/Enhancer Elements of Three Developmentally Regulated Genes of *D. Melanogaster*. Cell

46: 521-530.
Graham, C. F. and Morgan, R. W. (1966). Changes in the Cell Cycle During Early Amphibian Development. Dev. Biol. 14: 439-460.
Groen, A. K., Van der Meer, R., Westerhoff, H. V., Wanders, R. J. A., Akerboom, T. P. M. and Tager, J. M. (1982). Control of Metabolic Fluxes. In: Metabolic Compartmentation (Sies, H., ed.). Academic Press, New York, pp. 9-37.
Han, M., Chang, M., Kim U. -J. and Grunstein, M. (1987). Histone H2B Repression Causes Cell-Cycle-Specific Arrest in Yeast: Effects on Chromosomal Segregation, Replication and Transcription. Cell 48: 589-597.
Harborne, N. and Allan, J. (1986). The Resolution of Five Linker Histone Subtypes From Chicken Erythrocytes. FEBS Lett. 194: 67-72.
Heintz, N., Sire, H. and Roeder, R. G. (1983). Regulation of Human Histone Gene Expression: Kinetics of Accumulation and Changes in the Rate of Synthesis and in the Half-lives of Individual Histone mRNAs During the HeLa Cell Cycle. Mol. Cell. Biol. 3: 539-550.
Hentschel, C. C. and Birnstiel, M. L. (1981). The Organization and Expression of Histone Gene Families. Cell 25: 301-313.
Isenberg, I. (1979). Histones. Annu. Rev. Biochem. 48: 159-191.
Jeffreys, A. J. (1982). Evolution of Globin Genes. In: Genome Evolution (Dover G.A. and Flavell, R.B., eds.). Academic Press, New York, pp. 157-176.
Jin, Y. and Cole, R. D. (1986). H1 Histone Exchange is Limited to Particular Regions of Chromatin that Differ in Aggregation Properties. J. Biol. Chem. 261: 3420-3427.
Kaumeyer, J. F. and Weinberg, E. S. (1986). Sequence, Organization and Expression of Late Embryonic H3 and H4 Histone Genes from the Sea Urchin, *Strongylocentrotus Purpuratus*. Nucl. Acids Res. 14: 4557-4576.
Kedes, H. (1979). Histone Genes and Histone Messengers. Annu. Rev. Biochem. 48: 837-870.
Kimelman, D., Kirschner, M. and Scherson, T. (1987). The Events of the Midblastula Transition in *Xenopus* are Regulated by Changes in the Cell Cycle. Cell 48: 399-407.
Knowles, J. A. and Childs, G. J. (1984). Temporal Expression of Late Histone Messenger RNA in the Sea Urchin *Lytechinus Pictus*. Proc. Natl Acad. Sci. USA 81: 2411-2415.
Koster, J. G. (1987). Histone Gene Expression in *Xenopus Laevis*. Academisch Proefschrift, University of Amsterdam.
Koster, J. G., Kasinsky, H. E. and Destrée, O. H. J. (1979). Newly Synthesized Histones in Chromatin of Early Embryos of *Xenopus Laevis*. Cell Diff. 8: 93-104.
Koster, J. G., Destrée, O. H. J., Raat, N. J. and Westerhoff, H. V. (1990). Histones in *Xenopus Laevis'* Early Development: The Race Against Time. Biomed. Biochim. Acta. (in press).
Koster, J. G., Destrée, O. H. J. and Westerhoff, H. V. (1988). Kinetics of Histone Gene Expression During Early Development in *Xenopus Laevis*. J. theor. Biol. 135: 139-167.
Krieg, P. A., Robins, A. J., D'Andrea, R., Wells, J. R. E. (1983). The Chicken H5 Gene is Unlinked to Core and H1 Histone Genes. Nucl. Acids Res. 11: 619-627.
Laskey, R. A. (1986). Prospects for Reassembling the Cell Nucleus. J. Cell. Sci. Suppl. 4: 1-9.
Lennox, R. W. and Cohen, L. H. (1984). The H1 Subtypes of Mammals: Metabolic Characteristics and Tissue Distribution In: Histone genes: Structure, organization and regulation (Stein, G. S., Stein, J. L. and Marzluff, W. F., eds.) Wiley and Sons, New York, pp. 373-397.
Lewis, C. D., Lebrowsky, J. S., Daly, A. and Laemmli, U. K. (1984). Interphase Nuclear Matrix and Metaphase Scaffolding Structures: A Comparison of Protein Components. J. Cell Sci. Suppl. 1: 103-127.
Lifton, R. O., Goldberg, M. L., Karp, R. W. and Hogness, D. S. (1978). The Organization of the Histone Genes in *Drosophila Melanogaster*: Functional and Evolutionary Implications. Cold Spring Harbor

Symp. Quant. Biol. 42: 1047-1051.
Matsumoto, S. and Yanagida, M. (1986). Histone Gene Organization of Fission Yeast: A Common Upstream Sequence. Nucl. Acids Res. 14: 635-644.
Maxson, R. E. and Wilt, F. H. (1982). Accumulation of the Early Histone Messenger RNAs During the Development of *Strongylocentrotus Purpuratus*. Dev. Biol. 94: 435-440.
Maxson, R., Cohn, R., Kedes, L. and Mohun, T. (1983a). Expression and Organization of Histone Genes. Ann. Rev. Genet. 17: 239-277.
Maxson, R., Mohun, T., Gormezano, G., Childs, G. and Kedes, L. (1983b). Distinct Organizations and Patterns of Expression of Early and Late Histone Gene sets in the Sea Urchin. Nature 301: 120-125.
Maxson, R., Mohun, T. and Kedes, L. (1983c). Histone Genes. In: Eukaryotic genes: Their Structure, Activity and Regulation (Maclean, N., Gregory, S. and Flavell, R., eds.). Butterworths, London, pp. 277-298.
McGhee, J. D. and Felsenfeld, G. (1980). Nucleosome Structure. Ann. Rev. Biochem. 49: 1115-1156.
Mirkovitch, J., Mirault, M. E. and Laemmli, U. K. (1984). Organization of the Higher-order Chromatin Loop: Specific DNA Attachment Sites on Nuclear Scaffold. Cell 39: 223-232.
Neelin, J. M., Callahan, P. X., Lamb, D. C., and Murray, K. (1964). The Histones of Chicken Erythrocyte Nuclei. Can. J. Biochem. 42: 1743-1752.
Newport, J. and Kirschner, M. (1982). A Major Developmental Transition in Early *Xenopus* Embryos: I. Characterization and Timing of Cellular Changes at the Midlblastula Stage. Cell 30: 675-686.
Newrock, K. M., Cohen, L. H., Hendricks, M. B., Donnelly, R. F. and Weinberg, E. S. (1978). Stage Specific mRNAs Coding for Subtypes of H2A and H2B Histones in the Sea Urchin Embryo. Cell 14: 327-336.
Old, R. W., Woodland, H. R., Ballantine, J. E. M., Aldridge, T. C., Newton, C. A., Bains, W.A. and Turner, P.C. (1982). Organization and Expression of Cloned Histone Gene Clusters From *Xenopus Laevis* and *Xenopus Borealis*. Nucl. Acids Res. 10: 7561-7580.
Pehrson, J. R. and Cole, R. D. (1981). Bovine $H1^0$ Histone Subfractions Contain an Invariant Sequence which Matches Histone H5 Rather than H1. Biochemistry 20: 2298-2301.
Perry, M., Thomsen, G. H. and Roeder, R. G. (1985). Genomic Organization and Nucleotide Sequence of Two Distinct Histone Gene Clusters from *Xenopus Laevis*. (Identification of Novel Conserved Upstream Sequence Elements.) J. Mol. Biol. 185: 479-499.
Perry, M., Thomsen, G. H. and Roeder, R. G. (1986). Major Transitions in Histone Gene Expression do not Occur During Development in *Xenopus Laevis*. Devel. Biol. 116: 532-538.
Ramsay, N. (1986). Deletion Analysis of a DNA Sequence that Positions Itself Precisely on the Nuclesome Core. J. Mol. Biol. 189: 179-188.
Richmond, T. J., Finch, J. T., Rushton, B., Rhodes, D. and Klug, A. (1984). Structure of the Nucleosome Core Particle at 7 Å Resolution. Nature 311: 532-537.
Risley, M. S. and Eckhardt, R. A. (1981). H1 Histone Variants in *Xenopus Laevis*. Dev. Biol. 84: 79-87.
Roche, J., Gorka, A., Goeltz, P. and Lawrence, J. J. (1985). Association of Histone $H1^0$ with a Gene Repressed During Liver Development. Nature 314: 197-198.
Ruderman, J. V., Woodland, H. R. and Sturgess, E. A. (1979). Modulations of Histone mRNA During the Early Development of *Xenopus Laevis*. Dev. Biol. 71: 71-82.
Scherrer, K. (1980). Cascade Regulation. A Model of Integrative Control of Gene Expression in Eukaryotic Cells and Organisms. In: Eukaryotic Gene Regulation, Vol. I (Kolodny, G.M., ed.). CRC Press, Boca Raton, Florida, USA, pp. 57-129.
Schümperli, D. (1986). Cell-cycle Regulation of Histone Gene Expression. Cell 45: 471-472.
Seyedin, S. M. and Kistler, W. S. (1980). Isolation and Characterization of Rat Testis H1t (An H1 Histone Variant Associated with Spermatogenesis). J. Biol. Chem. 55: 5949-5954.

Simon, R. H., Camerini-Otero, R. D. and Felsenfeld, G. (1978). An Octamer of Histones H3 and H4 Forms a Compact Complex with DNA of Nucleosome Size. Nucl. Acids Res. 5: 4805-4818.
Simpson, R. T. (1978). Structure of the Chromatosome, a Chromatin Particle Containing 160 Base Pairs of DNA and all the Histones. Biochemistry 17: 5524-5531.
Simpson, R. T. (1981). Modulation of Nucleosome Structure by Histone Subtypes in Sea Urchin Embryos. Proc. Natl. Acad. Sci. USA 78: 6803-6807.
Simpson, R. T., and Stafford, D. W. (1983). Structural Features of a Phased Nucleosome Core Particle. Proc. Natl. Acad. Sci. USA 80: 51-55.
Smith, M. M. (1984). The Organization of Yeast Histone Genes. In: Histone Genes (Stein, G. S., Stein, J. L. and Marzluff, W. F., eds.). Wiley and Sons, New York, pp. 3-33.
Stedman, E. and Stedman, E. (1950). Cell Specificity of Histones. Nature 166: 780-781.
Stephenson, E., Erba, H. and Gall, J. (1981). Histone Gene Clusters of the Newt *Notophtalamus* are Separated by Long Tracts of Satellite DNA. Cell 24: 639-647.
Stockley, P. G. and Thomas, J. O. (1979). A Nucleosome-like Particle Containing an Octamer of the Arginine-rich Histones H3 and H4. FEBS Lett. 99: 129-135.
Thoma, F., and Simpson, R. T. (1985). Local Protein-DNA Interactions May Determine Nucleosome Positions on Yeast Plasmids. Nature 315: 250-252.
Thoma, F. A., Koller, T. H. and Klug, A. (1979). Involvement of Histone H1 in the Organization of the Nucleosome and of the Salt-dependent Superstructures of the Chromatin. J. Cell Biol. 83: 403-427.
Thomas, J. O., Rees, C. and Pearson, E. C. (1985). Histone H5 Promotes the Association of Condensed Chromatin Fragments to Give Pseudo-Higher-Order Structures. Eur. J. Biochem. 147: 143-151.
Tripputi, P., Emanuel, B. S., Croce, C. M., Green, L. G., Stein, G. S. and Stein, J. L. (1986). Human Histone Genes Map to Multiple Chromosomes. Proc. Natl. Acad. Sci. USA 83: 3185-3188.
Turner, P. C. and Woodland, H. R. (1983). Histone Gene Number and Organization in *Xenopus*: *Xenopus Borealis* has a Homogeneous Major Cluster. Nucl. Acids. Res. 11: 971-986.
Turner, P. C., Aldridge, T. C., Woodland, H. R. and Old, R. W. (1983). Nucleotide Sequences of H1 Histone Genes from *Xenopus Laevis*. A Recently Diverged Pair of H1 Genes and an Unusual H1 Pseudogene. Nucl. Acids. Res. 11: 4093-4107.
Urban, M. K. and Zweidler, A. (1983). Changes in Nucleosomal Core Histone Variants During Chicken Development and Maturation. Dev. Biol. 95: 421-428.
Van Dam, K. (1986). Biochemistry is a Quantitative Science. Trends Biochem Sc. 11: 13-14.
Van Dongen, W., De Laaf, L., Zaal, R., Moorman, A. and Destrée, O. H. J. (1981). The Organization of the Histone Genes in the Genome of *Xenopus Laevis*. Nucl. Acids Res. 9: 2297-2311.
Weintraub, H. (1978). The Nucleosome Repeat Length Increases During Erythropoiesis in the Chick. Nucl. Acids Res. 5: 1179-1188.
Westerhoff, H. V. and Van Dam, K. (1987). Thermodynamics and Control of Biological Free-energy Transduction, Elsevier, Amsterdam.
Widom, J. and Klug, A. (1985). Structure of the 300 Å Chromatin Filament: X-ray Diffraction from Oriented Samples. Cell 43: 207-213.
Williams, S. P., Athey, B. D., Muglia, L. J., Schappe, R. S., Gough, A. H. and Langmore, J. P. (1986). Chromatin Fibers are Left-handed Double Helices with Diameter and Mass Per Unit Length that Depend on Linker Length. Biophys. J. 49: 233-248.
Woodland, H. R. (1980). Histone Synthesis During the Development of *Xenopus*. FEBS Lett. 121: 1-17.
Woodland, H. R. (1982). The Translational Control Phase of Early Development. Biosc. Rep. 2: 471-491.
Woodland, H. R. and Wilt, F. H. (1980a). The Functional Stability of Sea Urchin Histone mRNA Injected into Oocytes of *Xenopus Laevis*. Dev. Biol. 75: 199-213.
Woodland, H. R. and Wilt, F. H. (1980b). The Stability and Translation of Sea Urchin Histone Messenger RNA Molecules Injected into *Xenopus Laevis* Eggs and Developing Embryos. Dev. Biol. 75: 214-

221.

Wu, R. S. and Bonner, W. M. (1981). Separation of Basal Histone Synthesis from S-phase Histone Synthesis in Dividing Cells. Cell 27: 321-330.

Wu, R. S., Panusz, H. T., Hatch, C. L. and Bonner, W. M. (1986). Histones and their Modifications. CRC Crit. Rev. Biochem. 20: 201-263.

Wu, R. S., West, M. H. P. and Bonner, W. M. (1984). Patterns of Histone Gene Expression in Human and Other Mammalian Cells. In: Histone Genes (Stein, G.S., Stein, J. L. and Marzluff, W. F. eds.). Wiley and Sons, New York, pp. 457-483.

Zernik, M., Heintz, N. Boime, J. and Roeder, R. G. (1980). *Xenopus Laevis* Histone Genes: Variant H1 Genes are Present in Different Clusters. Cell 22: 807-815.

Zweidler, A. (1984). Core Histone Variants of the Mouse: Primary Structure and Differential Expression. In: Histone Genes (Stein, G.S., Stein, J.L. and Marzluff, W.F., eds.). Wiley and Sons, New York, pp. 339-374.

Chapter 15

HOW FUNDAMENTAL KNOWLEDGE CAN HELP SOLVE PRACTICAL PROBLEMS IN TOXICOLOGY

Robert A. Neal

15.1. INTRODUCTION

Toxicology is usually defined simply as the science of poisons. In its broader sense, the discipline of toxicology is the study of the adverse effects of chemicals on living organisms. It is, thus, the science that defines limits of safety of chemical agents. From this definition it follows that a toxicologist is someone who generates data, using suitable models, on the toxicity of chemicals to living organisms. In a modern sense, toxicologists do more than this. They must be able, based on knowledge of the mechanisms by which chemicals produce their effects, to make predictions on the potential hazards of these chemicals to humans and the environment. According to these definitions, the discipline of toxicology encompasses both the plant and animal kingdom. However, in practice, the most frequent application of toxicology is a study of the toxic effects or potential toxic effects of chemicals on higher animals, especially humans. I will restrict my comments to the discipline of toxicology as applied to predicting the toxicity or lack of toxicity of chemicals to humans.

At this colloquium we have been discussing those biochemical factors and spatial arrangements of macromolecule which define a living cell. We have also discussed those alterations in biochemistry or spatial arrangements which lead to cell death. It would be useful to examine which of the toxic effects of chemicals seen in animals are caused by cell death and those which are not caused by the death of the cell, but, rather, by an alteration in the morphology or biochemistry of the cell.

The toxic effects of chemicals which one might reasonably attribute, at least under some conditions, to cell death include necrosis in various organs, impaired neurological function caused by the death of cells in the central or peripheral nervous system, fibrosis (particular in the lung), hormonal imbalance caused by chemically induced death of the cells which produce the hormone, sterility due to chemical killing of germ cells, and aplastic anemia due to chemically mediated death of stem cells in the bone marrow. This list is meant to be representative rather than exhaustive.

Other toxic effects of chemicals can be attributed to a permanently altered phenotypic expression or to a temporary alteration in the biochemistry or morphology of the cell, leading to a change in the normal functioning of the cell. Cancer and mutagenesis are examples of a chemically induced change in the phenotype of a cell or cells. Impaired neurological function could occur not by cell death but by a chemically induced alteration in the functioning of cells by, for example, inhibiting nerve transmission. The organophosphate, carbamate, and chlorinated hydrocarbon insecticides are examples of such chemicals. Chemically induced teratogenesis or birth defects are in my opinion more likely to result from an alteration in the differentiation of a group of cells rather than from cell death, although in some cases selective cell death may lead to a birth defect. Toxic effects of chemicals on the immune systems are also more likely to occur as a result of inhibition of the normal functioning of particular cells in the immune system or an inhibition or alteration in the differentiation of precursor cells than to death of cells in the immune system, although an altered immune competence is one result of aplastic anemia which more often than not is the result of cell death. Anemias other than aplastic anemia are more often the result from an alteration in function of differentiation of cells rather than cell death. A hormonal imbalance can be caused not by cell death but by chemically induced changes in the normal functioning of cells which produce the hormone or those which react to the hormone. Finally, reproductive dysfunction can result from functionally altered germ cells rather than the death of these cells.

What emerges from the above considerations is the perception that an alteration in the normal functioning of the cell is more often the cause of chemical toxicity than is the death of cells, although both effects are important events in toxicology.

In studies of the chronic toxicity of chemicals in experimental animals, one occasionally observes the early and/or more frequent appearance of a disease state commonly seen in that species or strain of experimental animals such as, for example,

Fundamental Knowledge in Toxicology

kidney disease, benign liver tumors, or testicular atrophy. The symptoms and the pathology of the chemically induced disease is sometimes indistinguishable from that which occurs normally but at a later time or less frequently. A possible mechanism by which this chemically induced disease is occurring is by accelerated senescence of the cell. Theories of cellular aging can be divided into two broad but overlapping categories, those that view all cell death as an actively programmed developmental process, and second, those which consider cell death as resulting from the passive accumulation of errors in informational and functional macromolecules. Of the two general theories, the accumulation of errors in macromolecule as a result of chronic exposure to a chemical is the most logical hypothesis of chemically induced senescence, if in fact such a phenomena does occur.

The general functional characteristics of senescence cells include: impaired transcription, altered translation, altered DNA repair, decreased proteolysis, decreased responsiveness to hormones, appearance of abnormal proteins, increased number of lysosomes, and increased number of chromosomal abnormalities.

With the exception of the appearance of abnormal proteins, these various effects could be caused by mutational events or as a result of chronic exposure of the cell to a chemical. The appearance of proteins with an altered primary sequence would require mutational events, although proteins with an altered, secondary or tertiary structure could result from the interaction of the chemical directly with the protein. I will not dwell on this topic any longer; however, accelerated senescence is a reasonable hypothesis on how some chemicals cause chronic disease in animals.

In practice, the most important problems in toxicology come down to these five issues: (i) understanding of mechanisms of chemical toxicity, (ii) prediction of responses at low doses, (iii) validity of animal models in predicting human risk, (iv) need for improved (more predictive, cheaper, faster) methods to assess potential toxicity of chemicals, and (v) measurement of chemical exposure in humans. It is valuable to discuss each of these from the stand-point of fundamental knowledge at the cellular level.

15.2. MECHANISMS OF TOXICITY

More than any other single factor, the future of toxicology as a predictive science depends on an understanding of the mechanisms by which chemicals cause toxic effects in animals. The solution to all of the other items on my list of practical problems will be heavily dependent on an understanding of the mechanism by which chemicals or classes of chemicals cause toxic effects in animals.

One has a choice of studying the mechanisms of toxicity at a variety of different levels of organization within animals, for example, in the whole animal, in isolated organ systems, at the cellular or at the subcellular level. In my opinion, experiments using isolated cells are likely to yield the most useful information on mechanisms of chemical toxicity for most compounds. The whole animal or whole organ will more often than not be much too complex a level of organization with too many modifying factors to yield the most useful information about mechanisms of toxicity. Studies with particulate or soluble portions of cells or purified macromolecules from cells lack the control factors and spatial arrangements operative in the whole cell and are likewise not likely to yield the most useful information on mechanisms of toxicity for most compounds. Once hypotheses on the mechanism of toxicity are formulated from studies at the cellular level, they can be examined at the whole organ or whole animal level to ascertain if they may account for the toxic effects seen in the whole animal or at the subcelllular level to determine if the target molecules or macromolecule have been correctly identified.

15.3. TOXIC RESPONSES AT LOW DOSES

Most studies of the potential toxicity of chemicals are carried out at high doses in experimental animals and the effect of the lower dose levels to which humans are usually exposed are estimated using the dose responses seen in animal studies and mathematical models in the case of cancer causing and mutagenic chemicals or, for toxic effects other than cancer or mutation, by applying safety factors to no-effect levels of administered compounds determined in studies in animals. Considerable scientific uncertainty exists about the accuracy of the predictions of potential cancer incidence determined using the mathematical modeling procedures. The scientific validity of the safety factors applied to no-effects level for toxic compounds observed in experimental animals is also open to question. There is a crucial need to improve our low-dose extrapolation procedures. One way to improve these procedures is to use target site dose rather than the administered dose in the mathematical extrapolations or in establishing no-effects levels for toxic effects other than cancer or mutagenesis. That is to measure the level of the chemical in the target cell or better yet, at the target site in the target cell after administration of the compound by the appropriate route and use this as the measure of exposure in the low-dose extrapolation. Knowledge of the mechanisms of toxicity will help in identifying the target site within cells and facilitate the determination of these more appropriate measures of exposure.

Fundamental Knowledge in Toxicology

15.4. VALIDITY OF ANIMAL MODELS

Perhaps the most vexing problem in toxicology is uncertainty about the validity of our animal models in predicting the response of humans exposed to toxic chemicals. In other words, will humans exposed to a particular chemical develop the same qualitative symptoms of toxicity as experimental animals? Even more uncertain is whether humans will exhibit the same toxic effects at the same or similar doses as those seen in experimental animals. Less uncertainty exists in predicting in a qualitative manner, acute responses in humans based on results in experimental animals. A compound which is acutely toxic in the rat, mouse, or dog will likely be acutely toxic in humans although the dose required in the various species to produce acute toxicity may be quite different. A much greater uncertainty exists in the comparative toxic effects in experimental animals and humans to be expected after exposure to nonacutely toxic doses of a compound for a major portion of the lifetime of each species. This is particularly the case for compounds which are "weak" carcinogens. The uncertainty about the relative dose-response relationships in experimental animals and humans is again greater than the uncertainty about the qualitative responses. Again, knowledge of the mechanisms of toxicity of the chemical will be helpful in reducing this uncertainty. Knowledge of mechanisms of toxicity together with comparative studies using human and experimental animal tissues or studies in experimental animals and exposed human populations, if available, will be essential in addressing this problem.

15.5. IMPROVED METHODOLOGY

We also need to develop better (more predictive, faster, cheaper) methods to assess the potential toxicity of chemicals to humans. Again, knowledge of mechanisms of toxicity of chemicals or classes of chemicals will greatly facilitate the development of these new procedures. With the knowledge of the mechanism by which a chemical, acting at the cellular level, can cause toxic effects in a whole animal, we can develop specific tests to determine if other chemicals have the ability to affect these cells in the same way and have the potential to cause the same toxic effects in the whole animal including humans.

Examples of these newer methods could be the ability to measure protooncogene to oncogene conversion, improved ability to measure genetic damage in living animals (including humans), increased use of human cells in culture in the measurement of potential toxicity, and measurement of covalent binding of a specific nature to macromolecule in target cells.

15.6. MEASUREMENTS OF HUMAN EXPOSURE

Finally, we need to develop noninvasive procedures for detecting in human populations evidence of excessive exposure to potentially toxic chemicals. Some techniques for measuring exposure to chemical carcinogens currently under development are: (i) chromosomal aberrations, (ii) sister chromatid exchanges, (iii) micronucleus, (iv) DNA repair, (v) metabolites in urine and feces, and (vi) biochemical measurements such as HGPRT, Hb variants, Hb alkylation and DNA adducts. Each of these techniques, in their current state of development, have limitations which strongly limit their usefulness. Nevertheless, they are a start and more research in this area must continue.

Let us examine each of these briefly. The measurement of chromosomal aberrations in peripheral blood lymphocytes is the most well developed method for monitoring humans for exposure to genotoxic chemicals. Although chromosomal aberrations are not specific for distinct chemicals, they reasonably represent a measure of the total clastogenic effect of exposures to chemicals in the environment (including food, water, drugs) and in the workplace. The major disadvantage of the method is that the scoring for chromosomal aberrations is laborious and subjective. Additionally, the number of aberrations is influenced by age, personal habits such as smoking and viral infections. Sister chromatid exchanges in circulating peripheral lymphocytes can also be a possible measure of exposure to genotoxic chemicals. However, the relationship of sister chromatid exchange to DNA damage is not well known and the relationship to genotoxicity and clastogenicity (chromosomal aberrations) is not well established.

Micronuclei are thought to originate from aberrant cell division and unequal chromosome distribution. The method can be applied to a number of human tissues and is not unduly difficult or time-consuming. However, it has not been validated for biomonitoring purposes. The main advantage of this method is that it may be specific for clastogenic effects and that it measures effects directly in the target tissue of exposed persons. However, the target tissue in humans for the chemical under study is frequently either unknown or unavailable for analysis.

Induction of DNA repair in somatic cells in humans with high cancer risks have been measured. In principle, the method can be applied to cells of various human target organs and/or tissues. However, the method lacks sensitivity and specificity and its application to somatic cells of humans exposed to mutagens and/or carcinogens needs validation.

Urine and feces of exposed human populations can be analyzed for the presence of mutagenic agents or mutagenic metabolites by various short-term tests such as mutagenicity assays in bacteria or in mammalian cells. However, because of the mutagenic differences in urine and feces that result from diet variation and personal habits, the method is not expected to be useful in biomonitoring populations occupationally exposed to genotoxic agents.

Some biochemical measurements of exposure to genotoxic chemicals are available. One of these is the measurement of the activity of the enzyme hypoxanthine-guanine phosphoribosyltransferase (HGPRT) in white blood cells of exposed populations. This test is based on the detection of a mutation in a structural gene located on the x-chromosome that codes for this enzyme. The major disadvantage of this technique is that the procedure is cumbersome for application to large populations and it is nonspecific for the chemical under study.

Another method is to measure mutation rates in the hemoglobin (Hb) gene in human red blood cells. The altered hemoglobin molecules are detected with the help of monospecific antibodies that recognize changes in certain amino acid sequences in the hemoglobin molecule. The great advantage of this method is that it measures human mutation induction directly. However, the technique is time-consuming and laborious and is not currently suitable for large-scale population monitoring.

Another technique is to assess the alkylation of hemoglobin as a measure of exposure to alkylating agents. This technique has been extensively used to measure exposure to ethylene oxide. The advantage of the method is that it is specific for alkylating agents and makes it a valuable tool for biomonitoring exposure to these agents. The disadvantage of the method applied to ethylene oxide is the unexpectedly high background fluctuation of hemoglobin alkylation in control populations. This lowers the method of sensitivity for establishing low external exposure to this alkylating agent in human populations.

As we noted previously, the most direct way to measure the primary effect of a compound is to detect and measure its reaction with the intracellular target. For mutagens and for genotoxic carcinogens, the target has been identified as DNA. Recently, immunochemical methods for detecting specific DNA adducts in blood cells have been developed. These methods when fully developed may be sensitive enough for biomonitoring of the exposed human populations and it seems possible to detect those DNA adducts that are indicative not only of exposure but also possibly of adverse health effects of concern.

15.7. CONCLUDING REMARKS

I have considered above some of the practical problems in toxicology that directly affect human lives. Unlike numerous problems tackled and solved successfully by physicists and chemists in the past, the toxicological and environmental health problems facing humankind are immensely more complex. To solve such problems of unprecedented complexity, it may not be sufficient for biologists and environmental health scientists just to assimilate and apply existing knowledge derived from established sciences such as physics and chemistry to their problems but it may be mandatory for them to develop fundamentally new scientific approaches, both experimental and theoretical, taking into account the rich experimental data provided by biological observations.

The history of science is replete with examples of major breakthroughs in human knowledge resulting from the need to solve practical problems. The development of *thermodynamics* was spurred by the desire to improve the efficiency of the steam engine invented in the 18th century, the birth of *quantum mechanics* in the early decades of this century was foreshadowed by practical applications of electromagnetism and synthetic chemistry in industry in the 19th century, and the development of *information (or communication) theory* in the mid-twentieth century was closely associated with the inventions of telegraphy and telephony and the need for accurate and rapid computations in weapons development and other military applications. In a similar vein, it seems plausible that the current global awareness of the environmental health problems occasioned by excessive human exposures to artificial chemicals originating from industry, agriculture, and automobiles, will serve as a powerful driving force that will eventually catapult the birth of a new scientific discipline of the 21st century. This science may be characterized by an integration of existing sciences (thermodynamics, quantum mechanics, information theory, and cosmology) into a *unified science of life*. It is possible that, only with the development of such a novel science of life, can we be equal to the difficult and challenging task of solving practical problems in toxicology and environmental health sciences in the coming decades and beyond.

GLOSSARY

algebra The branch of mathematics that uses positive and negative numbers, letters, and other systematized symbols to express and analyze the relationship between concepts of quantity in terms of formulas, equations, etc.; generalized arithmetic.

algebra of machines The branch of algebra developed to describe the behavior of machines and their interactions with environment using algebraic expressions.

algorythm 1) An explicit or effective procedure for producing something. 2) The act of calculating with any species of notation.

apoenzyme The protein portion of an enzyme as contrasted with the nonprotein portion, or coenzyme, or prosthetic portion (if present).

apoptosis Controlled cell deletion complementary to mitosis in the regulation of animal cell populations. Controlled cell death. Apoptosis = Greek, meaning "dropping off" or "falling off" of petals from flowers or leaves from trees. Coined by J. F. R. Kerr, A.H. Wyllie and A.R. Currie in 1972 (Brit. J. Cancer 26, 239 (1972)).

arithmetic The science or art of computing by positive real numbers.

automata theory A branch of mathematics investigating the behavior of various automata. Automata theories have been applied to modeling the development of plants and animals (A. Lindenmayer, J. Theor. Biol. 54, 3 (1975)).

automaton 1) Anything that can move or act of itself; 2) A machine or controlling mechanism designed to follow automatically a predetermined sequence of operations or respond to encoded instructions and correct errors or deviations occurring during operation.

Belousov-Zhabotinskii reaction The oxidation of citric acid by $KBrO_3$ catalyzed by the ceric-cerous ion couple (Ce^{+4} - Ce^{+3}). This reaction has the unusual property that the progression of the reaction can lead to a temporal oscillation of the concentrations of certain chemicals participating in the reaction (e.g. Br^- and the Ce^{+4}/Ce^{+3} ratio) and a spatial pattern formation (e.g. chemical waves in a petridish or colored bands along the length of a test tube). The spatiotemporal organizations last as long as the chemical reaction proceeds. Discovered by B.B. Belousov in 1958 and later extended by A.M. Zhabotinskii (see G. Nicolis and I. Prigogine, Self Organization in Non-Equilibrium systems, Wiley-InterScience Publication, New York, 1977, pp 339-353).

biological reactive intermediates Stable chemicals that are converted into unstable, reactive intermediates catalyzed by enzymes in living tissues (e.g. the trichloromethyl radical from carbon tetrachloride catalyzed by enzymes located in the endoplasmic reticulum of the cell).

Bhopalator A hypothetical model of the living cell that is based on the postulate that the self-replicating properties of the cell derive from the coupling among three major classes of the cellular constituents—information-storing nucleic acids, energy-transducing enzymes, and intracellular gradients of chemical species (called *intracellular dissipative structures*, IDS). The primary role of nucleic acids is to store genetic information in a stable molecular form for transmission from one generation to another, the primary role of enzymes is to utilize the free energy released from chemical reactions to produce far-from-equilibrium distributions of matter inside the cell (i.e. IDS), and the primary role of IDS is to exchange information with, and to do work on, its extracellular environment. First proposed in Bhopal in 1983 during an international meeting (S. Ji., J. theor. Biol. 116, 399 (1985)).

Big Bang theory A theory in cosmology postulating that at some time about $10\text{-}15 \times 10^9$ years ago all the matter of the universe was packed into a super dense small agglomeration which was subsequently hurled in all directions at enormous speeds by a cataclysmic explosion.

boson Any elementary particle having integral spin (e.g. photons, mesons). Bosons obey Bose-Einstein statistics. The Pauli exclusion principle is not obeyed by bosons. All particles are either bosons or fermions (elementary particles having half integer spin).

Brusselator A theoretical model of a chemical reaction proposed by I. Prigogine and R. Lefever in Brussels (J. Chem. Phys. 48, 1965 (1968)) that has self-organizing properties. Also called the "trimolecular model" because of the presence of a termolecular collision step in the mechanism;

$$A \rightleftharpoons X$$
$$B + X \rightleftharpoons Y + D$$
$$2X + Y \rightleftharpoons 3X$$
$$X \rightleftharpoons E$$
$$A + B \rightleftharpoons D + E$$

where A,B = reactants; D,E = products; X,Y = reaction intermediates.

Ca^{++} homeostasis hypothesis The hypothesis that, when the intracellular concentration of Ca^{++} rises beyond a certain critical level (0.1 to 5×10^{-6} M), normal cell functions are severely perturbed, leading to cell death. For recent reviews, see S. Orrenius, D. Reed and J. Farber in the November 6, 1990 issue of Chemical Research in Toxicology.

catalase The enzyme that catalyzes the dismutation of H_2O_2 into O_2 and water.

cell A small unit or protoplasm (i.e., a semifluid, viscous, translucent colloid), usually with a nucleus and an enclosing membrane: all plants and animals are made up of one or more cells.

cell force The totality of the chemical and physical processes that go on inside the cell to effectuate gene-directed cellular behaviors.

Glossary

cell water — The structure of water inside the living cell may be highly organized and not randomly mobile.

cellular automata — The mathematical systems constructed from many identical components (cells), each simple, but together capable of generating complex behavior. A one-dimensional cellular automaton consists of a line of sites, with each site carrying a value of 0 or 1. The value of a site at position 1 is updated in discrete time steps according to an identical deterministic rule depending on a neighborhood of sites around it.

cellular protection mechanisms — Living cells have evolved to possess various mechanisms by which they can protect themselves against injurious chemical species such as reactive oxygen species and other biological reactive intermediates (e.g. GSH as a scavenger of free radicals and electrophiles).

chaos — 1) The disorder of formless matter and infinite space, supposed to have existed before the ordered universe. 2) An irregular motion which is described by deterministic equations.

chemical oscillations — Under appropriate conditions, the concentrations of certain chemical species participating in self-organizing chemical reactions can oscillate in time (e.g. the Br^- concentration in the Belousov-Zhabotinskii reaction).

chemical potential — The change in the free energy of the system with a change in the chemical composition due to chemical reaction or due to matter transport. Under the condition of constant temperature, any change in the system proceeds from a state of higher chemical potential to a state of lower chemical potential. At constant temperature and pressure, the chemical potential is given by Gibbs free energy (G); and under constant temperature and varying pressure, the chemical potential of the system is given by Helmholtz free energy (H).

chemical waves — During certain self-organizing chemical reactions proceeding under appropriate conditions, the concentrations of some chemicals can vary in space, forming wave-like or spiral-like patterns.

chemiluminescence — The emission of photons from energy-releasing chemical reactions.

chromatin — The genetic material of the nucleus consisting of deoxyribonucleo-protein. During mitotic division the chromatin condenses to form chromosomes.

chromosome — One of the bodies (normally 46 in man) in the cell nucleus that is the bearer of genes.

coenzymes — Substances that are necessary for the catalytic action of enzymes. Coenzymes are of smaller molecular size than the enzymes themselves (e.g. several vitamins such as thiamine, pyridoxal, nicotinamide).

cofactor	A prosthetic group such as heme, coenzymes, and inorganic ions such as magnesium ion, essential for enzyme action.
configuration	Unlike conformation, the configuration of a molecule cannot be changed without breaking and forming at least one covalent bond.
conformation	The 3-dimensional arrangement of atoms in a molecule that can be changed without breaking or forming any covalent bonds. The isotopically labeled ethane, CHDT-CHDT, can exist in three stable conformations, which differ from one another by a rotation around the C-C bond.
conformons	Hypothetical mobile energy packets stored in the form of conformational strains of biopolymers (nucleic acids and enzymes). Conformons carry potential energy (due to strains) and information (in the form of the unique spatiotemporal alignments of catalytic amino acid residues in enzymic active sites or nucleotide bases in functional sites of nucleic acids). Because of the availability of both free energy (to do work) and information (to control work), conformons can drive controlled molecular work processes in enzymes and nucleic acids, leading to goal-oriented catalysis, molecular recognition, and gene expression (S. Ji, Ann. N.Y. Acad. Sci. 227, 211-226 and 419-437 (1974); Asian J. Exp. Sci. 1, 1 (1985); J. theor. Biol. 116, 399 (1985)).
conjugation reactions	Cells remove foreign compounds (xenobiotics) by sulfation or glucuronidation, or by adding the glutathionyl group. These reactions are called conjugation reactions, or phase II reactions.
control	Influencing a device or machine in such a way as to achieve a specific goal.
covalent binding hypothesis	Many chemicals injure cells by being converted into biological reactive intermediates (BRI) which subsequently bind covalently to essential cellular macromolecules like enzymes, phospholipids, and nucleic acids. Formulated in the early 1970's by Mitchell, Jollow, Gillette, Potter, Davis and Brodie.
creation and annihilation operators	A creation operator, a_i^+, is defined in such a way that, when acting on a basis vector, adds a particle to it with quantum number i. Similarly an annihilation operator, a_i, will remove a particle with quantum number i upon acting on a basis vector.
curvature	The rate of deviation of a curve or curved surface from a straight line or plane surface tangent to it.
cytochromes P-450	A group of enzymes that have a heme and an iron ion at the active site and absorb maximally at 450 nm when reduced in the presence of carbon monoxide. These enzymes participate in mono-oxygenation reactions of various chemicals, both endogenous and exogenous.
cytons	The totality of the intracellular chemical and physical processes that are postulated to

Glossary

	mediate cellular influences, namely the cell force.
cytoskeletons	A group of filamentous protein molecules (microfillaments; microtubles) that provide highly organized structures to the cytoplasm of the living cell. Also called the cytomatrix.
cytosociology	The concept that the various enzymes in the living cell are not randomly distributed but form a coherently interacting organized system in the cytoplasm. (R. Welch and T. Keleti, J. theor. Biol. 93, 701 (1981)).
dissipative structures	Any distribution of matter in space and time that is maintained far from equilibrium though dissipation of free energy (e.g. the flame of a candle; the K^+ and Na^+ ion gradients across the cell membrane; self-organizing chemical reactions such as the Belousov-Zhabotinskii reaction).
DNA gyrase	A bacterial enzyme that uses the energy of ATP hydrolysis to pump supercoils continuously into the DNA, thereby maintaining conformational strains in localized regions of DNA (B. Alberts et al. The Cell, Garland Publishing Inc., New York, 1983, p. 446).
DNA supercoils	The helical turns of a double-stranded DNA molecule generated when the DNA double helix is rotated around the helical axis. The potential energy stored in one supercoil is sufficient to break the hydrogen bonds of 10 base pairs.
electrical potential	The electric potential at a point in an electric field is the work required to bring unit positive electric charge from infinity to the point. It is measured in volts. If work of 1 joule is required to move a charge of 1 coulomb to the point its potential is 1 volt.
electrophiles	Any chemical species having a great tendency to react with a electron-rich center (e.g. carbonium ions, electron-deficient carbon centers such as the carbonyl carbon).
emergent property	A property of a complex system which is not contained in its parts (e.g. rigidity, superconductivity, consciousness).
endergonic	Free-energy consuming (as in muscle contraction, active transport, etc.).
energy	The property of a system that is a measure of its capacity for doing work. The energy of an isolated system (e.g. the universe) remains constant (the First Law of Thermodynamics). Energy and mass are interconvertible according to Einstein's law, $E = mc^2$.
enthalpy	A thermodynamic function (i.e. the value of the function depends only on the initial and final states of the system and independent on the path followed) defined as $E + PV$, where E is the internal energy of the system, P the pressure, and V the volume.

entropy	The thermodynamic function of a system that is a measure of the degradation of the system (e.g. when a crystal dissolves in water to form a solution; when the human body decays due to starvation). A measure of randomness or uncertainty. Whenever irreversible processes occur, the entropy of the universe must increase (the Second Law of Thermodynamics). In addition to the "thermodynamic" entropy defined above, the term "information-theoretic" entropy is frequently used which is equivalent to the information defined by Shannon's formula (see information). The information-theoretic entropy of a message can be regarded as a measure of the uncertainty that is removed as a consequence of receiving the information.
enzymes	Protein molecules that can accelerate the rate of chemical reactions which proceed very slowly or not at all in their absence. Most, if not all, chemical reactions occurring in the living cell are enzyme catalyzed (presumably so that these reactions can be controlled by the cell). Certain RNA molecules can act as enzymes (ribozymes).
equilibrium structures	Distributions of matter in space that persist without the need of dissipating free energy (e.g., crystals, lipid bilayers, etc).
excited state	When a molecule absorbs energy, it can be excited to a higher energy state in terms of electronic, vibrational, or rotational motions.
exergonic	Free-energy releasing (as in $ATP + H_2O \rightarrow ADP + P_i + H^+$).
extracellular matrix	Proteins and other materials that form a deformable structural framework outside the cell.
Fenton Reaction	The electron transfer reaction from reduced Fe^{+2} or Cu^+ to H_2O_2 to generate $HO\cdot$ and HO^-.
fermion	Any elementary particles having half-integer spin (e.g. electrons). Fermions obey Fermi-Dirac statistics and the Pauli exclusion principle.
fluctuations	Random thermal (or Brownian) motions of molecules and intramolecular segments of polymers.
Franck-Condon principle	1) The absorption of visible or ultraviolet light by a molecule occurs by a process which substitutes a new electronic structure for an old electronic structure so rapidly that there is no significant change in internuclear distance during the course of the transition. This is due to the fact that electrons have velocities about 100 times (or more) greater than that of vibrating nuclei (typically 10^8 cm/sec vs. 10^6 cm/sec.). First proposed by J. Franck in 1925 (Trans. Faraday Soc. 21, 536 (1925)) and later put on a theoretical basis by E.U. Condon in 1928 (Phys. Rev. 32, 858 (1928)) in order to account for the electronic and vibrational spectral data of simple molecules. 2) W.F. Libby (J. Phys. Chem. 56, 863 (1952)) extended the Franck-Condon principle to include inorganic electron transfer reactions with the conclusion that inorganic electron transfer processes in aqueous media (e.g. from Fe^{+2} to Fe^{+3}) must be preceded by the reorganization of the primary

Glossary 487

hydration shells around the electron donor (Fe^{+2}) and acceptor (Fe^{+3}). 3) In 1974, the Franck-Condon principle was applied for the first time to the molecular mechanism of enzymic catalysis (S. Ji, Ann. N.Y. Acad. Sci. 227, 432 (1974)). According to the so-called generalized Franck-Condon principle, the rate of the electronic rearrangement of a substrate bound to an active site of an enzyme (i.e. catalysis) is determined by the rate with which the catalytic residues of that active site can rearrange to a conformational state complementary to the altered substrate structure which is intermediated between the molecular shape of the bound reactant and that of the bound product. Since it is reasonable to think that the rate of conformational rearrangements of an enzyme would depend on the amino acid sequence of the protein, it follows that the rate of enzymic catalysis can be encoded in the primary sequence of an enzyme. Therefore, the enzymic theory based on the Franck-Condon principle provides a possible molecular mechanism to link the genetic information of DNA to specific rate constants of elementary chemical reaction steps catalyzed by an enzyme. The strained conformational state of the catalytic residues that is localized at the active site and precedes an electronic rearrangement of a bound substrate is named the Franck-Condon conformon (S. Ji, J. Theor. Biol. 116, 399 (1985)).

free energy A thermodynamic functions (a function of a system that depends only upon the initial and final states of the system and is independent of the path followed) that gives the amount of work available when a system undergoes some specified change.

free radicals Reactive chemical species that possess an unpaired (or odd) electron (e.g. $\cdot CCl_3$, $O_2^{\cdot -}$ etc.). Free radicals can be divided into anion free radicals (odd electron + negative charge), cation free radicals (odd electron + positive charge), and neutral free radical (odd electron with no electrical charge).

frustrations When a system consists of numerous interacting components (e.g. electron spins in spin glass), it is often impossible to arrange all the components in such a way that all of them exist in minimum energy states. For example, if one considers a system of three electrons in which each electron can exist with its spin directed either up (u) or down (d); and if one further assumes that the antiparallel spin arrangements (du or ud) is of a lower energy content than the parallel arrangement (uu or dd), then there are two possible arrangements of 3 electrons, namely duu and udd, that contain one unfavorable interaction between neighboring electrons that cannot be avoided no matter how one arranges the middle electron. This is an example of frustration. Certain conformational strains of biopolymers can be regarded as equivalent to frustrations in spin glass.

gene The functional unit of heredity. Each gene occupies a specific place or locus on a chromosome, is capable of reproducing itself exactly, at each cell division, and is capable of directing or controlling the formation of an enzyme or other proteins.

Gibbs free energy Defined by the relation,

$$G = E + PV - TS$$

where G = Gibbs free energy, E = internal energy, P = pressure, V = volume, T = temperature, and S = entropy. The Gibbs free energy of a system approaches zero as the system becomes equilibrated under constant T and P.

glutathione peroxidase The enzyme that catalyzes the oxidation of reduced glutathione (GSH) by H_2O_2; i.e., $2GSH + H_2O_2 \rightarrow GSSG + 2H_2O$.

gnergy A new term formed from the Greek roots, "gne-" (to know) and "-ergy" (to work), to indicate a physical entity composed of energy (to do work) and information (to control work) thought to be essential for performing goal-oriented or goal-seeking work processes (S. Ji, J. theor. Biol. 116, 399 (1985)). The discrete units of gnergy are called "gnergons." The gnergons essential for driving the goal-oriented behaviors of biopolymers in the living cell (e.g. enzymic catalysis, active transport, contraction of the actomyosin system, etc.) are identified with conformons (conformational strains of biopolymers carrying free energy and genetic information), and the gnergons essential for the living cell to communicate with its environment (e.g. chemotaxis, mitosis, secretion of hormones, etc.) are identified with intracellular dissipative structures (IDS).

Harber-Weiss reaction The electron transfer reaction from O_2^- to H_2O_2 to generate $HO\cdot$; $O_2^- + H_2O_2 \rightarrow O_2 + HO\cdot + HO^-$. This reaction is catalyzed by Cu or Fe ions.

holoenzyme The complete enzyme, i.e., apoenzyme + coenzyme.

hydrogen bond Hydrogen atoms in the groups -O-H and -N-H can form non-covalent bonds with another N or O atom. The bonding is ionic and weak with a bond energy of about 5 Kcal/mole and a bond length of about 2 Å.

information "A measure of one's freedom of choice when one selects a message" (W. Weaver, in The Mathematical Theory of Communication by C.E. Shannon and W. Weaver, University of Illinois Press, Urbana, 1949, p. 9). The smallest amount of information is 1 bit (binary digit) which is equivalent to the amount of information required to correctly predict the outcome of one toss of a coin. In general, the amount of the information carried by a message can be calculated by using the simplified form of Shannon's formula:

$$I = \log_2 \frac{Wo}{W}$$

where I = the amount of information expressed in bits, Wo = the total number of equally possible messages (e.g. all possible DNA molecules 10^9 nucleotide long), W = the number of messages actually selected (e.g. all possible 10^9 nucleotide—long DNA molecules that carry genetic information). When W = 1, I is maximal. There is another aspect to information; i.e., the meaning or the semantics of information (see M.V. Volkenstein, Foundations of Physics 7:97 (1984)) which is independent of the amount of information (i.e. of the immense number of 10^9 nucleotide-long DNA molecules, only a very small fraction will carry biologically meaningful genetic information).

irreversible change Changes in which the system is not in equilibrium at all instants during the change. All practical processes involve irreversible changes. Associated with all irreversible changes is a net increase in the entropy of the universe.

Glossary

kinetic energy
: The energy possessed by virtue of motion, equal to the work that would be required to bring the body to rest.

language
: 1) The medium through which we code our feelings, thoughts, ideas and experiences. 2) the concept of a language can be generalized to include any logical system of symbols that are used to communicate information. The generalized language can be represented as a 5-tuple, L = L(A, V, G, D, C), where A is the alphabet, V is the vocabulary, G is the grammar, D is the dictionary, and C is the historical or cultural contexts within which L has meanings. So defined, the generalized language can include, in addition to natural languages, various organized systems of knowledge in chemistry, physics and biology.

ligand
: Anything that binds to a receptor. It can be a metal ion, organic ion, neutral molecule, or radical. Ligands can be small molecular weight species, proteins, or nucleic acids.

limit cycle
: 1) Special solutions of differential equations that can be represented by closed curves in the phase plane (N. Minorsky, Non-Linear Oscillations, R.E. Krieger Publ. Co., Huntington, N.Y., 1974). 2) If one plots the concentrations of the substrate and the product of an oscillating chemical reaction on the x- and y-axes, respectively (i.e. a "phase plane"), the time evolution of the reaction system tends toward a closed loop or circle in the phase plane. This stable trajectory is called the "limit cycle," term first used by H. Poincaré in 1881.

lipid peroxidation
: The oxidation of lipids leading to the formation of peroxides, namely compounds containing the peroxy group, -O-O-.

looped domains
: Segments of DNA double helix forming a loop, 10^2 - 10^3 nucleotide long.

mechanism
: 1) A philosophical doctrine that maintains that all events or phenomena, no matter how complex, can be ultimately understood in a mechanical framework; as opposed to dualism, idealism, mentalism, and vitalism. 2) A set of physical or chemical processes on the microscopic level that are postulated to underlie observed macroscopic phenomena.

membrane potential
: The electrical potential difference that exists between the cytoplasm and the extracellular medium of the living cell. Always smaller than 100 mV. The membrane potential is generated as a result of the balance between the active transport of ions across biomembranes catalyzed by energy-driven pumps located in the plasma membrane and the passive diffusion of ions through channels spanning the membrane.

mitochondrion
: The subcellular organelle responsible for coupling the oxidation of substrates to the chondrion synthesis of ATP (oxidative phosphorylation), the universal currency of free energy in the living cell. Thousands of mitochondria are contained in each cell.

mitosis
: The usual process of cell reproduction consisting of a sequence of modifications of the nucleus (prophase, prometaphase, metaphase, anaphase, telophase) that results in the formation of two daughter cells with exactly the same chromosome and DNA content as

that of the original cell.

molecular machine A biopolymer or a system of biopolymers (e.g. enzymes, DNA, RNA) that can utilize chemical or light energy to perform work on its environment, including transport of mass in space (e.g. active transport, muscle contraction, chromosomal rearrangements, biosynthesis). Unlike macroscopic machines, a molecular (or microscopic) machine must be able to undergo thermal fluctuations (i.e. conformationally dynamic) so that it can receive energy from an external source, store the energy as conformational strains for times longer than characteristic times of thermal motions, and utilize it to perform catalysis or mechanical work.

multienzyme complexes Systems of enzymes organized to accomplish some biological functions. Substrates and products involved in catalyses may be channeled between appropriate enzymes and not allowed to freely diffuse away into the cytosol before catalyses are completed.

neural network A system of structural and functional interconnections among neurons.

non-linearity 1) Differential equations containing terms other than the differential terms raised to the 1st power. 2) Any observables whose magnitude is a non-linear function of independent variables (e.g. enzymic rate is a non-linear function of substrate concentrations in Michaelis-Menten kinetics).

nucleophiles Any chemical species that have a great tendency to react with an electron-deficient center (and hence exposed nucleus of an atom) (e.g. carbanion, HO^-, $O_2^{\cdot -}$).

Oregonator A kinetic model of the self-organizing Belousov-Zhabotinskii reaction proposed by J. Field and R.M. Noyes in Oregon (J. Chem. Phys. <u>60</u>, 1877 (1974)).

organization Whenever a particular pattern of distribution of matter in space and/or time is selected out of a group of alternative choices, the selected distribution of matter is said to possess an increased amount of information as defined by Shannon's formula and a higher degree of organization. In this sense, the terms organization, structure, and information can be treated as equivalent (see J. Rothstein, In: <u>The Maximum Entropy Formalism</u>, ed. by R.D. Levine and M. Tribus, The MIT Press, Cambridge, 1979, pp. 423-468.)

origin of life Many believe that the first self-replicating molecular system (e.g. self-replicating RNA strands) originated on the surface of the earth spontaneously about 2-3 billion years ago. The first Homo sapiens is believed to have originated from ape-like creatures about 0.3 to 3×10^6 years ago. The universe itself originated about 15 billion years ago.

phase I and II reactions Cells have the capacity to remove hydrophobic xenobiotics by first hydroxylating them through the cytochrome P-450-mediated mono-oxygenation reaction (phase I reaction) and then adding polar groups to the resulting OH group through glucuronidation or sultation (phase II reactions).

phase space A multidimensional space in which the coordinates represent the variables required to specify the state of the system. In particular, a six-dimensional space incorporating three

Glossary

	dimensions of position and three of momentum.
photon	The quantum of electromagnetic radiation.
potential	Electrostatic, magnetostatic, and gravitational potentials, at a point in the field: the work done in bringing unit positive charge, unit positive pole, or unit mass, respectively, from infinity (i.e. a place infinitely distant from the causes of the field) to the point.
potential energy	The energy possessed by a body or system by virtue of position, equal to the work done in changing the system from some standard configuration to this existing state.
potential gradient	The rate of change of potential, V, at a point with respect to distance x, measured in the direction in which the variation is a maximum. It is measured in volts per meter (e.g. the membrane potential of normal cells = 60 mV/60Å = 10^5 V/cm).
prostaglandins (PG)	A group of lipid-soluble compounds first isolated from the prostate gland in 1973 by U.S. von Euler. PG's have a wide variety of pharmacological effects on living tissues and are all derived from the 20-carbon fatty acid, arachidonic acid, released from biomembranes by the action of phospholipases.
protein dynamics	Enzymes under physiological conditions are constantly undergoing rapid fluctuating motions. Such fluctuations are thought by many to be essential for initiating catalysis.
quantum	The smallest discrete unit of energy in nature. First invoked by M. Planck in 1900 in order to explain the phenomenon of the "ultraviolet catastrophe" in black-body radiation. Planck postulated that an oscillator could acquire energy only in discrete units called quanta. In 1905, Einstein extended the concept of quanta to explain the photoelectric effect by assuming that light was radiated in quanta (photons).
quantum field theory	A field theory in which all the physical observables of a system are represented by appropriate operators which obey certain commutation relations. The quantized field can be considered as an assembly of particles each of which is characterized by its own energy, momentum, charge, etc., the total energy, momentum, etc., of the field being built up additively from the individual contributions of the particles present. Any particle may thus be considered as a "quantum" of a corresponding field.
reactive oxygen species	The oxygen molecule, O_2, can be reduced by adding 1, 2, 3 or 4 electrons to generate O_2^- (superoxide anion), H_2O_2 (hydrogen peroxide), $HO\cdot$ (hydroxyl radical) + HO^- (hydroxide anion), or $2HO^-$ respectively. Of these, O_2^-, H_2O_2 and $HO\cdot$ are chemically reactive and hence are called reactive oxygen species, reactive oxygen intermediates, or reactive oxygen metabolites. In addition, the electronically excited oxygen molecule called the "singlet" oxygen that can be generated under appropriate conditions is also included as a member of reactive oxygen species.
receptors	Protein molecules that possess a high binding affinity for a select molecule (or ligand) and exclude others. Receptors can be located in biomembranes, the cytosol, or the nucleus. Receptors can be distinguished from enzymes because receptors in general do not catalyze

chemical reactions. Nucleic acids can also act as receptors.

redox cycling   The cyclical reduction and oxidation reaction catalyzed by certain chemicals which have the ability to receive one electron to form a metastable anion radical which in turn donates one electron to an acceptor (e.g. O_2) to regenerate the original molecule. Quinones are well-known examples of redox cyclers.

reification   1) To treat an abstraction as substantially existing, or as a concrete material object; 2) The conversion of virtual particles of vacuum in relativistic quantum mechanics into real particles through input of energy.

relativistic quantum theory   The quantum theory of particles which is consistent with the special theory of relativity, and thus can describe particles moving arbitrarily close to the speed of light.

reversible change   A change that is carried out in such a way that the system is in equilibrium at any instant and that a slight decrease in the factor affecting the change causes every feature of the forward process to be completely reversed.

second quantization   Quantization of fields. Introduced to describe many-body identical particle systems. It is the process by which a classical field may be considered as an assembly of particles. Wave functions in the traditional quantum mechanics derived from the first quantization are replaced by field operators, and these and all other physical operators act on the Fock space, whose basis vectors consist of occupation numbers of various quantum states.

simplex   The simplest polyhedron in a given geometric space. The simplex of the 2-dimensional space is a triangle; the simplex of the 3-dimensional space is a tetrahedron. In general, the n-dimensional simplex has (n+1) vertices.

solitons   1) The term was coined in 1965 by N.J. Zabusky and M.D. Kruskal (Phys. Rev. Lett. $\underline{15}$, 240 (1965)) to indicate a special solution to a non-linear differential equation which was developed in 1885 by Kortweg and de Vries to describe the solitary wave phenomenon (i.e. solitary water waves that travel long distances without dispersion) first observed by J.S. Russel in 1844. This mathematical equation is known as the Kortweg-de Vries equation. 2) The generic term denoting all examples of dynamic, localized structural deformations (e.g. water waves, protein conformational strains, etc.) that carry energy (A.C. Scott, Comments Mol. Cell. Biophys. $\underline{3}$(1), 15 (1985)). 3) Conformons (conformational strains of biopolymers selected by evolution to perform biological functions) can be regarded as solitons of biopolymers endowed with some useful biological functions. Conformons are solitons, but not all solitons of biopolymers are conformons (S. Ji, in The Living State - II (R.K. Mishra, ed.), World Scientific Publishing Co., Singapore, 1985, pp. 563-574.)

spatio-temporal organization   Organization of matter both in space and in time (e.g. Belousov-organization Zhabotinski reactions).

Glossary

superoxide dismutase (SOD)	The enzyme that catalyzes the dismutation of superoxide anion into hydrogen peroxide and oxygen.

$$2O_2^- \rightarrow H_2O_2 + O_2$$

Discovered by J.M. McCord and I. Fridovich in 1969.

symbiosis — Any intimate association between two species.

symmetry — 1) Equality or correspondence in form of parts distributed around center or an axis, at the two extremities or poles, or on the two opposite sides of any body. 2) Any property of an object, material or mathematical, that remains constant or invariant upon changing the object in certain ways.

symmetry breaking — A process by which the degree of the symmetry of an object, entity, or process is reduced (e.g. thermal gradient-driven convective flow of a fluid, ATP-driven active transport of cations across biomembranes).

teleonomy — The doctrine that life is characterized by endowment with a project or purpose; i.e. the existence in an organism of a structure or function implies that it has had evolutionary survival value.

thermal energy — The energy associated with a system of molecules or atoms undergoing vibrational, rotational or translational motions at a given temperature. Thermal energies can be harnessed to perform useful work only during a heat flow along a temperature gradient.

thiols — Organic molecules containing the sulfhydryl group, SH.

topology — A branch of mathematics that deals with the properties of a geometric figure which do not vary when the figure is transformed in certain ways.

toxicology — 1) The science of poisons—their source, chemical composition, action, tests, and antidotes. 2) The science aimed at elucidating molecular mechanisms of interactions between chemicals and living systems (cells, organs, organisms, and societies) that lead to undesirable effects on human health.

useful numbers —
1) h (Planck's constant) = 6.626×10^{-27} erg·sec
   = 1.58×10^{-34} cal·sec
2) 1 eV = 23 kcal/mole
3) N (Avogadro's number) = 6×10^{23}
4) 1 cm^{-1} = 2.85 cal/mole
5) ΔG_{ATP} (free energy of hydrolysis of ATP under physiological conditions) = 16 kcal/mole
6) the vibrational quantum of the amide-I bond (i.e. the stretching motion of the C = O bond) = 1660 cm^{-1}
   = 4.73 kcal/mole (or einstein).

virtual particles	Because of the uncertainty principle it is possible for the law of conservation of mass and energy to be broken by an amount ΔE providing this only occurs for a time Δt such that $\Delta E \cdot \Delta t \leq h/2\pi$, where h = Planck constant. This makes it possible for particles to be created for short periods of time where their creation would normally violate conservation of energy. These particles are called virtual particles. The electrostatic force between charged particles may be described in terms of the emission and absorption of virtual photons by the particles.
wave function	Symbol: ψ. A mathematical function appearing in the Schrödinger wave equation and describing the behavior of a particle according to wave mechanics. The particle is thought of as a wave and ψ is the displacement or amplitude of this wave with respect to position. The square of the amplitude of the matter wave, $\|\psi\|^2$, is proportional to the probability of finding the particle in a specified position.
wave number	Symbol ν. The number of waves per unit path length (units = cm^{-1}, m^{-1}, etc.). The reciprocal of the wavelength.
zymons	Transient correlated motions of groups of particles in enzymes. Thought to be involved in the energization of enzyme active sites (B. Somogyi, G.R. Welch and S. Damjanovich, Biochim. Biophys. Acta. 768, 81 (1984)). Similar to conformons.

INDEX

acetaminophen, 227, 387-388, 390; -protein (58kD) adduct, 390; -induced cell necrosis, 394; -induced DNA fragmentation, 394.

acetanilide, 264.

active transport, 58, 59.

action, quantum of, 4, 30, 61.

action at a distance, 95, 114, 283.

adiabatic approximation, 270.

adjacency matrix, 262.

aging, molecular mechanisms of, 200-205; and free radicals, 201, 204; and diseases, 200; and cancer, 195-199.

algebra, 481; of machines, 15-17, 250-262, 481.

alkylation, 391; of genomic DNA, 393.

amide-I band, 265; the red shift of, 265.

analogy, 6, 9; table, 9, 10; F and U vectors in an analogy table, 10.

analogical thinking, 297.

Anderson, P. W., and a model of origin of biological information, 39-40, 224-225.

Anderson conformon, 41.

animal models, 477; validity of, 477.

ansatz, Davydov's, 267, 269, 279; Bethe, 275.

anthropic principle, 154, 236, 319.

antioxidants, 406-408.

aperiodic solids, 238, 240; and information, 240.

apoptosis, 170, 481.

Aristotle and DNA, 138; and unmoved mover, 138.

ATP, 32, 60, 264; hydrolysis and DNA supercoils, 38; synthesis in mitochondria, 56-61; usages through soliton generation, 264-265; usages through generation of Green-Ji-Davydov conformons, 42, 59, 280; usages through generation of excimer states, 280; in hypoxic cell injury, 418, 425.

ators and self-organizing systems, 26.

attractors, strange, 367.

autocatalysis, 245, 248.

automata, see automaton; theory, 481.

automaton, 13-17, 347-350, 481; finite-state, 347.

ß-approach, 290.

ß-decay, 244.

bandwidth, 125, 130; and channel capacity, 125,

Belousov-Zhabotinskii (BZ) reaction, 27, 69, 481.

Bérnard instability, 68, 241.

Bethesda-ator, 28.

Bhopalator, 28, 61-139, 482.

bifurcation, 72.

Big Bang, 154, 158, 233-234, 482.

binding energy of enzymes for substrates, 42.

biochemistry, 44-52, 114-116.

biocommunication system, 122-124.

biocybernetics, 5, 138, 210.

biodesic path, 306, 310.

biological communication, quantum of, 122.

biological reactive intermediates (BRI), 481.

biological uncertainty principle (BUP), 118.

biology, definition of, 102-106; in

relation to physics, 109-118; as the center of science, 163-164; functional, 104; evolutionary, 104; user, 105; sender, 105; macroevolutionary, 104; microevolutionary, 104; molecular, 40-41; and its unification with physics, 319.
bios, 320.
biotic, 320.
biotonic laws, 341.
blebs, see surface blebs.
Bohm, D., 235, 319.
Bohr, N., 4, 235; atomic model, 63; and complementarity principle, 4; see the Bohr-Elsasser theorem.
Bohr-Elsasser theorem, see Bohr-Elsasser incompleteness theorem.
Bohr-Elsasser incompleteness theorem, 230; of physicochemical explanation of life, 230.
Boltzmann constant, 120.
bosons, 482; gauge, 94, 112-113.
boundary conditions, 334.
breaks, 233; Bohr-Elsasser, 233-234; Einstein-Bohm-Wheeler, 233-234; Hawking-Penrose-Guth, 233-234; Homo abstractus, 233-234; Planck-Heisenberg, 233-234; Wilson, 233-234;
Brownian motions, 59; see thermal fluctuations or motions.
Brusselator, 27, 482.
Ca^{++}, intracellular, 391; extracellular, 391; in the nucleus, 394; mitochondrial, 392; microsomal, 392; in cell injury, 373, 378; accumulation in the nucleus, 394; gradient across the nuclear membrane, 394.
Ca^{++}-calmodulin blockers, 391-392; protection of cells by, 392.
Ca^{++}-dependent degradative processes, 380, 382.
Ca^{++}-dependent endonucleases, 393.
Ca^{++} (homeostasis) hypothesis, xxviii, 391, 482.
Ca^{++} ionophore, A23187, 380, 381.
calmodulin, 378, 391.
cancer, 195-199; and aging, 198; mechanisms of, 195-199; micro cancer state, 439.
carbonyls, electronically excited, 401, 402, 410.
carcinogenesis, 195-199; multistage, 199; initiation, 196-197; promotion, 196-197; progression, 197-199.
catalase, 482.
catalysis, and curved chemistry, 116; and conformons, 41-42; information and free energy requirements in, 31-35; and thermal fluctuations, 35-37, 136.
causes, material, 14, 231; efficient, 14, 210, 231; formal, 14, 232; final, 14, 232; external, 147; internal, 137.
causality, 10.
cell, 482; as the atom of biology, 62-67; and cytology, 52-56; and the Bhopalator, 28, 62-141; and cytosociology, 284-290; and biochemistry, 250-263; force, 7-8, 95-96, 98-118; death, mechanisms of, 168-173.

Index

cell death, mechanisms of, 168-173; kinetics of, 171-173, 225-277; program, 170.

cell force, 7-8, 95-118, 482; and cytons, see cytons; in analogy to the gravitational force, 114-118; in analogy to the strong force, 110-114.

cell injuries, compartmental, 384-391.

cell life program, 170.

cellular automata, 483.

cellular communication system, 122-139.

cellular communication theory, 122-139; and biocybernetics, 138-139.

cellular metabolism, see metabolism.

cellular protection mechanisms, 483.

cellular uncertainty principle (CUP), 118-122.

channel, communication, 122-124; capacity, 125-126, 222-224.

chaodynamics, 139-141.

chaomachine, 140.

chaos, 139, 483; deterministic, 139, 242; and information theory, 140-141.

chemical oscillations, 483.

chemical potential, 483.

chemical waves, 483.

chemiluminescence, 402, 403, 483; low-level, 403..

chemiosmotic hypothesis, 60, 173; and proton gradient, 173; vs. the conformon hypothesis, 174-176.

chemistry, 238; encapsulates irreversible time into matter, 240; of free radicals, see free radicals; and electron pairs, 228.

chiral asymmetry, 243-247.

chiral symmetry breaking, see symmetry breaking.

chirality, homo-, 243; biomolecular, 243-248.

Christoffel symbol, 299.

chromodynamics, 7.

chromatin, 483.

chunk, 55, 56.

chunked, see chunk.

chunking, see chunk.

Circe effect, 228.

collective phenomena, 361.

clock, chemical, 241.

code, 361.

coenzymes, 255, 483.

cofactor, 252, 483.

coherence, 239, 241; long-range, 242.

communication, 122; molecular, 31; quantum of molecular, 61; biological, or bio-, 122, 124; cellular, 122; system, 122-124; theory, see information theory.

communication channel, see channel.

complementarity principle, 3-4; generalized complementarity, 341.

complexes I, III, and IV, 58.

computation, 339.

computer, 81; cell as a molecular, 81-83; and mind, 205-209.

concatenation, N-dimensional, 20-21.

conduction band in proteins, 293.

configuration, 484; vs. conformation, 484.

conformation, 484; of a protein, 32; and conformons, see conformons; conformational strains, 32, 136-137; conformational level, 254.

conformons, 31-35, 280, 315, 484; and solitons, 33, 280; and frustrations, 225; Anderson, 41; Gedda, 42-43; Green-Ji-Davydov, 42; Franck-Condon, 42; Frauenfelder-Lumry, 41; Kehkls, 42; Klonowski-Klonowska, 42; Volkenstein-Jencks, 42; in proteins, 222; in DNA, 129, 222-224; generases, 129; generators, 129; free energy content of, 31-34, 222; information content of, 34, 222; quanta of molecular communication, 61; virtual, 136; and hydrogen-bond phonon field, 315.

conjugation reaction, 484.

connection, 94-95; affine, 299, 302; and fiber bundle, 92.

control, 484; parameter, 139.

convection, 68, 239.

cooperative effects, in neural nets, 361, 369.

coupling, electron-nuclear, 50; respiration-phosphorylation, see oxidative phosphorylation; respiration-glycolysis, see Pasteur effect; hormone-second messenger, 47, 130; electron-phonon, 293.

covalent binding hypothesis, 484.

Crabtree effect, 47.

creation and annihilation operators, 484.

curvature, 484; of spacetime, 94, 282; of metabolic space-time, 310.

curved chemistry, 114-117.

cybernetics, 13-17; bio-, see biocybernetics.

cyclophorase, 52.

cytochrome oxidase, 417; see also complex IV, 58.

cytochrome P-450, 178, 228, 484; structural gene for, 182; receptor site of, 180; effector site of, 180; substrate binding in, 180; catalytic site of, 180; coupled state of, 179; uncoupled state of, 180.

cytolytic reaction, 432; and switching parameter, 432.

cytons, 8, 9-96, 484; mediators of the cell force, see cell force; and conformons, 96; and IDS, 96.

cytoskeletons, 85, 484.

cytosociological factors, 288.

cytosociology, 49, 284-290, 484.

Davydov's soliton, see solitons.

death, cell, see cell death; consciousness, 152.

decode, 361.

decoder, 123.

Delbrück, M., 138; and Aristotle's unmoved mover, 138.

delta approach, 289.

deterministic chaos, see chaos.

diffusion coefficient matrix, 308, 310.

diffusion Lagrangian, 308.

dimol reaction, 403.

diseases development, 191; frustrated self-defense mechanisms and, 191-195.

disorder, 240; entropy and, 240.

dissipation function, 306.

dissipative processes, 240; since Shrödinger's time.

dissipative structures, 67-76, 485; intracellular (IDS), 69-73; and 4-

Index

structures, 68-69; and far-from-equilibrium systems, 67; and non-equilibrium thermodynamics, 240; and dissipative symmetry-breaking, 158.

DNA, 78-80; transcription, 78-80; replication, 78-80; gyrase, 27, 37, 485; topoisomerase, 37; helicase, 37; supercoils, 485; single strand breaks, 422; and Aristotle's unmoved mover, 138; conformons in, 222; fragmentation, 394; damage by singlet oxygen, 408-409.

DT diaphorase, see quinone oxidoreductase.

duality, wave-particle, 3; information-energy, xxiv; generalization to tetrahedrality, 231, xxiv.

Edinburghator, 28.

effector, 123; in biocommunication system, 123; of cytochrome P-450, see cytochrome P-450.

Einstein, A., 235-236, 328; Einstein's equation, 11; Einsteinian questions, 237.

Einstein-Podolski-Rosen (EPR) paradox, 340.

Einsteinian questions, 237.

electrical potential, 485.

electron-phonon coupling, see coupling.

electron transfer complex (ETC), 59.

electronic-nuclear interactions, 51.

electronic semiconduction, 293.

electronically excited states, in biological molecules, 401; lipid peroxy radicals as, 404.

electrophiles, 485; GSH defense against, 374.

elongation complexes, 137.

emergent properties, 20, 361, 485; principle of, 20, 356; 371.

enantiomer, 245.

encapsulation of irreversible time into matter, 240.

encoder, 122-124.

encoding, 252.

endergonic, 485.

energy, 31, 485; free, 31-34, 487; potential, 490; binding energy of Davydov-like solitons, 265; thermal, 30, 136, 292, 492; biological, storage and transport of, 264; vibrational, self-trapping of, 265.

enfolding, 235.

enthalpy, 485.

entropy, 239, 485; production, internal, 239; negative entropy flow, 239; thermodynamic, 123; information-theoretic, 123; informational, 123; Shannon, 123; physical, 123; generalized, 341, 357; in relation to order and disorder, 240; and the second law, 238-242; and irreversible processes, 238-242.

environmental health sciences, 480.

enzyme, 35-37, 285-294, 485; control net, 255; catalysis and conformons, 40-43; catalysis, 285; catalytic vs. recognition functions of, 43; holoenzymes, 488; coenzymes, see coenzymes.

equilibrium fluctuations, 35-36.

equilibrium structures, 486.
ergons, 152, 160; and gnergons, 152; and gnons, 152; and conformons, 152; and IDS, 152; and cytons, 152.
event, 105; set of past events and evolution, 105; set of present events and phenotypes, 106.
evolution, 5; biological, 3, 183-184; Darwinian, 183-184; function-evolution duality, 3; cosmological, 156; of enzymes, 288, 290; of organisms, 158; punctuated equilibria and, 184; and molecular biology, 104.
excimer, 280.
excited carbonyls, 410.
excited states, 486.
exergonic, 486.
extracellular matrix, 133, 486.
far-from-equilibrium thermodyanmics, see thermodynamics.
Fenton reaction, 486.
fermion, 487.
fiber bundle, 92; theory, 92-96; and connections, 93, 95; and fiber space, 93-94; and genetic information, 94.
fidelity in communication, 130; and redundancy, 130; and multiple channels, 198.
field, 83; 294-296; and particles, 83, 283, 294; intracellular metabolic velocity (IMVF), 84; intracellular chemical concentration (ICCF), 85-86; biopolymeric force (BFF), 84-85; gauge, 311; and long-range interactions in cell metabolism, 283; and theories of living state, 283; field-effect element, 290.
fims, functionally important motions, 35-36; and conformons, 36.
firing, in metabolic machines, 257.
fitness value, 165-166.
fluctuations, 487.
fluctuation-dissipation theorem, 50-51, 311; and enzymic catalysis, 51
Fokker-Planck equation, 308; as applied to protein dynamics, 309.
folding of proteins, 47-48.
force, 101; cell, see cell force; electromagnetic, 7, 11, 21; electroweak, 7, 21; in Newtonian mechanics, 101, 114; in general relativity theory, 114-115; gravitational, 114-115; society, 151-152; proton-motive, 173; strong, 110-113.
fractals, 92; biological machines as, 92; in medicine, 92.
Franck-Condon principle, 50-53, 343, 486; generalized, 50, 53; and electronic-nuclear coupling, 50.
Franck-Condon factor, 268-269.
free energy, 309; barrier, 287, 288, 290; intrinsic free-energy transition-state barrier, 288; extrinsic factors affecting free-energy barrier, 288; complementarity, 315.
free radicals, 191-193, 487; in aging, 200-205; in xenobiotic metabolism, 180, 228-229; in cancer, 191-195; controlled vs. uncontrolled, 191-195.
free will, in relation to Newton's third

Index

law, 138.
frustrations, 39-40, 487; and conformons, 39; and Anderson conformons, 41, 225; and origin of biological information, 39-40, 225.
function space, 93.
gauge, 97, 312; field, 311; bosons, 315; theories, 96-98; symmetry, 97; principle, 98; local gauge symmetry, 97; local gauge principle and new force, 98.
gene, 487; cell-death, 170; cell-life, 170; cellular adaptive, 170; cellular defense, 170; spatiotemporal, 177; structural, 177.
genetic information, 34, 121; and aperiodic solids, 238; in conformons, 34.
genotype and phenotype, 101-102.
geodesic, 282-283, 310; paths in the living cell, 298.
geometrization, 302.
geometry, affine, 298; curved, 114-116; connected, 297; Euclidean, 115; flat, 115; Riemannian, 115.
Gibbs free energy, 487.
gluons, 7, 94.
glycolysis, 46; and respiration, 47.
glutathione, 375; disulphide, 374; peroxidase, 375, 487.
gnergons, 1, 152, 210; and conformons, 152; and IDS, 152; and gnergy, 1, 152.
gnergy, 1, 152, 487; tetrahedron, 231, 234; tetrahedrality of, 231.
gnons, 1, 152, 160; and conformons, 152; and IDS, 152; and DNA sequence information, 160, 222.

Gödel, K., 356.
Gödel's incompleteness theorem, 356.
gravitons, 94; and gravitational force, 94.
gravity, 114; and general relativity theory, 114.
h-particles, 86, 165.
hadrons, 111, 113; and strong force, 113.
Hamiltonian, 271, 272.
Hamilton's principle, 97-98.
Hanoverator, 28, 193-194.
Harber-Weiss reaction, 488.
Heisenberg ferromagnet, 316.
Heisenberg uncertainty principle, 65.
helicase, 37.
hepatocyte, 225, 376; cell-death kinetics of, 225-227; cytoskeletal structures, 378.
hepatocellular necrosis, 384.
Higgs field, 155; and Shannon information, 155.
histone, 444; genes, 443-444; kinetics of synthesis of, 454; control vs. regulation of gene expression of, 459-466.
historicity, 10; and the common descent of light and life, 13.
Holcombe-Welch machine, see machines.
Homo abstractus, 233-234.
Homo sapiens, 185; evolution of, 185-186.
Hofstadter cubes, 208.
Hofstadter images, 208.
human body, 141; the Piscatawaytor model of, 141-147; and exposure to toxic chemicals, 478.

hydrogen bond, 488; networks, 291, 293; role of, in soliton formation, 266; structures in proteins, 293; phonons, 315.
hypoxia, 414, 415; and cell death, 414.
hypoxic cell injury and ATP, see ATP.
hysteresis, 371.
immortality program, 170.
immune responses, 430.
immune parameters, 430-440.
implicate order, , 235, 283; and the Shillongator, 235.
inflammation, 186-191; and inflammatory response, 187; and the Londonator, 188-191.
information, 10, 488; amount of, 10, 222; complexity and, 123; cosmological, 154; genetic, 34, 120, 176; minimum free energy requirement for transmitting, 222; primeval, 159-160; post-Big Bang, 155, 158; Shannon, 160; and entropy of message source, 123-125; source of, 123-125; theory, 242; theory and physics, 242; and organization, 356; and thermodynamic entropy, 123; and generalized entropy, 341; transmission, 123; transmission in time, 128; from the law of physics and chemistry to, 242; theory and interpretation of natural laws, 242; storage, global, among cells, 371.
inputs, 23.
intelligent behaviors, 166; cognitive and effector functions in, 166; and cellular intelligence (CI), 166-167, 170, 172.
intermediate form of free energy (IFFE), 175; in oxidative phosphorylation, 175; in muscle contraction, 175; and conformons.
internal charges, 294.
internal coordinates, 94.
internal degrees of freedom, 95.
internal space, 95.
internal states of machines, 15, 24-25.
intracellular dissipative structures (IDS), 69-73, 253.
irreversible change, 488.
kinetic energy, 488.
kinetic power, 307.
Krebs cycle, 58; as a metabolic machine, 44, 251-252.
Krebs machine, 47, 252.
L-space, 121, 165-166; subspaces of, 166.
l-particles, 86, 165.
lactate dehydrogenase (LDH) leakage, 419.
Lagrangian, 295; method, 305; diffusion, 308; metabolic field, 305; density, 304.
language, 488; and emergent properties, 20; as a metaphor of biological complexity, 17-20; formal specification, 260.
Lamarckism, 178; lamarckian processes on the cellular level, 178.
law of requisite variety (LRV), 221; and human evolution, 185-186; and the inflammatory response, 189-190; and the complexity of enzymes, 35-37; and the

Index

complexity of the cellular communication system, 129-131.
least action, principle of, 97, 295.
Leibniz's model of the brain, 326-327.
leptons, 94-95, 158.
life, 164-168; origin of, 224-225; and the Bohr-Elsasser incompleteness theorem, xxix; 230; and its irreducibility to chemistry and physics, xxix, 230; and non-equilibrium, 324.
ligand, 488; ligand-operated molecular vending machines (LOMVM), 135-136.
light, 12; and life from gnergy, 11-13.
limit cycle, 371, 488.
lipid peroxidation, 389, 402, 407, 423, 424, 425, 488.
liver, perfused, 74.
liver cells, isolated, 225-226.
local equilibrium assumption, 305.
logistic equation, 139.
Londonator, 28, 186-191; and inflammation, see inflammation.
McClare limit, 318.
McClare machine, see machine.
machines, 13-17; algebra of, 15, 250-262; artificial vs. biological, 17; and cellular automata, 351; finite-sate, 15, 251, 347; general abstract, 259-260; Holcombe-Welch, 44-45, 47; Krebs, 47; McClare, 29-35; macroscopic vs. microscopic, 32; Mealy, 348; molecular, 32, 318, 489; Moore, 348; Pasteur-Crabtree, 47; self-replicating, 353; Turing, see Turing machine; X-, 259-260.

macroscopic vs. microscopic, 32.
macrophages, activated, 431.
Madisonator, 28, 56-61; and oxidative phosphorylation, 56, 59.
manifold, 90-92; metabolic, 284-285.
Markov chain, 241.
matter and energy, 231; interconversion through Einstein's equation, 11; self-activity of matter, 301, 310.
Matisse, H., 167.
measurement, 341; in relation to communication, 352; in relation to generalized entropy, 341.
mechanics, Newtonian, 32; quantum, 32; of biopolymers, 4-43.
mechanisms, 489; of toxicity, 475-476.
membrane potential, 489.
memory, content-addressable (CAM), 371.
menadione, 377, 380, 405.
messengers, different classes of, 13.
meta-automatonistic, 326.
metabolism, cellular, 282; computer modeling of, 250: and non-linear models, 285; holistic description of, 282; interconnectiveness of, 282; and channeling, 285; and compartmentation, 45-49, 386; and field theory, see metabolic field theory.
metabolic field theory, 300, 308.
metabolic gauge field, 313.
metabolic infrastructure, 283.
metabolic metric, 283.
metabolic non-commutativity, 73.
metabolic pathways, 45-49;

organizational hierarchy of, 45-49; as machines, 251.
metabolic space, 303; curvature of, 303.
meta-chemical, 167.
metalanguages, 327.
metamolecular, 327.
metaquantal, 327.
metric tensor, 298.
microscopic reversibility, 136.
mind, in relation to computers, 205-209.
misonidazole, 422; mechanism of toxicity of, 422.
mitochondrial electron transfer chain, 58-59.
mitochondrion, 56-61, 489.
mitosis, 489.
molecular biology, 40.
molecular communication theory (MCT), 31.
molecular machines, see machines.
molecular theories, xiii.
molecularization of macroscopic machines, 32-35; and the thermal barrier, 30.
monomol reaction, 403.
morphology and morphogenesis, 52-53.
multiple channel hypothesis of bio-communication, 130, 198.
multienzyme complexes, 489.
natural killer (NK) cells, 431.
netlets, 363-364.
neural network, 361-369, 489.
Newbrunswickator, 28, 149-152.
Newtonian mechanics, see mechanics; the third law of, 138; and force, 113-114; and action at a distance, 113, 283; the second law of, 101.
Noether's theorem, 312.
nonequilibrium, constructive role of, 240; as the source of order, 242.
non-commutativity, in quantum mechanics, 72; in organ metabolism, 72.
non-linearity, 139, 489; and deterministic chaos, 138-139.
normoxia, 415.
nucleons, 166.
nucleophiles, 489.
nucleus as a critical site of cell injury, 392-395.
Oklahomacityator, 28; and xenobiotic metabolism, 181-183, 228-229.
one-dimensional, aperiodic solids, 240.
ontogeny vs. phylogeny, 41.
operator, 294; Hamiltonian, 332; see also Hamiltonian; Laplacian, 304.
order, 240; role of entropy in, 240.
Oregonator, 27, 489.
organisms, 25-26.
organization, 167, 489; and generalized entropy, 341; and Shannon information, 160; and life, 167; and beauty, 167; spatiotemporal, 67, 492; self-, see self-organization.
origin of life, 224-225, 242, 490; and molecular biology, 41.
origin of biological information, 224-225.
origin of the universe, 157-163, 230-237.
outputs, 24-25; of frustrated self-

defense mechanisms, 193.
oxidative phosphorylation, 58-61; and chemiosmotic hypothesis, 60-61; and the Madisonator, 56, 59; and conformons, 59; and mitochondria, 58; and intermediate form of free energy (IFFE), 175; and intramembrane protons of Williams, 174.
oxygen uptake, 74, 418, 419.
oxygen tensions (or partial pressures), tissue, 416.
parallel processing, in brain, 370.
parameter, nonlinear, 139.
particles, l-, 86, 165; h-, 86, 165; virtual, 493.
parity non-conserving interactions, 247.
Pasteur effect, 47.
Perutz, M. F., 238.
phase I and phase II reactions, 490.
phase space, see space.
phenotype, 41, 101.
Philadelphiator, 28.
phonons, optical, 269; acoustic, 269, 315; condensation, 293; hydrogen-bond, 315.
photodynamic action, of endogenous aromatic compounds, 226; and free radicals, 226; and light-induced cell death, 226-227.
photoemission, 403.
photons, 81, 490.
phylogeny, 41, 101.
physics, 242; and information, 155, 242; vs. biology, 2-3, 4, 319.
Piscatawaytor, 28; and the human body, 141-147.

places, in automata theory, 255-256.
Planck constant, 65.
Poisson law, 241.
Poisson equation, 304.
potential, 490; chemical, 301; electromagnetic, 94-95; gravitational, 94; energy, 490; gradient, 490.
primordial soup, 225.
principle of emergent properties, see emergent properties.
principle of geometrization of physics, 302.
principle of self-organization, see organization.
Princetonator, 28; and the origin of life, 224-225.
programmed cell death, see apoptosis.
programmed organismic death, 200.
prooxidant state, 192.
prostaglandins, 188, 490.
protein dynamics, 490.
protein kinases, 47.
proteins, α-helical, 33, 265; conformational strains in, 32-33, 58-59.
proteins thiols, in cell injury, 376.
protochemical, 291, 293; events, 293.
protons, transmembrane, 56-59, 61; intramembrane, 58-59, 61; Williams, 61; Mitchellian, 61; high-energy, 293.
proton-motive force, see force.
proton transfer complex (PTC), 58-59.
quantum, 490; virtual vs. real, 136.
quantum of action, 4, 30.
quantum of biological communication, 122.

quantum of molecular communication, 61.
quantum chromodynamics (QCD), 7.
quantum electrodynamics (QED), 7.
quantum field theory, 490; relativistic, 136, 491.
quarks, 7; and the strong force, 112.
quinone oxidoreductase, 405.
radiation, mitogenic, 410.
reactive oxygen species (ROS), 228, 229, 401-402, 404, 422, 491.
receptors, 131, 132, 491; osmo-, 132.
redox cycling, 404-406, 491.
reification, 491.
relativity theory, special, 167; general, 166; analogical application to biology, 114-116, 296-311.
replisomes, 137.
reversible change, 491.
requisite variety, law of, see law of requisite variety (LRV).
scalar field, 84.
Schrödinger, E., 238; and biology, 238-242; and wave mechanics, 332.
second law of thermodynamics, 239; and internal entropy production, 239.
second law of Newtonian mechanics, see Newtonian mechanics.
self-activity of matter, 301, 310.
self-defense mechanisms, 170, 187-190; frustrated, 191, 195; and cancer, 195-199; and inflammation, 186-191; and cellular intelligence (CI), 170.
self-organization, 159, 243, 366; principle of, in cosmology, 160; free energy and information requirements for, 159-160; and far-from-equilibrium thermodynamics, 67; gnergy principle of, 160; in neural networks, 363; and the orign of life, see the Princetonator; and the cell, see the Bhopalator; and the universe, see the Shillongator; and the self-defense mechanisms, see the Hanoverator; and the human body, see the Piscatawaytor; and the human society, see the Newbrunswickator; and oxidative phosphorylation, see the Madisonator; and the inflammatory response, see the Londonator; and xenobiotic metabolism, see the Oklahomacityator.
self-similarity, 92.
self-trapping, 268-280.
sender, 122; biology, see biology.
Shannon's inequality, 125.
Shannon information, see information.
Shannon communication theory, 122-126.
Shillongator, 28; and the origin of the universe, 28; the concrete version of, 156-163; the abstract version of, 230-237.
signal transduction, cellular, 130.
simplex, 491; of 3-dimensional space, 144.
singlet oxygen, 402, 408-410; DNA damage by, 409.
social factors in enzymic catalysis, 288.
sociobiology, 148; in relation to the Piscatawaytor, 147-149.

Index

solitons, 33, 265-268, 491; and conformons, 33, 280; and excimer, 280; Davydov's, 265-268; and enzymic catalysis, 280; quantum mechanics of, 265-279.

source of information, 122.

space, Eigen, 120, 166; fiber, 93; fiber bundle, 93; function, 93; genetic information, 120; internal coordinate, 94; L-, 120, 166; mathematical, of biology, 94; Mayr, 166; total, 93; basespace, 93; phase, 490.

spacetime, biological, 296.

spatiotemporal organization, see organization.

splicesomes, 137.

statistical mechanics, 355-356; generalizing, 356.

strange attractors, see attractors.

strong force, see force; the cell force in analogy to, 110-114.

sulfhydryl group, 378-379.

superoxide anion (radical), 405; from one-electron reduction of redox cyclers, 405.

superoxide dismutase, 492.

superstrings, 157-158; and Shannon information, 155; and the Big Bang, 158.

surface blebs, 376-377; and cortical microfilaments, 376.

symmetry, 492; chiral, 244; between the living and the nonliving, 159.

symmetry breaking, 492; chiral, 244-247; equilibrium, 158; dissipative, 158; in space, 49; in spacetime, 49; in time, 49; spontaneous, 244.

symmetry principle of bioenergetics, 174.

symmetry principle of bioenergetic coupling mechanisms, 175.

systems, isolated, 239; open, 239; and subsystems, 324, 357; dynamical, 13; general, 13.

table symmetry, 12.

table theory, 9.

TCCD, 393, 396.

teleonomy, 492.

termite society, nest building behavior of, 209; in relation to human rationality, 209.

tetrahedrality of gnergy, 231, 234; and energy, matter, life and information, 231, 234; as an extension of the particle-wave duality of light, 231.

thermal barrier, 29, 31.

thermal energy, see energy.

thermal motions, 30.

thermal rigidity, 18, 30.

thermal flexibility, 18, 30.

thermodiffusion, 240.

thermodynamics, second law of, 14, 30, 239; and living systems, 66.

thiols, 492; protein, modification of, 376.

tokens in metabolic machines, 257.

tocopherol, 402.

topoisomerase, 37.

topology, 492; of death, 261-262.

Torontoator, 28.

toxic responses at low doses, 476.

toxicology, xxiii, 473, 492.

transcription, 25, 37, 42, 45, 47.

transition state, 288.

translation, 47-48.
transmitter, 123.
trypan blue uptake, 382, 419.
tumors, 430; regression, 430; progression, 430; sneaking-through phenomena, 430; dormancy, 440; immunostimulation of tumor growth, 440; see also cancer.
Turing machine, 82, 208, 347, 350; as a model of human rationality, 205-207.
uncertainty principle, Heisenberg, 8; biological, 118-120; cellular, 118-120.
uncertainty principle of proteins, 318.
unfolding, 158, 235.
unification of biology and physics, 319; through the Shillongator, 237.
unmoved mover, Aristotle's, 138; and Newton's third law, 138; and internal causes, 137; thermal stroke and, 138.
user, 122, 124; biology, 105.
Valhallator, 28; and carcinogenesis, 195-199.
variational calculus, 298.
variety, 221; the law of requisite, see law.
vector, 298; field, 84; F and U vectors in table theory, 9.
virtual quanta, see quanta.
Volkenstein-Jencks conformons, see conformons.
wave function, 332, 493.
wave number, 493.
wave-particle duality, see duality; and its generalization to the tetrahedrality of gnergy, 231, 234.

waves, intracellular chemical (ICW), 49; chemical concentration, 26; intracellular Ca^{++}, 7; and phase angle, 11.
wave packet reduction, 344.
weak interactions, 248; and the origin of biomolecular chirality, 243.
well-informed heat engine (WHE), 137, 352-353; a molecular realization of, 137.
wholeness of quantum systems, 344.
xenogenes, 229.
Yang-Mills gauge theory, 111.
Xeonopus laevis, 443, 448; genome, 451; histone synthesis in, 454; control and regulation of histone synthesis in, 459.